Papers contributed to the international symposium on
foundations of plasticity

Monographs and textbooks on mechanics of solids and fluids

editor-in-chief G. Æ. Oravas

Mechanics of plastic solids

editor J. Schroeder

Volume 1

Foundations of plasticity

editor A. Sawczuk

Volume 2

Problems of plasticity

editor A. Sawczuk

Problems of Plasticity

Papers contributed to the international symposium on
foundations of plasticity

Warsaw, August 30 – September 2, 1972

Edited by

A. SAWCZUK

Polish Academy of Sciences, Warsaw

NOORDHOFF INTERNATIONAL PUBLISHING
LEYDEN

© 1974 Noordhoff International Publishing
Softcover reprint of the hardcover 1st edition 1974
A division of A. W. Sijthoff International Publishing Company B.V., Leyden,
The Netherlands

ISBN-13: 978-94-010-2313-9   e-ISBN-13: 978-94-010-2311-5
DOI: 10.1007/978-94-010-2311-5

Library of Congress Catalog Card Number 73-85410

List of contents

Invited lectures

List of contents

Discussion notes

List of contents

List of contents

List of contents

Preface

The theory of plasticity has a broad range of applications in industry and engineering. Its methods permit to devise metal processing, allow to assess safety of structures against collapse, make possible to estimate flow of bulk solids. The technological progress calls both for deeper understanding of sources of plastic deformation and for more rigorous description of mechanisms of plastic motion.

It was felt at the Institute of Fundamental Technological Research in Warsaw several years ago that an international meeting bringing together experts in various domains of plasticity and studying plastic behavior from different positions could be of use for further development of the discipline itself and for its service to society. The appropriate occasion for such a meeting sponsored by the Polish Academy of Sciences was provided by the 70th anniversary of Professor Wacław Olszak.

The idea of a fundamentals oriented symposium was favorably received by the scientific community over the world. Eminent scientists generously advised me as to the scope of the meeting and helped in shaping the program. It was eventually decided to concentrate the debates on a) physical foundations of plasticity, b) thermodynamics of plastic and visco-plastic deformation, c) experiments regarding plastic behavior, d) mathematical methods in plasticity, e) computer techniques suitable for solving important technological problems. The selection of papers was made accordingly.

The symposium contributed papers were published in the volume entitled 'Foundations of plasticity' and in the 'Archives of Mechanics' No. 3, 1972. Several related papers were printed in the 'Archives of Mechanical Engineering' No. 1, 1973 and in other specialized journals.

The present volume entitled 'Problems of plasticity' contains the 'Invited lectures' and discussions to the contributed papers as well as a number of brief communications presented in the discussions.

It is a pleasure to express my gratitude to the Members of the International Scientific Committee and to the Referees for the advise and to the Polish Academy of Sciences for the support. The ready cooperation of Noordhoff International Publishing is gratefully acknowledged.

I wish to express my warm thanks and appreciation to Dr. Janas for his

Preface

help and enthusiasm. Thanks are due to the Members of the Organizing Committee and to my Collegues and Associates here for their sustained assistance during the years of preparation of the Symposium.

Antoni Sawczuk

INTERNATIONAL SYMPOSIUM
ON FOUNDATIONS OF PLASTICITY

Warsaw, August 30 – September 2, 1972

SYMPOSIUM ORGANIZED BY
THE POLISH ACADEMY OF SCIENCES

INTERNATIONAL SCIENTIFIC COMMITTEE

J. M. Alexander, London

D. C. Drucker, Urbana

L. Finzi, Milan

A. E. Green, Oxford

P. G. Hodge, Jr., Minneapolis

H. G. Hopkins, Manchester

A. A. Ilyushin, Moscow

W. T. Koiter, Delft

J. Mandel, Paris

P. M. Naghdi, Berkeley

W. Nowacki, Warsaw

F. K. G. Odqvist, Djursholm

E. T. Onat, New Haven

W. Olszak, Udine

W. Prager, Providence

Yu. N. Rabotnov, Moscow

M. Reiner, Haifa

L. I. Sedov, Moscow

V. V. Sokolovsky, Moscow

T. Y. Thomas, Los Angeles

C. Truesdell, Baltimore

H. Ziegler, Zürich

ORGANIZING COMMITTEE

S. Kajfasz, R. Kowalczyk, M. Kwieciński, P. Perzyna, S. Zahorski,
M. Życzkowski

TECHNICAL COMMITTEE

J. Bauer, L. Dietrich, A. Drescher, J. A. König, K. Turski, T. Wierzbicki,
J. Zawidzki

LADIES COMMITTEE

E. Drescher, M. K. Duszek, W. Grabczyńska, H. Kajfasz, J. Ostrowska-
Maciejewska

SYMPOSIUM CHAIRMAN
A. Sawczuk

SYMPOSIUM SECRETARY
M. Janas

Dedication

Symposium organized in connection with Professor Wacław Olszak's 70th anniversary.

Address of Welcome

by W. Nowacki

Vice-President of the Polish Academy of Sciences

It is, indeed, a pleasure to be present at this meeting and to welcome on behalf of the Polish Academy of Sciences the participants of the Symposium on Problems of Plasticity.

Much importance is attached in this country to the development of basic research. The Polish Academy of Sciences has given a great deal of careful thought and of resolute effort to the promotion of research in many branches of natural science. We set up the Institute of Fundamental Technological Research which eventually became the most important Polish center in the domain of mechanics.

Efforts have been made to carry on work of mechanics on a broad front, such as providing a forum for discussion in the form of annual scientific conferences, symposia, summer courses, and in addition carrying out extensive publishing activities.

The Theory of Plasticity has its old traditions in Poland, dating back to the first years of Professor Huber's activities. The greatest strides, however, have been witnessed during the last twenty years. Professor Olszak is the one to whom we owe valuable and lasting achievements. His creative work set an example and encouraged a group of his young co-workers to dedicate themselves to the pursuit and to the expansion of that discipline of science.

Professor Olszak's concepts in the field of anisotropy and non-homogeneity in plasticity represent a notable contribution to the progress in the Theory of Plasticity in our country.

The Symposium is devoted to the foundations of the Theory of Plasticity. It is interesting to note that after many years of research in mechanics, we reflect again on the basic notions, such as elasticity, plasticity, viscosity, brittleness, and we examine thermodynamic fundamentals of plasticity, trying to get closer to the physical reality of material behaviour.

I think it appropriate to conclude my short address by wishing you all most successful debates.

Session Program

Wednesday, August 30, 1972

OPENING SESSION

Address of Welcome: W. NOWACKI, Vice-President of the Polish Academy of Sciences

Inauguration Concert: H. CZERNY-STEFANSKA playing Chopin

SESSION 1 Invited Lecture

Chairman: W. NOWACKI, Warsaw

W. PRAGER, Providence, R.I.: Limit analysis: the development of a concept

SESSION 2A Constitutive Relations

Chairman: W. T. KOITER, Delft

T. TOKUOKA, Kyoto: Fundamental relations of plasticity as derived from hypo-elasticity

J. KRATOCHVÍL, Prague: Finite-strain theory of inelastic behavior of crystalline solids

Y. HORIE, Raleigh, N.C.: On the thermodynamic states of plastically deformed solids

Discussion: S. Nemat-Nasser, Evanston, Ill.; P. Germain, Paris; E. H. Lee, Stanford, Ca.; B. R. Seth, Ranchi, Bihar; J. T. Fong, Washington D.C.; E. T. Onat, New Haven, Conn.; D. R. Owen, Pittsburgh, Penn.; F. A. McClintock, Cambridge, Mass.;

SESSION 2B Experimental Plasticity

Chairman: F. K. G. ODQVIST, Stockholm

E. SHIRATORI, K. IKEGAMI, K. KANEKO, Tokyo: Subsequent yield surfaces in consideration of the Bauschinger effect

W. SZCZEPIŃSKI, Warsaw: On the concept of surfaces of influence of plastic prestraining on the mechanical performance of metals

N. Como, Naples: A theoretical evaluation of the offset sensitivity of experimental yield surfaces

O. A. Shishmarev, Riazan: Experimental study on one type of plastic anisotropy forms not considered in simplified flow theories (presented by title)

Discussion: H. Lippmann, Karlsruhe; O. A. Shishmarev, Riazan; E. Shiratori, Tokyo; B. Gowda, Oxford; M. Save, Mons

Session 3 Invited Lectures

Chairman: W. Olszak, Udine and Warsaw

D. C. Drucker, Urbana, Ill.: Some questions continuum plasticity can answer at the microscopic level in metals

E. Kröner, Stuttgart, C. Teodosiu, Bucharest: Lattice defect approach to plasticity and viscoplasticity

Discussion: M. Wnuk, Brookings, S.D.; J. Kosiński, Warsaw; H. Lippmann, Karlsruhe; J. A. Simmons, Washington, D.C.; E. H. Lee, Stanford, Ca.; D. R. Owen, Pittsburgh, Penn.; S. Nemat-Nasser, Evanston, Ill.; K. C. Valanis, Hoboken, N.J.; P. Perzyna, Warsaw

Session 4A Applied Plasticity

Chairman: G. Backhaus, Dresden

V. Nagpal, F. A. McClintock, C. A. Berg, M. Subudhi, Cambridge, Mass.: Traction-displacement boundary conditions for plastic fracture by hole growth

J. L. Duncan, Hamilton, Ont.: Superplasticity: constitutive relations

A. R. Ragab, J. L. Duncan, Hamilton, Ont.: Superplasticity: forming problems

Discussion: M. Wnuk, Brookings, S.D.; H. G. Hopkins, Manchester; D. C. Drucker, Urbana, Ill.; W. Szczepiński, Warsaw; B. Gowda, Oxford; V. Guttmann, Petten

Session 4B Numerical Methods

Chairman: N. Jones, Cambridge, Mass.

O. C. Zienkiewicz, G. C. Nayak, D. R. J. Owen, Swansea: Composite and 'overlay' models in numerical analysis of elasto-plastic continua

A. H. SHABAIK, E. G. THOMSEN, Berkeley, Ca.: Computer aided visioplasticity solution of some deformation problems

A. G. UGODCHIKOV, YU. G. KOROTKIKH, Gorki: Constitutive equations of non-isothermal elastic-plastic deformation and methods of their practical application (presented by title)

Discussion: M. Kleiber, Warsaw

Thursday, August 31, 1972

SESSION 5 Invited Lecture

Chairman: M. SAVE, Mons

E. T. ONAT, F. FARDSHISHEH, New Haven, Conn.: On the state variable representation of mechanical behavior of elastic-plastic solids

Discussion: P. Podio Guidugli, Pisa; D. C. Stouffer, Cincinnati, Ohio; J. Kratochvíl, Prague; P. Perzyna, Warsaw; L. H. N. Lee, Notre Dame, Ind.; E. H. Lee, Stanford, Ca.; S. Nemat-Nasser, Evanston, Ill.; D. C. Drucker, Urbana, Ill.; W. Koiter, Delft; Th. Lehmann, Bochum; E. Kröner, Stuttgart; F. A. McClintock, Cambridge, Mass.; J. J. Gilman, Morristown, N.J.; F. Sidoroff, Paris.

SESSION 6A Wave Propagation Problems

Chairman: B. R. SETH, Ranchi, Bihar

P. GERMAIN, Paris, E. H. LEE, Stanford, Ca.: Plane waves in elastic-plastic media

H. FUKUOKA, Osaka: A note on the strength of the combined tension-torsion waves of elasto-plastic tubes

T. C. T. TING, Chicago, Ill.: A unified theory of elastic-plastic wave propagation of combined stress

Discussion: J. Najar, Warsaw

SESSION 6B Applied Plasticity

Chairman: S. KALISZKY, Budapest

G. MAIER, Milan: A shakedown matrix theory allowing for workhardening and second-order geometric effects

G. I. N. Rozvany, S. R. Adidam, Clayton, Victoria: Recent advances in optimal plastic design

G. Augusti, A. Baratta, Naples: Theory of probability and limit analysis of structures under multi-parameter loading

D. A. Gokhfeld, O. F. Cherniavski, Chelabinsk: Methods of solving problems in the shakedown theory of continua (presented by title)

Discussion: M. Życzkowski, Cracow; J. A. König, Warsaw; W. Szczepiński, Warsaw; S. Kaliszky, Budapest; Z. Mróz, Warsaw; N. C. Lind, Waterloo, Ont.; J. Murzewski, Cracow.

Session 7 Invited Lectures

Chairman: M. Reiner, Haifa

J. Mandel, Paris: Relations de comportement des milieux élastiques-plastiques et élastiques-viscoplastiques. Notion de repère directeur

P. Perzyna, Warsaw: Internal variable description of plasticity

Discussion: H. Lippmann, Karlsruhe; E. Kröner, Stuttgart; Z. Bychawski, Cracow; P. E. Pobedria, Moscow; A. Baltov, Sofia; F. Sidoroff, Paris; P. Germain, Paris.

Session 8A Mathematical and Physical Problems

Chairman: D. R. Owen, Pittsburgh, Penn.

J. T. Fong, J. A. Simmons, Washington, D.C.: A non-equilibrium thermodynamic theory of simple materials based on a single-integral entropic functional

V. Kafka, Prague: Theory of slow elastic-plastic deformation of polycrystalline metals with micro-stresses as latent variables

T. Mura, Evanston, Ill.: Semi-microscopic plastic distortions and disclinations.

R. de Wit, Washington, D.C.: Relation between continuous distributions and discrete disclinations

P. Villaggio, Pisa: An inequality formulation of some homogeneous thermodynamic processes

R. Takserman-Krozer, Haifa: The yield value of the Reiner fluid

F. A. Leckie, A. R. S. Ponter, Leicester: Theoretical and experimental investigation of the relationship between the plastic and creep deformation of structures

A. BERIO, L. BORTOLETTI, P. MANCA, A. PAGLIETTI, Cagliari: On the plastic behavior of time-dependent materials. Theoretical and experimental investigations.

N. FOX, Sheffield: Some problems of finite plastic deformation (presented by title)

SESSION 8B Applied Plasticity

Chairman: G. SACCHI, Pavia

N. N. MALININ, Moscow: Creep in metal processing

S. VALIAPPAN, Kensington, NSW: Elasto-plastic analysis of anisotropic work-hardening materials

G. GUDEHUS, Karlsruhe: Elastic-plastic constitutive equations for dry sand

J. BACKLUND, Gothenburg: Mixed finite element analysis of elasto-plastic plates in bending

J. BEJDA, Warsaw: Propagation of two-dimensional strong discontinuity waves in elastic-viscoplastic medium

Z. WASZCZYSZYN, Cracow: Calculation of sandwich shells of revolution of large elastic-plastic deflections

M. I. ERKHOV, Moscow: Extremum principles in the dynamics of rigid-plastic bodies and mathematical programming

T. Z. BLAZYNSKI, Leeds: Underwater explosive forming of rectangular sheets

R. N. YONG, E. McKYES, V. SILVESTRI, Montreal, Que.: Yield and failure of clays (presented by title)

B. A. GORDIENKO, Khabarovsk: Buckling of inelastic cylindrical shell under axial impact (presented by title)

Friday, September 1, 1972

SESSION 9 Invited Lectures

Chairman: S. MURAKAMI, Nagoya

H. G. HOPKINS, Manchester: Mathematical methods in plasticity

E. H. LEE, Stanford, Ca., P. GERMAIN, Paris: Elastic-plastic theory at finite strain

Discussion: L. H. N. Lee, Notre Dame, Ind.; W. J. Morales, Tampa, Fla.; S. Nemat-Nasser, Evanston, Ill.; F. A. McClintock, Cambridge, Mass.;

G. Gudehus, Karısruhe; H. F. Bueckner, Schenectady, N.Y.; D. R. Owen, Pittsburgh, Penn.; K. C. Valanis, Hoboken, N.J.

SESSION 10A Constitutive Equations

Chairman: S. NEMAT-NASSER, Evanston, Ill.

TH. LEHMANN, Bochum: On large elastic-plastic deformations

Z. MRÓZ, Warsaw: A description of workhardening of metals with application to variable loading

K. C. VALANIS, Hoboken, N.J.: Observed plastic behavior of metals vis-a-vis the endochronic theory of plasticity

Discussion: E. H. Lee, Stanford, Ca.; O. Bruhns, Bohum; K. S. Havner, Raleigh, N.C.; V. Kafka, Prague; M. Reiner, Haifa; P. Perzyna, Warsaw; F. Sidoroff, Paris; E. T. Onat, New Haven, Conn.; E. H. Lee, Stanford, Ca.; E. Kröner, Stuttgart; Y. Horie, Raleigh, N.C.; D. R. Owen, Pittsburgh, Penn.

SESSION 10B Experimental Plasticity

Chairman: A. BALTOV, Sofia

P. K. FUNG, D. J. BURNS, N. C. LIND, Waterloo, Ont.: Yield under high hydrostatic pressure

J. KLEPACZKO, Warsaw: Some experimental investigations of the elastic-plastic waves in bars

N. JONES, Cambridge, Mass.: Some remarks on the strain-rate sensitive behavior of shells

A. A. LEBEDYEV, N. V. NOVIKOV, Kiev: Deformation and fracture of structural metals under complex stress at low temperatures (presented by title)

Discussion: J. Litoński, Warsaw; H. Fukuoka, Osaka

SESSION 11 Invited Lectures

Chairman: Z. MARCINIAK, Warsaw

A. PHILLIPS, New Haven, Conn.: Experimental plasticity

N. CRISTESCU, Bucharest: On dynamic plasticity

Discussion: J. Zarka, Paris; G. Gudehus, Karlsruhe; E. T. Onat, New Haven, Conn.; T. H. Lin, Los Angeles, Ca.; J. L. Duncan, Hamilton, Ont.; H. Fukuoka, Osaka; J. Gilman, Morristown, N.J.

SESSION 12A Mathematical Methods

Chairman: H. LIPPMANN, Karlsruhe

J. SIDOROFF, Paris: The geometrical concept of intermediate configuration and the elastic-plastic finite strain

A. J. M. SPENCER, J. E. FERRIER, Nottingham: Some solutions for a class of elastic-plastic solids

H. STRIFORS, B. STORÅKERS, Stockholm: The initiation of buckling of rigid-plastic thick-walled cylindrical vessels under hydrostatic pressure

T. W. TING, Urbana, Ill.: Elasto-plastic torsion of solid bars (presented by title)

Discussion: H. Lippmann, Karlsruhe; E. H. Lee, Stanford, Ca.; R. N. Dubey, Waterloo, Ont.; M. Kleiber, Warsaw; P. Germain, Paris.

SESSION 12B Physical Foundations

Chairman: H. ZORSKI, Warsaw

H. D. BUI, A. ZAOUI, J. ZARKA, Paris: Sur le comportement élasto-plastique et viscoplastique des monocristaux et des polycristaux métalliques de structure cubique à faces centrées.

T. H. LIN, Los Angeles, Ca.: Microstress fields of slip bands and ninhomogeneity of plastic deformation of metals

K. S. HAVNER, Raleigh, N.C.: An analytical model of large deformation effects in crystalline aggregates

Discussion: M. Wnuk, Brookings, S.D.; E. Kröner, Stuttgart; H. Zorski, Warsaw; F. A. McClintock, Cambridge, Mass.; K. S. Havner, Raleigh, N.C.; T. H. Lin, Los Angeles, Ca.; J. Zarka, Paris

Saturday, September 2, 1972

SESSION 13 General Lectures

Chairman: W. GUTKOWSKI, Warsaw

PH. G. HODGE, Jr., Minneapolis, Minn.: Computer solutions of plasticity problems

D. RADENKOVIC, Paris: Constitutive laws for granular media

Discussion: O. C. Zienkiewicz, Swansea; J. Zarka, Paris; G. Maier, Milan; W. Szczepiński, Warsaw; G. Gudehus, Karslruhe; A. Drescher, Warsaw; J. Salençon, Paris; J. Graham, Kingston, Ont.; T. J. Chung, Huntsville, Ala.; F. A. Leckie, Leicester; S. Nemat-Nasser, Evanston, Ill.; E. T. Onat, New Haven, Conn.; F. A. McClintock, Cambridge, Mass.

SESSION 14A Numerical Methods

Chairman: M. ŻYCZKOWSKI, Cracow

J. H. ARGYRIS, A. S. L. CHAN, London: Static and dynamic elastic-plastic analysis by the method of finite elements in space and time

N. V. BANICHUK, F. L. CHERNOUS'KO, Moscow: A method of local variations for numerical solutions for elastic-plastic problems

J. S. GUNASEKERA, J. M. ALEXANDER, London: Matrix analysis of the large deformation of an elasto-plastic axially symmetric continuum

Discussion: E. H. Lee, Stanford, Ca.; S. Valliappan, Kensington, NSW

SESSION 14B Mechanics of Granular Media

Chairman: J. KRAVTCHENKO, Grenoble

P. STUTZ, Grenoble: Comportement élasto-plastique des milieux granulaires

T. HUECKEL, Warsaw: Some remarks on granular hardening-softening media

V. N. NIKOLAYEVSKII, Moscow: Continuum theory of plastic deformation of a granular medium (presented by title)

I. A. BEREZHNOI, D. D. IVLEV, V. B. TCHADOV, Moscow and Kuibyshev: On constructing models of cohesionless media by specifying the dissipation function (presented by title)

Discussion: G. Gudehus, Karlsruhe; D. C. Drucker, Urbana, Ill.; Z. Mróz, Warsaw; D. Radenkovic, Paris; A. Sawczuk, Warsaw

CLOSING SESSION

Chairman: A. SAWCZUK, Warsaw

J. J. GILMAN, Morristown, N.J.: The microdynamics of plastic flow

Discussion: C. Teodosiu, Bucharest; J. Kratochvíl, Prague; F. A. McClintock, Cambridge, Mass.

Session Program

Film: 'Dynamics of dislocations' commented by V. GUTTMANN, Petten, Netherlands

Free Discussion: S. Nemat-Nasser, Evanston, Ill.; K. C. Valanis, Hoboken, N.J.; F. A. McClintock, Cambridge, Mass.

Closure: A. SAWCZUK

Invited lectures

Limit analysis: the development of a concept

William Prager

Brown University, Providence, R.I., USA

The paper uses examples to trace the slow development of the concept of limit analysis in the 18th and early 19th centuries, its temporary eclipse by the emerging theory of elasticity, and its firm establishment in the 20th century. Early work is discussed in greater detail than recent developments. A possible moral relevant to current trends in solid mechanics is pointed out.

1. Introduction

Linear theory of elasticity is now used so extensively in engineering stress analysis that we are apt to forget that it is a comparatively recent addition to the analytical tools of the engineer. It is only 150 years that Cauchy presented the foundations of this theory to the French Academy of Sciences (Sept. 30, 1822) [1]. Many famous structures were therefore designed without the benefit of this theory.

The methods used before the advent of the theory of elasticity may be classified as early forms of limit analysis. The earliest example is due to Galileo. In his Discorsi [2], which were published in Leyden in 1638 and contain the first printed discussion of problems in strength of materials, he considers the failure of a cantilever beam (Fig. 1). Stated in modern terms, his assumption is that failure occurs by rigid-body rotation of the cantilever about the lower edge of the rectangular root section. According to this picture of the process of failure, all fibers would fail in tension, and the resulting moment of the uniformly distributed rupture stresses at the root section with respect to the axis of rotation would have to equal the moment of the failure load with respect to this axis.

It is hard to imagine that Galileo did not realize that his rupture stresses had a horizontal resultant that was not balanced by horizontal loads acting on the beam. We must therefore assume that, though using an equilibrium

3

Fig. 1. Galileo's estimate of strength of beam (1638).

condition to determine the failure load, he was deliberately concentrating on the kinematics rather than the statics of the failure process. In modern terminology, he was using the kinematic method of limit analysis. What he did not realize was that this method furnishes only an upper bound to the failure load and not necessarily this load itself.

2. Stability of masonry vaults and domes

Masonry vaults and domes provide further examples of early forms of limit analysis. In considering these examples, we must keep in mind the different roles structural weight plays in these masonry structures and in modern reinforced concrete shells. The dome of St. Peter's, for instance, has a diameter of 40 m and a thickness of 3 m. Compared to the weight per unit area of the median surface of this kind of massive shell, other loads such as wind pressure are very small. Moreover, the compressive stresses caused by the structural weight are only a few percent of the crushing strength of the stone. Accordingly, failure can only be due to the inability of masonry to support significant tensile stresses.

In his mechanics treatise, de LaHire [3] considers a circular voussoir arch of constant thickness. He asks what weights the voussoirs must have if the forces adjacent voussoirs transmit on each other are to be normal to the joint and to pass through its center. He solves this problem in an elegant way by extending the radial joints to their intersections *K*, *L*, *M*, *N* with the horizontal through the crown of the arch (Fig. 2). He finds that the weights of the voussoirs must be proportional to the lengths of the segments *KL*, *LM*, *MN* and that the normal pressures between adjacent voussoirs are proportional to *OL*, *OM*, *ON*. In modern terms, when the figure *OKLMN* is rotated by a right angle in the clockwise sense, it becomes the force diagram for the weights and pressures, and *ABCD* is the corresponding funicular polygon.

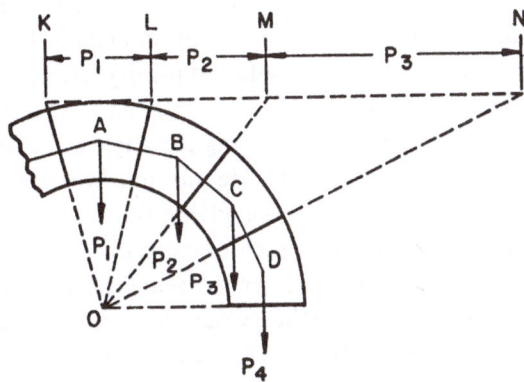

Fig. 2. De la Hire's attempt to construct statically admissible stress field for arch (1695).

De LaHire notes that the voussoir at the horizontal base of the arch would have to have an infinite weight. He concludes from this that a dry voussoir arch of constant thickness would not be stable if the voussoirs were perfectly smooth. His analysis may be viewed as an attempt to construct a statically admissible stress field. On account of the constraint of vanishing bending moment, which he imposes on this field, his failure to find a statically admissible stress field does not prove that the arch will collapse. In fact, de LaHire's conclusion becomes valid only if the ratio of the constant thickness to the radius of the semicircular arch reaches a critical lower bound.

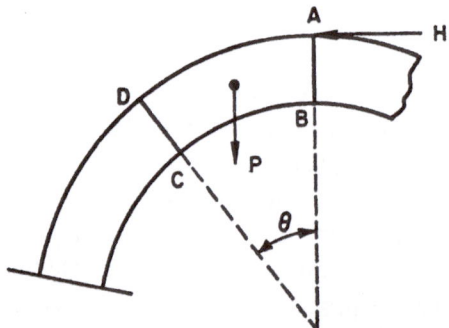

Fig. 3. Coulomb's analysis of arch (1773).

Coulomb took a closer look at ways in which arches may fail [4]. For an assumed failure joint CD forming the angle θ with the vertical (Fig. 3), he determined the greatest value H_1 that the horizontal thrust H at the crown A of the arch may have if the part $ABCD$ is not to slide up along the joint CD or to rotate about the corner D under the joint influence of its weight P and the thrust H. In this computation, he allowed for dry friction along the joint. He also found the smallest value H_2 of H if the part $ABCD$ is not to slide down along DC or rotate about C. Varying the angle θ, he then determined min H_1 and max H_2 as bounds for H. He did not, however, use his results to derive a design rule, for instance one giving the smallest allowable ratio of thickness to radius for a circular arch of constant thickness. (Using limit analysis, Heyman [5] finds this ratio for the semicircular arch to be 0.106.) From his examination of typical arches, Coulomb concludes that the slipping mode of failure may usually be disregarded so that only the hinging of neighboring parts of the arch needs to be considered.

Lamé and Clapeyron [6] discuss this type of failure in greater detail assuming that the arch breaks into four parts as shown in Fig. 4. They observe that the fracture line DF usually starts normal to the inner contour of the arch but deviates from this direction before it reaches the outer contour. While these authors use statical arguments, their results is more readily obtained by the following kinematic reasoning. Let W be the weight of the left half of the arch (Fig. 5a) and X its moment arm with respect to A. Similarly, let w be the weight of the part between the fracture line and the axis of symmetry, and x its moment arm with respect to D. Finally, let H and h be the elevations of the crown B of the arch above the horizontals through A and D. As far as the left of the arch is concerned, the failure shown in Fig. 4 may be viewed as resulting from the superposition of a small counterclockwise rotation Ω of the *entire left half* about A and a small clockwise rotation ω of the upper part with respect to the lower part about D. If the arch is to be stable, the work of the weights of the parts of the arch in the assumed mode of failure must be negative, or

$$\Omega XW - \omega xw > 0. \tag{1}$$

Now, the condition that the point B moves along the axis of symmetry is

$$\Omega H - \omega h = 0. \tag{2}$$

Elimination of the ratio Ω/ω from (1) and (2) yields the condition

$$(WX/H) - wx/h > 0, \tag{3}$$

which would assure stability if rupture could only occur along the considered line.

Actually, condition (3) must be fulfilled for any assumed rupture line. Since the first term on the left of (3) is fixed, the arch will be stable if (3) is fulfilled even if the rupture line is chosen to maximize the absolute value of the second term. To simplify the discussion, Lamé and Clapeyron restrict it to vertical rupture lines. In Fig. 5b, two neighboring positions of the rupture line are considered; the one shown in full line is supposed to yield the maximum of wx/h. To obtain the value of this expression for the dotted rupture line, we must take the resulting moment of the weight w and the weight dw of the part between the two rupture lines with respect to a point on the dotted rupture line and divide this moment by $h + dh$. Since the moment of dw is small of higher order, it may be neglected. Now, the second term on the left of (3) is stationary in the neighborhood of its maximum. Accordingly, $wx/h = w(x + dx)/(h + dh)$ or

$$dh/dx = h/x. \tag{4}$$

This relation shows that the rupture line furnishing the maximum of wx/h meets the inner contour of the arch at a point such that the tangent of this contour at this point and the horizontal through the crown B intersect on the line of action of the weight w of the part of the arc between the vertical through D and the axis of symmetry.

Lamé and Clapeyron extend their discussion to the failure of a hemispherical masonry dome of constant thickness that is supported by a cylindrical wall. Mentioning fissures observed at St. Peter's, they assume

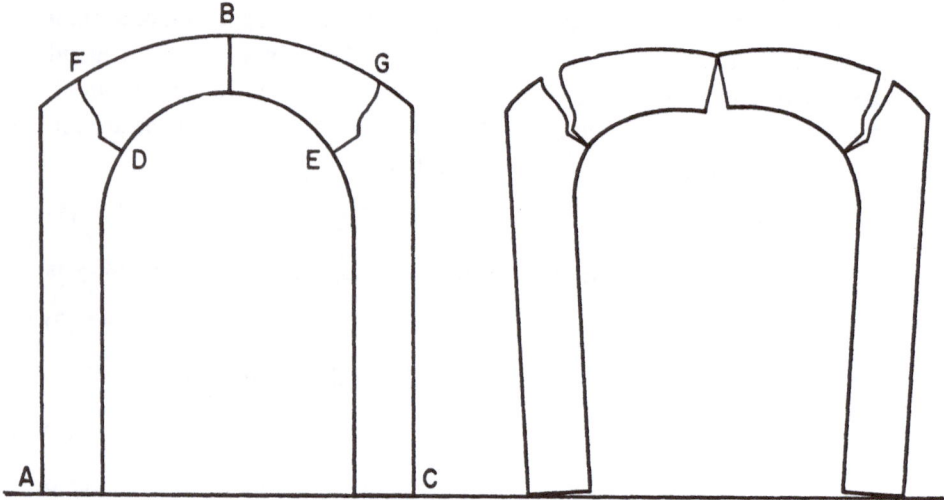

Fig. 4. Failure mode envisaged by Lamé and Clapeyron (1823).

that fissures along meridional planes will divide the structure into sectors, each of which fails in essentially the same way as the arch in Fig. 4, except that the weight per unit length of the centerline of a sector is not constant but decreases towards the crown. This means that the hemispherical dome of constant thickness will be stable for a smaller ratio of thickness to radius than the semicircular arch of constant thickness. (Heyman [7] gives the critical ratio for the dome as 0.042, which is less than half the value of the critical ratio for the semicircular arch).

It appears worthwhile to rephrase the reasoning of Lamé and Clapeyron in the terms of limit analysis. To make limit analysis applicable to masonry, we must make the following idealizing assumptions (see [5], [8]–[10]): masonry has no tensile strength but infinite compressive strength, and sliding failure cannot occur.

Any line of thrust (funicular polygon) constructed for the weights of the voussoirs will then represent a state of stress that satisfies the equilibrium conditions for the arch that carries only its own weight. This statically admissible state of stress will be safe (i.e., will satisfy the yield condition) if the line of thrust nowhere leaves the masonry. The tangency condition established by Lamé and Clapeyron for the position of the critical section (Fig. 5b) assures that the line of thrust does not leave the masonry in the neighborhood of this section. The inequality (3) then assures that this thrust line intersects the footing of the arch within the masonry. Interpreted

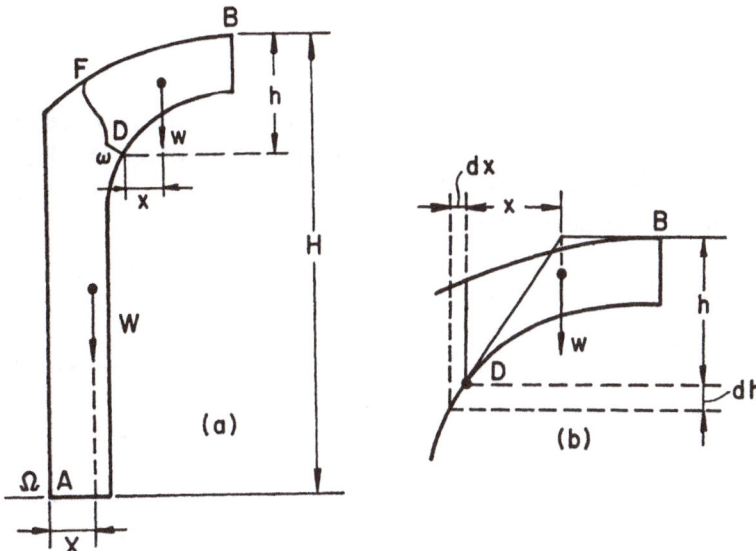

Fig. 5. Determination of position of hinge *D* according to Lamé and Clapeyron (1823).

in this manner, the analysis of Lamé and Clapeyron furnishes conditions for the existence of a safe, statically admissible stress field. If these conditions are satisfied, it follows from the static theorem of limit analysis that the arch cannot collapse even though the considered thrust line may not correspond to the actual stress resultants in the arch. When this thrust line intersects the footing at *A* (Fig. 5a), the considered statically admissible stress field is compatible with the collapse mechanism of Fig. 4. This indicates that the arch is ready to collapse under its own weight.

Heyman [5] pointed out an important consequence of the static theorem: a small settlement of the footing cannot cause the collapse of an otherwise stable arch because the thrust line that represented a safe, statically admissible stress field before the settlement will continue to represent such a field after the settlement provided that this does not significantly change the geometry of the arch.

We have specified a stress field by a thrust line. A more conventional way is to use the stress resultants: the bending moment *M*, the axial force *N*, and the shear force *S*. Only the first two are *generalized stresses* in the sense of limit analysis; the shear force is a reaction to the kinematic constraint of zero slippage between adjacent voussoirs. The generalized stresses will be treated as positive if they act as in Fig. 6a. Since tensile stresses are ruled

9

out,

$$N \geqq 0. \tag{5}$$

The stress resultants M and N are equipollent to an excentric force that has the magnitude and direction of N and an excentricity whose absolute value cannot exceed $t/2$, where t is the thickness of the arch. Accordingly, admissible combinations of M and N are restricted by (5) and

$$|M| \leqq Nt/2. \tag{6}$$

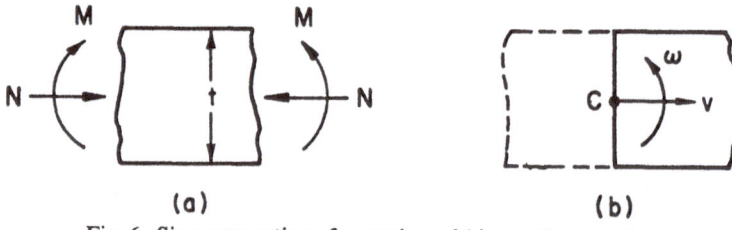

(a)   (b)

Fig. 6. Sign conventions for static and kinematic variables.

Any deformation of the arch with rigid voussoirs results from the hinging of neighboring voussoirs about a common edge (Fig. 6b). The resulting state of relative motion of the right with respect to the left voussoir may be specified by a rightward translational velocity v and an angular velocity about the center C of the left face of the right voussoir. Note that $v \geqq 0$ regardless of whether hinging occurs about the upper or the lower common edge of the two voussoirs. Because a compressive force N is treated as positive, the ‚associated strain rate' is

$$\alpha = -v = -|\omega|t/2. \tag{7}$$

Note that the bending moment M and the angular velocity ω must have the same sign:

$$M\omega \geqq 0. \tag{8}$$

The inequalities (5) and (6) show that, in a stress plane with rectangular coordinates M and $Nt/2$, admissible states of stress are represented by points of the shaded convex domain in Fig. 7. If a strain rate plane with rectangular coordinates ω and $2\alpha/t$ is superimposed on the stress plane, the relations (7) and (8) show that the strain rate vector for a given limiting state of stress is normal to the boundary of the convex admissible domain at the point that

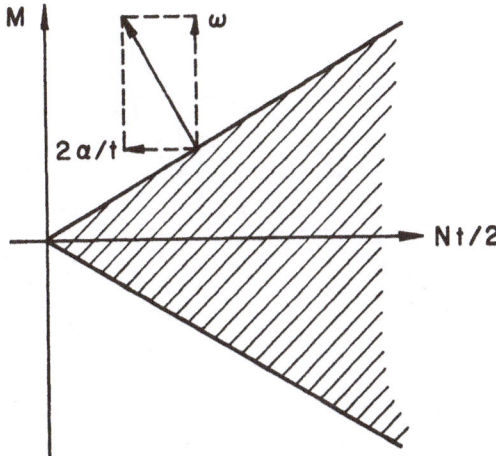

Fig. 7. Admissible domain in stress plane.

represents this state of stress. This is a special case of the *normality condition* that plays an essential role in the theory of limit analysis.

For usual values of the ratio of thickness to radius, the stresses in masonry vaults and domes are very small compared to the crushing strength of the stone. Accordingly, the problem is one of *stability* rather than strength. The analysis of Lamé and Clapeyron realistically accounts for this fact by treating the voussoirs as rigid and regarding the hinging of adjacent voussoirs as the only possible mechanism of deformation. It seems a pity that, under the influence of the developing theory of elasticity, later authors lost sight of these features of masonry construction.

Navier [11], Bresse [12], and Grashof [13] extended Euler's theory of elastic bending of bars [14] to bars with initial curvature. Their methods were increasingly applied to the design of arches and vaults, and the sound arguments of Lamé and Clapeyron were muddled by the introduction of irrelevant elastic considerations. Culmann [15] stipulated that the thrust line defined by the axial force and the bending moment of the arch should remain within the central third of the thickness. Winkler [16] suggested that, if the axis of the arch was a statically admissible thrust line, the actual thrust line of the elastic arch would practically coincide with the axis. Engesser [17] treated cylindrical vaults under loads that vary in the direction of the cylinder axis (Fig. 8). Assuming that plane material cross sections remain plane, he stipulated linear relations between the six degrees of

11

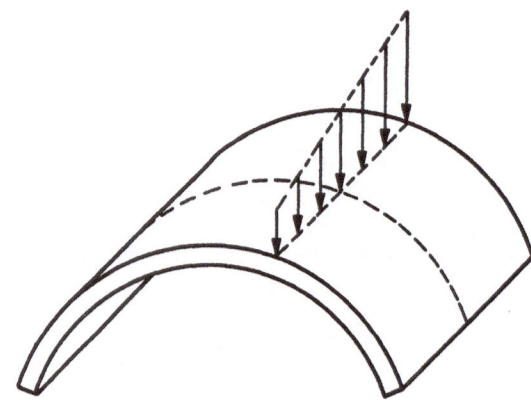

Fig. 8. Engesser's treatment of obliquely loaded vault (1909).

freedom of the cross section and the corresponding stress resultants. This arbitrary quasi-elastic treatment represents a retrograde step from the well founded analysis of Lamé and Clapeyron. Engesser's approach, however, was soon superseded by the developing theory of elastic shells.

Before leaving the subject of masonry arches, we mention Pauker's contribution of 1849. The following brief account of Pauker's work is based on a historical paper by Bernstein [18].

Pauker resumes Coulomb's approach and investigates sliding as well as hinging at some cross section CD (Fig. 3). By assuming that the thrust H across the section of symmetry acts at the upper edge A of this section, Coulomb reduced the arch to a simply redundant structure. Pauker instead considers a doubly redundant structure by replacing the thrust across the symmetry section by two equipollent horizontal forces u and v acting at A and B (Fig. 3). He establishes the inequalities that exclude hinging about C or D or sliding along CD in one or the other direction. Using u and v as rectangular coordinates in a stress plane, he remarks that, for the assumed failure section CD, the four inequalities specify a domain of admissible values of u and v, and he states that the arch must fail if the admissible domains corresponding to all possible failure sections have no points in common. If, on the other hand, there are common points, the arch will be stable. From the value pairs u, v that correspond to common points, he then chooses the one that minimizes $u + v$, that is the resulting thrust across AB.

Pauker apparently failed to recognize that the admissible domain for any assumed failure section is convex, and that the intersection of these convex domains is also convex; otherwise, his analysis could be described as an early example of convex programming.

3. Stability of retaining walls

Other early examples of limit analysis are found in soil mechanics. Here again Coulomb [4] laid the foundations for a rational analysis. For horizontal terrain and a retaining wall with a prefectly smooth, vertical rear surface (Fig. 9a), he investigates the force P that a unit length of the retaining wall would have to exert on the soil to prevent sliding along a line BC that forms the angle α with the horizontal. The equilibrium conditions for the horizontal force P, the weight W of the prism with the cross section ABC, and the reaction Q along BC furnish

$$P = W \tan (\alpha - \phi). \tag{9}$$

where ϕ is the angle of friction. Substituting $W = (\rho g h^2 \cot \alpha)/2$, where ρ is the density of the soil, and determining the value of α that furnishes the greatest P, he finds

$$\alpha = \frac{\pi}{4} + \frac{\phi}{2}, \quad B_{max} = \tfrac{1}{2}\rho g h^2 \, \frac{1 - \sin \phi}{1 + \sin \phi}. \tag{10}$$

Coulomb's use of the maximum of P is obviously motivated by the thought that the wall will be stable if it can withstand this greatest force.

Beginning with Poncelet [19], numerous authors have expressed Coulomb's results in geometric terms. We mention only Rebhann's theorem [20], which

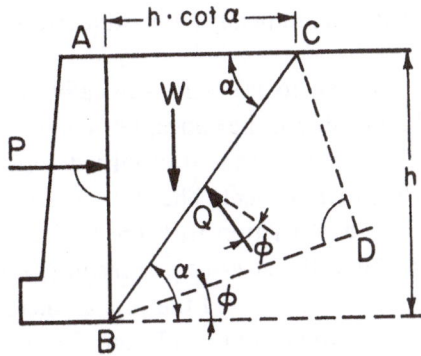

Fig. 9a. Coulomb's analysis of retaining wall (1773).

also applies to a rough wall and sloping terrain (Fig. 9b). If the line *BD* is drawn through the foot of the wall at the friction angle ϕ against the horizontal, and if the line *DC* forms the same angle β with *BD* as the pressure *P* forms with the vertical, the triangles *BAC* and *BDC* have the same area when the angle α between *BC* and the horizontal is chosen to maximize *P*. (In the case of Fig. 9b, we have $\beta = \pi/2$, and the triangle *BDC* is obtained from *BAC* by reflection on *BC*. Accordingly, $\alpha - \phi = (\pi/2) - \alpha$, which yields Coulomb's value of α.)

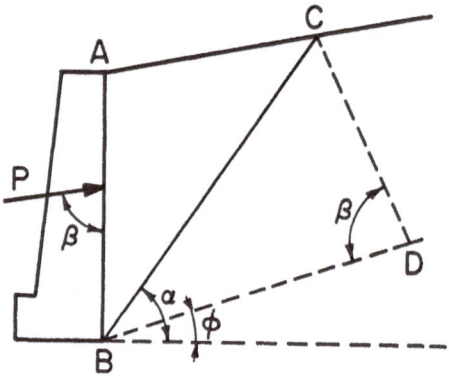

Fig. 9b. Rebhann's theorem (1871).

From the point of view of limit analysis, Coulomb's approach must be classified as an incomplete static approach. He only uses global equilibrium conditions in conjunction with the yield condition, which specifies the direction of the force *Q* transmitted across *BC* (Fig. 9a). He then treats his value of the pressure *P* for an assumed angle α as if it were a lower bound derived from a complete safe, statically admissible stress field, and maximizes *P* with respect to α.

Rankine [21] was the first to use safe, statically admissible stress fields in soil mechanics. He introduced the concept of conjugate line elements at a point in the plane of stress: the stress transmitted across each of these elements is parallel to the other element. The stress p' transmitted to a vertical wall makes the angle of wall friction with the horizontal (Fig. 10a). If the terrain has the same slope, the stresses p transmitted across line elements parallel to the terrain are vertical, and the rate of increase of p with depth follows from the vertical equilibrium condition. For a smooth vertical wall and horizontal terrain (Fig. 10b), p' and p are the principal stresses σ_x and σ_y

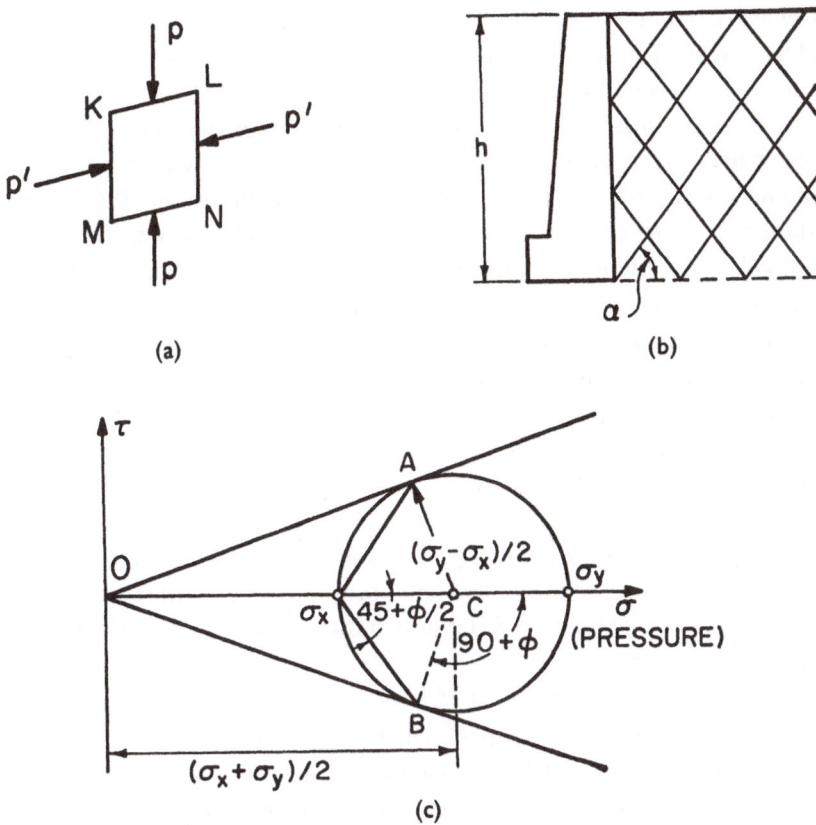

Fig. 10. Rankine's analysis of retaining wall (1857).
  (a) conjugate line elements
  (b) slip line field
  (c) Mohr circle for limit state of stress.

in the horizontal and vertical directions, and $\sigma_y = \rho g y$, where y is the depth below the surface of the ground. The yield condition then shows that the greatest possible angle between line element and stress vector, i.e. the angle ϕ of internal friction, is reached for the line elements that form the angles $\pm \alpha$ with the horizontal, where α is given by (10). This is readily seen from Fig. 10c, which uses a concept not known to Rankine, the representation of the yield condition by the envelope AOB of the stress circles for limiting states of stress (Mohr, [22]). Figure 10b shows the slip line field, which yields

$$\sigma_x = \rho g y \, \frac{1 - \sin \phi}{1 + \sin \phi} \tag{11}$$

15

and hence Coulomb's value (10) for the resultant pressure P on the wall. Rankine invoked a special principle in asserting that this value was the actual earth pressure. This principle was stipulated by Moseley [23] in the following vague form: 'If there be a system of pressures in equilibrium, among which are a given number of resistances, then is each of these a minimum subject to the conditions imposed by the equilibrium of the whole'. Rankine's version is: 'The forces which balance each other in or upon a given body or structure being distinguished into two systems, called respectively *active* and *passive*, which stand to each other in the relation of cause and effect, then will the passive forces be the least which are capable of balancing the active forces, consistently with the physical condition of the body or structure'. Applying this principle, he continues as follows: 'In a mass of earth, the active forces are the vertical pressures produced by the gravitation of its parts, the passive forces are the pressures conjugate to these vertical pressures by which the earth is prevented from spreading. The pressures conjugate to the vertical pressures will therefore be the least which are consistent with the conditions of equilibrium ... and with the conditions of stability at each point ...'. (The dots indicate references to earlier parts of the paper.)

Whereas any similarity between Moseley's principle and the fundamental theorems of limit analysis is superficial, Kötter [24] to some extent anticipated the static theorem of limit analysis. Discussing the pressure of sand on the bottom of a vertical cylindrical container, he divides the conditions to which the stress field in the sand is subject into purely statical conditions and physical conditions representing the investigator's concept of the nature of the material. As examples of conditions of the second kind, he mentions that the stress transmitted across a surface element should act along the normal to this element in an ideal fluid and not deviate from this normal by more than the angle of friction in a granular material. He then defines an admissible stress field as one that satisfies both statical and physical conditions. Visualizing the bottom of a container as separated from the cylindrical well and subject to a given upward vertical force, he asks whether the bottom will remain at rest or begin to move up or down. He states that the bottom will remain at rest as long as there exists an admissible stress field. This is the static approach of limit analysis supported by a postulate rather than a theorem.

Rankine's solution in Fig. 10b has straight slip lines (i.e. lines for which the stress transmitted across each line element forms the maximum allowable angle with the normal to the element). Lévy [25] pointed out that the slip lines, which he calls sometimes 'lignes de glissement' and sometimes 'lignes de rupture', will in general be curved. Kötter [26] derived the equations that link the increments dp and dθ of the mean pressure p and the inclination θ of a slip line as one proceeds along this line. By far the most complete discussion of slip line fields in soils prior to the development of general slip line theory is due to Massau [27].

Massau establishes the characteristic equations along slip lines when the weight of the soil is negligible (p. 151) and when it must be taken into account (p. 154). He also investigates the change of the radii of curvature of the slip lines of one family as one progresses along a slip line of the other family (p. 162). He discusses relations along an envelope of slip lines (p. 163) using the case of a completely rough retaining wall as example (p. 166). He treats lines of stress discontinuity (p. 267), and extends his results to soils with cohesion (p. 309). Since a soil with cohesion and zero angle of friction is a perfectly plastic material, Massau's results include those of Hencky [28] and Prandtl [29] as special cases.

Massau discusses many special geometries of terrain and retaining wall. From the point of view of limit analysis, his solutions are *incomplete* safe, statically admissible stress fields, because he does not investigate whether these fields can be suitably continued into the rigid regions.

It is interesting to compare the early development of concepts of limit analysis in the two fields considered above. For an arch, the static and kinematic approaches are linked by the concept of the thrust line. A thrust line for the given loads that remains within the thickness of the arch represents a safe, statically admissible stress field, and a thrust line that does not leave the thickness of the arch but is alternatingly tangent to the inner and outer contours of the arch at a sufficient number of points (5 for symmetric collapse) indicates an unsafe, kinematically admissible velocity field. In soil mechanics, there is no obvious connection of this kind between the static and kinematic approaches. The concept of the plastic potential, which provides this connection, was introduced by v. Mises [30], who also pointed out that it implied maximum local plastic dissipation. Melan [31] and Prager [32] recognized its importance in the discussion of uniqueness, Koiter [33] extended it so singular yield conditions, of which the Coulomb condition is an example, and Drucker and Prager [34], [35] applied it to soils. Shield [36] gave a useful interpretation of the kinematic consequences of the theory of the plastic potential for a Coulomb material in plane strain: because the

17

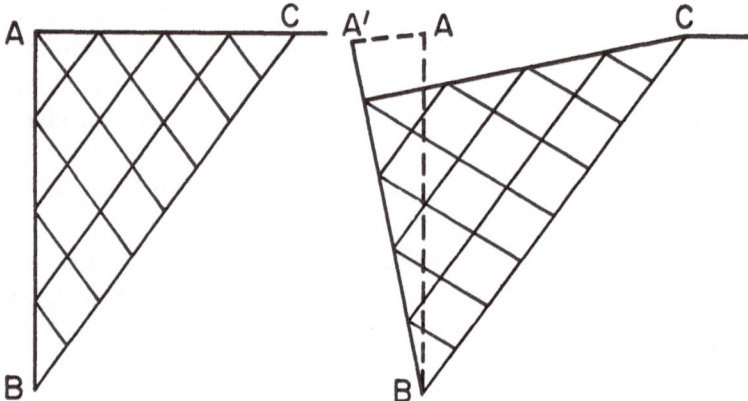

Fig. 11. Deformation of slip net according to Shield (1953).

rate of extension along slip lines is zero, the slip line net deforms as if it consisted of rigid elements hinged together at the nodes. Figure 11 shows this kind of deformation for the slip line net of Fig. 10b if the retaining wall tilts about its lower edge. The deformation is associated with a volume increase. It can, in fact, be shown [35] that any planar deformation of a Coulomb material that obeys the theory of the plastic potential is accompanied by a mean rate of extension $(\dot{\varepsilon}_1 + \dot{\varepsilon}_2)/2$ that equals the product of the maximum shear rate $|\dot{\varepsilon}_1 - \dot{\varepsilon}_2|/2$ by the sine of the angle of friction. Since this amount of dilatation exceeds that observed with granular soils, modifications of Coulomb's yield condition have been proposed (see, for instance, [37], [38]).

4. Load-carrying capacity of steel structures

While methods used in early work on arches and retaining walls may be retrospectively classified as limit analysis, the formal development of this branch of structural theory started with the attempt to predict the collapse loads of steel structures. In 1914, Kazinczy [39], a young engineer in the employ of the city of Budapest, concluded from his experiments on clamped steel beams that the theory of elasticity could not provide a realistic estimate of the load-carrying capacity. He pointed out that a section in which the yield stress is reached acts as a hinge as far as further load increases are concerned, and that at least three hinges of this kind must form in a clamped beam before catastrophic deflections can develop.

Kazinczy's paper unfortunately failed to receive the attention it merited, partly because it was written in a not widely read language, and partly because the war closed normal channels for the international exchange of technical information. A few years later, Kist [40] independently presented similar ideas in his inaugural lecture at the Technical University of Delft.

The habit of relying on elastic analysis, however, proved hard to overcome, and it took numerous investigations spread over the next 30 years before the new point of view was generally accepted. Maier-Leibnitz [41], for example, states that designers of steel structures had been hesitant to use continuous beams ever since Mohr warned of the influence of support settlements. He stresses the fact that small settlements do not change the load-carrying capacity of a continuous beam of mild steel. References [41] through [50] list representative papers from this 30-year period. The discussion originally centered on beams [41]–[47], where the locations of the yield hinges were intuitively clear. When frames [48], [49] and plates [50] began to be considered, the need for a more systematic approach became obvious. This was developed by Gvozdev [51]. As Kazinczy's paper in 1914, Gvozdev's paper in 1938 was published on the eve of a world war and did not attract attention until more than a decade later, when its results had been independently established by Greenberg, Prager, and Drucker [52], [53]. Because the Proceedings of the conference on plastic deformation at which Gvozdev presented his results are not readily accessible, it appears worthwhile to review the main part of this paper here.

Gvozdev states that the determination of the load-carrying capacity of a structure, i.e. the determination of the common factor that must be applied to given safe loads before these will cause failure, is one of the most important tasks of structural analysis. He points out that the theory of elasticity can furnish this factor only when the structural material is brittle, but not when it is capable of yielding under constant stress. For a structure made of the latter kind of material, he defines failure as an increase of the deformation of the structure that is exclusively due to the yielding of its elements.

Instead of deriving the plastic behavior of a structural element from assumptions regarding the behavior of its material, Gvozdev proposes directly to define the behavior of the element. To this end, he introduces generalized stresses and plastic strains, s_i and e_i ($i = 1, 2, ..., n$), defined in such a manner that $\sum_i e_i s_i$ is the energy dissipated in a plastic deformation e_i that occurs under the constant stress s_i. He then assumes that the plastic deformation e_i uniquely determines the stress s_i. Accordingly, the dissipated energy may be written as a function of the plastic strains:

$$\sum_i e_i s_i = f(e_i). \tag{12}$$

Regarding the quantities e_i ($i = 1, 2, ..., n$), as the components of a strain vector with respect to rectangular axes in a Euclidean n-space, and the quantities s_i as the coordinates of a stress point with respect to the same axes, Gvozdev remarks that, for a fixed strain vector, the stress points satisfying (12) fill a hyperplane that is normal to the strain vector. He then *tacitly* assumes that stress points outside the half-space bounded by this hyperplane and containing the origin represent states of stress that cannot be attained in the considered structural element, and that the only attainable stress points are the points of the convex intersection of the half-spaces corresponding to all possible strain vectors. The principle of maximum local plastic dissipation [30], [54], [55] follows from this tacit assumption. Gvozdev then uses this principle in the now customary manner to prove the fundamental theorems of limit analysis.

Once the fundamental principles were clearly understood, the field of application of limit analysis expanded rapidly within a few years. As a discussion of the numerous contributions from this period, approximately from 1950 to 1958, would transcend the scope of this paper (for a partial list, see [56]), we mention only the application of the basic theorems of limit analysis to the minimum-weight design of structures for a given load factor at plastic collapse [57], the development of the theory of ideal looking materials as a dual to the theory of limit analysis [58], and, in view of the important role of the normality of the strain rate vector to the yield locus and the convexity of this locus, the attempts to derive these properties from first principles [59]. (A recent survey of these attempts is found in [60].)

5. Concluding remark

In conclusion, it appears worthwhile to stress the fact that, by the middle of the last century, the time was ripe for the development of the theory of limit analysis. It is extremely likely that this development would have taken place but for the strong preoccupation of workers in solid mechanics with the emerging theory of elasticity. If the development had occurred then, the long period could have been avoided, in which the ingrained habit of thinking exclusively in terms of the elastic behavior of structures was gradually discarded.

There is perhaps a moral for the present day in this. We are experiencing an extremely rapid, not to say hectic, development of new theories in solid mechanics. While one or the other of these may become as important to engineering design as the theory of elasticity, we should be careful not to abandon old, and perhaps incompletely developed, theories that may yet have to play a vital role.

Acknowledgment

The author gratefully acknowledges the most valuable assistance of Dr. Jan A. König of Warsaw in the evaluation of the work of Pauker and Gvozdev, and of Mrs. A. Królikowska in obtaining copies of Polish and Russian papers on the subject.

References

[1] Cauchy, A. L., *Exercices de mathématiques*, vol. 2, 1827, pp. 1, 42, 60, 108; vol. 3, 1828, pp. 160, 188, 213, 237.
[2] Galilei, G., Discorsi e dimostrazioni matematiche intorno à due nuove scienze, Leyden, 1638, p. 157.
[3] de LaHire, P., Traité de Mécanique, Paris, 1695; see also: Sur la construction des voutes dans les édifices, Histoire de l'Académie Royala des Sciences, Année 1712, p. 70.
[4] Coulomb, C. A., Essai sur une application des règles de maximis et minimis à quelques problèmes de statique relatifs à l'architecture, Mémoires de math. et de phys. prés. à l'Ac. Royale des Sciences par divers savants, Année 1773, p. 343.
[5] Heyman, J., The safety of masonry arches, *Int. J. Mech. Scis. 11* (1969) p. 363.
[6] Lamé, G. and E. Clapeyron, Sur la stabilité des voutes, *Annales des Mines, 8* (1823) 789.
[7] Heyman, J., On shell solutions for masonry domes, *Int. J. Mech. Scis. 3* (1967) p. 227.
[8] Kooharian, A., Limit analysis of voussoir (segmental) and concrete arches, *J. Amer. Concrete Inst. 24* (1952) 317.
[9] Prager, W., Théorie générale des états limites d'équilibre, *J. Math Pures et Appl. 34* (1955) 395.
[10[Heyman, J., The stone skeleton, *Int. J. Solids Structs. 2* (1966) 249.
[11] Navier, L. M. H., Leçons de Mécanique à l'Ecole des Ponts et Chaussées, Paris, 1826.
[12] Bresse, J. A. C., Recherches Analytiques sur la Flexion et la Résistance des Pièces Courbes, Paris, 1854.

[13] Grashof, F., Theorie der Elasticität und Festigkeit, Berlin, 1866.

[14] Euler, L., Methodus inveniendi lineas curvas maximi minimave proprietates gaudentes, Lausanne, 1744.

[15] Culmann, K., Graphische Statik, Zürich, 1866.

[16] Winkler, E., Die Lage der Stützlinie im Gewölbe, *Deutsche Bauzeitung 13* (1879) 117.

[17] Engesser, F., Das elastische Tonnengewölbe als räumliches System betrachtet, *Zeitschrift f. Bauwesen* (1909 107.

[18] Bernstein, S. A., Collected Papers, Moscow, 1961, p. 368.

[19] Poncelet, J. V., Mémoire sur la stabilité des revêtements et de leurs fondations, *Mémoires de l'Officier du Génie 13* (1840) 7.

[20] Rebhann, G., Theorie des Erddrucks und der Futtermauern mit besonderer Rücksicht auf das Bauwesen, Vienna, 1871.

[21] Rankine, W. J. M., On the stability of loose earth, *Phil. Trans. Roy. Soc. 147* (1859) 9.

[22] Mohr, O., Welche Umstände bedingen die Elastizitätsgrenze und den Bruch eines Materials, *ZS. d. Vereins Deutscher Ingenieure 44* (1900) pp. 1524 and 1572.

[23] Moseley, H., The Mechanical Principles of Engineering and Architecture, London, 1843, Part IV.

[24] Kötter, F., Der Bodendruck von Sand in verticalen, cylindrischen Gefässen, *J. reine u. angew. Math. 120* (1899) 189.

[25] Lévy, M., Sur une théorie rationnelle de l'equilibre des terres fraîchement remuées et ses applications au calcul des murs de soutenement, *J. Math. Pures et Appl. (2) 18* (1873) 241.

[26] Kötter, F., Die Entwickelung der Lehre vom Erddruck, *Jahresbericht d. Deutschen Mathematiker-Vereinigung 2* (1891–92) 78.

[27] Massau, J., Mémoire sur l'intégration des équations aux dérivées partielles, Ghent, 1899; reprinted as Edition du Centenaire, Mons, 1952. (The page references in the text are to the latter edition.)

[28] Hencky, H., Über einige statisch bestimmte Fälle des Gleichgewichts in plastischen Körpern, *ZS angew. Math. und Mech. 3* (1923) 241.

[29] Prandtl, L., Anwendungsbeispiele zu einem Henckyschen Satz über das plastische Gleichgewicht, *ZS angew. Math. und Mech. 3* (1923) 401.

[30] Mises, R. v., Mechanik der plastischen Formänderung von Kristallen, *ZS angew. Math. und Mech. 8* (1928) 161.

[31] Melan, E., Zur Plastizität des räumlichen Kontinuums, *Ing.-Arch. 9* (1938) 116.

[32] Prager, W., Recent developments in the mathematical theory of plasticity, *J. Appl. Phys. 20* (1949) p. 235; see also R. Hill, On the problem of uniqueness in the theory of a rigid-plastic solid, *J. Mech. and Phys. Solids 4* (1956) p. 247, and *5* (1957) pp. 1, 153, 302; J. F. W. Bishop, A. P. Green, and R. Hill, A note on the deformable region in a rigid-plastic body, *J. Mech. and Phys. Solids 4* (1956) p. 256; R. M. Haythornthwaite and R. T. Shield, A note on the deformable region in a rigid-plastic structure, *J. Mech. and Phys. Solids 6* (1958) 127.

[33] Koiter, W. T., Stress-strain relations, uniqueness and variational theorems for elastic-plastic materials with a singular yield surface, *Quart. Appl. Math. 11* (1953) 350.

[34] Drucker, D. C. and W. Prager, Soil mechanics and plastic analysis or limit design, *Quart. Appl. Math. 10* (1952) 157.

[35] Prager, W., On the kinematics of soils, Mémoires des Sciences, *Ac. Roy. Belgique, 28* (1954) fasc. 6, p. 3.

[36] Shield R. T., Mixed boundary value problems in soil mechanics, *Quart. Appl. Math. 11* (1953) 61.

[37] Ziegler, H., Zum plastischen Potential in der Bodenmechanik, *J. Appl. Math. und Phys. 20* (1969) 659.

[38] DiMaggio, F. L. and I. S. Sandler, Material model for granular soil, *Proc. ASCE 97* (1971) EM3, p. 935.

[39] Kazinczy, G., Experiments with clamped beams (in Hungarian), *Betonszemele 2* (1914), pp. 68, 83 and 101. For an English translation of part of this paper, see N. J. Hoff's contribution to the discussion of B. G. Johnston, C. H. Wang, and L. S. Beedle, An evaluation of plastic analysis as applied to structural design, *The Welding J. 33* (1954) p. 14-s. This contribution also lists further publications by Kazinczy on limit analysis.

[40] Kist, N. C., Does a strength analysis based on the proportionality of force and deformation lead to a good design of steel bridges and buildings? (in Dutch), *De Ingenieur 4* (1917) p. 743; see also N. C. Kist, Die Zähigkeit des Materials als Grundlage für die Berechnung von Brücken, Hochbauten und ähnlichen Konstruktionen aus Flusseisen, *Der Eisenbau 11* (1920) 425.

[41] Maier-Leibnitz, H., Beitrag zur Frage der tatsächlichen Tragfähigkeit einfacher und durchlaufender Balkenträger aus St37 und aus Holz, *Bautechnik 6* (1928) 11 und 27; Versuchte mit eingespannten und einfachen Balken von I-Form aus St37, *Bautechnik 7* (1929) 313 und 366.

[42] Fritsche, J., Tragfähigkeit von Balken aus Stahl bei Berücksichtiging des plastischen Verformungsvermögens, *Bauingenieur 4* (1930).

[43] Patton, E. and B. Gorbunov, Carrying capacity of welded beams under plastic deformation (in Russian).

[44] Van den Broek, J. A., Theory of limit design, *Trans. ASCE 105* (1940) 638.

[45] Luxion, W. W. and B. G. Johnston, Plastic behaviour of wide flange beams, *The Welding J. 27* (1948) 538-s.

[46] Prager, W., Über das Verhalten statisch unbestimmter Konstruktionen aus Stahl nach Überschreitung der Elastizitätsgrenze, *Bauingenieur 14* (1933) 65.

[47] Hrennikoff, A., Theory of inelastic bending with reference to limit design, *Trans. ASCE 113* (1948) 213.

[48] Girkmann, K., Bemessung von Rahmentragwerken unter Zugrundelegung eines ideal plastischen Stahls, *Sitz. ber. Ak. Wiss. Wien (II a) 140* (1931) 679.

[49] Baker, J. F., The rational design of steel building frames, *J. Inst. Civ. Engrs. 3* (1935–36) 127; A review of recent investigations into the behavior of steel frames in the plastic range, *J. Inst. Civ. Engrs. 31* (1949) 188; The design of steel frames, *The Structl. Engr. 27* (1949) 397.

[50] Johansen, K. W., Brudlinieteorier, Copenhagen, 1943; an English translation (Yield-line theory) was published by the Cement and Concrete Assoc., London, in 1962.

[51] Gvozdev, A. A., The determination of the value of the collapse load of statically indeterminate systems undergoing plastic deformation (in Russian), Proc. Conf. Plastic Deformation, Akad. Nauk USSR, 1938, p. 19; see also Calculation of the load-carrying capacity of structures by the method of limit equilibrium, Moscow, 1949, Chap. 6.

[52] Greenberg, H. J. and W. Prager, Limit design of beams and frames, Brown University Technical Report A18–1, Providence, R. I., 1949 (printed with discussion in *Trans. ASCE 117* (1952) 447).

[53] Drucker, D. C., H. J. Greenberg, and W. Prager, The safety factor of an elastic-plastic body in plane stress, *J. Appl. Mech. 18* (1951) 371; D. C. Drucker, W. Prager, and H. J. Greenberg, Extended limit design theorems for continuous media, *Quart. Appl. Math. 9* (1952) 381; W. Prager, General theory of limit design, Invited sectional lecture, 8th Internatl. Congr. Appl. Mech., Istanbul, 1952, (Proceedings., vol. 2, Istanbul, 1956, p. 65).

[54] Taylor, G. I., A connection between the criterion of yield and the stress-strain relationship in plastic solids, Proc. Roy. Soc. (A) 191 (1947) 441.

[55] Hill, R., A variational principle of maximum plastic work in classical plasticity, *Quart. J. Mech. and Appl. Math. 1* (1948) 18.

[56] Prager, W., An introduction to plasticity, Reading, Mass., 1959, references at end of Chap. 3.

[57] Drucker, D. C. and R. T. Shield, Design for minimum weight, Proc. 9th Internatl. Congr. Appl. Mech., Brussels, 1956, vol. 5, p. 212.

[58] Prager, W., On ideal locking materials, *Trans. Soc. Rheology 1* (1957) 169.

[59] Drucker, D. C., A more fundamental approach to plastic stress-strain relations, Proc. 1st U.S. Nat. Congr. Appl. Mech., ASME 1951, p. 487.

[60] Ziegler, H., Thermomechanics, *Quart. Appl. Math. 30* (1972) 91.

On plastic analysis of the microstructure
of metallic alloys

D. C. Drucker

University of Illinois, Urbana, Ill. USA

Plasticity theory, along with the other subdisciplines of mechanics, permits the rational analysis and design of devices, machines, and structures. It is a time-independent idealization which takes into account the salient features of ductile behavior (for states of stress and strain homogeneous on the macroscale) and then predicts reasonably well the local and global ductile behavior of actual structures. Equal success of the unmodified continuum theory in application to the analysis and design of the microstructure of ductile metallic alloys is not to be expected when atomic scale effects intervene. Yet much has been learned and far more can be learned from the simple continuum approach, as is illustrated by a summary of earlier results on sintered tungsten carbide-cobalt, aluminum-silicon, and carbon steel in pearlitic and spheroidized form. Several unifying concepts are merged to form some tentative conclusions and basic questions about the design of suitable microstructures for desired flow and fracture properties.

1. Introduction

The design and the analysis of the microstructure of metallic alloys to obtain desired flow and fracture properties requires a much clearer understanding of the relation between the desired properties and the structure than is now available. Macroscopic properties are governed by details observable with the unaided eye, with an optical microscope, and with an electron microscope, as well as by atomic structure and substructure. Different properties, mechanical and other, are controlled by different scale ranges from the subatomic to meters. Some of the scales of importance in fracture and in flow must be related because flow precedes fracture in a metal. However, the dominant or governing scale for each is likely to be very different.

This paper is concerned mainly with the question of flow stress or yield strength and emphasizes the probable lack of direct significance for many

25

alloys of the structural details observable in the ordinary optical microscope. A more recent comparison of pearlitic and spheroidized plain carbon steel by Butler [1] is added to an earlier analysis of an aluminum-silicon system with a small volume fraction of inclusions and of sintered WC–Co with moderate to very large volume fractions of tungsten carbide [2]. The steels were selected, not for their practical importance alone but because their ferrite matrix is bcc as contrasted with fcc for aluminum-silicon. Also, the platelike and spheroidal cementite shapes are dramatically different. Furthermore, the pearlite inclusions occupy an appreciable to a very large volume fraction of a plain carbon structural steel while, on the contrary, the cementite in a spheroidized steel occupies a rather small volume fraction, even at high carbon content.

In the sections which follow, the essential features and results of a straightforward and quite elementary continuum approach are outlined. Modifications are described which take a few aspects of atomistic behavior into account very crudely yet permit decisions to be made on the mechanisms which govern yield and flow strengths. The continuum analyses of the several microstructures are compared with experiments to demonstrate rather conclusively that the structural features which dominate the picture in the optical microscope are mainly spectators rather than governing participants in the determination of yield and flow stress. The largest volume fraction of pearlite (100 %), for example, produced less than a 25 % increase in flow stress over the value for a spheroidized steel of the same composition and grain size in the experiments of Rinebolt [3] and Rinebolt and Harris [4].

However, once the flow properties of the ductile matrix are determined it is likely that structure visible in the optical microscope will govern the remaining and possibly most important fracture properties of the structural alloys in common use.

2. The role of continuum mechanics at the microstructural level

As a first approximation, scale can be ignored entirely, the pictures obtained in the microscope looked at as macro-elements of elastic, plastic, and/or viscous material. The detailed mechanics of such inhomogeneous assemblages of material are messy but well-understood in principle until fracture intervenes. Broad, general, and useful principles are available for the calculation of the overall response of the composite which can do much better than the 'rule' of mixtures. These include, among many others [5], the minimum potential energy and minimum complementary energy theorems of linear and non-linear elasticity [6], the limit theorems of perfect plasticity [7], and the bounding theorems for linear or nonlinear creep or viscous response [8]. Much more can be done if desired. Stress and strain distributions in all the components can be computed with reasonable accuracy, given sufficient patience and time on a large computer.

It is rare, at present, in the analysis or design of ordinary structures or component parts to require the precision of local calculation which the theory is able to provide. Drastic idealizations are made. This is not a handicap. Quite the contrary, the analyst and even more the designer are served best by the strongest of idealizations and simplifications which still contain the essence of behavior for the problem to be solved.

The application of mechanics to the microscale for the determination of flow and fracture properties is at a primitive stage and calls for the most transparent of procedures and answers. Consequently, the general (averaging) theorems are the most useful for elastic, plastic, or viscous response to the extent that continuum mechanics is relevant on the microscale. Quasi-static elastic responses (elastic moduli of the composite assemblage) then are given by the energy theorems. Plastic responses, the main subject of this paper, then can be determined adequately by the plastic limit theorems of perfect plasticity. Work-hardening may be included closely enough by choosing σ_0, the yield stress for the perfectly plastic idealization, or the equivalent stress for multiaxial states of stress, in accord with the strain levels reached [2].

Prime results of interest here are the gross responses of models of alloys containing one or more of a wide variety of shapes and volume fractions of inclusions in an isotropic ductile matrix. The stress-strain characteristics have been reported previously and are summarized in Fig. 1 and Table I for fully bonded inclusions. Plastic deformation takes place only in the ductile

Table I

Plastic constraint factor in three dimensions (see Fig. 1b)

f	0.9	0.8	0.7	0.6	0.5	0.4	0.3	0.0
σ_0^c/σ_0	5.9	3.1	2.1	1.7	1.4	1.2	1.05	1.00
a/h	16.2	7.4	4.6	3.1	2.2	1.6	1.2	0

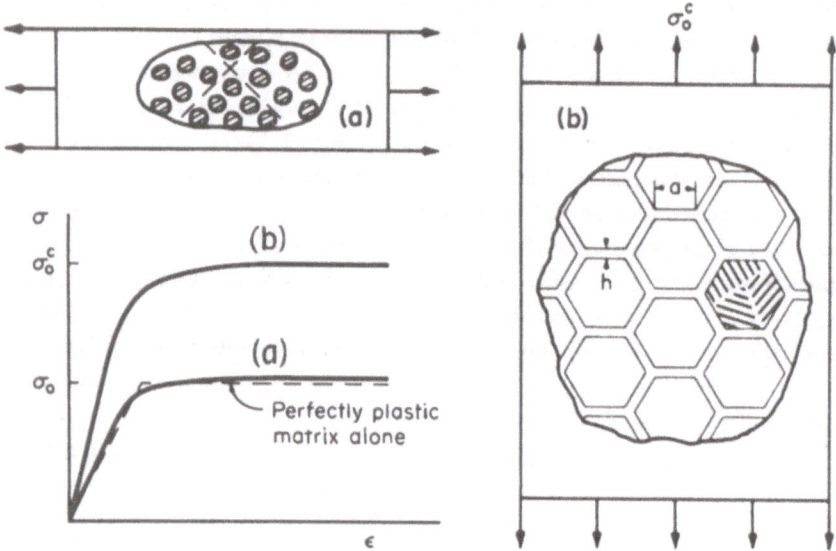

Fig. 1. Plastic constraint requires large volume fraction of rigid (strong) bonded inclusions [1, 2].

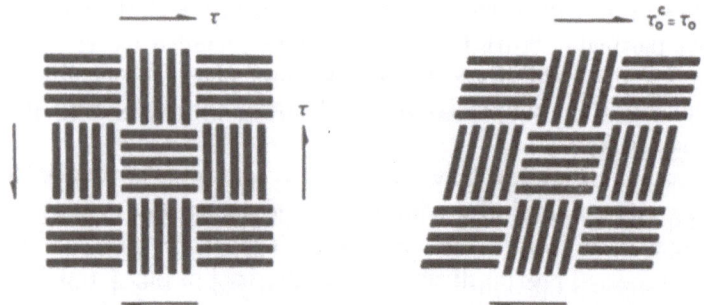

Fig. 2. No need for clear slip path (yield strength in shear of the composite oriented as shown is the yield strength in shear of the perfectly plastic matrix) [2].

matrix when the inclusions are strong enough to behave as elastic or rigid.

The striking features are that inclusion shape is relatively unimportant, and that until volume fractions, f, exceed $\frac{1}{3}$, the plastic constraint factor (ratio of yield strength σ_0^c of composite to yield strength σ_0 of matrix) introduced by any set of inclusions is little or negligibly greater than unity (no effect at all). As illustrated in Fig. 2, the absence of a clear slip path does not necessarily cause any constraint. Any initial or subsequent anisotropy in the plastic range could be taken into account properly. However, except in extreme cases, the main conclusions would be unaffected.

Interpretation on the microscale of these unambiguous and simple results of macroscopic continuum plasticity theory is direct for inclusion dimensions and interparticle spacings so large that atomic dimensions and dislocation considerations are washed out. In ductile metals, features discernible with an ordinary magnifying glass (100 microns or more) would seem to be in this category. Dimensions made visible by the optical microscope (micron or more) fall into a more problematic range. Results previously reported [1, 2] show that they too may be treated by ordinary continuum plasticity to a good first approximation. Continuum *elasticity* is appropriate at this scale, and well below, because near neighbor atomic interactions govern and a micron covers 3000 atomic spacings. However, as the dimensions of interest get smaller, it is necessary to modify the continuum plasticity approach to include some aspects of the atomic and dislocation scale. Some earlier work along these lines [9] is extended and sharpened in the next section.

3. Continuum approach including a few atomic and dislocation concepts

The modified continuum analysis to be described here follows the approach outlined by Wu and Drucker [9]. Physical information on metals and alloys at the atomic and dislocation scale is averaged qualitatively to provide a basis for the continuum analysis. Continuum concepts in turn are employed to help distinguish among competing local atomic and dislocation explanations. The results then are fed back to the continuum model in an iterative process. A bare beginning only has been made; much more remains to be done on this extremely complex and somewhat controversial procedure. The iterative process starts from the well-known fact that single crystals of pure ductile metals deform extensively in simple shear at very low stress levels. Typical shear stress values for this region of easy glide are in the range of 0.1 kg/mm² (140 psi) to 5% strain or more when specimens are held properly [10]. This level of Peierls force or frictional drag on dislocation motion is negligible in comparison with the 20 kg/mm² range of flow strengths in shear for the common rather course-grained plain carbon steels and structural aluminum alloys. (Flow stress or large-offset yield strength is considered to avoid the difficult issues associated with the upper and the lower yield point of steel.)

In a continuum analysis of a ductile structural metal or alloy, therefore, portions of crystals or grains which are pure and in perfect array may be considered to offer essentially no resistance to plastic deformation (to the passage of the many mobile dislocations present).

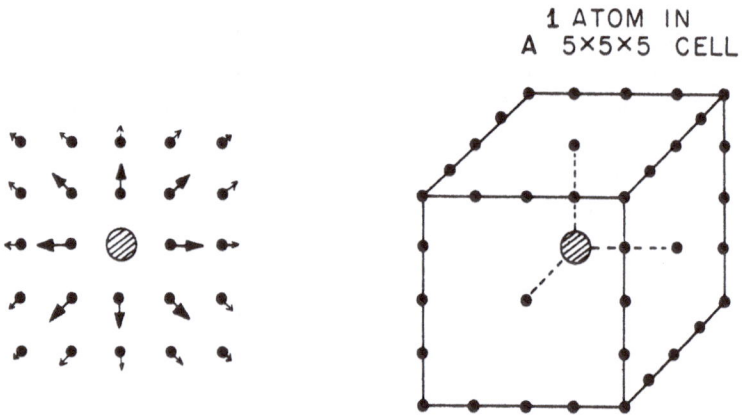

1 ATOM IN
A 5×5×5 CELL

Fig. 3. Solution hardening explanation requires a look at the atomic scale. The position of each atom is disturbed appreciably by a 1 per cent solution [2].

Common structural metals are far from simple alloys, but some of their microconstituents may be. Solute atoms, substitutional or interstitial, and vacancies warp the possible slip planes in their vicinity (a few atomic distances away), Fig. 3. All areas of all planes can be affected noticeably, therefore, when the atoms in solution constitute 1 % or more of all atoms in the region [2]. The greater the atomic fraction of solute atoms and the larger the difference in diameter from the surrounding atoms, the larger the curvatures induced and the higher the shear stress needed to move dislocations through the solution-hardened matrix. However, the greater the mismatch in atomic size, the lower the solubility [11]. Consequently, although solution hardening of ductile metals is significant and the appropriate flow strength of the matrix alloy should be included in a continuum approach, it does not produce the major portion of a 20 kg/mm^2 or so flow strength in shear [12, 13] of a ductile alloy. Quite a bit more than 10 kg/mm^2 must be accounted for by other mechanisms of hardening.

Coherent and non-coherent particles, inclusions and precipitates in the range covered by the electron microscope, are the next larger scale features to be examined. Coherent particles, in the 100 Å range of size [13], cause elastic distortion somewhat as solute atoms do. Elastic strains are as noticeable over a 500 Å region as atomic displacements are in a 15 Å domain for the atom. However, the mismatch accommodated at the interface of particle and matrix is of atomic dimension not particle dimension. (There is no distinction between the two for the solute atom.) Consequently the curvature of crystallographic planes is large in the atomic sense only at the interface and a few atomic layers away. The relative atomic arrangement of all but a very small volume fraction of the matrix is essentially unaffected. Non-coherent particles have a disordered matrix in the immediate vicinity of their surface. Despite their much larger size, they produce even less curvature of atomic planes just a few atomic distances away than do coherent particles.

In a continuum calculation, both classes of particles would be considered as immersed in a solution-hardened matrix whose stress-strain behavior is unaffected by the particles. Nevertheless, the degree of coherency does play an important role in a continuum approach to both flow and fracture. It determines the type and magnitude of stress concentrations and the permissibility of relative displacements between particle and matrix.

The most straightforward and simplest of continuum calculations is the most instructive in the next step of sorting out competing dislocation explanations of strain hardening. Over any plane region of total stress-carrying area A and corresponding particle area A_p, subjected to an average shear stress τ, the average shear stress on the particles τ_p is given by

$$\tau A = \tau_p A_p + \tau_s (A - A_p) \tag{1}$$

where τ_s is the shear stress carried by the solution-hardened matrix.

$$\tau_p - \tau_s = (\tau - \tau_s) \frac{A}{A_p} \tag{2}$$

The area ratio A/A_p is the same as the volumetric ratio V/V_p for a random array of particles [14], or $1/f$ where f is the volume fraction of particles.

This picture of particles acting as keys holding the matrix from slipping corresponds to Orowan's picture of dislocation lines pinned by the particles [15], exhibited clearly in electron microscopy, and to the subsequent loop stage described by Fisher, Hart, and Pry [16]. If it were the governing

31

mechanism at flow strengths of structural metals, $\tau = \tau_s$ would be greater than 10 kg/mm^2; and if f were 0.05 or less, τ_p would exceed 200 kg/mm^2 (280,000 psi).

An average shear stress of 200 kg/mm^2 is extremely high, far greater than can be carried by non-metallics in bulk. It is impossibly high for the most perfect of small inclusions which respond elastically, because it then must be multiplied by a large stress concentration factor for a particle resisting the effort to introduce an abrupt shear displacement discontinuity (or equivalently a dislocation pile up). Consequently, coherent particles would (and do) shear long before such high levels of stress could be reached. Noncoherent particles of several microns in size or larger have bulk properties and would fracture very early or separate from the matrix, or both.

Perhaps this statement on fracture is debatable, but more to the point, it is irrelevant for strong ductile materials like structural carbon steels or aluminum alloys. Suppose the noncoherent particles to be infinitely strong, able to withstand any shear stress. The matrix-particle interface then would decohere at moderate average applied stress and slip would occur on a wide variety of planes in the matrix. Mobile dislocations are present in vast number in such structural metals. Also, the interface region is one of disorder or defects. Additional dislocations can be created as needed by the high strain concentrations around the particles. Slip planes would shoot out tangentially from the particle-matrix interface [9]. Continuum theory here must abandon the simple isotropic model and limit the plastic flow to bands of discrete sets of planes and directions governed by the crystallographic structure and dislocation motion. The beginnings of an intersecting slip network would be set up long before the particle itself would fracture, even if it did have only the strength of the same compound in bulk.

As the applied stress is increased, any bonded large weak particles would be likely to fracture, while the smaller strong non-coherent ones would remain intact although progressively less well bonded to the matrix as slip is triggered off. More and more slip planes would be activated and the intersecting network would become more and more dense. The crystallographic planes in the slip bands are not perfect [17]. These imperfections serve as impediments to further intersecting slip, and the material continues to work-harden. Planes or bands always are in the way of any additional slip; they cannot be avoided as most of the particles are. They provide 'line' intersection rather than 'point' intersection, but point by point are not nearly as strong barriers as the small particles. They make up in quantity (really almost in order of infinitely, line vs. point) what they lack in quality. Grain boundaries, in the absence of particle segregation there, have this same characteristic of relatively weak impediment but omnipresence. In addition, they multiply the pattern of intersection because they cut across possible slip planes at a variety of angles, and so serve to reflect and refract the planes of slip which run up against them. In fcc or bcc crystals, there are so many slip systems available that a crystal orientation change alone is not a major block to slip.

In this modified continuum result, the particles in common structural metals act as triggers for slip much more than as keys against slip. Their role as keys loses importance well down in the working range of stress. The picture which emerges for most ductile alloys is that of a rather weak (solution-hardened) matrix divided into cells formed by the intersection pattern. Particles will no longer play an important role at this stage if the network is in stable chemical and physical equilibrium or if rate processes are slow enough to be neglected. The individual cell boundaries are weak and so must be very closely spaced to be effective on a force divided by equivalent effective area basis. The dislocation picture which corresponds is the resistance offered by dislocation cells or tangles to being cut through by other dislocations which move at low stress within the cell until piled up against the barrier.

A continuum analysis involving microstructural dimensions large compared with cell dimensions would simply average the matrix and cell-wall properties into the stress-strain relation for the work-hardening matrix as determined by test. A continuum analysis on a microstructural scale comparable to cell dimensions would have to treat matrix and cell boundaries as separate constituents.

A further refinement of the continuum picture is required when cell dimensions or interparticle distances become so small that very few mobile dislocations are available or can be generated. At dislocation densities of 10^{10} to 10^{12} per cm^2 on the average, regions of 1 to 0.1 micron dimension, respectively, just begin to fall into this category. At 0.1 to 0.01 micron, respectively, dislocations are few and relatively far between. A crude but useful approach is to consider the shear flow stress of the metal, τ_0, in such a region to be increased in inverse proportion to the small dimension [2];

$$\tau_0 = \tau_0^i \left(1 + \frac{c}{h} \right).$$
(3)

33

where τ_0^i is the bulk flow stress of the material and c is a very small dimension at which the flow stress is doubled.

A comparison with Ashby's description [18] of cells or tangles of dislocations produced by particles is appropriate here because of a fundamental difference despite some similarity. In an earlier paper [2], his pattern was termed 'passive' to distinguish it from the 'active' triggering off of dislocations and sending them long distances to pile up eventually against another particle or obstacle. The term 'passive' involved no adverse comment on merit. It was employed as a direct carry-over from soil mechanics where it denotes the soil pressure induced by an object (particle) forced against or through the soil (metal). The motion and the force exerted by the soil are opposed. 'Active' denotes pressure exerted by the soil on the object, when the object is pushed along by the soil. The force has a positive component in the direction of the motion. Punching out or shear motion [18] described by Ashby is passive in the sense that these dislocations oppose the motion and would collapse into the hole and disappear if the particle could be removed.

Ashby's pictures are real. They do explain well the intersecting slip hardening induced by thermal stresses or other local residual stresses in cooling down to room temperature. Such patterns also are likely to be observed on the release of applied stress. In the very different active picture of intersecting slip proposed earlier and in this section, the dislocations move far off in the directions imposed by the applied shear stress. Removal of the originating particle then would have little effect; the intersection pattern dominates.

4. Brief summary of earlier results for sintered WC-Co, aluminum-silicon, and carbon steel (pearlitic and spheroidized)

Useful and interesting results have been obtained by applying conventional continuum mechanics to microstructure. In an earlier analysis of sintered WC–Co (see [2]), the tungsten carbide particles were treated as brittle and the cobalt (which contains small but significant amounts of tungsten and carbon) as perfectly plastic. Most of the peculiar behavior observed by Gurland and others in the interesting and practical range of low cobalt content [19, 20] was clarified through the plastic constraint factors, Table I, for a hexagon model of the system, Fig. 1b. In particular, the essential ductility explained

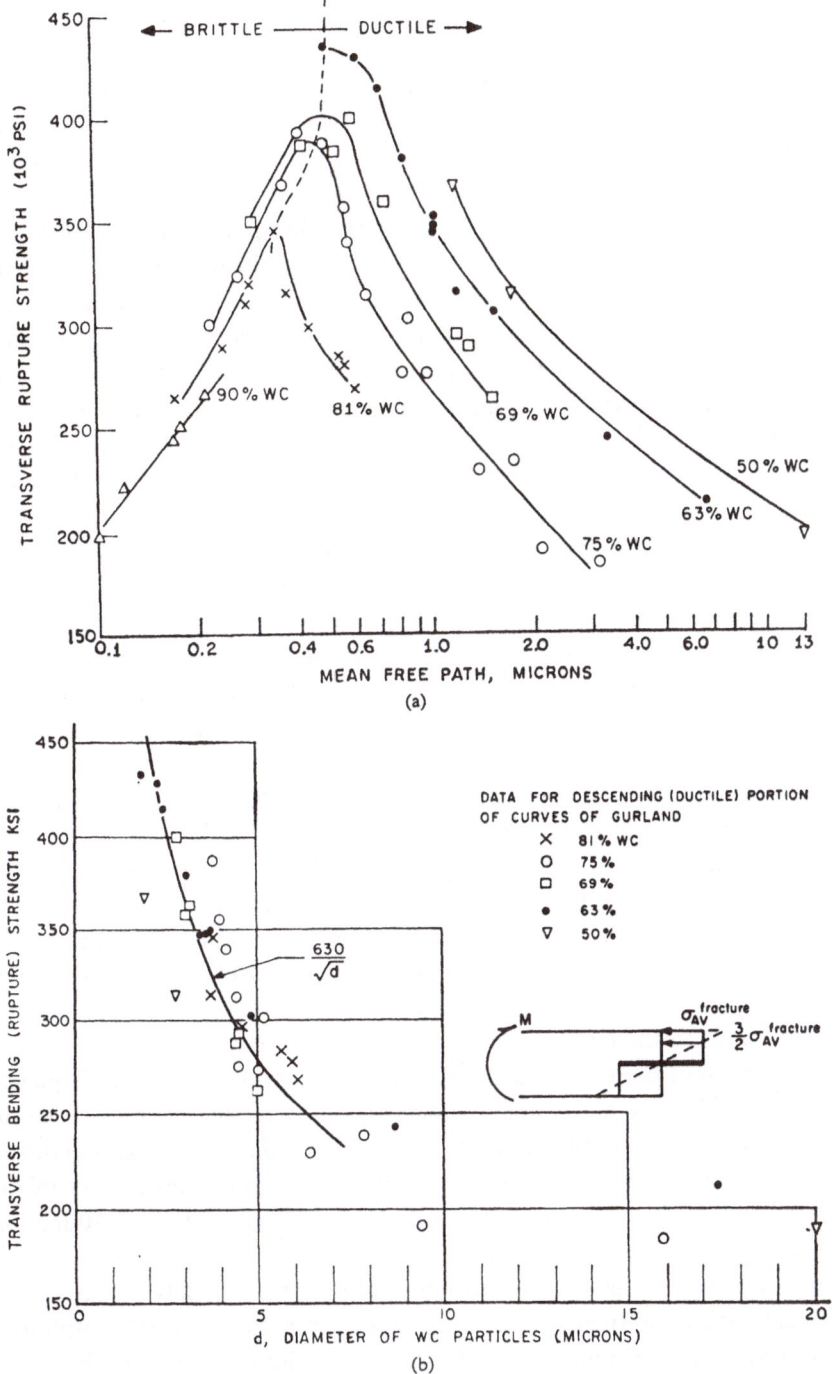

Fig. 4. Strength of sintered WC-Co. (a) Brittle to ductile behavior of cobalt. (b) For ductile behavior of cobalt actual tensile strength is approximately two-thirds of nominal transverse bending strength plotted.

the approximately 50% higher bending 'strength' than tensile strength and the relatively small scatter in test results, Fig. 4, despite the apparently extremely brittle fractures. Particle and bond strength limitations coupled with the high plastic constraint factors explained the peaking of strength at an intermediate particle size for a given volume fraction of cobalt, the increase in tensile or compressive strength with an increase in ductility [21] on the brittle side of the peak of strength, and the effect of temperature on strength [22]. Quantitative agreement at small particle sizes and associated small thickness h of the cobalt layers separating the particles required the ad hoc assumption (Eq. 3) of an increase in yield strength with the inverse of h. Except for this indirect and qualitative use of a dislocation density concept, no attempt was made to include atomic or dislocation level values.

The next application was to the very different aluminum-silicon system, studied by Gurland and his students [23], which is ductile on the macro-scale as well as the microscale. Their observations for silicon contents ranging up to 13.2% can be summarized as:

1. Silicon inclusions of the order of 5 to 10 microns in length and several microns in diameter were scattered reasonably uniformly in the matrix. They were the main, and just about the only feature observed in the optical microscope.

2. The presence of small amounts of silicon raised the yield strength of pure aluminum by a large factor.

3. The higher the volume fraction of silicon, the greater the number of inclusions seen and the higher the flow strengths.

4. As each alloy was subjected to increasing strain, more and more of the silicon inclusions fractured in tension but the stress continued to rise. Tensile fractures in the silicon often were associated with well-defined slip bands in the matrix.

The largest of these volume fractions of silicon produces negligible plastic constraint, Table I and Fig. 1a, and little increase in the rate of strain hardening over that of the matrix by itself. Reconciliation of this conclusion with the experimental observation 3 (which attests to the importance of the silicon) and 4 (which confirms the lack of importance of the 5 to 10 micron particles) was not immediate [23, 2].

Solution hardening does contribute more proportionately to this lower flow strength alloy than it does to the high strength structural aluminums described in the previous section. It cannot and does not account, however, for the continuing increase in yield and flow strengths with increasing volume fraction of silicon far beyond the saturation limit. The explanation had to lie in strengthening by particles not discernible in the optical microscope but on a scale well above the atomic. Following Kelly and Nicholson's discussion of coherent and non-coherent particles [13], the proposal was made [2, 9] that (then invisible) non-coherent particles of the order of 500 Å in diameter existed in enormous number, and that the intersecting slip they would trigger off could account for the observed stress-strain behavior. These crucial small particles would occupy an extremely small volume fraction despite their great number; the 5 to 10 micron particles would account for practically all of the silicon not in solution.

Particles in this size range were observed later in the electron microscope by Prince and Richman [24] in strong association with dislocation lines. Enough seem to be present to account qualitatively for the validity of the continuum conclusion that an alloy with 5 micron inclusions would have closely the same stress-strain curve as the matrix by itself.

It would then appear true that, with the usual techniques of specimen manufacture, as the silicon content is increased, more large inclusions would be present and more small inclusions would be present as well. In this sense, the observations in the optical microscope on interparticle distances, etc., lead to the correct conclusions on silicon strengthening but with the wrong reasoning.

The success of continuum theory in its application to WC–Co with the crudest of modifications for scale and in its much more fundamental approach to aluminum-silicon lacked sufficient generality. Therefore, the iron-carbon system was examined from these same points of view. Not only is there a basic change from fcc aluminum to bcc iron but both pearlitic and spheroidized compositions can be obtained and compared with each other experimentally and theoretically.

The solubility of carbon in α-iron [25] is negligible at room temperature but ferrite does contain extremely fine carbide dispersions [26] [27]. Pearlitic clumps, of many extremely thin platelets of cementite (Fe_3C) of high strengths separated by ductile layers of ferrite several times thicker, aggregate as shown very schematically in Fig. 2 and in one of the hexagons of Fig. 1b. At eutectoid composition, the entire volume is filled by pearlite. A little carbon goes a long way because the weight of cementite is 16 times the weight of the carbon, and the volume fraction of pearlite is 8 times the volume fraction of cementite or about 120 times the weight fraction of carbon. Consequently the familiar 1020 (0.20% carbon) plain carbon steel

37

will have a cementite volume fraction slightly over 0.03 and a pearlite volume fraction of almost 0.25. For a 1040 steel, these numbers double. In the spheroidal form, the volume fraction of cementite is the same as in pearlitic steel of the same carbon content but the separate cementite particles are immersed in a ferrite matrix with considerable tendency for segregation at the original grain boundaries [28] much as the silicon in the aluminum-silicon system, Fig. 1a. Interparticle distances of the order of 10 microns and particle diameters of one to three microns are reported [28]. Hypereutectoid spheroidized steels still show appreciable ratios of interparticle spacings to particle diameters and may be studied along with the hypoeutectoid.

In summary, the following reasoning and results were reported in Ref. [1]. The greater the number of dispersed submicroscopic carbide particles per unit volume, the higher the local flow stress. The shorter the time available at elevated temperature for the migration and coalescence of the sub-microscopic particles to microscopic size, and the larger the mean free path in the ferrite between carbide particles visible in the microscope, the more the number of submicroscopic particles which remain dispersed. Time available for the migration at elevated temperature should be reflected by grain size. Consequently, at a given low or medium carbon content, the ferrite separating the pearlite clumps in a pearlitic steel should have a some-what heavier loading of dispersions than the ferrite in a spheroidized steel of the same grain size and composition with its more evenly distributed cementite. On the other hand, far fewer dispersed particles should be left in the ferrite which lies between the closely spaced cementite plates of the pearlite. The associated decrease of flow stress which might be expected is offset, however, by an appreciable increase associated with smallness of absolute dimension, Eq. (3). The flow stress of the ferrite in the pearlite will depend upon the fineness of the pearlite but should not vary much with the carbon content of the steel as the process of formation sets the local environment.

Fig. 5 has a pictorial representation of these statements for flow stress in the ferrite. The change with carbon content is moderate for the ferrite matrix because carbide dispersions in very large number are already present as the low carbon content of 0.013 %. Fig. 5 also translates the curves for the ferrite matrix to flow stress or limits on flow stress for steel. For the pearlitic steel, the volume fraction of its pearlitic clumps becomes appreciable rather soon because it is 8 times the cementite volume fraction. Both the flow stress for the pearlite and the plastic constraint exerted by the pearlite on the free

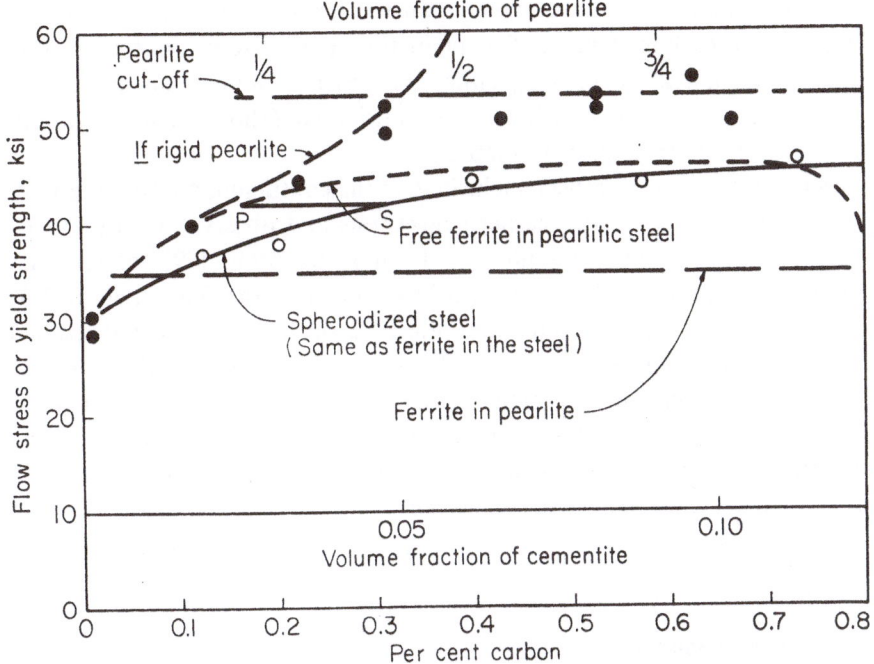

Fig. 5. Yield strengths for pearlitic and spheroidized steels and ferrite (1)

○ Rinebolt, 1954 [3] Spheroidized, ASTM #8
● Rinebolt and Harris, 1951 [4] Pearlitic, ASTM #8

Note: Curve for free ferrite in pearlitic steel is the curve for spheroidized steel at twice the carbon content as indicated by the horizontal line P–S.

ferrite must be estimated and taken into account properly. On the contrary, the volume fraction of the cementite in the spheroidized steel is small, and the interparticle distances observed in the microscope are of the order of several microns. Therefore, the flow stress of the spheroidized steel is indistinguishable from the flow stress of its ferrite. The pearlitic steel curve is somewhat arbitrarily taken as the spheroidized steel curve at twice the carbon content. A smooth transition is described in Ref. [1] between the two upper bound curves of Fig. 5, one for rigid pearlite in a ferrite matrix and the other for 100% pearlite. It clearly would agree well with the data points of Rinebolt and Harris [4] and for clarity is not repeated in Fig. 5.

The flow stress of the ferrite in the pearlite is about 53 000/1.5 or 35 000 psi, while the yield strength of the ferrite between the pearlite clumps never

39

gets more than about 10% higher than the ferrite in a spheroidized steel of the same carbon content. Except in the very narrow range of carbon content (from 0.25 to 0.40%), either the pearlite clumps can be ignored (below 0.25%) or the pearlite alone considered (above 0.40%) without introducing an error of more than 5%.

Many of these qualitative conclusions of the continuum analysis are well-known and in accord with earlier suggestions and observations by others. Experimental evidence presented by Danko and Stout [29] showed that pearlite clumps do not deform nearly as much as the free ferrite in low carbon steels. At 25% overall strain of a 1025 steel, the pearlite clumps had only undergone a strain of about 5%. Petch [30] suggested that most of the flow took place in the ferrite rather than in the pearlite and that the ferrite in a 1025 steel work hardens at a higher rate than does ferrite without the pearlite clumps. Experimental data of Irvine and Pickering [31] for low carbon steels show little dependence of the yield strength on the volume fraction of pearlite.

5. The conclusion for flow

The consistent picture which was emerged from the earlier studies of fcc aluminum-silicon, and the more recent bcc pearlitic and spheroidized forms of plain carbon steel, is that the flow stress of the ductile matrix of structural metals is determined primarily by microstructural features not visible in the optical microscope. Those features a few microns in size and larger such as the visible cementite particles in spheroidized steel and the silicon inclusions in the aluminum, the associated interparticle distances, the grain boundaries (as distinct from the submicroscopic particles and defects which may gather there), and the mean free paths between visible particles or between pearlite clumps in a low and medium carbon pearlitic steel have no appreciable direct influence on yield strength. Dispersion hardening by submicroscopic precipitates through an intersecting network of slip seems the dominant mechanism. It is, of course, superposed on solution hardening.

Yet excellent correlations of yield strength data have been obtained with measured interparticle distances or grain sizes as seen in the optical microscope. The conclusion is that this success is coincidental in principle for ductile structural metals and alloys. It can be achieved only when the visible is a scaled image of the salient features of the optically invisible. For example, longer time at elevated temperature leads to larger grain size and at the same time to agglomeration of ultrafine particles with a consequent denuding of the matrix. Therefore, the larger the grain size the lower the yield strength, but without the direct causal relation which does exist for ionic crystals and for bicrystals or polycrystals of very pure metals.

6. The contrary conclusion for fracture by void coalescence

A complementary result holds for fracture of an alloy composed of a small volume fraction of larger than micron size inclusions in a fully ductile matrix. As already described, the properties of the matrix, which are determined by features on the atomic and electron microscope scale, give the plastic response of the alloy directly and without modification. Fracture of such an alloy, on the contrary, is governed by the size, shape, and distribution of the inclusions visible in the optical microscope. Fracture in this

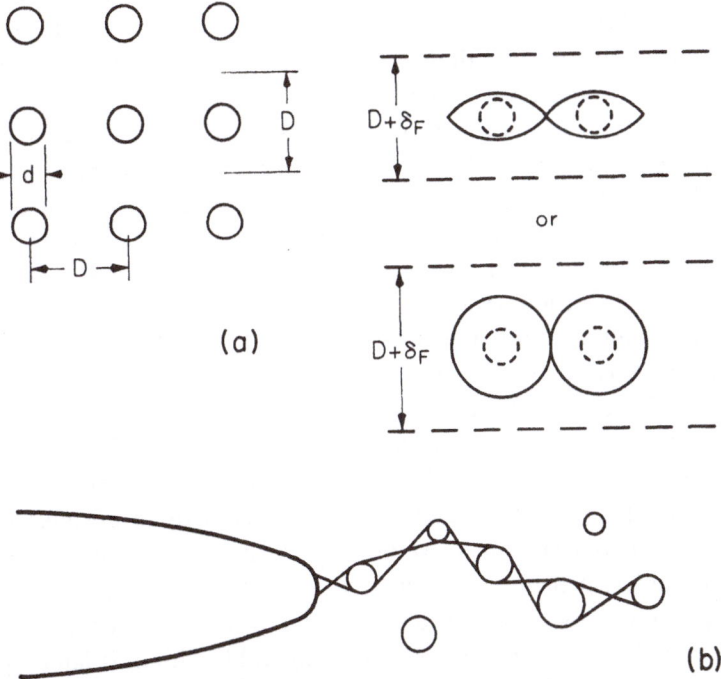

Fig. 6. Void coalescence along fracture surface (shown schematically in two dimensions but much more complex in three dimensions).

simplest and common situation takes place by void coalescence. The voids originate through breaking of the inclusions or their decoherence from the matrix. Local elongation to fracture (δ_F) in the region of void coalescence will depend upon the properties of the matrix (primarily the plastic properties) and the geometry of the microstructure.

Consideration of the drastic idealization in Fig. 6a, with uniform spherical holes of diameter d in a regular cubic array of spacing D replacing the inclusions, clarifies some questions and raises others. The volume fraction of holes determines the ratio d/D, which in turn determines the local average stress level required for void coalescence. Ratios δ_F/d or δ_F/D will be a function of d/D, σ_0/E, and other dimensionless characterizations of the stress-strain relation in the plastic range. A real material has a distribution of shapes, sizes, and locations of inclusions which crack and decohere. It will have an equivalent d/D for the fracture path which is considerably smaller than that calculated from the volume fraction on the basis of a regular cubic array.

The elongation to fracture, δ_F, which determines the crack opening displacement and the fracture toughness or critical stress intensity factors such as K_{Ic}, Fig. 6b, increases with the size of the holes at a given volume fraction of holes or d/D equivalent. If inclusions of micron size and larger must be present in the amount of a percent or so, the larger the particles the better from this point of view. However, the statement of the previous section still holds. Without special treatment, the microstructure visible in the optical microscope and that visible in the electron microscope tend to relate strongly to each other. Large inclusions at large spacings would likely indicate a matrix of low yield and flow strength.

An important question remains unanswered for arrays of holes or inclusions in a three dimensional array. In going from a composition with an effective d/D of, say, $\frac{1}{3}$ to a somewhat purer material of $\frac{1}{4}$, for example, is there much increase in δ_F when the inclusion size is kept the same and the spacing is increased? The simple picture of a large spherical expansion of the holes until the plastic links between them neck in an unstable manner like ordinary tensile specimens would indicate that δ_F is comparable with D. The quite different picture of primarily lateral expansion of the holes with an early takeover by plastic instability [32], a picture *I* prefer for this range of d/D, would indicate that δ_F remains comparable with d. If lateral growth prevails, there would be little change in local elongation for a marked decrease in d/D at constant d; if spherical expansion dominates, the elongation would go up appreciably. Regardless of the validity of either answer, however, the conclusion remains that in void coalescence the optical microscope scale and ordinary continuum mechanics determines the fracture behavior once the properties of the matrix are given, [33].

None of this implies that properties governed by the submicroscopic scale cannot control some important aspects of fracture. Cleavage, separation at grain boundaries in creep, and stress corrosion cracking are so troublesome precisely because they operate on a very small scale and may not permit the void coalescence on the micron and larger scale which is so much more desirable.

References

[1] Butler, T. W. and D. C. Drucker, Yield strength and microstructural scale: a continuum study of pearlitic vs. spheroidized steel, *J. of Applied Mechanics, Trans. ASME*, in press.

[2] Drucker, D. C., The continuum theory of plasticity on the macroscale and the microscale, *J. Materials, 1* (1966) 873–910.

[3] Rinebolt, J. A., Effect of metallurgical structures on the impact properties of steels, *ASTM STP No. 158* (1954) 203–214.

[4] Rinebolt, J. A. and W. A. Harris, Effect of alloying elements on notch toughness of pearlite steels, *Trans. ASM, 43* (1951) 1175–1201.

[5] Koiter, W. T., General theorems for elastic-plastic solids, in *Progress in Solid Mechanics, 1*, edited by I. N. Sneddon and R. Hill, North-Holland Publishing Co., Amsterdam, 1960, 165–222.

[6] Langhaar, H. L., *Energy methods in applied mechanics*, John Wiley and Sons, Inc., New York, 1962.

[7] Drucker, D. C., W. Prager and H. J. Greenberg, Extended limit design theorems for continuous media, *Quarterly of Applied Mathematics., 9* (1952) 381–389.

[8] Hoff, N. J., Approximate analysis of structures in the presence of moderately large creep deformations, *Quarterly of Applied Mathematics, 12* (1954) 49–55.

[9] Wu, T. T. and D. C. Drucker, Continuum plasticity theory in relation to solid solution, dispersion, and precipitation hardening, *J. of Applied Mechanics, 34, Trans. ASME, 70* (1967) 195–199.

[10] Davis, R. S., R. L. Fleisher, L. D. Livingston and B. Chalmers, Effects of orientation on the plastic deformation of aluminum single crystals and bicrystals, *J. of Metals, 9* (1957) 136–140.

[11] Hume-Rothery, W., G. W. Mabbott and K. M. Channel-Evans, The freezing points, melting points, and solid solubility limits of the alloys of silver and copper with the elements of the B sub-groups, *Phil. Trans. of Roy. Soc. 233[A]* (1934) 1–97.

[12] Irvine, K. J. and F. B. Pickering, Low-carbon steels with ferrite-pearlite structures, *JISI, 201* (1963) 944.

[13] Kelly, A. and R. B. Nicholson, Precipitation hardening, *Progress in Materials Science*, edited by B. Chalmers, Pergamon-Macmillan, *10*, No. 3, (1963) 149–391.

[14] Rostoker, W. and J. R. Dvorak, *Interpretation of metallographic structures*, Academic Press, New York, 1965.

[15] Orowan, E., *Symposium on internal stresses in metals and alloys*, Institute of Metals, 1948, 451–484.

[16] Fisher, J. C., E. W. Hart and R. H. Pry, The hardening of metal crystals by precipitate particles, *ACTA Metallurgica, 1* (1953) 336–339.

[17] Johnson, W. G. and J. J. Gilman, Dislocation multiplication in lithium fluoride crystals, *J. Appl. Phys. 31* (1960) 632–643.

[18] Ashby, M. F., The deformation of plastically non-homogeneous materials, *Phil. Mag., 21* (1970) 399–424.

[19] Nishimatsu, C. and J. Gurland, Experimental survey of the deformation of the hard-ductile two-phase alloy system WC-CO., *Trans. ASM, 52* (1960) 469–484.

[20] Gurland, J., The fracture strength of sintered tungsten carbide-cobalt alloys in relation to composition and particle spacing, *Trans. Met. Soc. AIME, 227* (1963) 1146–1150.

[21] Suzuki, H., Variations in some properties of sintered tungsten carbide-cobalt alloys with particle size and binder composition, *Trans. Jap. Inst. Metals, 7* (1966) 112–117.

[22] Betser, A. A. and J. Gurland, Some effects of temperatures and heat treatment on the strength of sintered WC-Co alloys, *ASME preprint 66-MD-17.*

[23] Gangulee, A. and J. Gurland, On the fracture of silicon particles in aluminum-silicon alloys, *Trans. Met. Soc. AIME, 239* (1967) 269–272.

[24] Prince, K. D. and M. H. Richman, Direct observation of the effect of particle size on dispersion hardening, *J. Materials, 4* (1969) 145–158.

[25] Wert, C. A., Solid solubility of cementite in alpha iron, *J. Metals, 188* (1950) 1242–1244.

[26] Allen, N. P., W. P. Rees, B. E. Hopkins and H. R. Tipler, Tensile and impact properties of high purity iron-carbon and iron-carbon-manganese alloys of low carbon content, *J. of Iron and Steel Inst., 174* (1953) 108–120.

[27] Brick, R. M., R. B. Gordon and A. Phillips, *Structure and properties of alloys*, Third Edition, McGraw Hill Book Company, New York, 1965, 225–227.

[28] Liu, C. T., *An experimental study of the strengthening mechanism and fracture behavior of spheroidized carbon steel*, Brown University Report, August 1967.

[29] Danko, J. C. and R. D. Stout, The effect of microstructure on the morphology of fracture, Part II, *Weld. J., 35* (1956) 775.

[30] Petch, N. J., The cleavage strength of polycrystals, *J.I.S.I., 174* (1953) 25.

[31] Irvine, K. J. and F. B. Pickering, Low-carbon steels with ferrite-pearlite structures, *J.I.S.I., 201* (1963) 944.

[32] Drucker, D. C., Macroscopic fundamentals in brittle fracture, in *Fracture*, Vol. 1, Ed. H. Liebowitz, Academic Press, 1968, 473–531.

[33] Nagpal, V., F. A. McClintock, C. A. Berg and M. Subudhi, Traction displacement boundary conditions for plastic fracture by hole growth, in *Foundations of Plasticity* (Warsaw 1972), A. Sawczuk, editor, Noordhoff Int. Publ., Leyden, 1973, 365–386.

Lattice defect approach to plasticity and viscoplasticity

E. Kröner

University of Stuttgart, Stuttgart, Germany

C. Teodosiu

Center of Solid Mechanics, Bucharest, Rumania

The plasticity and viscoplasticity of crystalline materials is discussed in terms of lattice defects, in particular of dislocations. It is argued that physical observations on a microscopic scale can provide some guidance for macroscopic theories of the deformation behaviour of such material. This should apply in particular to the description of the internal mechanical state and its variation during deformation. Hence viscoplastic deformation is related to the dislocation motion. A theory is proposed that provides a general framework for microscopic analysis of viscoplastic deformation of a single crystal.

1. Introduction

Half a century of physical research has shown that plasticity and viscoplasticity are typical properties of *crystalline* materials and that the *defects* in the crystalline structure are the main factors which give rise to the elementary processes which are observed macroscopically as the permanent deformation.[1]) The defects which play an important role in this are above all the dislocations, the point defects and the grain and phase boundaries

The dislocations act as the carriers of plastic deformation and produce most of the remarkable changes of the internal mechanical state during the deformation. The point defects, such as vacancies and solute atoms of a

---

[1]) This statement is also meant to explain the sense in which the terms plastic and viscoplastic are understood in this lecture. Unfortunately the nomenclature of the deformation properties of solids is not very uniform. The plasticity and viscoplasticity of crystalline materials always include a certain amount of elasticity, too. Therefore, it would be more accurate to speak of elasto-plastic and elasto-viscoplastic solids. We renounce this for reasons of simplicity.

45

different species usually introduce a strong viscous component into the deformation behaviour. Finally, the grain and phase boundaries not only can increase the resistance against plastic flow but they can also be additional sources of viscous behaviour.

A number of factors such as temperature, flow velocity, concentration of solute atoms or second phases, mechanical and heat treatment, path of loading (unidirectional, cyclic, statistical) etc. influence strongly the elementary processes and thus the phenomenological behaviour. It is therefore clear that a theory which describes the viscoplastic behaviour of materials taking into account all these factors must necessarily be extremely complex. As a consequence, the numerous physical theories in which macroscopic equations are derived by applying certain averaging procedures involving selections of the elementary processes are, without exception, good approximations only in limited ranges of the above influence quantities.

In this lecture we shall try to answer the question how far the results of solid state physics can provide some sort of guidance for the development of theories which describe the deformation of crystalline materials.

2. History type theories

Leaving aside for the moment the thermodynamical aspects of the deformation there appears to be an agreement that the main problem left for solution is that of the constitutive equations. In particular we think of the task of giving explicit forms of such equations which allow the solution of actual problems in the desired approximation.

The existing approaches can be classified essentially as two types of theories which we like to call *history type* and (*internal*) *state type theories*. We postpone the discussion of the state type theories until section 5 because for this purpose we need the knowledge of some physical results presented in sections 3 and 4. In the history type theories one describes the proceeding deformation in terms of very few phenomenological variables – mainly by the measures of plastic and elastic strain, perhaps including strain gradients. The material, for simplicity assumed to be uniform, is then specified by a response functional which is usually taken as a local time functional of the strain of a priori unknown form. The term local implies that a small piece of material already shows all the properties of the whole material.

The description of materials in terms of time functionals has been very successful in the case of viscoelastic bodies. If plasticity, i.e. a rearrangement of lattice defects, is involved, an essential difficulty arises. In order to obtain the unknown functional from experiments one must apply in these experiments all possible load histories. This is the situation surmised by Drucker [1] who wrote in 1960: 'In the strict sense it is not possible to predict the strain components which will be found for a given stress history; the experiment itself must be run to get the answer'.

The situation would be better if plastic materials behaved like simple bodies. Then only experiments with homogeneous stress and strain would suffice to determine the complete time functional. In other words: If plastic or viscoplastic bodies behave simple it is possible to predict the outcome of experiments with *inhomogeneous* stress and strain provided the behaviour under *homogeneous* stress and strain has been measured for all possible load histories. In this case the theory allows genuine predictions, i.e. predictions on experiments which have not been performed before.

We shall argue later that crystalline plastic materials are not simple bodies for basic physical reasons. To these materials Drucker's surmise applies in its full import. This fact can be expressed by saying: In a strict sense a genuine phenomenological theory of plasticity, i.e. a theory of plasticity which allows genuine predictions on a phenomenological basis, cannot exist. Of course, this statement does not exclude the possibility of assuming a priori or by interpolation of experimental data certain forms of the time functional which then lead to reasonable or even excellent predictions in certain areas of application.

3. Direct observation of the dislocation state

The physical research on plastic and viscoplastic, i.e. crystalline, materials has revealed the existence of an *internal mechanical state* which is the *lattice defect state*. The quantities used for the description of this state are called *internal* quantities. From the phenomenological standpoint they are *hidden* quantities because devices on a microscopic scale are necessary to make these quantities visible.

We call internal state *variables* those quantities which vary during the experiments, in contrast to internal state *parameters* which also specify the internal state but remain constant during the deformation. An often encountered situation on which we concentrate in the following is that the arrangement of point defects and grain or phase boundaries is unchanged whereas the distribution of the dislocations varies during the experiment. This means that the internal state variables of the theory should be those which specify the dislocation arrangement.

There exist several well developed techniques which allow the direct observation of dislocations. Transmission electron microscopy has proved to be particularly successful. Except for Fig. 1 all pictures shown in the forthcoming figures have been obtained with this technique. We present these pictures in order to give a visual impression of the dislocation state and its complexity which is responsible for the similarly complex pheno-menological behaviour of crystalline materials. We mention that the pre-paration as well as the quantitative interpretation of such micrographs requires considerable experimental and theoretical experience. So particular devices must be applied in order to ensure that the dislocation arrangements seen are those really present in the interior of a deformed specimen. For details we refer to the work of Mughrabi [2].

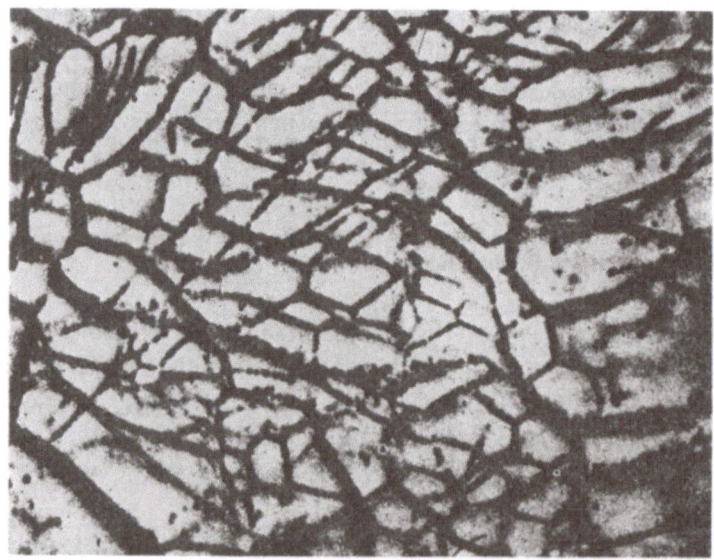

Fig. 1. A three-dimensional network of dislocations below the surface of a single crystal of AgBr. The network becomes visible in the microscope by the precipitation of silver along the dislocation lines. 1000 ×. After J. M. Hedges and J. E. Mitchell [3].

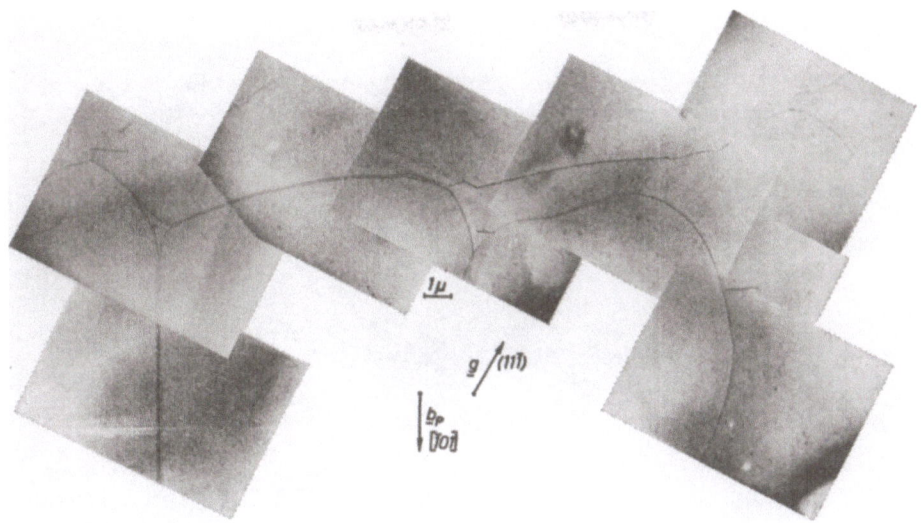

Fig. 2.

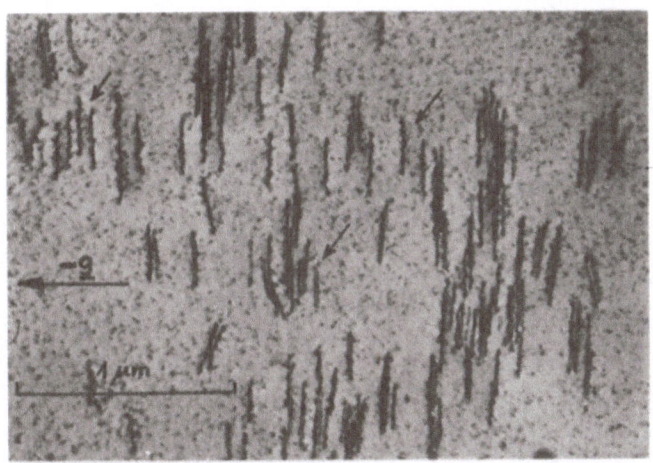

Fig. 3.

Fig. 2–5. Dislocation arrangements in slightly deformed (5 %) copper single crystals (stage I). Thin lines represent single dislocation lines. For details of figs. 2, 3 see H. Mughrabi [2], of figs. 4, 5 see U. Eßmann [4].

49

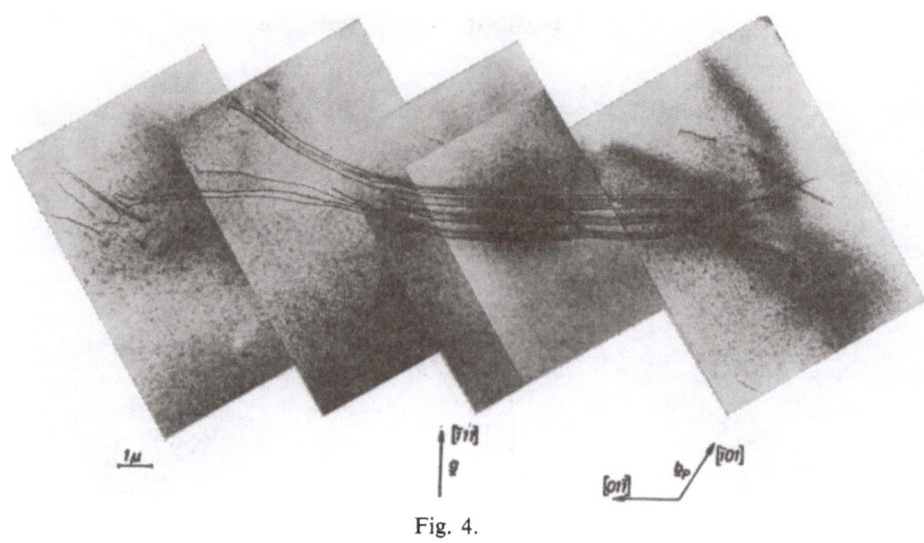

Fig. 4.

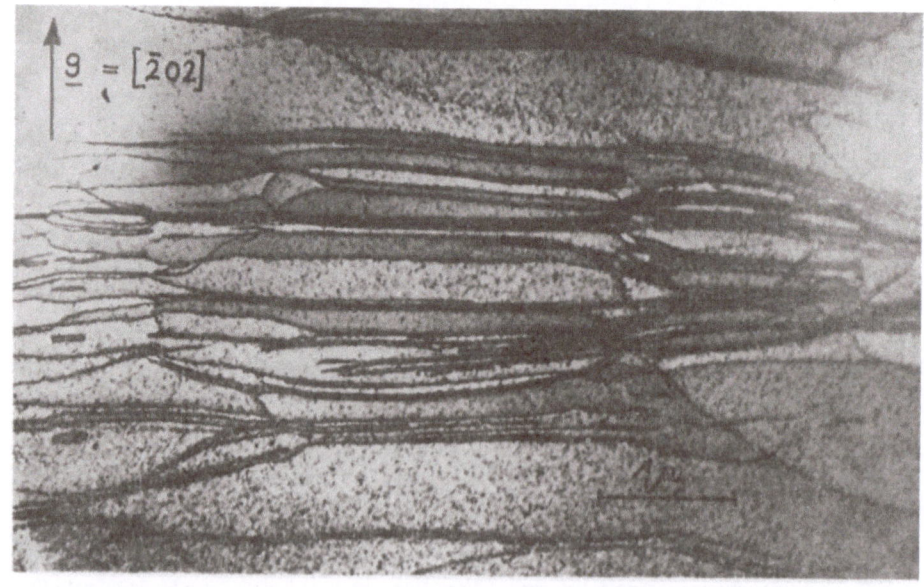

Fig. 5.

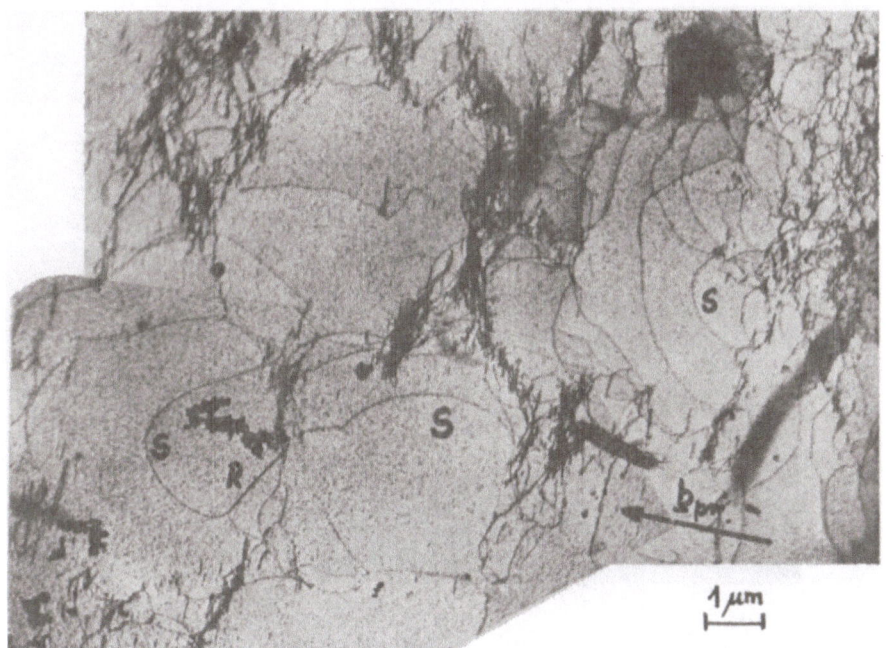

Fig. 6.

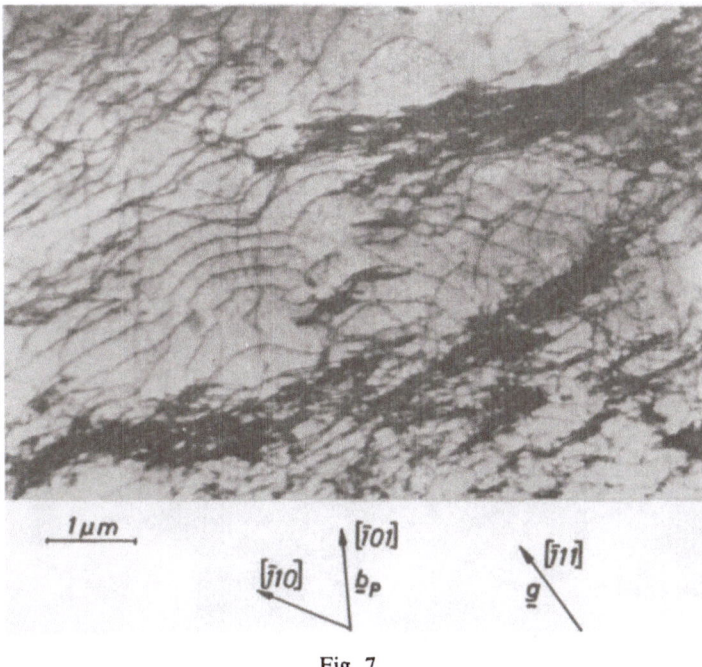

Fig. 7.

Fig. 6–9. Dislocation arrangements in copper single crystals after a tensile strain of 0.17 (stage II). The pictures resemble those obtained after cyclic deformation. The reason is that in both cases (in contrast to stage I) the glide occurs in more than one glide system. After U. Eßmann [5] and H. Mughrabi [2].

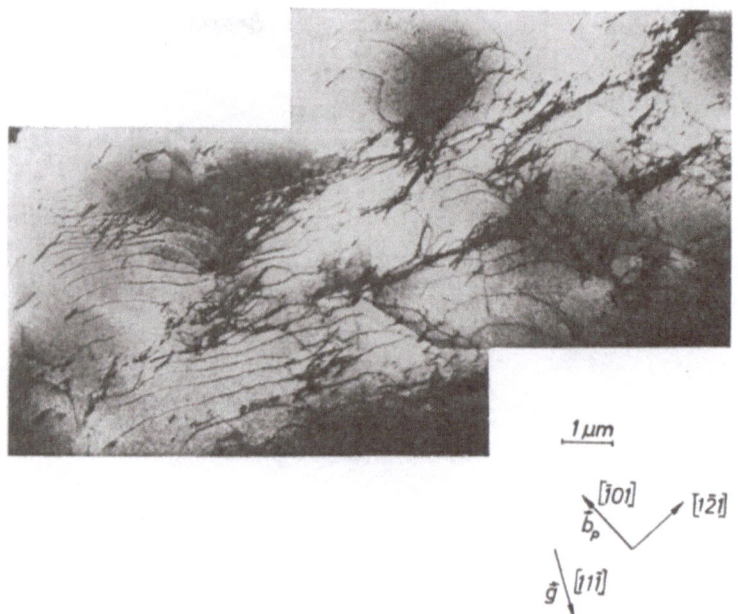

1 µm

$[\bar{1}01]$

$\vec{b}_P$

$[1\bar{2}1]$

$\hat{g}$ $[11\bar{1}]$

Fig. 8.

b_P

1 µm

Fig. 9.

Fig. 10.

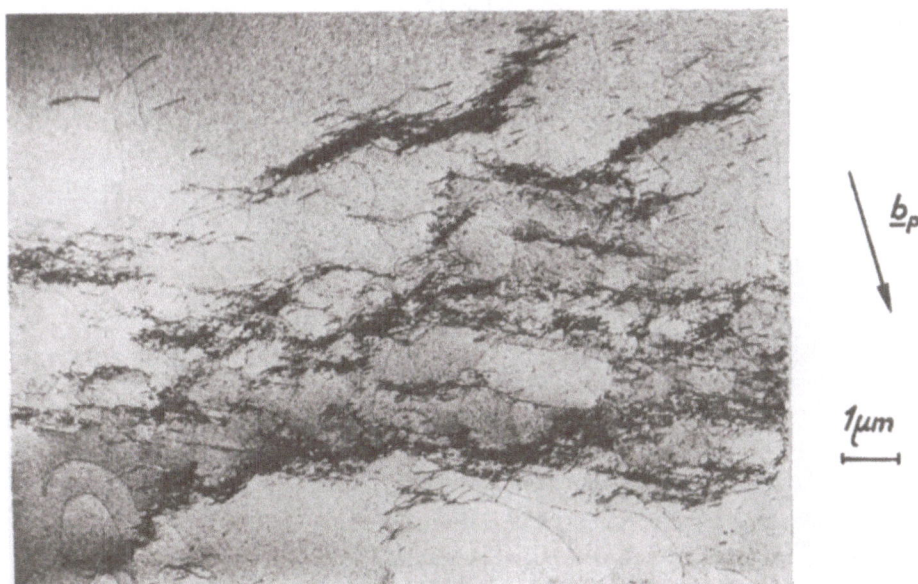

Fig. 11.

Fig. 10–13. Dislocation arrangements in copper single crystals which have undergone 15 cycles in a tension-compression test of constant strain amplitude $\varepsilon = \pm 0.0065$ at room temperature. Typical is the formation of dislocation multipoles and the coexistence of free single dislocation lines. After H. Mughrabi [7].

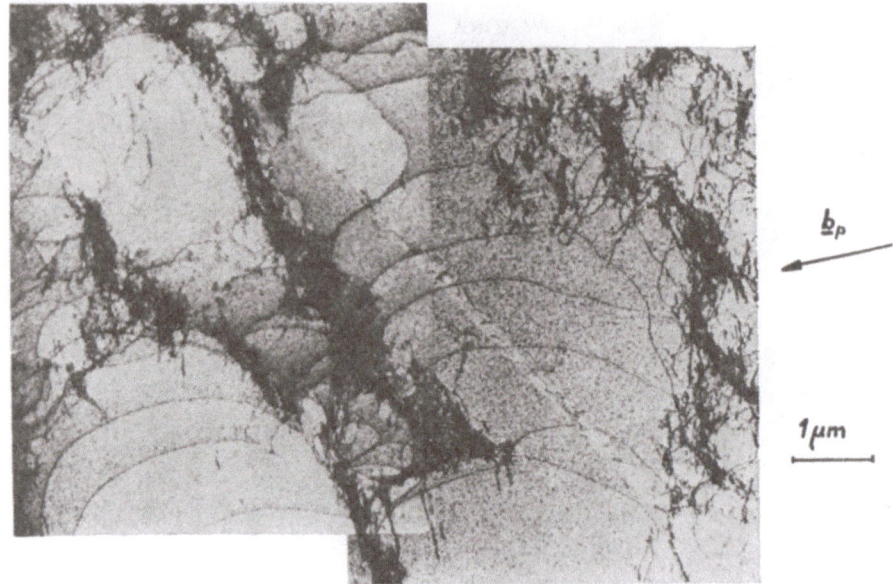

Fig. 12.

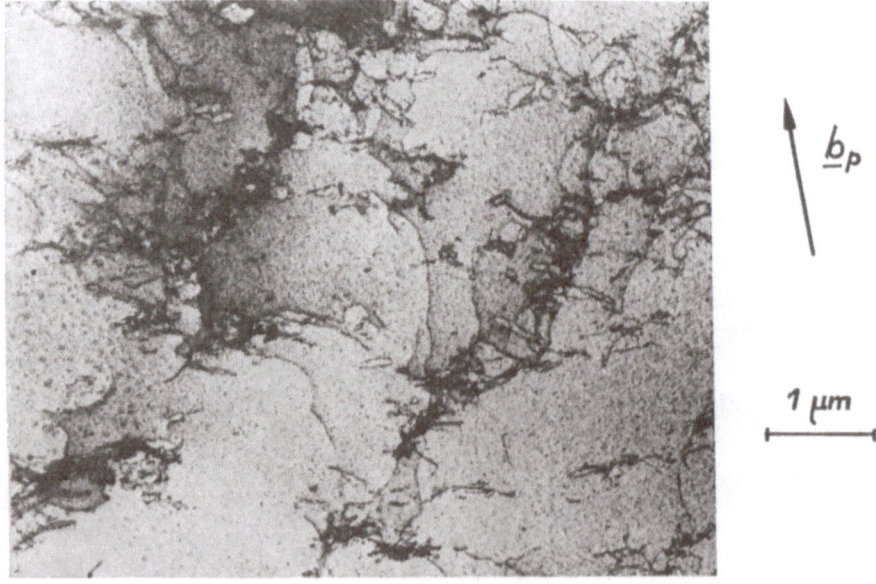

Fig. 13.

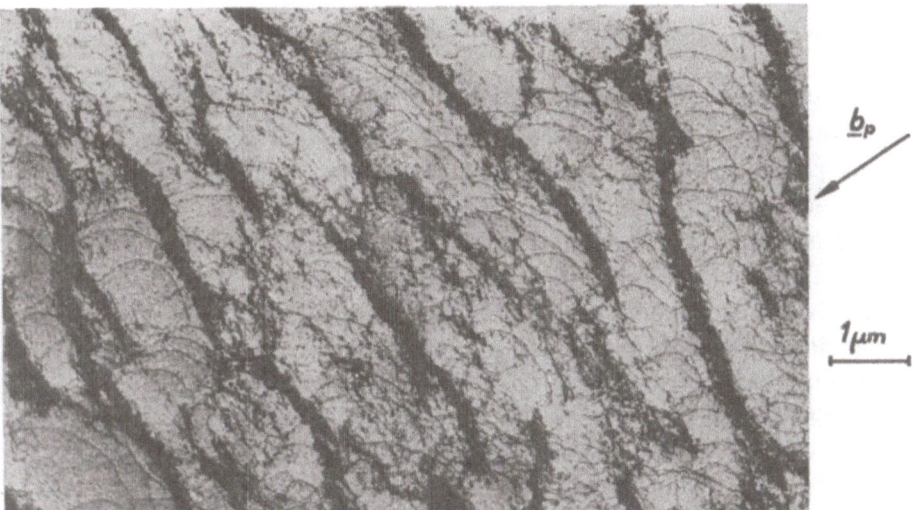

Fig. 14. As in figs. 10–13, but after 300 cycles. After H. Mughrabi [7].

The already historical picture (1953) of Hedges and Mitchell,[1]) [3], in Fig. 1 shows a so-called three-dimensional network of dislocations which was found in an undeformed crystal of AgBr. According to Frank's theory of crystal growth such networks are formed during the crystallization process.

The next pictures show the dislocation state after a small unidirectional elongation (stage I, Figs. 2–5), after a somewhat larger unidirectional elongation (stage II, Figs. 6–9) and after a cyclic tensile deformation (Figs. 10–14) of a copper single crystal. A closer inspection reveals, that thin lines are single dislocations, thicker lines are bundles of several dislocations which are formed in increasing number during the deformation. Since the dislocations imply disturbances of the regular lattice we see that these disturbances increase in the course of the deformation. Beside this we observe that the dislocations arrange themselves neither completely regularly nor completely irregularly: one sees partially ordered distributions. This is an important hint to the statistical nature of plastic deformation on which we shall comment shortly.

[1]) We would like to thank all authors the original figures of whom were used to illustrate the present text.

4. The physical basis of plasticity and viscoplasticity

If dislocations move in a glide plane they produce a permanent relative displacement of the size of one lattice spacing of the two adjacent atomic lattice planes. If billions of dislocations move one observes a deformation on a macroscopic scale.

The motion of a dislocation implies a cooperative movement of all the atoms along the line. This motion is *athermal* because *thermal* activation can assist the cooperative motion of particles only within a volume of linear dimensions of a few lattice spacings. Such an activated motion is typical for point defects. Thermally activated processes lead to strongly temperature and time dependent behaviour. This observation explains that dislocations make a crystal plastic whereas point defects make it viscous.[1]) It is in agreement with the well-known fact that plastic deformation changes the internal mechanical state of a crystal whereas no such change is observed in pure viscous deformation.

Strictly speaking, one always has simultaneous plastic and viscous deformation. In fact, in nature there exists neither a purely plastic nor a purely viscous crystalline material: one always finds the two properties combined in the form of the viscoplastic body. This statement is underlined by the fact that dislocations and certain types of point defects, the socalled eigen-defects, are mutually convertible.

Doubtless, plasticity is the more complex and more interesting of the two properties at least in the case of crystalline matter. In order to understand it one must understand the variation of the internal mechanical state during the deformation. This is the fundamental problem not only on the microscopic scale but also in the macroscopic constitutive laws.

Most important for the development of the dislocation state is the interaction between dislocations which takes place via the elastic stress and strain fields by which the dislocations surround themselves. The mathematical theory of this interaction is well developed. As we have seen in the pictures, the number of dislocations increases with deformation. Hence the mutual distances decrease and the interaction becomes stronger. Since the

---

[1]) This statement is somewhat simplified. Also dislocations undergo thermally activated processes, for instance when two dislocations intersect during their motion. Such processes occur within very small volumes and are usually accompanied by the production of special point-like defects such as jogs, kinks, vacancies.

interaction above all produces a mutual hindering we understand roughly that the resistance against plastic flow increases with the deformation (work-hardening).

More quantitatively, one studies primarily the interaction of two dislocations which, for simplicity, are often taken to be straight parallel lines. Due to the special structure of elasticity theory (biharmonic rather than harmonic) and, simultaneously, due to the fact that straight dislocations glide only in certain selected planes (their glide planes) there exist stable configurations of two dislocations in finite distances (typically e.g. 0.1 μm). If the two dislocations have opposite signs one speaks of dislocation dipoles. Such dipoles can possess a considerable mechanical stability and form obstacles to the motion of other dislocations. Fig. 15 shows how the interaction energy E of two dislocations of opposite sign changes with their distance x.

In a similar manner one can calculate the existence of stable groups of 3, 4 or more dislocations of one or two signs. These are the dislocation bundles, also called dislocation multipoles, which we had seen in our pictures. The degree of stability of such a multipole depends strongly on the mutual

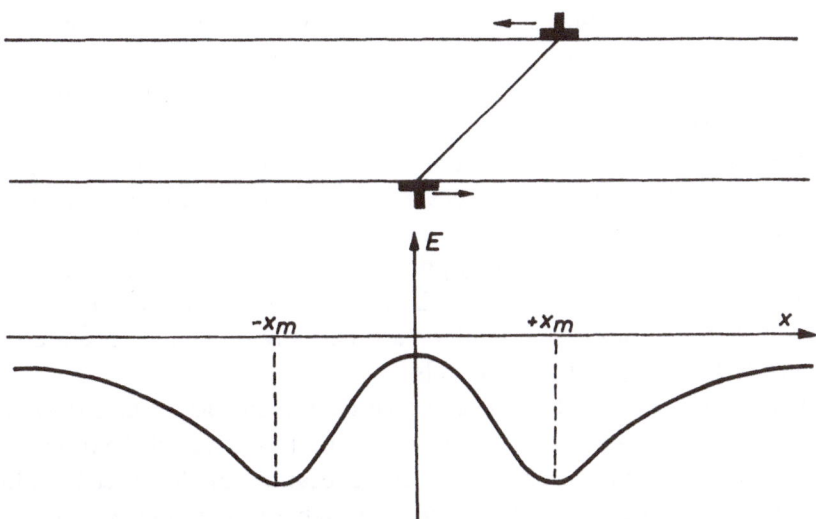

Fig. 15. Two edge dislocations of opposite sign in dipole configuration. Below: interaction energy E as a function of the horizontal relative distance of the dislocations (think lower dislocation fixed at $x = 0$).

position of the dislocations. Roughly speaking, the tendency is towards the formation of multipoles of increasing order during the proceeding deformation. This applies to unidirectional as well as to cyclic deformation.

We mention in passing that the dislocation arrangements built up during cyclic deformation are often only little stable. At a certain point they can decay into other more stable arrangements so giving rise to sudden larger deformations, so-called strain bursts (Neumann [8]). The tendency is towards a more and more pronounced cell structure.

5. Statistical and state type theories

Theories which use internal variables have been considered by many workers in the field. References are given further below. In our opinion a theory which claims to be a *physical* theory must not leave open the physical meaning of the internal variables. It is above all the physical identification of these variables which fills the frame of the theory with physical life.

When dealing with plasticity the variable part of the internal mechanical state is the dislocation state. Therefore we need variables which describe this state on a macroscopic scale. On the other hand, we have seen that devices on a microscopic scale are necessary for a direct observation of the dislocation state (or of any other imaginable hidden variable). For this reason we hesitate to call a theory of plasticity phenomenological if it uses internal variables.

We shall now argue that the number of internal state variables is infinite, a result likewise surmised by Drucker in 1960. A convincing way to do this is to construct a complete set of those variables. There are various possibilities out of which we prefer the description in terms of correlation functions of order n ($n = 1, 2, ..., \infty$). This method is suggested by the previous observation that dislocation distributions are partially ordered, hence are conveniently handled by statistical tools.

Let us first recall that in the so-called continuum theory of dislocations the internal state is specified by the tensor function $\alpha_{ij}(x)$ of the dislocation density which counts the strength of the dislocations of glide vector ($=$ Burgers vector) of direction j which pierce through a macroscopic area element oriented perpendicular to direction i. However, the observation of dislocation arrangements shows that dislocations do not form good line densities but much more irregular distributions. This is the main reason why the continuum theory of dislocations in its presently developed form can never become a good macroscopic theory of plasticity. It nevertheless is the fundamental physical theory for the phenomenon of plasticity because it describes correctly the physical situation on the microscopic scale, i.e. if the dislocation density tensor is specified in the form of delta functions along the dislocation lines. In this way $\alpha_{ij}(x)$ becomes a random function of position.[1])

The physical equations of this dislocation theory must now be combined with the mathematical equations of probability theory. This situation is similar to that encountered in the statistical theory of turbulence where the physical Navier-Stokes equations are combined with the probability theory. Using statistics means taking averages. For this we need the concept of the statistical ensemble, ergodic hypothesis etc. There is no space here for going into the details. We only recall two important results of the probability theory, namely: (i) the ensemble average of a random function of position (and/or time) is a macroscopic function of position (and/or time) and (ii) the complete macroscopic description of a random function of position (and/or time) requires the infinite set of n-point correlation functions of position (and/or time). These theorems prove our former statement on the number of internal state variables.

Let us now consider the function with $n = 1$, i.e. the average $\langle \alpha_{ij}(x) \rangle$ of the dislocation density. It is, in fact, this quantity which was introduced by Nye in 1953 as the tensor of dislocation density. Clearly the macroscopic description of a dislocation state by this tensor is very poor. For instance, $\langle \alpha_{ij}(x) \rangle = 0$ if dislocations of opposite sign in equal number pierce through the macroscopic area element at x. For evident reason of symmetry this is the case in the tension tests of which we have shown the electron micrographs. In order to specify the clearly visible dislocation arrangements one has to consider at least the 2-point correlation functions.

The average dislocation density tensor $\langle \alpha_{ij}(x) \rangle$ is generally not zero if the plastic strain is macroscopically inhomogeneous. This observation proves our previous statement that for physical reasons crystalline plastic materials are never simple bodies. In fact, since dislocation arrangements with $\langle \alpha_{ij}(x) \rangle \neq 0$ do not occur in experiments with homogeneous strain one cannot predict from such experiments the response of the body in cases where $\langle \alpha_{ij}(x) \rangle \neq 0$. The present result does not exclude, of course, that in

---

[1]) In this description the crystal is still treated as a material continuum!

certain situations the treatment of a plastic material as a simple body may be a reasonable or even excellent approximation.

Of much more practical importance than the average dislocation density tensor is the 2-point correlation function $\langle \alpha_{ij}(x_1)\alpha_{kl}(x_2)\rangle$. This is a 4th rank tensor which contains such important information as the total length of all dislocations in a volume element, densities of dislocation dipoles and also partial information about dislocation multipoles [9]. This observation shows that it is worthwhile to attempt a theory which includes the 2-point correlation tensor.

We mention briefly that the dislocation correlation functions can be replaced by the correlation functions of stress and elastic strain produced by the dislocations. This possibility reflects the observation that the internal mechanical state is determined by the dislocations but, equivalently, also by the fluctuations of stress and elastic strain produced by the dislocations. The fundamental fact that the *fluctuating* stresses and elastic strains determine the deformation behaviour of crystalline matter is not mentioned in many texts on plasticity.

As we have seen, a rigorous macroscopic theory of plasticity and viscoplasticity requires the inclusion of an infinite number of state variables. Such a theory would not be manageable anymore; therefore one has to look for simplifications. The most suggestive approximation consists in the restriction to a small number of variables which are best adapted to the kind of problems to be solved. It is the program of achieving such approximations where in my opinion the discoveries of solid state physics can be particularly helpful. That there is a sensible basis for this statement will show itself in the following brief report on macroscopic theories of plasticity which use concepts of dislocation theory.

6. Kinematics of the elastic-viscoplastic deformation

Consider a single crystal $\mathscr{C}$ at time t_0, free of any surface tractions and body forces, at a uniform absolute temperature θ_0, and choose this configuration, say (k_0), as *reference configuration* of $\mathscr{C}$.

Assume that the crystal is imperfect. Then, a global natural configuration, i.e., a stress-free configuration of the whole body does not exist. Let X be an arbitrary particle of $\mathscr{C}$ and let $N(X)$ be a material neighbourhood of X whose dimensions in (k_0) are small with respect to those of $\mathscr{C}$ but large in comparison with the mean separation distance between crystal defects. To determine the residual macroscopic elastic deformation of $N(X)$ in (k_0) we may, at least in principle, cut out of the body this neighbourhood and release it under constant positions of all dislocations.[1]) Let $(\tilde{k}_0)$ denote the configuration of $N(X)$ as obtained immediately after cutting out, i.e., before any time-dependent strain recovery could have taken place. We then call the deformation of $N(X)$ from $(\tilde{k}_0)$ to (k_0) the *residual elastic deformation* of $N(X)$ in the unloaded reference configuration (k_0).

Assume now that the crystal undergoes an elastic-viscoplastic deformation under the action of the external loads and of an inhomogeneous temperature field. Let (k) denote its *current configuration* at time t, and let x_0 and x be the position vectors of the material particle X in the configurations (k_0) and (k), respectively. The *motion* of the crystal is given by the one-parameter family of mappings

$$x = x(x_0, t) \tag{6.1}$$

and the particle *velocity* field by

$$\dot{x} = \frac{\partial}{\partial t} x(x_0, t). \tag{6.2}$$

The *deformation gradient* associated with the motion (6.1) is

$$F = \frac{\partial}{\partial x_0} x(x_0, t). \tag{6.3}$$

To determine the thermoelastic deformation undergone by $N(X)$ at time t we may repeat the cutting out procedure used at time t_0. Let $(\tilde{k})$ be the configuration of $N(X)$ obtained by cutting it out at time t, suddenly reducing its temperature to θ_0 and instantaneously releasing it from the constraints exerted by the remaining part of the crystal. Then, the deformation of $N(X)$ from $(\tilde{k})$ to (k) will be called the *thermoelastic deformation* of $N(X)$ at time t.

Like $(\tilde{k}_0)$, the local configuration $(\tilde{k})$ is a natural state only for the macroscopic stress produced by the external loads and the inhomogeneous temperature field. To remove also the microscopic stress produced by dislocations it would be necessary to cut the crystal into even smaller pieces

---

[1]) For thin specimens, this requirement has been pratically fulfilled by irradiating the crystal with fast neutrons before load-removal and thus pinning the dislocations in their under-load positions (see Mughrabi [2], [6]).

with diameters comparable to the mean separation distance between the dislocation lines. According to our definition of the configurations $(\tilde{k}_0)$ and $(\tilde{k})$ we shall always understand by stresses and strains the corresponding macroscopic quantities, i.e., mean values calculated over macroscopic surface and volume elements which, therefore, do not contain microscopic fluctuations. The only exception will be made in Sect. 8, where the microdynamics of viscoplastic flow will be analyzed by taking into account the action upon dislocations of macro- and microstresses as well.

The configurations $(\tilde{k}_0)$ and $(\tilde{k})$ were obviously defined to within a rigid-body rotation. We now remove this indeterminancy, by requiring that the mean lattice orientation, as defined at points far from crystal defects, be the same throughout the motion and for all particles $X \in \mathscr{C}$. According to this convention, the glide directions and planes in the configuration $(\tilde{k})$ will be parallel to those in the configuration (k_0) for any X and t. Then the deformation of $N(X)$ from $(\tilde{k}_0)$ to $(\tilde{k})$ is called the *viscoplastic deformation* of $N(X)$ at time t.

Let now Y be another particle of $N(X)$, and let dx, dx_0, $d\tilde{x}$, $d\tilde{x}_0$ denote the position vectors of Y with respect to X in the configurations (k), (k_0), $(\tilde{k})$, and $(\tilde{k}_0)$ (see Fig. 16). We define the *thermoelastic distortion*, A, the *residual elastic distortion*, A_0, and the *viscoplastic distortion*, P, by the relations

$$dx = A d\tilde{x}, \quad d\tilde{x}_0 = A_0 d\tilde{x}_0, \quad d\tilde{x} = P d\tilde{x}_0. \tag{6.4}$$

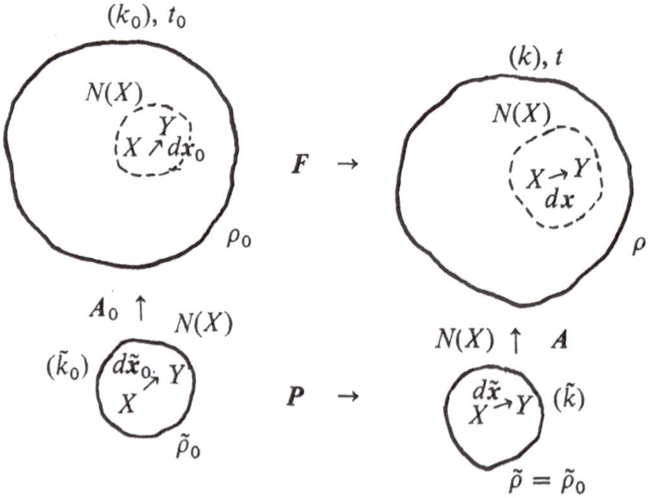

Fig. 16. Configurations of a single crystal during the elastic viscoplastic deformation.

We also assume that, for sufficiently small neighbourhoods $N(X)$, the so-defined values of the distortions at X do not depend on the choice of the neighbourhood $N(X)$ and of the particle $Y \in N(X)$. Consequently, by repeating the same procedure for all particles $X \in \mathscr{C}$ and times t, we may define the fields $A(x, t)$, $A_0(x_0)$, and $P(x, t)$.

Suppose that the tensor fields F, A, and P are continuously differentiable and admit *for every fixed time t* the inverses F^{-1}, A^{-1}, and P^{-1}. Since, by (6.3),

$$dx = F dx_0, \tag{6.5}$$

it follows from (6.4) that[1)]

$$F = APA_0^{-1}. \tag{6.6}$$

From (6.2) and (6.3) we deduce that

$$\dot{F} = \frac{\partial \dot{x}}{\partial x_0} = (\text{grad } \dot{x})F,$$

where a superposed dot denotes as usual the material time derivative, i.e., the time derivative for $x_0 = \text{const}$, and grad denotes $\partial/\partial x$ for $t = \text{const}$. The last equation may be written in the equivalent form

$$\text{grad } \dot{x} = \dot{F}F^{-1}, \tag{6.7}$$

from which, by making use of (6.6), we obtain

$$\text{grad } \dot{x} = \dot{A}A^{-1} + A\dot{P}P^{-1}A^{-1}. \tag{6.8}$$

[1)] Eckart [10] was the first to use time-dependent, local natural configurations, like $(\mathring{k})$- in order to separate the elastic from the inelastic part of the total deformation. The non, linear composition rule (6.6) of the distortions was independently introduced by Lee and Liu [11] and Fox [12], [13] in the special case $A_0 = I$, and by Teodosiu [14] and Rice [15] in the general case. The special choice of the orientation of the configuration $(\mathring{k})$ made in this paper was first used by Teodosiu [14], and independently considered by Rice [15] and Mandel [16]. As pointed out by Lee [17] and Mandel [16], the configuration $(\mathring{k})$ may be uniquely defined only if the macroscopic stress tensor T is uniquely determined by the current values of A and θ, being thus independent of the defect content of the crystal. This simplifying assumption is based upon the low sensitivity of the elastic characteristics to the viscoplastic deformation, which has been revealed by various experiments. It may be mathematically expressed by the additive decomposition of the free energy density into a thermoelastic and a dislocation-dependent part (see Eq. (8.6)$_1$ below).

This relation shows that in the nonlinear case the velocity gradient cannot be decomposed into an elastic and a plastic part unless the elastic strains and rotations are small enough to allow the replacement of A by 1 in the last term of Eq. (6.8).

Eqs. (6.6) and (6.8) are the basic kinematic equations. In the next section we shall study the dependence on the dislocation motion of the macroscopic quantities entering Eq. (6.8).

7. The relation between the viscoplastic deformation and the dislocation motion

Let n be the number of the (potential) glide systems of the crystal, and denote by $\tilde{g}^{(r)}$ and $\tilde{n}^{(r)}$, $r = 1, ..., n$, the unit vectors of the glide direction and of the normal to the associated glide plane of the rth glide system in the configuration $(\tilde{k})$, and by $g^{(r)}$ and $n^{(r)}$ the unit vectors of the same directions in the configuration $(\mathring{k})$.

If only the rth glide system is active, then

$$\dot{\overline{\text{d}\tilde{x}}} = (\mathring{a}^{(r)}\tilde{g}^{(r)} \otimes \tilde{n}^{(r)})\text{d}\tilde{x},$$ (7.1)

where $\mathring{a}^{(r)}$ is the rate of the viscoplastic shear in the glide system (r). In the general case, due to the vectorial additivity of the velocities, we have

$$\dot{\overline{\text{d}\tilde{x}}} = (\sum_r \mathring{a}^{(r)}\tilde{g}^{(r)} \otimes \tilde{n}^{(r)})\text{d}\tilde{x},$$ (7.2)

where the sum extends over all active glide systems. On the other hand, by differentiating (6.4)$_3$ with respect to t, we obtain

$$\dot{\overline{\text{d}\tilde{x}}} = \dot{P}\text{d}x_0 = \dot{P}P^{-1}\text{d}\tilde{x}.$$ (7.3)

By comparing (7.2) with (7.3), we deduce

$$\dot{P}P^{-1} = \sum_r \mathring{a}^{(r)}\tilde{g}^{(r)} \otimes \tilde{n}^{(r)}.$$ (7.4)

It should be noted that the current value of P is *not* uniquely determined by the current values of $\mathring{a}^{(r)}$, $r = 1, ..., n$. To obtain P it would be necessary to integrate the system of differential equations

$$\dot{P} = (\sum_r \mathring{a}^{(r)}\tilde{g}^{(r)} \otimes \tilde{n}^{(r)})P$$ (7.5)

for given time-dependences of $\mathring{a}^{(r)}$, $r = 1, ..., n$, with the initial condition $P(t_0) = 1$.

Consider now the expansion of a dislocation loop in the glide plane of unit normal $\tilde{n}$ in the configuration $(\tilde{k})$. Let $\vec{b}$ denote the Burgers vector of the loop calculated by choosing the positive direction of the loop clockwise when looking down along $\tilde{n}$. It may be shown ([14], p. 852) that the viscoplastic distortion produced by the motion of the dislocation loop is

$$P = 1 + \delta(\tilde{S})\vec{b} \otimes \tilde{n}, \tag{7.6}$$

where $\tilde{S}$ is the surface swept out by the dislocation during the time interval $[t_0, t]$. The distribution $\delta(\tilde{S})$ is defined by the equation

$$\int_V \varphi(x)\delta(\tilde{S})\mathrm{d}V = \int_{\tilde{S}} \varphi(x)\mathrm{d}S, \tag{7.7}$$

where V is any three-dimensional region containing $\tilde{S}$. From (7.6) it follows that the viscoplastic shear associated with the dislocation motion is

$$\tilde{a} = \vec{b}\delta(\tilde{S}), \tag{7.8}$$

where $\vec{b}$ is the magnitude of the Burgers vector. Assume now that many dislocation loops with the same Burgers vector $\vec{b}$ are expanding over the glide plane of the system (r). According to (7.8), the mean value of the corresponding viscoplastic shear is

$$\tilde{a}^{(r)} = \frac{\vec{b}}{\Delta\tilde{V}} \int_{\Delta\tilde{V}} \delta(\tilde{S}^{(r)})\mathrm{d}V = \frac{\vec{b}}{\Delta\tilde{V}} \int_{\tilde{S}^{(r)}} \mathrm{d}S = \frac{\vec{b}\tilde{S}^{(r)}}{\Delta\tilde{V}}, \tag{7.9}$$

where $\Delta\tilde{V}$ denotes the region occupied by the material neighbourhood $N(X)$ in the configuration $(\tilde{k})$ and its volume, and $\tilde{S}^{(r)}$ is the area swept out by all dislocation loops of the system (r) in the time interval $[t_0, t]$. Differentiating (7.9) with respect to t and assuming that plastic flow produces no volume changes ($\Delta\tilde{V} = $ const) we obtain

$$\dot{\tilde{a}}^{(r)} = \frac{\vec{b}\dot{\tilde{S}}^{(r)}}{\Delta\tilde{V}}, \tag{7.10}$$

and hence (7.4) becomes

$$\dot{P}P^{-1} = \frac{1}{\Delta\tilde{V}}\sum_r \dot{\tilde{S}}^{(r)}\vec{b}^{(r)} \otimes \tilde{n}^{(r)}. \tag{7.11}$$

To calculate the term $A\dot{P}P^{-1}A^{-1}$ we now make use of Nanson's formula

$$\tilde{n}\mathrm{d}\tilde{S} = jA^T n\mathrm{d}S, \tag{7.12}$$

65

Title: *E. Kröner and C. Teodosiu*

where

$$j = \frac{\rho}{\tilde{\rho}} = \frac{\Delta \tilde{V}}{\Delta V}, \tag{7.13}$$

ρ and $\tilde{\rho}$ are, respectively, the mass densities in the configurations (k) and $(\tilde{k})$, and ΔV and dS are the volume and surface elements corresponding to $\Delta \tilde{V}$ and $d\tilde{S}$ in the configuration (k). Replacing $\tilde{S}$, $\tilde{n}$, and n in (7.12) by $\tilde{S}^{(r)}$, $\tilde{n}^{(r)}$, and $n^{(r)}$, respectively, and dividing both sides of this equation by dt, we infer

$$\tilde{n}^{(r)} \dot{\tilde{S}}^{(r)} = \frac{\Delta \tilde{V}}{\Delta V} A^T n^{(r)} \dot{S}^{(r)},$$

or

$$\frac{1}{\Delta \tilde{V}} \dot{\tilde{S}}^{(r)}(A^{-1})^T \tilde{n}^{(r)} = \frac{1}{\Delta V} \dot{S}^{(r)} n. \tag{7.14}$$

Next, we may write

$$A\tilde{b}^{(r)} = b^{(r)}, \tag{7.15}$$

where $b^{(r)}$ is the deformed Burgers vector of the dislocation loops.

From (7.11), (7.14), and (7.15) it follows that

$$A\dot{P}P^{-1}A^{-1} = \frac{1}{\Delta \tilde{V}} \sum_r \{ \dot{\tilde{S}}^{(r)}(A\tilde{b}^{(r)}) \otimes [(A^{-1})^T \tilde{n}^{(r)}] \},$$

whence

$$A\dot{P}P^{-1}A^{-1} = \frac{1}{\Delta V} \sum_r \dot{S}^{(r)} b^{(r)} \otimes n^{(r)}. \tag{7.16}$$

We note that

$$\dot{S}^{(r)} = \int_{l^{(r)}} v dl = v^{(r)} l^{(r)}, \tag{7.17}$$

where v is the expansion velocity of the loops, i.e., the projection of the dislocation velocity on the outward normal to the loop[1]), $l^{(r)}$ is the total

---

[1]) In general, v varies from one loop to the other and also along the same dislocation loop. When it is difficult to recognize the loop to whom a given dislocation segment dl belongs, v is to be calculated as projection of the dislocation velocity on the vector $n \times dl$.

length of the mobile dislocation lines at time t, and $v^{(r)}$ is the *mean expansion velocity*, which is given by

$$v^{(r)} = \frac{1}{l^{(r)}} \int_{l^{(r)}} v\,dl. \tag{7.18}$$

Finally, by defining the *mobile dislocation density*, α_M, as total length of the mobile dislocation lines per unit volume in the configuration (k), i.e.,

$$\alpha_M^{(r)} = l^{(r)}/\Delta V \tag{7.19}$$

we obtain from (7.16) and (7.17)

$$A\dot{P}P^{-1}A^{-1} = \sum_r \alpha_M^{(r)} v^{(r)} b^{(r)} \otimes n^{(r)}, \tag{7.20}$$

and hence Eq. (6.8) becomes

$$\text{grad }\dot{x} = \dot{A}A^{-1} + \sum_r \alpha_M^{(r)} v^{(r)} b^{(r)} \otimes n^{(r)}. \tag{7.21}$$

Note that Eq. (7.20) may be also written as

$$A\dot{P}P^{-1}A^{-1} = \sum_r \dot{a}^{(r)} g^{(r)} \otimes n^{(r)}, \tag{7.22}$$

where

$$\dot{a}^{(r)} = b^{(r)} \alpha_M^{(r)} v^{(r)} \tag{7.23}$$

is the rate of the viscoplastic shear in the glide system (r), measured with respect to the vectors $g^{(r)}$ and $n^{(r)}$. Equation (7.23) may be viewed as a generalization of Orowan's relation [18] to the case of the finite elastic-viscoplastic deformation.

8. Thermodynamics of the elastic-viscoplastic deformation

Before passing to this section a word on the thermodynamic aspects of plasticity may be appropriate. As is well-known, the entropy, as a measure of the degree of order, plays a decisive role in thermodynamic considerations. In our problem the interest is not only in the degree of order as determined by the heat vibrations, but also in the degree of order of the dislocation arrangement, which is responsible to a very large extent for the instantaneous state of the body.

Whereas the first part can be accounted for by one function (the entropy production) one needs an infinite number of functions in order to describe the dislocation state. It therefore seems that the development of a good method for the description of the internal mechanical state and its variation has to be more complex than the application of classical thermodynamic considerations which usually lead to certain restrictions on the form of the functions or functionals which characterize the material properties. We mention only briefly that the plastic state is usually far away from thermal equilibrium, a fact which may lead to further complications. However, due to the high complexity of the problem, we confine ourselves in the following with describing the thermodynamics of the macroscopic deformation.

The thermodynamic process must be compatible with the laws of *balance of linear and angular momenta*

$$\operatorname{div} T + \rho f = \rho \ddot{x}, \quad T = T^T, \tag{8.1}$$

and the law of *balance of energy*

$$\rho \dot{\varepsilon} = T \cdot \operatorname{grad} \dot{x} + \operatorname{div} q + \rho r. \tag{8.2}$$

Here T is the stress tensor, f is the specific *body force* per unit mass, ε is the *internal energy density* per unit mass, q is the *heat flux* vector, and r is the *heat supply* per unit mass and unit time.

Along with Eq. (8.2), which expresses the first principle of thermodynamics, we shall use the second principle in the form of the *Clausius-Duhem inequality*

$$\rho \dot{\eta} \geqq \operatorname{div}(q/\theta) + \rho r/\theta, \tag{8.3}$$

where η is the *entropy density* per unit mass.

Eliminating r between Eqs. (8.2) and (8.3), and taking into account (6.8), we obtain

$$\rho \dot{\eta} \geqq \rho \dot{\varepsilon} - (\dot{A}A^{-1} + A\dot{P}P^{-1}A^{-1}) \cdot T - \frac{1}{\theta} q \cdot \operatorname{grad} \theta \geqq 0.$$

Finally, by introducing the *free energy density*

$$\psi = \varepsilon - \eta\theta, \tag{8.4}$$

the last inequality becomes

$$-\rho \dot{\psi} - \rho \eta \dot{\theta} + (\dot{A}A^{-1} + A\dot{P}P^{-1}A^{-1}) \cdot T - \frac{1}{\theta} q \cdot \operatorname{grad} \theta \geqq 0. \tag{8.5}$$

We adopt now a set of constitutive equations, by taking as independent variables besides the classical thermoelastic variables A, θ, and grad θ, a set of internal (or structural) state variables, namely the dislocation densities $\alpha^{(1)}, ..., \alpha^{(n)}$, and the concentration c of point defects per unit mass[1]). These constitutive equations are:

$$
\begin{aligned}
\psi &= \hat{\psi}(A, \theta) + \bar{\psi}(S), \\
\eta &= \hat{\eta}(A, \theta, S), \\
T &= \hat{T}(A, \theta, S), \\
q &= \hat{q}(A, \theta, \text{grad } \theta, S),
\end{aligned}
\right\}
\tag{8.6}
$$

where

$$
S = \{\alpha^{(1)}, ..., \alpha^{(n)}, c\}
\tag{8.7}
$$

denotes the vector of the structural state variables.

The particular additive form adopted for ψ assures the independence of the elastic constants on the lattice defect content of the crystal (cf. footnote p. 63). $\bar{\psi}(S)$ may be interpreted as residual free energy in the configuration (k) [15].

Including grad θ as independent variable into the first three Eqs. (8.6) proves to be incompatible with the Clausius-Duhem inequality and, therefore, it has been omitted from the very beginning.

Let us now investigate the restriction imposed by the Clausius-Duhem inequality on the possible form of the constitutive equations. From (8.6)$_1$ we find

$$
\dot{\psi} = \frac{\partial \hat{\psi}}{\partial A} \cdot \dot{A} + \frac{\partial \hat{\psi}}{\partial \theta} \dot{\theta} + \frac{\partial \bar{\psi}}{\partial S} \cdot \dot{S}.
$$

Introducing this expression into (8.5) gives after some manipulation

$$
\left(T(A^{-1})^T - \rho \frac{\partial \hat{\psi}}{\partial A} \right) \cdot \dot{A} - \rho \left(\eta + \frac{\partial \hat{\psi}}{\partial \theta} \right) \dot{\theta} - \rho \frac{\partial \bar{\psi}}{\partial S} \cdot \dot{S} +
$$

$$
+ (A\dot{P}P^{-1}A^{-1}) \cdot T + \frac{1}{\theta} q \cdot \text{grad } \theta \geq 0.
\tag{8.8}
$$

---

[1]) There exists an extensive literature concerning the thermodynamics of materials with internal state variables (see, e.g., [14, 16, 19–36]). Some of the papers on this topics discuss the physical nature of the internal state variables for the viscoplastic deformation, too [29–36].

It may be shown that there always exists a thermodynamic process for which A and θ take arbitrarily prescribed values at a given time. Therefore, in order for the Clausius-Duhem inequality to be satisfied by all thermodynamic processes compatible with the balance laws, it is necessary that the cofactors of A and θ in the above inequality vanish, whence

$$T = \rho \, \frac{\partial \hat{\psi}(A, \theta)}{\partial A} \, A^T, \quad \eta = - \frac{\partial \hat{\psi}(A, \theta)}{\partial \theta}. \tag{8.9}$$

The possible form of the constitutive equations $(8.6)_1$, $(8.6)_4$ and (8.9) may be further restricted by requiring their form invariance under superimposed rigid-body motions. It results that:

$$\left. \begin{aligned} \psi &= \check{\psi}(E, \theta) + \bar{\psi}(S), \\[4pt] T &= \rho A \, \frac{\partial \check{\psi}(E, \theta)}{\partial E} \, A^T, \quad \eta = - \frac{\partial \check{\psi}(E, \theta)}{\partial \theta}, \\[4pt] q &= A\check{q}(E, \theta, A, \operatorname{grad} \theta, S), \end{aligned} \right\} \tag{8.10}$$

where

$$E = \tfrac{1}{2}(A^T A - 1) \tag{8.11}$$

is the elastic strain tensor.

Equation (6.8) suggests that the term $(A\dot{P}P^{-1}A^{-1}) \cdot T$ in (8.5) is connected with the mechanical power expended in viscoplastic flow. Indeed, from (7.20) it follows that

$$(A\dot{P}P^{-1}A^{-1}) \cdot T = \sum_r \dot{a}^{(r)} \tau^{(r)}, \tag{8.12}$$

where

$$\tau^{(r)} = (g^{(r)} \otimes n^{(r)}) \cdot T = (Tn^{(r)}) \cdot g^{(r)} \tag{8.13}$$

is the *resolved shear stress*, in the glide system (r), i.e., the component of the stress tensor acting on the glide plane and in the glide direction of the system (r).

Taking into account (8.4), (8.7) and $(8.10)_2$, the balance law of energy (8.2) and the Clausius-Duhem inequality (8.8) assume the following forms:

$$\rho \theta \dot{\eta} + \rho \, \frac{\partial \bar{\psi}}{\partial S} \cdot \dot{S} = (A\dot{P}P^{-1}A^{-1}) \cdot T + \operatorname{div} q + \rho r, \tag{8.14}$$

$$-\rho \, \frac{\partial \bar{\psi}}{\partial S} \cdot \dot{S} + (A\dot{P}P^{-1}A^{-1}) \cdot T + \frac{1}{\theta} \, q \cdot \operatorname{grad} \theta \geqq 0. \tag{8.15}$$

When no dislocation motion takes place, these relations take the familiar forms corresponding to the purely thermoelastic deformation:

$$\rho\theta\dot{\eta} = \text{div } q + \rho r, \tag{8.16}$$

$$q \cdot \text{grad } \theta \geqq 0. \tag{8.17}$$

Inspection of Eqs. (8.15) and (8.17) reveals that the non-thermal part of the dissipation is entirely due to the dislocation motion.

9. Microdynamics of viscoplastic flow

In this section we shall analize the relations between the state variables on the microscale, in order to determine the possible form of the macroscopic flow laws. For the sake of simplicity we consider only the case of the single glide.

a) *Quasistationary motion of dislocations*

The notion of quasistationarity of the dislocation motion has been independently introduced by Mecking [38] and by de Rosset and Granato [39], being further discussed by Mecking and Lücke [40], Neuhäuser et al. [41], and others.

Suppose that, at a given time t, the mobile dislocation density in the active glide system is α_M. From these glide dislocations a small fraction, say α_F, are moving between obstacles with the flight velocity, v_F. The other glide dislocations (density α_W) are waiting at each obstacle (such as forest dislocations, point defects etc.) a mean waiting time t_W before overcoming it by thermal activation. According to Eq. (7.23) we may write

$$\dot{a} = b\alpha_F v_F. \tag{9.1}$$

The flight motion is said to be *quasistationary* if

$$\alpha_F \approx \text{const.} \tag{9.2}$$

during the mean flight time t_F. This requirement is fulfilled if no essential changes in the values of the resolved shear stress τ, temperature θ, and structure S take place during the time t_F, i.e.,

$$\dot{\tau}t_F \ll \tau, \quad \dot{\theta}t_F \ll \theta, \quad \dot{c}t_F \ll c, \quad \dot{\alpha}t_F \ll \alpha. \tag{9.3}$$

Since t_F is as small as 10^{-6} sec, [40], conditions (9.3) are certainly satisfied for all macroscopic, currently performed, tests.

Differentiating (9.2) with respect to t gives

$$\dot{\alpha}_F = \dot{\alpha}_F^+ - \dot{\alpha}_F^- = \alpha_W/t_W - \alpha_F/t_F = 0, \tag{9.4}$$

where $\dot{\alpha}_F^+$ is the mobilization rate of the waiting dislocations, and $\dot{\alpha}_F^-$ is the immobilization rate of the flying dislocations. Since $\alpha_W + \alpha_F = \alpha_M$, Eq. (9.4) yields

$$\alpha_F/t_F = \alpha_W/t_W = \alpha_M/(t_W + t_F), \tag{9.5}$$

and hence

$$\alpha_F v_F = \alpha_M l_F/(t_W + t_F) = \alpha_M v, \tag{9.6}$$

where $l_F = v_F t_F$ is the mean distance between obstacles and

$$v = l_F/(t_W + t_F). \tag{9.7}$$

From (9.1) and (9.6) it follows that

$$\dot{a} = b\alpha_M v, \tag{9.8}$$

where now α_M means the *total* mobile dislocation density, which includes flying and waiting dislocations as well.

Assume now that during the viscoplastic deformation stronger obstacles (such as Lomer-Cottrell barriers) are generated, which can no longer be overcome by thermal activation. Let us denote by $L \gg l_F$ the mean separation distance between these obstacles[1]), i.e., the mean free path of dislocations before their final immobilization, and by t_L the mean life time of a mobile dislocation. The entire dislocation motion is *quasistationary* if

$$\alpha_M \approx \text{const} \tag{9.9}$$

during the life time t_L. This condition is satisfied if the parameters τ, θ and S do not significantly change within the time t_L, i.e.,

$$tt_L \ll \tau, \quad \dot{\theta}t_L \ll \theta, \quad \dot{c}t_L \ll c, \quad \dot{\alpha}t_L \ll \alpha. \tag{9.10}$$

As t_L is of the order of a few seconds, condition (9.10) is always fulfilled for moderate strain rates. On the contrary, sudden changes of the tempera-

---

[1]) In most theories of work-hardening (see, e.g., [42]–[45], this free path is assumed to decrease with increasing plastic deformation, thus producing the increase of the flow stress.

ture or strain rate cause a transitory behaviour, until a new quasistationary state is attained [38, 40].

Condition (9.9) may be rewritten as

$$\dot{\alpha}_M = \dot{\alpha}_M^+ - \dot{\alpha}_M^- = \dot{\alpha}_M^+ - \alpha_M/t_L = \dot{\alpha}_M^+ - \alpha_M v/L = 0, \tag{9.11}$$

whence

$$\alpha_M v = \dot{\alpha}_M^+ L, \tag{9.12}$$

where $\dot{\alpha}_M^+$ is the production rate of the mobile dislocation density. Finally, if the immobilization of the dislocations by strong barriers is definitive, then the production rate of the mobile dislocations equals the rate of the total dislocation density, i.e., $\dot{\alpha}_M^+ = \dot{\alpha}$, and hence

$$\alpha_M v = \dot{\alpha} L. \tag{9.13}$$

Introducing now (9.13) into (9.8) leads to

$$\dot{a} = b\dot{\alpha}L. \tag{9.14}$$

This relation has been repeatedly used in the literature [42–45]. If the dependence of L on α is known, then Eq. (9.14) expresses the viscoplastic shear rate in terms of quantities characterizing the dislocation multiplication and motion.

b) *Thermally activated motion of dislocations*

In this subsection we summarize the present knowledge concerning the thermally activated dislocation motion. For details on this subject we refer to the review articles by Seeger [43], Conrad [46], Argon [47], Evans and Rawlings [48], and to Gibbs [49].

The obstacles impeding the dislocation motion may be conveniently divided into two groups which will be referred to as extended and non-extended obstacles.

Extended obstacles generate long-range stress fields, which vary only slowly with the position of the dislocation on the glide plane and cannot be overcome by thermal activation. Therefore, such obstacles are also termed *athermal*. To this group belong second phase particles, dislocation pile-ups lying in the same or in parallel glide planes etc. The resolved shear stress generated by extended obstacles is commonly denoted by τ_μ, in order to recall that it depends on temperature only through the temperature-dependence of the elastic constants, e.g., the shear modulus, μ; it is sometimes called athermal stress.

Non-extended (or local) obstacles generate short-range stress fields, which are effective only over a few atomic distances. Thermal fluctuations can assist the applied stress in overcoming such obstacles, which are also named *thermal* obstacles. To this group belong impurity atoms, forest dislocations, precipitates, jogs in the glide dislocations etc.

Assume that the resolved shear stress τ produced by the external loads exceeds the athermal stress τ_μ, so that the extended obstacles can be overcome. Then the dislocations move from sources to strong, insurmountable barriers with a mean velocity given by Eq. (9.7).

It may be shown that the mean waiting time of a glide dislocation in front of a non-extended obstacle is given by

$$1/t_W = v^+ - v^-, \tag{9.15}$$

where

$$
\left.
\begin{aligned}
v^+ &= v_0 \exp\left[-\frac{\Delta G(S, \tau^*, \theta)}{k\theta} \right] \\
v^- &= v_0 \exp\left[-\frac{\Delta G(S, -\tau^*, \theta)}{k\theta} \right],
\end{aligned}
\right\}
\tag{9.16}
$$

are, respectively, mean values of the forward and backward jumps of a dislocations over the obstacles, v_0 is the attempt frequency of a dislocation for overcoming an obstacle, $\Delta G(S, \tau^*, \theta)$ is the free enthalpy which must be supplied by thermal lattice vibrations in order to push a dislocation over an obstacle, $k = 1{,}38 . 10^{-25}$ J/mol°K is Boltzmann's constant, and τ^* is the effective shear stress acting on the dislocation, which is defined as difference between the resolved stress produced by the external loads and the athermal stress τ_μ, i.e.,

$$\tau^* = \tau - \tau_\mu(\alpha^{(1)}, ..., \alpha^{(n)}). \tag{9.17}$$

The attempt frequency v_0 is approximately given [50] by the formula $v_0 = bv_D/l$, where l denotes the mean separation between neighbouring local obstacles along the dislocation line and $v_D = 8 . 10^{12}$ sec is the Debye frequency of thermal lattice vibrations.

The free enthalpy ΔG depends on the interaction free energy between a glide dislocation and an obstacle. A typical plot of ΔG against τ^* for given temperature and structure is shown in Fig. 17, where $\Delta G_0 \equiv \Delta G(S, 0, \theta)$, and the stress τ_0^* is defined by $\Delta G(S, \tau_0^*, \theta) = 0$. The temperature-depen-

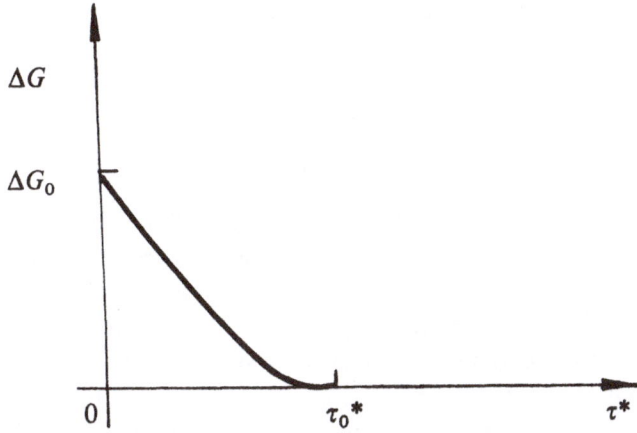

Fig. 17. Typical plot of the free enthalpy of activation against the effective stress.

dence exhibited by ΔG is weak and usually ascribed to the temperature-dependence of the elastic constants, too.

According to Schoeck [51], the free enthalpy may be written in the form

$$\Delta G = \Delta g - \tau^* b l \Delta x, \qquad (9.18)$$

where Δx is the so-called activation distance, and Δg and $b l \Delta x$ mean the work done over Δx against the dislocation obstacle interaction force and by the effective stress τ^*, respectively. By making use of (9.16) and (9.18), Eq. (9.15) becomes

$$1/t_W = (2bv_D/l) \exp\left(-\frac{\Delta g}{k\theta}\right) \sinh \frac{\tau^* b l \Delta x}{k\theta}. \qquad (9.19)$$

On the other hand, the flight velocity, v_F, is determined by the drag force acting on the dislocation and may be expressed as

$$v_F = b\tau^*/B, \qquad (9.20)$$

where B is the drag constant. Hence

$$t_F = l_F/v_F = l_F B/(b\tau^*). \qquad (9.21)$$

Introducing now (9.19) and (9.21) into (9.7), and assuming for the sake of simplicity that $l_F \approx l$, we find

$$v = \frac{b}{\exp\left(\dfrac{\Delta g}{k\theta}\right)\left(2v_D \sinh \dfrac{\tau^* b l \Delta x}{k\theta}\right)^{-1} + \dfrac{B}{\tau^*}}. \qquad (9.22)$$

75

Finally, by substituting (9.22) into (9.8), we obtain

$$\dot{a} = \frac{b^2 \alpha_M}{\exp\left(\dfrac{\Delta g}{k\theta}\right)\left(2\nu_D \sinh \dfrac{\tau^* b l \Delta x}{k\theta}\right)^{-1} + \dfrac{B}{\tau^*}}. \tag{9.23}$$

It should be noted that passing from (9.22) to (9.23) requires averaging for all dislocation segments moving within the macroscopic volume element. Therefore, when using Eq. (9.23) for the interpretation of macroscopic experiments, the quantities Δg, τ^*, l, Δx, and α_M should be regarded as phenomenological parameters, which correspond to those denoted by the same symbols in Eq. (9.22) only in special circumstances [49, 52]. However, Eq. (9.23) has the merit of suggesting the right macroscopic variables to be included in the continuum theory, and will be used to this aim in the following two sections. Before closing this section, we mention that for $\tau^* \geqq \tau_0^*$ the attempt frequency equals the effective frequency of overcoming local obstacles. Then the mean waiting time vanishes, i.e., the local obstacles become penetrable to the applied shear stress without thermal assistance, and the dislocation motion is only drag-controlled. Hence Eq. (9.23) is to be replaced for $\tau^* \geqq \tau_0^*$ by

$$\dot{a} = b^2 \alpha_M \tau^* / B. \tag{9.24}$$

10. Rate-independent plasticity

The numerical analysis of Eq. (9.23) for current values of the parameters involved shows that for sufficiently small shear rates at a given temperature, or for sufficiently high temperatures at a given plastic shear rate, the effective stress τ^*, implicitly defined by (9.23), is approximately zero. Then the flow condition becomes

$$\tau = \tau_\mu(\alpha^{(1)}, \ldots, \alpha^{(n)}), \tag{10.1}$$

and hence the deformation proceeds athermal and rate-independent.

The critical values of $\dot{a}$ and θ corresponding to the boundary between athermal and thermally activated processes may be found by neglecting in Eq. (9.23) the terms that account for back jumps and drag, and setting $\tau^* = 0$ in the remaining equation. It results that the plastic flow may be considered athermal if, for a given temperature θ^1),

$$\dot{a} < \dot{a}_{cr} = b^2 \alpha_M v_D \exp\left(-\frac{\Delta G_0}{k\theta}\right) \tag{10.2}$$

or, alternatively, for a given plastic shear rate,

$$\theta > \theta_{cr} = \frac{\Delta G_0}{k \ln(b^2 v_D \alpha_M / \dot{a})}. \tag{10.3}$$

Suppose that an isothermal elastic-plastic deformation proceeds at sufficiently small plastic shear rates such that condition (10.2) be satisfied for all active glide systems, i.e.,

$$\dot{a}^{(r)} < \dot{a}^{(r)}_{cr} = b^2 \alpha_M^{(r)} v_D \exp\left(-\frac{\Delta G_0^{(r)}}{k\theta}\right)$$

for $r = 1, ..., p$. Then, since $\tau^{*(r)} \approx 0$, in order for the glide system (r) to be active it is necessary that the corresponding resolved shear stress satisfies the condition

$$\tau^{(r)} = \tau_\mu^{(r)}(\alpha^{(1)}, ..., \alpha^{(n)}), \tag{10.4}$$

where $\tau^{(r)}$ is the athermal shear stress in the glide system (r).

If plastic flow takes place, say on the first p glide systems, then Eq. (10.4) is identically satisfied for $r = 1, ..., p$. Differentiating the corresponding equations with respect to t yields

$$\dot{\tau}^{(r)} = \frac{\partial \tau_\mu^{(r)}}{\partial \alpha^{(1)}} \dot{\alpha}^{(1)} + ... + \frac{\partial \tau_\mu^{(r)}}{\partial \alpha^{(p)}} \dot{\alpha}^{(p)}, \quad r = 1, ..., p. \tag{10.5}$$

By solving this system with respect to $\dot{\alpha}^{(1)}, ..., \dot{\alpha}^{(p)}$, we obtain

$$\dot{\alpha}^{(r)} = \sum_{s=1}^{p} A^{(rs)}(\alpha^{(1)}, ..., \alpha^{(n)}) \dot{\tau}^{(s)}, \tag{10.6}$$

where $A^{(rs)}$ are known functions of their arguments.

Assuming now that the entire dislocation motion proceeds quasi-

---

[1]) Combining this result with Eq. (9.24), we deduce that the range of the plastic shear rate for which dislocation motion needs thermal activation is

$$\dot{a}_{cr} < \dot{a} < b^2 \alpha_M \tau_0^* / B.$$

stationary, we may apply Eq. (9.14) for every active glide system, thus obtaining

$$\dot{a}^{(r)} = b^{(r)} L^{(r)}(\alpha^{(1)}, \ldots, \alpha^{(n)}) \sum_{s=1}^{p} A^{(rs)}(\alpha^{(1)}, \ldots, \alpha^{(n)}) \dot{t}^{(s)}, \tag{10.7}$$

where $L^{(r)}(\alpha^{(1)}, \ldots, \alpha^{(n)})$ is the mean free path of the dislocations in the glide system (r). Finally, substituting (10.7) into (7.22) yields

$$A \dot{P} P^{-1} A^{-1} = \sum_{r,s=1}^{p} A^{(rs)} L^{(r)} \dot{t}^{(s)} b^{(r)} \otimes n^{(r)}, \tag{10.8}$$

where the arguments of $A^{(rs)}$ and $L^{(r)}$ have been omitted for conciseness. This equation[1]) is valid if at least one glide system is active. If $\tau^{(r)} < \tau_\mu^{(r)}$ for any r, then all rates $\dot{\alpha}^{(r)}$, $\dot{a}^{(r)}$ vanish, and hence $A \dot{P} P^{-1} A^{-1} = 0$, i.e., the deformation proceeds elastically.

11. Rate-dependent plasticity (viscoplasticity)

For $\theta < \theta_{cr}$ or $\dot{a} > \dot{a}_{cr}$ the influence of the thermal activation and/or that of the viscous drag can no longer be disregarded, and the plastic behaviour exhibits temperature and rate effects.

A necessary condition for the activation of an arbitrary glide system (r) is that the resolved shear stress exceeds the athermal stress[2]), i.e.,

$$\tau^{*(r)} = \tau^{(r)} - \tau_\mu^{(r)}(\alpha^{(1)}, \ldots, \alpha^{(n)}) > 0. \tag{11.1}$$

Assume that the first p systems are active. The contribution of the glide system (r) to the viscoplastic flow is characterized by the total dislocation density, $\alpha^{(r)}$, the mobile dislocation density, $\alpha_M^{(r)}$, the mean dislocation velocity $v^{(r)}$, and the mean free path, $L^{(r)}$, of the dislocations belonging to that system. Indeed, by (7.20), we have

$$A \dot{P} P^{-1} A^{-1} = \sum_{r=1}^{p} \alpha_M^{(r)} v^{(r)} b^{(r)} \otimes n^{(r)}. \tag{11.2}$$

The microscopic analysis done in Sect. 9, as well as the reasoning which

---

[1]) The linear form of Eq. (10.8) has been discussed by Zarka [30, 36] for a special choice of the functions $L^{(r)}$. The present form of this equation has been independently derived by Teodosiu [31].

[2]) For $\tau^{*(r)} = 0$, Eq. (9.23) yields $\dot{a}^{(r)} = 0$.

led to Eq. (10.6), suggests the adoption of the following system of algebraic and differential equations for the evolution of the parameters listed above:

$$v^{(r)} = \begin{cases} \vartheta^{(r)}(\tau^{(r)} - \tau_\mu^{(r)}, S, \theta) & \text{for} \quad r = 1, ..., p, \\ 0 & \text{for} \quad r = p+1, ..., n, \end{cases}$$

$$\dot{\alpha}^{(r)} = \begin{cases} \sum_{s=1}^{p} B^{(rs)}(S, \theta)t^{(s)} & \text{for} \quad r = 1, ..., p, \\ 0 & \text{for} \quad r = p+1, ..., n, \end{cases} \qquad (11.3)$$

$$\dot{\alpha}_M^{(r)} = \dot{\alpha}^{(r)} - \alpha_M^{(r)} v^{(r)}/L^{(r)},$$

$$L^{(r)} = \hat{L}^{(r)}(S, \theta).$$

This system does no longer make use of the quasistationarity of the entire dislocation motion during the life time t_L. The functions $\tau_\mu^{(r)}$, $\hat{L}^{(r)}$, $\vartheta^{(r)}$, $B^{(rs)}$ are characteristics of the given crystal. Their form may be theoretically deduced from microscopic models, as has been done under certain simplifying hypotheses by Zarka [30], or may be chosen on empirical grounds.

We complete the system (11.3) by adopting for the concentration c of the point defects the evolution equation

$$\dot{c} = c_1(S, \theta) \, tr(\text{grad} \, \dot{x}) + c_2(S, \theta)\dot{\theta} + \sum_r \bar{c}_r(S, \theta)\dot{\alpha}_M^{(r)}. \qquad (11.4)$$

The first term in the right-hand side accounts for the coupling between the volume dilatation and the concentration, and the second one takes into consideration the influence of the temperature on the equilibrium concentration of point defects.

12. Discussion of the theory proposed

For convenience we collect here the basic field equations, namely the *kinematic equations*

$$F = APA_0^{-1}, \quad E = \tfrac{1}{2}(A^T A - 1), \qquad (12.1)$$

$$\text{grad} \, \dot{x} = \dot{A}A^{-1} + A\dot{P}P^{-1}A^{-1}, \qquad (12.2)$$

the *balance laws*

$$\text{div} \, T + \rho f = \rho\ddot{x}, \quad T = T^T, \qquad (12.3)$$

$$\rho\theta\dot{\eta} + \rho \frac{\partial\bar{\psi}}{\partial S} \cdot \dot{S} = (A\dot{P}P^{-1}A^{-1}) \cdot T + \text{div} \, q + \rho r, \qquad (12.4)$$

79

the *thermoelastic constitutive equations*

$$\psi = \check{\psi}(E, \theta) + \overline{\psi}(S),$$ (12.5)

$$T = \rho A \frac{\partial \check{\psi}(E, \theta)}{\partial E} A^T, \quad \eta = -\frac{\partial \check{\psi}(E, \theta)}{\partial \theta},$$ (12.6)

$$q = A\check{q}(E, \theta, A \text{ grad } \theta, S),$$ (12.7)

which have to be associated with the *condition of activation* and with the *inelastic constitutive equations*, i.e., (10.4), (10.6) and (10.8) for the rate-independent deformation, and (11.1), (11.3) and (11.4) for the viscoplastic deformation. To complete the formulation of the boundary-value problem we have to add to the above equations the thermomechanical boundary conditions. In the most common case of the isothermal deformation, this can be done by prescribing the loading program, i.e., the dependence of the surface traction t on place and time, and by using the boundary condition

$$Tn = t \text{ for } x \in \mathscr{S},$$ (12.8)

where $\mathscr{S}$ is the surface bounding the single crystal.

To test the completeness of the above system of equations we shall use a time discretization. Let us assume that we know at a given time t the configuration (k), and all thermoelastic and internal variables, as well as the values of $\dot{T}$ and $\dot{\theta}$ as functions of x. We may then determine the active glide planes at each place x by using the condition (11.1). Next, by using Eqs. (11.3), (11.4) we find the rates $\dot{\alpha}^{(r)}$, $\dot{\alpha}_M^{(r)}$, and $\dot{c}$, as well as the quantities $L^{(r)}$ and $v^{(r)}$. The determination of $\dot{x}$, A, θ, and T from the system (12.1)–(12.7) with prescribed boundary conditions, and with the last term of Eq. (12.2) known by (11.2), is now a boundary-value problem of the thermoelasticity. By considering that the solution of this problem is valid during the time interval $(t, t + \Delta t)$, where Δt is small enough, we may calculate $\dot{T}(t + \Delta t)$, $\dot{\theta}(t + \Delta t)$, and then determine the values of the internal state variables at time $t + \Delta t$ by

$$S(t + \Delta t) \approx S(t) + \dot{S}(t)\Delta t.$$ (12.9)

We see that we know at time $t + \Delta t$ the same data as those assumed as given at time t, and thus the solving procedure may be in principle continued in the same way.

We conclude that the theory proposed provides a sufficiently general framework for the macroscopic analysis of viscoplastic deformation of a single crystal.

References

[1] Drucker, D. C., *Structural Mechanics*, Pergamon, 1960, 407.

[2] Mughrabi, H., *Phil. Mag. 23* (1971) 897, 931.

[3] Hedges, J. M. and J. W. Mitchell, *Phil. Mag. 44* (1953) 223.

[4] Essmann, U., *phys. stat. sol. 12* (1965) 707, *Acta Met. 12* (1964) 1468.

[5] Essmann, U., *phys. stat. sol. 12* (1965) 723.

[6] Mughrabi, H., *Phil. Mag. 18* (1968) 1211.

[7] Mughrabi, H., *Proc. 3rd Inf. Conf. on Strength of Metals and Alloys*, Cambridge, England 1973 (to appear) and unpublished results.

[8] Neumann, P., *Acta Met. 18* (1969) 1219.

[9] Kröner, E., in *Inelastic behaviours of solids*, M. F. Kanninen, W. F. Adler, A. R. Rosenfield and R. I. Jaffee, Eds., Battelle Inst. Materials Science Colloquia 1969, McGraw Hill, 1970, 137.

[10] Eckart, C., *Phys. Rev. 73* (1948) 373.

[11] Lee, E. H. and D. T. Liu, *J. Appl. Phys. 38* (1967) 19.

[12] Fox, N., in *Mechanics of generalized continua*, IUTAM Symp. Freudenstadt–Stuttgart 1967, E. Kröner, Ed., Springer, 1968.

[13] Fox, N., *Q. J. Mech. Appl. Math. 21* (1968) 67.

[14] Teodosiu, C., *Proc. Conf. Fundamental Aspects of Dislocation Theory*, Washington 1969, J. A. Simmons, R. de Wit and R. Bullough, Eds., Nat. Bur. Stand. 1970, vol. 2, 837.

[15] Rice, J. R., *J. Mech. Phys. Sol. 19* (1971) 433.

[16] Mandel, J., *C. R. Acad. Sci. Paris (Ser. A) 272* (1971) 276, 1596, *273* (1971) 44.

[17] Lee, E. H., *J. Appl. Mech. Trans. ASME (Ser. E) 36* (1969) 1.

[18] Orowan, E., *Proc. Phys. (London) 52* (1940) 8.

[19] Kröner, E., *J. Math. Phys. 42* (1963) 27.

[20] Valanis, K. C., *J. Math. Phys. 45* (1966) 197.

[21] Valanis, K. C., *J. Math. Phys. 46* (1967) 164.

[22] Onat, E. T., *Irreversible Aspects of Continuum Mechanics*, IUTAM Symp. Vienna 1966, Springer 1967, 292.

[23] Coleman, B. D. and M. E. Gurtin, *J. Chem. Phys. 47* (1967) 597.

[24] Fox, N., in *Proc. Conf. Fundamental Aspects of Dislocation Theory*, Washington 1969, J. A. Simmons, R. de Wit and R. Bullough, Eds., Nat. Bur. Standards, 1970, vol. 2, 1041.

[25] Perzyna, P. and W. Wojno, *Arch. Mech. Stosowanej 20* (1968) 499.

[26] Kratochvil, J. and O. W. Dillon, Jr., *J. Appl. Phys. 40* (1969) 3207, *41* (1970) 1470.

[27] Lardner, R. W., *Int. J. Eng. Sci. 7* (1969) 417.

[28] Eisenberg, M. A., Office Naval Research, Contract No. N00014–68–A–0173–0002 (1969).

[29] Zarka, J., *Sur la viscoplasticité des métaux*, Thèse, Paris 1968.

[30] Zarka, J., *Mémorial de l'Artillérie Française*, 2ème fasc. 1970, 223.

[31] Teodosiu, C., Conf. on Mechanics of Solids, Oberwolfach 1971 (unpublished).

[32] Zarka, J., *Arch. Mech. Stosowanej 23* (1971) 369.

[33] Perzyna, P., *Bull. Acad. Polon. Sci. Sér. Sci. Tech. 19* (1971) 177.

[34] Perzyna, P., in *Advances in Applied Mechanics*, vol. 11, Academic 1971, 313.

[35] Perzyna, P., *J. Mécanique 10* (1971) 391.

[36] Zarka, J., *J. Mech. Phys. Solids 20* (1972) 179.

[37] Kratochvil, J., Acta Mechanica (1972), 307.

[38] Mecking, H., *Untersuchung der Plastizität von Silbereinkristallen durch Zugverformung bei konstanten und plötzlich wechselnden Versuchsbedingungen*, Thesis, TH Aachen, 1967.

[39] Rosset, W. S. de, and A. V. Granato, in *Proc. Conf. Fundamental Aspects of Dislocation Theory*, Washington 1969, J. A. Simmons, R. de Wit and R. Bullough, Eds., Nat. Bur. Stand. 1970, vol. 2, 1099.

[40] Mecking, H. and K. Lücke, *Scripta Met.* 4 (1970) 427.

[41] Neuhäuser, H., N. Himstedt and Ch. Schwink, *phys. stat. sol (a) 3* (1970) 585, 929.

[42] Seeger, A., in *Dislocations and Mechanical Properties of Crystals*, J. C. Fisher, W. G. Johnston, R. Thomas and T. Vreeland Jr., Eds., J. Willey & Sons 1957, 243.

[43] Seeger, A., in *Handbuch der Physik*, S. Flügge, Ed., Springer 1958, vol. VII/2, 1.

[44] Krommüller, H., in *Grundlagen des Festigkeitverhaltens von Metallen*, H. Klare et al., Eds., Akademie – Verlag 1965, 51.

[45] Kronmüller, H., in *Moderne Probleme der Metallphysik*, A. Seeger, Ed., Springer 1965, 126.

[46] Conrad, H., *J. Metals 16* (1964) 582.

[47] Argon, A. S., *Mater. Sci. Eng. 3* (1968/69) 24.

[48] Evans, A. G. and R. D. Rawlings, *phys. stat. sol. 34* (1969) 9.

[49] Gibbs, G. B., *Mat. Sci. Eng. 4* (1969) 313.

[50] Granato, A. V., K. Lücke, J. Schlipf and L. J. Teutonico, *J. Appl. Phys. 35* (1964) 2732.

[51] Schoeck, G., *phys. stat. sol. 8* (1965) 499.

[52] Hirth, J. P. and W. D. Nix, *phys. stat. sol. 35* (1969) 177.

DISCUSSION

John A. Simmons[1]). Professor Kröner and Dr. Teodosiu's approach provides an elegant theoretical framework relating dislocation theory to macroscopic plasticity, but it may be somewhat difficult to apply in practice. I would like to present a first-order theory which may be more directly applied to problems of polycrystalline plasticity. This work is being done in collaboration with Dr. Roger B. Clough, also of the National Bureau of Standards. It is based on the thermally-activated motion of dislocation segments on randomly oriented slip planes, and predicts most of the usual

[1]) National Bureau of Standards, Washington D.C., U.S.A.

assumptions of classical plasticity. These include conservation of volume during plastic flow, alignment of the principal stress and principal plastic strain rate axes, and pressure-independence of yielding, since the resultant strain rates are functions of the deviatoric stress. In addition, we have developed the concept of plastic power dissipation as a highly useful dynamic measure of multiaxial yielding.

The details of the theory summarized here are given elsewhere [1, 2]. Very briefly, the plastic strain rates are first calculated by assuming the thermally-activated motion of non-interacting, randomly oriented dislocation segments on randomly oriented slip planes with no back forces. At constant temperature, if the speed (v) of each segment is then only a function of the force per unit length of dislocation ($b_i \sigma_{ij} \mu_j$, where b_i is the Burgers vector, σ_{ij} is the applied stress, and μ_j is the slip plane normal), then the speed is explicitly

$$v = 2v_0 \exp(-G_0/kT) \cdot \sinh(\eta_i \sigma_{ij} \mu_j V^*/kT),$$

where v_0, G_0, and V^* are orientation-independent material parameters, kT is the thermal energy, and η_i is a unit Burgers vector. The macroscopic plastic strain rates are then obtained by integrating explicitly over all orientations:

$$\dot{\varepsilon}_{ij} = \int_\Omega \Lambda_{ij} bvn d\Omega,$$

where Λ_{ij} is a tensor containing the orientation dependence of the macroscopic strain rate on the slip plane distortion rate, and n is the mobile dislocation density per element of angular volume $d\Omega$.

In order to measure multiaxial yielding, we would desire some scalar physical quantity which among other things is constant over a physically reasonable yield surface. Plastic strain rate is a second order tensor, which then leads to unwieldy higher order tensors if derivatives are taken with respect to other tensor quantities. For example, the derivative with respect to stress of the logarithm of the plastic strain rate corresponds to a fourth order activation volume tensor. Moreover, by integrating the above equation we have found that the macroscopic plastic strain rate is not of an Arrhenius form (even though the dislocation speed is), but instead the macroscopic Arrhenius variable is the plastic power dissipation. For this and other reasons, some of which become apparent below, plastic power dissipation suggests itself as a natural and highly useful variable to characterize multi-

83

axial yielding. Using this, multiaxial yielding is characterized by surfaces of constant plastic power dissipation:

$$P = \sigma_{ij}\dot{\varepsilon}_{ij} = \sigma_1\dot{\varepsilon}_1 + \sigma_2\dot{\varepsilon}_2 + \sigma_3\dot{\varepsilon}_3 = P_0,$$

where σ_i and ε_i are the principal applied stresses and associated principal plastic strain rates (which were calculated above). P_0 is the uniaxial plastic power dissipation, which is proportional to $\exp(\frac{1}{2}B\sigma_0)$, where σ_0 is the uniaxial tensile yield stress[1]) and $B = V^*/kT$. The shape of the yield surface in stress space is then characterized by $B\sigma_0$, independently of σ_0, which acts as a scale factor. Biaxial yield surfaces in stress space, as shown in Fig. 1, in fact vary from a von Mises shape when $B\sigma_0 \ll 1$ (typical of

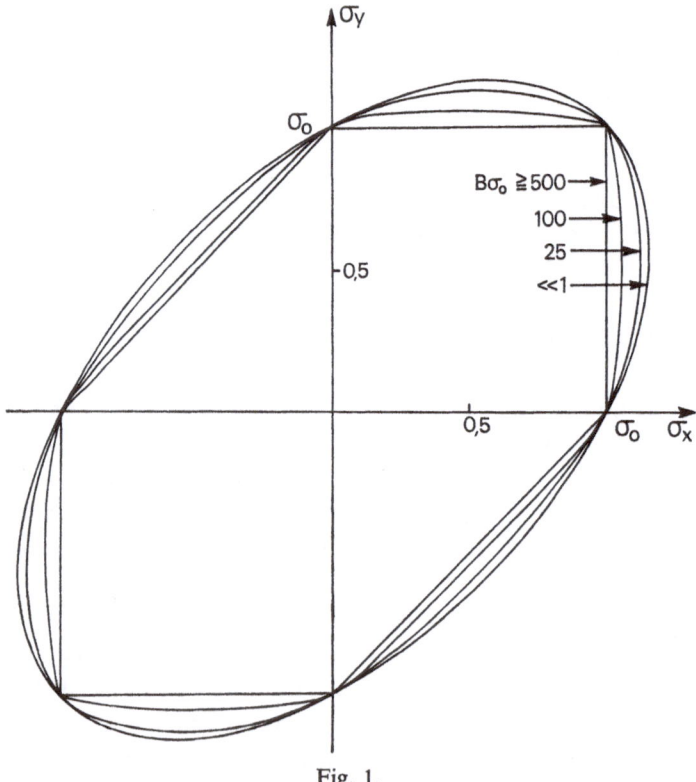

Fig. 1.

[1]) Uniaxial yielding is here defined as the uniaxial tensile stress corresponding to a prescribed value of plastic power dissipation. Direct measurement of it requires simultaneous measurement of stress and plastic strain rate.

high temperature plastic flow) to a Tresca shape when $B\sigma_0 \gtrsim 200$ (typical of low temperature flow). Also, calculations show that the plastic strain rate vector is normal to the yield surface, a verification that Drucker's stability postulate [3] holds for this type of yield surface.

The use of plastic power dissipation as a characterization of multiaxial yielding permits evaluation of effects of small temperature and multiaxial stress changes on the shape of the yield surface through the introduction of the second order activation volume tensor for plastic power dissipation. This is generally a deviatoric tensor which is diagonal in principal stress space, is defined as

$$kT\left(\frac{\partial \ln P}{\partial \sigma_{ij}}\right) = V_{ij},$$

and is a first-order coefficient of a Taylor expansion of $(\ln P)$ about temperature and miultaxial stress differentials which leads to the approximate Arrhenius expression for the plastic power dissipation:

$$P(1/kT, \sigma) \cong P' \exp\left[-\frac{Q}{k}\left(\frac{1}{T} - \frac{1}{T'}\right)\right] \cdot \exp\left[(V_{ij}/kT)(\sigma_{ij} - \sigma'_{ij})\right],$$

where $P' = P(1/kT', \sigma')$, $-Q = [(\partial \ln P/\partial(1/kT)]$, and the primed quantities refer to the reference state. This model assumed that the segments were non-interacting and that were no internal back forces. Internal back forces can be incorporated into the model, and at present the effects of internal stress and prestrain effects on the yield surface are being computed. In addition, an experimental program is being carried out to measure changes in the shape of the yield surface as a function of temperature.

References

[1] Clough, R. B. and J. A. Simmons, *A theory of multiaxial plasticity based on dislocation dynamics*, to be published.
[2] Clough, R. B. and J. A. Simmons, Thermodynamics of dislocation motion in multi-axial stress fields, to be published in *Proceedings of the J. E. Dorn memorial symposium on rate processes in plastic deformation*, Cleveland, Ohio ASM (1972).
[3] Drucker, D. C., *Proc. 1st U.S. Nat. Congr. Appl. Mech.* (1951) p. 487.

C. Teodosiu. The viscoplastic power dissipation has been already used to macroscopically characterized various types of loading-unloading processes in elastic-viscoplastic materials[1]). It is very instructive to see that this macroscopic variable can be shown to satisfy, under certain simplifying assumptions, an Arrhenius equation of the same type as that governing the motion of the glide dislocations.

Although Dr. Simmons' approach starts from microscopic considerations which are similar to those in our paper, it seems rather difficult to compare the results, because our analysis is mainly based on the introduction of an explicit set of internal variables – the densities of dislocations and the concentration of point defects – whose evolution and interaction are considered responsible for the viscoplastic behavior. On the other hand, our approach is intended to describe the inelastic behavior of single crystals, for which the anisotropy plays a decisive role and randomness hypotheses concerning the distribution of the glide planes could hardly be applied.

E. H. Lee[2]). You mention that constitutive relations based on continuum theory cannot be predictive, in contrast to the application of the principles of solid state physics. I presume you mean by this, for example, that postulating a yield surface in stress-space requires every point of the surface to be established experimentally. But continuity and smoothness concepts arise, so that a few experimental points may suffice to prescribe the surface, and the yield condition is then effectively predicted for other stress points.

E. Kröner. I hope that this question is answered to some extent in the written version of our lecture. The statement about predictability in the theory of plasticity was originally given by D. C. Drucker as a surmise. I agree that for determining a yield surface a few experimental points may suffice. However, the yield surface changes during the deformation depending on the particular loading path which not need be straight in the stress space. This means that different sequences of yield surfaces belong to different loading paths. Hence a number of yield surfaces which is not too small must be determined already for one loading path. To give a *full* account of the behaviour of the material one needs the sequences of yield surfaces for many loading paths. In other words, the complex space of loading histories is to be filled with experimental points in such a way that the particular loading path of our future experiment is nowhere too far away from other

[1]) See, e.g., Cristescu, N., *Dynamic plasticity*, North Holland, Amsterdam (1967) Chap. X.
[2]) Stanford University, Stanford, California, U.S.A.

already measured loading paths. In this way a huge interpolation scheme is built up. With its help effective predictions are possible, of course. However, the degree of predictability is quite below that of the theory of elasticity, for instance, where the predictions are not of an interpolation type.

W. Kosiński[1]). I have two questions in connection with the second part of the lecture. The first one regards the definition of the deformation tensor E. The second concerns the intermediate configuration. In your considerations the intermediate configuration is either a global configuration for the whole body or a local one for each particle of the body. I understand that the intermediate configuration is an unstressed state of an element of the body (i.e. a particle and its neighbourhood). In the case of a local configuration all stress free elements of the body no longer fit together to form a continuum which can be mapped continuously back to the reference configuration of the body. If F is the actual deformation gradient with respect to the reference configuration and F_p is the deformation gradient in the intermediate configuration (with respect to the reference configuration), then F_e is the deformation gradient in the actual configuration with respect to the intermediate configuration, such that

$$F = F_e F_p.$$

It is noticed that the quantities F_e and F_p are not composed of partial derivatives. On the other hand a global unstressed intermediate configuration is to be obtained in the case of a homogeneous deformation only.

C. Teodosiu. E is the elastic strain tensor defined by Eq. (8.11) in terms of the elastic distortion, A.

I agree completely with your second comment. The incompatibility of the elastic strains generated by an nonhomogeneous inelastic deformation has been often pointed out in the literature (see, e.g., Refs. [12]–[15]). Our Eq. (6.6) reduces in fact to your decomposition rule for the total deformation gradient, in case $A_0 = 1$. It should be mentioned, however, that when only local considerations are involved, as in the main part of our paper, the fact that the distortions A and P cannot be derived from displacement vector fields is immaterial.

[1]) Institute of Fundamental Technological Research, Warsaw, Poland.

H. Lippmann[1]).   The set of state variables considered, have been introduced by the authors by means of really convincing physical arguments. So I would like to pose two questions.

1) Has the theory been applied, or is it planned to be applied, to some nontrivial examples as for instance, torsion of cylindrical rods. I would be interested to know how the state variables are distributed, say, over a cross-sectional area, and what kind of information can be obtained on the distribution of dislocations.

2) Is it possible to determine, at least in principle or even practically, the distribution of the state variables by means of experiments so that the theory can be compared with the reality. Or are there other possibilities to check the theory experimentally.

C. Teodosiu.   The theory has been applied mainly to uniaxial deformation processes, including creep, relaxation, and cyclic experiments. In all these cases it has been possible to make quantitative predictions about and to measure experimentally the dislocation densities in various glide planes (see, e.g., Refs. [2]–[7], [38]–[44]). However, much more theoretical and experimental work has to be done before passing to the effective solving of boundary-value problems involving inhomogeneous deformations.

[1]) University of Karlsruhe, GFR.

Representation of elastoplastic behavior
by means of state variables

F. Fardshisheh and E. T. Onat

Yale University, New Haven, Connecticut, USA

Mechanical behavior of elastoplastic solids which develop *anisotropy* during the course of plastic flow is discussed in the case where deformations are *finite* and isothermal. The mode of description used is that of state variables and the resulting representation could, in principle, be constructed with the help of phenomenological experiments.

1. Introduction

This paper is concerned with the description of mechanical behavior of elastoplastic solids in the presence of *finite*, isothermal deformations.

This problem or its suitable generalizations have been the subject of numerous recent investigations. We mention amongst these the work of Green and Naghdi [1], Lee and Liu [2], Lee [3], Perzyna and Wojno [4], Willis [5], Owen [6], and Mandel [7].

The work of Lee and Liu [2] and Lee [3] has clarified several basic issues that arose when attempts were made to generalize the classical laws of elastoplasticity to the case of finite deformations. However, these authors treat only a restricted class of elastoplastic materials, namely the materials which remain isotropic in their unstressed state even after the occurrence of plastic flow.

In the present paper we consider materials which develop anisotropy during the course of plastic flow so that an unstressed element is no longer isotropic. (It may be pointed out that a reasonably general treatment of this case does not exist even for the case of small deformations).

The representation developed here is sufficiently general, and moreover it can be constructed, in principle, with the aid of experiments. Some of the above mentioned work suffer from the drawback that the constitutive relations developed therein cannot be established directly from experimental evidence.

The mode of description adopted here is that of state variables. We make extensive use of the results of [8], [9], and [10]. We show in Section 4 that the state of a stress free element can be represented, in a well-defined sense, by a number of *even* rank tensors. In Section 5 it is seen that the state of a deforming element is represented by these tensors (which define the state of the associated stress free element) and by a second rank symmetric tensor which measures deformations with respect to the stress free element. We discuss the concept of *elastic range* and introduce a *yield function*. In Section 7 we complete the representation by constructing growth laws for the state variables. The implications of certain energy considerations are then discussed and the constitutive relations are given in their final form.

The important case of metal like solids where large total deformations are accompanied by small 'elastic' deformations is considered in Section 10 as an application and illustration of the general theory.

In the last Section we discuss briefly the problem of determination of the number and the tensorial nature of state variables from phenomenological experiments.

2. Preliminary considerations

In order to clarify what is meant by representation of mechanical behavior we may consider the following set of thought experiments. We start at time $\tau = 0$ with a supply of identical homogeneous, virgin, isotropic test specimens. We imagine that a typical experiment consists of the application of a homogeneous[1]) time-dependent deformation to the specimen and the observation of the resulting stress. The applied deformation is measured in terms of deformation gradients $D_{ij}(\tau)$ defined as

$$x_i = D_{ij}(\tau)X_j$$

where X_j and x_i are initial and current coordinates of the material point X_j in a *fixed* rectangular cartesian coordinate frame. We denote by $\sigma_{ij}(\tau)$ the stress components observed in the same coordinate frame. A single test

[1]) Elastoplastic solids are 'simple' in the sense of Noll [11] so that it is appropriate to construct the constitutive relations by means of experiments involving homogeneous deformations.

involves two tensor histories

$$D(\tau), \sigma(\tau); \quad \tau \in [0, \infty)$$

where bold face latters constitute a short-hand notation for the corresponding components. We assume that $D(\tau)$ and $\sigma(\tau)$ are continuous and bounded in their components.

From a mathematical point of view a given solid acts (by means of such tests) as an operator F which assigns to each continuous tensor valued function $D(\tau)$ a function $\sigma(\tau)$:

$$\sigma(\tau) = F[D(\tau)], \ \tau \in [0, \infty)$$

Description of mechanical behaviour of a given solid is equivalent, in this isothermal setting, to stating the precise nature of F for this solid.

Of the methods available for the mathematical representation of F the one which is based on state variables is particularly suitable for elasto-plastic solids. The lack of 'analyticity' associated with the questions of loading and unloading is a characteristic feature of elastoplastic solids. It is this characteristic that makes integral representations unsuitable for the description of elastoplastic behavior [12].

It is shown in a series of papers [8], [9], [10] that, under certain require-ments of smoothness, we can construct a representation of the following type for F:

$$\sigma(t) = f(S(t)) \qquad \text{(a),}$$
$$S^{\cdot}(t) = g(S(t), V, \Omega) \qquad \text{(b).} \tag{1}$$

The first equation in (1) states that the current stress is a function of the current state and orientation $S(t)$ of the solid where $S(t)$ is a vector in the state and orientation space of the solid. The second equation is the law of growth or evolution for the state; here V and Ω are the tensors of current rate of deformation and rotation applied to the element. This equation asserts that the rate of change of state and orientation is a function of the current state and orientation and also of the current rate of deformation and rotation. In (1) f and g are subject to certain invariance requirements and S is composed of a number of 'tensors'.

It will be seen that elastoplastic solids admit a representation of type (1). In fact the definition of such solids will involve precise statements on the forms of f and g and on the nature of S.

3. Elastic range

A principal attribute of elastoplastic solids is the presence of *elastic range* in their behavior. To define the elastic range formally we consider a solid which has been subjected to a given deformation history on time interval $[0, t]$. This history will bring the solid to the state S at time t.

Now subject this solid to a future deformation on $[t, \infty)$. Let $D^*(\tau)$ be the future deformation measured with respect to the configuration at time t. Note that $D^*(t) = I$ where I is the identity tensor. Now the presence of elastic range implies the existence of a class, say $\mathscr{C}(S)$, of deformation histories $D^*(\tau)$ on $[t, \infty)$ such that

$$\sigma(\tau) = \hat{f}_S(D^*(\tau)),$$
$$\text{when} \quad D^*(\tau) \in \mathscr{C}(S), \ \tau \in [t, \infty), \tag{2}$$
$$S(\tau) = \hat{g}_S(D^*(\tau)),$$

where $\hat{f}_S$ and $\hat{g}_S$ are ordinary functions of their arguments.

The above relations state that for any $D^* \in \mathscr{C}(S)$ the solid[1] will behave 'elastically'. Namely the stress and the state of the element will depend only on the current value of $D^*(\tau)$ but not on its history on $[t, \tau)$ provided that $D^* \in \mathscr{C}(S)$. Note that the second equation in (2) states that the future behavior of the solid will not depend on the history of D^* on $[t, \tau)$ when $D^* \in \mathscr{C}(S)$. $\mathscr{C}(S)$ is subject to certain restrictions. One such restriction is the following one: if $D^* \in \mathscr{C}(S)$ then $RD^* \in \mathscr{C}(S)$ where R represents a time dependent rigid body rotation. This requirement is a consequence of the principle of objectivity and needs no further elaboration. Another requirement on $\mathscr{C}(S)$ is that it must contain D^* which take on the value I at times other than t. This restriction enables the material to return to its original state S by purely elastic deformations.

Let us now denote by $E(S)$ the totality of states which can be reached starting from S by the action of $D^* \in \mathscr{C}(S)$. It is easily seen from (2) that $E(S)$ is a nine-dimensional subset of the state space. We call $E(S)$ *the elastic range of S*. In view of above remarks concerning $\mathscr{C}(S)$ we can decide that the states belonging to $E(S)$ can be reached from each other by purely elastic deformations of the solid. This last remark suggests that it would be economical to write the 'elastic' relation (2) not for each state in $E(S)$ but rather for a definite reference state of the elastic range $E(S)$.

Can we find a natural reference state in the lastic range $E(S)$? In the case

[1] D^* is a short hand notation for $D^*(\tau)$, $\tau \in [0, \infty)$.

where the solid can always be brought to a stress free state by means of an elastic deformation the stress free state can be used as a natural reference state. We shall assume that our solid has the above property: namely $\mathscr{C}(S)$ contains deformation histories which bring an element of the solid to a stress free[1]) state. We shall use the stress free elements as reference states and configurations.

Once this decision has been made then it becomes necessary to study the totality of stress free elements that one can obtain by performing all tests of interest. Such a study should enable us to single out a given stress free element amongst many. This leads us to the question of the representation of stress free states.

4. Representation of internal structure of stress free elements

Each stress free element possesses an internal structure characterized by a particular distribution of dislocations, impurities, grains, etc. This internal structure will have come about as a result of the past deformation applied to a virgin element. Since elastoplastic solids are *time independent* the internal structure of a stress free element will not change in the absence of future deformations. A stress free element may be anisotropic: In such a case a rigid body rotation applied to the element will, in general, change the 'orientation' of the internal structure with respect to a fixed coordinate frame.

It is clear that the future behavior of a stress free element will depend on its present internal structure. If stress free elements are to be used as reference states and configurations then we must be able to describe mathematically the internal structure of a stress free element so that the influence of this structure on future behavior could be expressed in mathematical terms.

The problem of description of internal states has been studied in [9] and [10] from a phenomenological point of view. Here we summarize the results of these studies in a form appropriate to the present work.

We assert, following [10] that, the internal state of a (stress free) element may be represented by a point in a suitable linear space. We assume, for simplicity, that this space is finite dimensional, say, R_n. The coordinates of

---

[1]) If for a given solid this assumption does not hold, then a natural choice of reference states is more difficult, if not, impossible. In this case a reference state (or more correctly a reference orbit) may be assigned rather arbitrarily with due regard to smoothness and continuity.

the state point are the n-parameters which measure the salient and relevant features of the internal structure (e.g. distribution of dislocations, the orientation of this distribution, etc.) of a given stress free element.

Consider now a stress free element whose *state and orientation* is represented by a point s of R_n. Next *rotate* this element by an amount Q, where Q is a proper orthogonal 3×3 matrix:

$$QQ^T = I \quad \text{and} \quad \det Q = +1.$$

This rigid body rotation will change the orientation of the internal structure with respect to a fixed frame of reference. Thus the state and orientation of the rotated element, when studied in our fixed frame of reference, will in general be different from s. Let $P_Q s$ denote the state and orientation of the rotated element.

As s varies over all state points, $P_Q s$ represents (with an appropriate extension) a transformation of R_n into R_n.

On the other hand the set of transformations $\{P_Q\}$ where Q ranges over all rotations can be shown to constitute a group [10]. Thus the transformations $\{P_Q\}$ enjoy properties such as

$$P_Q P_R = P_{QR}.$$

Moreover by a suitable construction on the state and orientation space the P_Q's become *linear and orthogonal* [10].

The above remarks show that $\{P_Q\}$ is an n-dimensional representation of the proper orthogonal group $O^+(3)$.

Thus the state and orientation space gains the following structure ([13], p. 16).

(i) R_n is the direct sum of mutually orthogonal subspaces $M_1, ..., M_m$ ($m \leq n$) where

(ii) each of the M_i is of *odd* dimension and

(iii) each is invariant under the action of $\{P_Q\}$. (Namely if $s \in M_i$ then $P_Q s \in M_i$).

(iv) The action of the P_Q's on each M_i is again linear and orthogonal.

This means that each s in R_n may be expressed uniquely in the form

$$s = q_1 + ... + q_m, \quad q_i \in M_i \tag{3}$$

with

$$P_Q s = P_Q q_1 + ... + P_Q q_m \tag{4}$$

and

$$P_Q q_i \in M_i.$$

In mathematical language we say that the n-dimensional representation of $O^+(3)$ is decomposed into a direct sum of irreducible representations of $O^+(3)$.

Next we must specify the action of P_Q in the invariant subspaces M_i. As is well-known from Group Theory this can be done, clearly and economically, with the aid of tensors. Let us illustrate the use of tensors in the present context by an example. Suppose that one of the M_i's, say M_1, is 5-dimensional.

The space of all second rank tensors q_{ij} is a nine dimensional linear space and the space of all symmetric traceless ($q_{ij} = q_{ji}$, $q_{ii} = 0$) tensors is a 5-dimensional subspace of this space.

Since all linear spaces having the same finite dimension are essentially the same we may take M_1 to be the space of symmetric traceless tensors. The action of P_Q on M_1 is then given by the familiar tensor transformation:

$$P_Q r = Q r Q^T, \quad r \in M_1$$

or in component form

$$(P_Q r)_{ij} = Q_{ik} Q_{jl} r_{kl}.$$

Notice that $P_Q r$ is again a symmetric traceless tensor.

It can be shown that each M_i can be regarded as a space generated by an irreducible tensor (i.e. a tensor of certain symmetries) of *even* rank. The requirement of even rank is a consequence of the fact that the operator F in (1) involves tensors of *second* rank [10].

The action of P_Q on M_i is the ordinary tensor transformation appropriate to the rank of the irreducible tensor.

We now recapitulate the results announced in this section in a less formal manner:

The state of a stress free element is represented by means of *even* rank irreducible tensors ($q_1, \ldots, q_m$). A rigid body rotation Q of the element causes the state point ($q_1, \ldots, q_m$) to move to ($P_Q q_1, \ldots, P_Q q_m$) where $P_Q q_i$ has the meaning of ordinary tensor transformation.

The above tensors may be thought of as representing various relevant aspects of dislocation distribution within a stress free element (cf. Kröner [14]).

We shall find it convenient to denote the m-tuple $(q_1, ..., q_m)$ by a single symbol s. It must then be remembered that s and $P_Q s$ are related by (3) and (4) to q_i and $P_Q q_i$.

It is useful to refer here to previous work. Lee [3] has considered, in the language of the present paper, the class of materials satisfying

$$P_Q s = s$$

for any $Q \in O^+(3)$ and for any s. This means that for this material s is composed of scalars and therefore a stress free element is isotropic.

Mandel [7] has used a 'repère directeur' (essentially a second rank tensor) to describe the orientation of a stress free element. As pointed out by Mandel this is adequate for single crystals, but for polycrystals the use of 'repère directeur' may not be convenient. We base this claim on the results of the present section: As we have seen above, one can use a number of tensors (some of them higher than second rank) to describe the state and orientation of a stress free element without singling out a given state on an orbit as reference state.

5. The state of a deforming element

Consider an element which has been subjected to the deformation history $D(\tau)$ on $[0, t]$. Let the state and orientation of this element be denoted by S and let D stand for $D(t)$.

According to the assumption introduced in Section 3 this element may be brought into a stress free state by the application of some future 'elastic' deformation D^*, namely a $D^* \in \mathscr{C}(S)$. Consider now the resulting stress free element and let its deformation, measured with respect to the very initial configuration at $\tau = 0$, be denoted by D^P (cf. Fig. 1).

We assume that the shape of the unloaded element, namely $D^{P^T} D^P$, is uniquely determined by the previous deformation $D(\tau)$, $\tau \in [0, t]$ applied to the element. However the *orientation* of the unloaded element is not unique (cf. Fig. 1). Indeed the deformation $R(\tau)D^*(\tau)$ where $R(\tau)$ is a time dependent rigid body rotation will also belong to $\mathscr{C}(S)$ and it will result in the unloaded configuration RD^P where R is the terminal value of $R(\tau)$. Thus if s is the state and orientation of the configuration D^P then $P_R s$ will be the state and orientation of the element in configuration RD^P. The points obtained from s by the application of P_R as R ranges over $O^+(3)$ is called an *orbit* of the

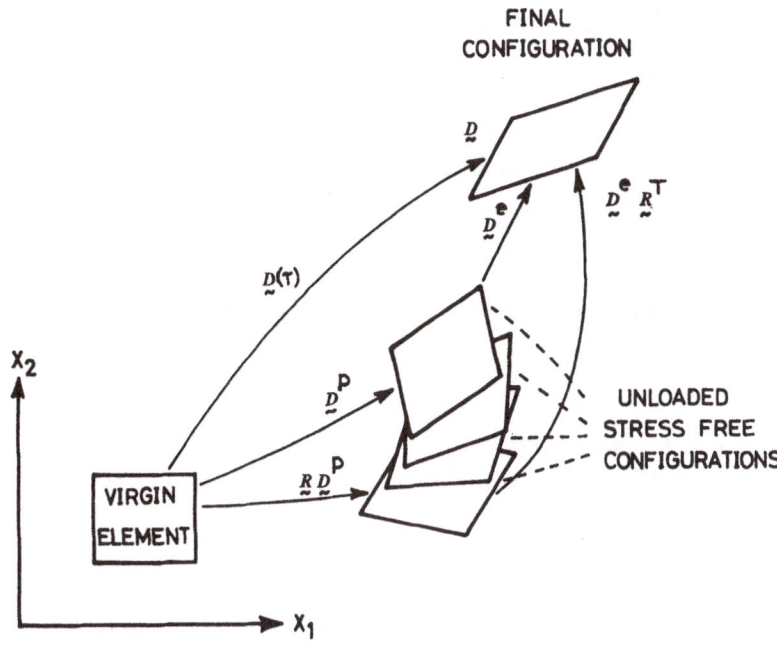

Fig. 1.

state and orientation space under $O^+(3)$. We see that by unloading from S we can reach all points of the orbit of stress free states associated with S.

Our aim in unloading was to create a reference configuration and state, but we see that there is an abundance of these configurations. Can we single out one of these configurations? The answer to this question is affirmative as can be seen from the following considerations.

We introduce the following decomposition of D which was used by Lee in [3]:

$$D = D^e D^P \qquad (5)$$

which as pointed above is not unique (cf. Fig. 1).

However we shall now show that the particular decomposition (5) which renders D^e symmetric is unique. Indeed we find from (5)

$$(D^e D^{e^T})^{-1} = D^{T^{-1}}(D^{P^T} D^P)D^{-1} \qquad (6)$$

which shows (since det $D^e > 0$ and $D^{P^T} D^P$ is known) that

$$D^e D^{e^T} \qquad (7)$$

97

is uniquely determined. Now consider the polar decomposition of D^e

$$D^e = UQ \tag{8}$$

where the symmetric tensor U is positive and $\det Q = +1$. Combinations of (6), (7) and (8) shows that U^2 is also uniquely determined:

$$(U^2)^{-1} = D^{T^{-1}}(D^{P^T}D^P)D^{-1}. \tag{9}$$

Since U is positive the knowledge of U^2 (cf. (9)) determines U uniquely. Thus the decomposition (5) which renders D^e symmetric is unique and is given by

$$D^e = U \quad \text{and} \quad D^P = U^{-1}D,$$

where U is determined by (9).

We reserve the symbol D^P for the unloaded configuration $U^{-1}D$ so that we have henceforward

$$D = UD^P \quad \text{where} \quad U = U^T. \tag{10}$$

Let the state of this uniquely determined stress free configuration be s.

The symmetric second rank tensor U accepts the following physical interpretation. Consider the unstressed configuration of the element. According to our remarks concerning $\mathscr{C}$ we can deform this element elastically so as to bring it to its previous shape D and to its previous state S. U is the terminal value of this deformation and it may be called the symmetric elastic deformation associated with the state S.

We shall assume that given the state S there exists a uniquely determined stress free state s and a symmetric elastic deformation U associated with this state.[1] On the other hand it should be clear from the above remarks that a given pair (s, U) defines the state S uniquely. Thus there exists a one-to-one correspondence between S and (s, U). Therefore we can take the pair (s, U) as defining the state of a deforming material element.

We can recapitulate these remarks in the following way. The state of a deforming element is characterized by the tensors

$$\underbrace{q_1, ..., q_n}_{s}; \quad U. \tag{11}$$

---

[1] Note that two elements which are in the same state at time t may have different final configurations $D(t)$ and different stress free configurations D^P. But by the assumption just introduced they give rise to the same stress free state s and the same U.

The first n tensors define the state of the associated stress free element, the last tensor measures the symmetric elastic deformation superimposed on the stress free state.

It might be noted here that an arbitrary pair (s, U) does not necessarily define a state. For a given s, an arbitrary $U(U = U^T)$ may not belong to a $D^* \in \mathscr{C}(s)$: If U is too 'large' plastic deformations would occur before D^* assumes the terminal value U. Thus for a given s we must specify those U which can be terminal values of $D^* \in \mathscr{C}(s)$. This will lead us to the study of yield condition in the next section.

Before going on to the next section we raise the following question.

Consider an element in the state (s, U). Apply to this element a rigid body rotation Q. What will be the state of the rotated element? We wish now to show that the state of the material element is

$$(P_Q s, Q U Q^T)$$

or equivalently with an obvious extension of notation

$$(P_Q s, P_Q U). \tag{12}$$

Let D be the total deformation of the element in the state (s, U). The total deformation of the rotated element is QD. In view of (10)

$$QD = Q U Q^T Q D^P. \tag{13}$$

Since superimposed rigid body rotations do not affect the shape of the associated stress free element, QD^P is an admissible residual deformation. Thus since $Q U Q^T$ is symmetric the assertion (12) follows from (13).

6. Yield condition

Consider an element which is in the stress free state s (or more formally in the state (s, I)) at time t. We wish now to define, in a more precise way, $\mathscr{C}(s)$, the class of future deformation histories which give rise to elastic behavior, and $E(s)$ the states which can be reached elastically from s. Let $D^*(\tau)$, $\tau \in [t, \infty)$ be a future deformation applied to the element s. Then the tensor $g^*(\tau) = D^{*^T}(\tau) D^*(\tau)$ measures *shape* changes with respect to the configuration at time t. We assume, following the tradition of plasticity that if $g^*(\tau)$, $\tau \in [t, \infty)$ falls within a closed surface in the g^*-space with coordinates g_{ij}^*, then the corresponding D^* will belong to $\mathscr{C}(s)$. This means that $D^* \in \mathscr{C}(s)$ if

$$Y_s(D^{*^T}(\tau)D^*(\tau)) \leqq 0, \qquad \tau \in [t, \infty).$$

It will be convenient to write the above inequality in the following form

$$Y(s, D^{*^T}(\tau)D^*(\tau)) \leqq 0, \qquad \tau \in [t, \infty), \tag{14}$$

where Y is a scalar valued function of both s and g^*. This inequality defines a surface and its interior in the g^*-space for each distinct stress free state s.

Now consider a state $(s, U) \in E(s)$. This state is reached by the application of a deformation $D^* \in \mathscr{C}(s)$ with the terminal value U ($U = U^T$) to the element in state (s, I). Then in view of (14) we have

$$Y(s, U^2) \leqq 0. \tag{15}$$

Thus $(s, U) \in E(s)$ satisfy the yield inequality (15).

If for a given (s, U), $Y(s, U^2) < 0$ then we say that the element in the state (s, U) is within an elastic range. If, on the other hand, $Y(s, U^2) = 0$ we say that the element is at *yield point*.

Since the superimposed rigid body rotations alter, by assumption, only the orientation of the internal structure of the element, it is natural to assume that an element at yield point will remain at yield point following the application of a rigid body rotation. Thus, in view of (12),

$$Y(s, U^2) = 0 \quad \text{must imply} \quad Y(P_Q s, P_Q U^2) = 0.$$

This invariance requirement for states at yield point can be extended, without loss of generality, to elements within an elastic range: Thus we impose the invariance requirement

$$Y(s, U^2) = Y(P_Q s, P_Q U^2) \tag{16}$$

for any state of interest. (16) means that Y is an invariant scalar function of the tensors $q_1, ..., q_n$ and U^2 under transformations belonging to $O^+(3)$.

7. Description of mechanical behavior

We have seen that the state of a deforming element is represented by the pair (s, U). According to (1) the stress σ carried by this element is a function of its state:

$$\sigma = f(s, U). \tag{17}$$

The above relationship is subject to the following invariance requirement

$$Q^T f(P_Q s, P_Q U)Q = f(s, U),$$

which emanates from (i) the assumption that a rigid body rotation Q applied to an element carrying the stress σ causes the stress to change to $Q\sigma Q^T$ and (ii) the observation associated with (12).

Energy consideration of the next section will show that f is derivable from a scalar potential Ψ.

Our next task is to write growth laws of the form (1b) which govern the evolution of the state (s, U) as a function of the deformation applied to the element.

Growth laws. We consider first an element which at time t is in the state $(s(t), U(t))$ satisfying the yield inequality

$$Y(s(t), U^2(t)) < 0.$$

Thus the state of the element lies within an elastic range. We apply next a deformation

$$D^*(\tau) \in \mathscr{C}(s(t), U(t)) \tag{18}$$

and we let $s(\tau)$ and $U(\tau)$ denote the resulting state of the material at time $\tau \in [t, \infty)$.

According to (18) no further plastic flow takes place during this deformation so that the state point will move within $E(s(t), U(t))$, or equivalently, within $E(s(t), I)$.

Furthermore in view of a previous remark $s(\tau)$ will remain on the orbit of $s(t)$ so that it will be related to $s(t)$ by

$$s(\tau) = P_{R(\tau)} s(t),$$

where $R(\tau)$ is a rigid-body rotation to be determined. We know from (12) that when $D^* = Q(\tau)$ where Q is a rotation then $R(\tau) = Q(\tau)$.

We see that the future deformation when studied with respect to the configuration of the element in the state $(s(t), I)$ will be observed as

$$D_s(\tau) = D^*(\tau)U(t). \tag{19}$$

With the help of the polar decomposition theorem $D_s(\tau)$ can be written as

$$D_s(\tau) = U(\tau)R(\tau), \tag{20}$$

where U is symmetric and R is orthogonal. We can easily show that the state of the element at time τ will be $s(\tau)$, $U(\tau)$ where

$$s(\tau) = P_{R(\tau)}s(t) \tag{21}$$

and $U(\tau)$ and $R(\tau)$ are given by (20). This observation follows from the fact that at time τ the state of the material is within $E(s(t), I)$ and therefore it can be reached uniquely starting from s by a rigid body rotation followed by a symmetric deformation.

Now combining (19) and (20) we obtain after some manipulation

$$U^2(\tau) = D^*(\tau)U^2(t)D^{*^T}(\tau),$$
$$\tau \geq t, \tag{22}$$
$$R(\tau) = U^{-1}(\tau)D^*(\tau)U(t),$$

which constitute together with (21) 'finite' growth laws in the elastic range. Of course, in view of (14), (22) will hold only if

$$Y(P_{R(\tau)}s(t), D^*(\tau)U^2(t)D^*(\tau)) \leq 0, \qquad \tau \geq t.$$

It will be desirable to write the above 'finite' growth laws in their 'incremental' form. For this purpose we take the time derivatives of (22) and (21) at time t. Note that

$$D^*(t) = I \quad \text{and} \quad \dot{D}^*(t) = V + \Omega, \tag{23}$$

where V and Ω are the symmetric and antisymmetric parts, respectively, of the velocity gradient tensor $\partial v_i / \partial x_j$ and are referred to as the tensors of rate deformation and rate of rotation.

In view of (23) we obtain from (22)

$$U(t)\dot{U}(t) + \dot{U}(t)U(t) = (V + \Omega)U^2(t) + U^2(t)(V - \Omega), \tag{24}$$
$$\dot{R}(t) = U^{-1}(t)[(V + \Omega)U(t) - \dot{U}(t)].$$

The above expressions can be written in a more elegant form by the use of the tensor c defined as

$$c = U^2. \tag{25}$$

Thus we obtain from (24)

$$\dot{c} = (V + \Omega)c + c(V - \Omega) \tag{26}$$

102

and after some manipulation

$$c\dot{R} - \dot{R}c = 2U^T V U - \dot{c}$$

or

$$c\dot{R} - \dot{R}c = 2U^T V U - (V + \Omega)c + c(V - \Omega), \tag{27}$$

where $\dot{R}$ is an antisymmetric second rank tensor $(R(t) = I)$. It is easily seen that (27) provides three independent equations for the three components of $\dot{R}$ and that the solution of these equations has the form

$$\dot{R} = \Omega + f(c, V), \tag{27'}$$

where $f(c, V)$ is antisymmetric. It can also be seen that $f(c, V)$ vanishes when $V = 0$ or when V has the principal directions of c, but otherwise it is generally non-zero. Thus even when $\Omega = 0$, $\dot{R}$ may be non-zero.

On the other hand the derivative of (21) provides

$$\dot{s}(t) = \frac{d}{d\tau} (P_{R(\tau)} s(t))_{\tau = t}. \tag{28}$$

It can be shown that (28) reduces to

$$\dot{s}(t) = T(\dot{R}(t)) s(t), \tag{29}$$

where T is a linear operator in R_n which depends linearly on $\dot{R}$. A more explicit form for T can be obtained by means of the decomposition (4):

$$P_{R(\tau)} s(t) = P_{R(\tau)} q_1(t) + \dots + P_{R(\tau)} q_m(t), \tag{30}$$

where $P_R q_i$ has the meaning of a tensor transformation.

Equations (28) and (30) yield

$$\dot{s}(t) = \sum_{i=1}^{m} T(\dot{R}) q_i(t)$$

with

$$T(\dot{R}) q_i(t) = \frac{d}{d\tau} (P_{R(\tau)} q_i(t))_{\tau = t}.$$

In order to obtain a clearer idea as to the form of $T(\dot{R}) q_i(t)$ consider the case where one of the q_i's, say q_1 is a second rank tensor. We have then

$$P_{R(\tau)} q_1(t) = R(\tau) q_1(t) R^T(\tau)$$

and

$$\frac{d}{d\tau}\left(P_{R(\tau)}q_1(t)\right)_{\tau=t} = \dot{R}(t)q_1(t) - q_1(t)\dot{R}(t),$$

hence

$$T(\dot{R})q_1 = \dot{R}(t)q_1(t) - q_1(t)\dot{R}(t).$$

Similar explicit expressions can be obtained for the action of $T(\dot{R})$ on higher rank tensors.

The growth laws (26), (27) and (29) apply to an element which is in the state[1]) (s, c) satisfying

$$Y(s, c) < 0.$$

Next we discuss the growth laws for an *element at yield point*, i.e. for an element whose state satisfies

$$Y(s, c) = 0.$$

In this case the growth laws must necessarily be incremental.

Let us apply to an element at yield point the rate of deformation V and the rate of rotation Ω. We expect that for certain combinations of V and Ω the resulting behavior will be elastic and the corresponding growth laws will have the previous form. But there will be other combinations of V and Ω which will cause plastic flow. Therefore we must first establish a criterion to distinguish between these two cases. In order to establish such a criterion we may find $\dot{c}$ and $\dot{s}$ from the 'elastic' growth laws (26–29) and then compute the resulting rate of change $\dot{Y}$ in the value of Y:

$$\dot{Y} = \frac{d}{d\tau}\left(Y(s(\tau), c(\tau))\right) = \frac{\partial Y}{\partial s}\cdot\dot{s} + \frac{\partial Y}{\partial c}\cdot\dot{c}, \qquad \tau = t, \tag{31}$$

where $(\partial Y/\partial s)\cdot\dot{s}$ stands for the scalar product in R_n of the vector $\partial Y/\partial s$ and $\dot{s}$. Similarly $(\partial Y/\partial c)\cdot\dot{c}$ represents a scalar product in the c-space. It can be shown using the invariance property (16) and the growth laws (26–29)

[1]) We shall often refer to the pair (s, U^2) or (s, c) as the state of the material. Since c and U are positive symmetric tensors the knowledge of one implies the unique knowledge of the other.

104

that (31) reduces[1]) to

$$\dot{Y} = \frac{\partial Y}{\partial c} \cdot 2 U^T V U. \tag{32}$$

Now we postulate that if

$$\frac{\partial Y}{\partial c} \cdot 2 U^T V U \leqq 0 \tag{33}$$

then the behavior is elastic and the growth laws (26–29) hold. Note that in this case $\dot{Y}$, when computed with 'elastic' growth laws, is non-positive.

On the other hand, when

$$\frac{\partial Y}{\partial c} \cdot 2 U^T V U > 0 \tag{34}$$

plastic flow must occur and a new set of growth laws must be introduced. We postulate that when plastic flow occurs $\dot{s}$ and $\dot{c}$ will depend only on s, c and V and Ω. We further require that the dependence on V is linear and that the growth laws appropriate to plastic flow must reduce to the previous ones as $(\partial Y/\partial c) \cdot 2 U^T V U \to 0$. These postulates are similar to the ones used by Handelman, Lin, and Prager [15].

It can be shown that under these requirements the new growth laws take the form

$$\dot{c} = (V+\Omega)c + c(V-\Omega) + \phi(s, c)\left(2\frac{\partial Y}{\partial c} \cdot U^T V U\right),$$

when $Y(s, c) = 0$, and $(\partial Y/\partial c) \cdot 2 U^T V U > 0$

$$\dot{s} = T(\dot{R})s + \psi(s, c)\left(2\frac{\partial Y}{\partial c} \cdot U^T V U\right) \tag{a} \tag{35}$$

where ϕ is a tensor valued function which is symmetric and second rank, and $\psi \in R_n$ and $\dot{R}$ is given as before;

$$c\dot{R} - \dot{R}c = 2 U^T V U - (V+\Omega)c - c(V-\Omega). \tag{b}$$

---

[1]) Another way of arriving at (32) is this: In view of (27)' we can choose Ω in such a way that $\dot{R} = 0$. Then $\dot{c} = 2 U^T V U$ and $T(\dot{R})s = 0$ and (32) follows. Since Ω is of no real consequence as far as physics of the situation is concerned we can expect that (32) will be valid when $\dot{R} \neq 0$.

A further consideration shows that ϕ and ψ are *form invariant* functions of their arguments under proper orthogonal transformations.

When plastic flow occurs at time t the state point at time $t + \Delta t$ must still be at yield point. Thus we must have

$$Y(s + \dot{s}\Delta t, c + \dot{c}\Delta t) = 0.$$

It can be shown using the growth laws (35) that the above expression leads to the consistency relation

$$\frac{\partial Y}{\partial s} \cdot \psi + \frac{\partial Y}{\partial c} \cdot \phi + 1 = 0, \tag{36}$$

which imposes a mild restriction on ψ and ϕ.

It is desirable to investigate whether other restrictions should be imposed on ψ and ϕ and also the function f in (17). Indeed we may try to generalize the normality rule of classical plasticity and the stability postulate of Drucker to the present case and to study restrictions that these generalizations may impose on the growth laws.

In the present paper we shall content ourselves with the study of a thermodynamic requirement.

8. Thermodynamic considerations

We now investigate the implications of certain thermodynamic considerations. Since we are interested only in isothermal deformations we need not use concepts such as entropy and free energy. Instead we can base our discussion on maximum recoverable work[1]) which is a purely mechanical concept.

Consider now an element in the state (s, c). Apply to this element a future deformation and determine the work required to perform this deformation. If this work is negative then we say that work is *extracted* from the element. We now introduce the postulate that the work extracted by each permissible future deformation is less than W_0 where $W_0 \geq 0$ and depends only on the state (s, c). (This postulate is related to Kelvin's Principle of classical thermodynamics). The postulate implies that for each state (s, c) there exists a least upper bound $\Psi(s, c)$ for the work which can be extracted from an element

---

[1]) For a discussion of maximum recoverable work see, Breuer and Onat [16] and Day [17].

of unit mass which is in the state (s, c). We assume, in addition, that there exists at least one future deformation that actually extracts the work $\Psi(s, c)$. It would then be permissible to refer to $\Psi(s, c)$ as the maximum recoverable work.

It is easily established from the definition of $\Psi(s, c)$ that the following inequality holds[1]) at any instant of interest

$$\rho\dot{\Psi} = \rho\left(\frac{\partial\Psi}{\partial s}\cdot\dot{s} + \frac{\partial\Psi}{\partial c}\cdot\dot{c}\right) \leq \sigma_{ij}V_{ij} = \sigma\cdot V \tag{37}$$

where ρ is the density which we shall assume to be a function of the state (s, c). The inequality (37) has some important consequences which we explore next. We first note that Ψ must possess the invariance requirement

$$\Psi(s, c) = \Psi(P_Q s, P_Q c). \tag{38}$$

Next we consider a state (s, c) satisfying

$$Y(s, c) < 0.$$

In view of (26–29) the inequality (37) becomes

$$\rho\frac{\partial\Psi}{\partial s}\cdot T(\dot{R})s + \frac{\partial\Psi}{\partial c}\cdot(2U^T VU + \dot{R}c - c\dot{R}) - \sigma\cdot V \leq 0.$$

But in view of (38), we have as before (cf. (31) and (32))

$$\frac{\partial\Psi}{\partial s}\cdot T(\dot{R})s + \frac{\partial\Psi}{\partial c}\cdot(\dot{R}c - c\dot{R}) = 0$$

so that (37) reduces to

$$2\rho\frac{\partial\Psi}{\partial c}\cdot U^T VU - \sigma\cdot V \leq 0$$

or equivalently

$$\left(2\rho\frac{\partial\Psi}{\partial c_{kl}}U_{ik}U_{jl} - \sigma_{ij}\right)V_{ij} \leq 0.$$

---

[1]) If the existence of the free energy Φ and the validity of Clausius-Duhem inequality are accepted then it can be shown that

$$\rho\dot{\Phi} \leq \sigma\cdot V.$$

Now for a given (s, c) satisfying (37) we can choose V arbitrarily (i.e. there is no restriction of the type (34) on V). In view of this observation the inequality demands that

$$\sigma = 2\rho U \frac{\partial \Psi}{\partial c} U^T. \tag{39}$$

Thus we established that the relationship (17) cannot be arbitrary but it must be related to the maximum recoverable work Ψ in the manner shown above.

Let us now examine the case of

$$Y(s, c) = 0 \tag{40}$$

and consider a rate of deformation satisfying (34). Then (37) becomes with the help of growth laws (35) and (39)

$$\rho \left[\frac{\partial \Psi}{\partial s} \cdot \psi + \frac{\partial \Psi}{\partial c} \cdot \phi \right] \left(2 \frac{\partial Y}{\partial c} \cdot U^T V U \right) \leqq 0.$$

Since $\rho > 0$ and $2(\partial Y/\partial c) \cdot U^T V U > 0$ we obtain

$$\frac{\partial \Psi}{\partial s} \cdot \psi + \frac{\partial \Psi}{\partial c} \cdot \phi \leqq 0 \quad \text{when} \quad Y(s, c) = 0. \tag{41}$$

The physical meaning of this inequality will be explored later.

It will be convenient to summarize here the results of this and previous sections. This would provide a compact account of the description of mechanical behavior of elastic-plastic solids.

State of deforming element : (s, c) or equivalently $(q_1, ..., q_m, U^2)$
Yield function : $Y(s, c)$
Maximum recoverable work: $\Psi(s, c)$
Tensors of rate of
deformation and rotation : V and Ω

$$\sigma = 2\rho U \frac{\partial \Psi}{\partial c} U^T$$

$$\dot{c} = \begin{cases} (V+\Omega)c+c(V-\Omega) & \text{when } e\colon Y(s,c) < 0 \text{ or} \\ & Y(s,c) = 0, \\ & \dfrac{\partial Y}{\partial c} \cdot U^T V U \leqq 0; \\[2ex] (V+\Omega)c+c(V-\Omega)+2\phi(s,c)\left(\dfrac{\partial Y}{\partial c} \cdot U^T V U\right) & \text{when } p\colon Y(s,c) = 0, \\ & \dfrac{\partial Y}{\partial c} \cdot U^T V U > 0; \end{cases}$$

(42)

$$\dot{s} = \begin{cases} T(\dot{R})s & \text{when } e, \\[1ex] T(\dot{R})s+2\psi(s,c)\left(\dfrac{\partial Y}{\partial c} \cdot U^T V U\right) & \text{when } p \end{cases}$$

where

$$c\dot{R} - \dot{R}c = 2U^T V U - [(V+\Omega)c+c(V-\Omega)]$$

and

$T(\dot{R})$ is defined below Eq. (30).

Y and Ψ are invariant, and ϕ and ψ are form invariant, under rigid body rotations of the element. There exists also the following restrictions

$$\frac{\partial Y}{\partial s} \cdot \psi + \frac{\partial Y}{\partial c} \cdot \phi + 1 = 0,$$

$$\frac{\partial \Psi}{\partial s} \cdot \psi + \frac{\partial \Psi}{\partial c} \cdot \phi \leqq 0 \quad \text{when} \quad Y(s,c) = 0.$$

Note that the representation has the general form of (1) but the growth laws are not analytic and have only the piece-wise linear dependence on V and Ω.

9. Plastic deformation and its rate of change

We now return briefly to the quantity D^p introduced in Section 5. We recall that $D^p(t)$ is defined by means of the relation

$$D(t) = U(t)D^p(t), \quad U^T = U.$$

In view of this equation and our previous comments D^p could be interpreted as the plastic deformation that accompanies the straining of the element. As Fig. 1 suggests D^p is a measurable quantity. As we shall see below it is useful to study the rate of change of D^p.

One finds easily from (10) that

$$\dot{D}^p D^{p-1} = U^{-1}[(U+\Omega)U - \dot{U}].$$

Now it is tempting to decompose $\dot{D}^p D^{p-1}$ into its symmetric and anti-symmetric parts. Let these parts be denoted by V^p and Ω^p so that

$$V^p + \Omega^p = U^{-1}[(V+\Omega)U - \dot{U}]. \tag{43}$$

When the instantaneous deformation is purely elastic (cf. when the case e occurs in (42)) we find from (24) that

$$V^p = 0 \quad \text{and} \quad \Omega^p = \dot{R} \quad \text{when } e \tag{44}$$

We can thus say that during elastic deformations the rate of plastic deformation V^p as defined above is zero and the rate of plastic rotation is identical with $\dot{R}$. Note from (27)' that Ω^p is not in general equal to Ω.

On the other hand when the instantaneous deformation is plastic we find from (35)

$$V^p = -U^{-1}\phi U^{-1}\left(\frac{\partial Y}{\partial c} \cdot U^T V U\right) \tag{45}$$

and

$$c\Omega^p - \Omega^p c = 2U^T V U - (\Omega + V)c - c(V - \Omega) - \left(\frac{\partial Y}{\partial c} \cdot U^T V U\right) \cdot$$
$$\cdot [2\phi - (U^{-1}\phi U + U\phi U^{-1})].$$

If we wanted to impose the restriction that plastic deformations are volume preserving then we would insist, in view of (45), that

$$\text{tr } U^{-1}\phi U^{-1} = 0 \tag{46}$$

which puts a mild restriction on ϕ. We shall presently see that the quantity V^p is important in the interpretation of the inequality (41).

10. The case of small elastic deformations

We wish now to discuss the form that the general representation (42) takes for a class of technically important materials. We consider materials such as ductile metals where large plastic deformations are accompanied by small 'elastic' deformations. Here by small elastic deformations we mean deformations for which the components of $c-I$ or equivalently $U-I$ are, in absolute value, much smaller than unity. Observe that in this case $\frac{1}{2}(c-I)$ may be interpreted as the classical infinitesimal strain tensor ε^e which measures the deformation with respect to the stress free configuration.

We next introduce the assumption that Ψ depends only upon c so that the stress free elements will have the same constant energy. That this assumption is a fairly reasonable one for metals is well-known (cf. Farren and Taylor [18]).

Another reasonable assumption is that the plastic deformations are volume preserving (cf. (46)).

In view of these assumptions the representation becomes to a first order of approximation and with the trivial change of notation

$$\varepsilon^e = \tfrac{1}{2}(c - I),$$

$$\sigma = 2\mu\varepsilon^e + \lambda I \operatorname{tr} \varepsilon^e$$

$$\dot{\varepsilon}^e = \begin{cases} V & \text{when } e: \ Y(s, \varepsilon^e) < 0 \text{ or} \\ & \qquad Y = 0 \text{ and } \dfrac{\partial Y}{\partial \varepsilon^e} \cdot V \leqq 0, \\[2em] V + \tfrac{1}{2}\phi\left[\dfrac{\partial Y}{\partial \varepsilon^e} \cdot V\right] & \text{when } p: \ Y(s, \varepsilon^e) = 0, \\ & \qquad \dfrac{\partial Y}{\partial \varepsilon^e} \cdot V > 0; \end{cases}$$

$$\hspace{10cm} (47)$$

$$\dot{s} = \begin{cases} T(\Omega)s & \text{when } e, \\[1em] T(\Omega)s + \psi(s, \varepsilon^e)\left(\dfrac{\partial Y}{\partial \varepsilon^e} \cdot V\right) & \text{when } p \end{cases}$$

where

$$\frac{\partial Y}{\partial s} \cdot \psi + \frac{1}{2} \frac{\partial Y}{\partial \varepsilon^e} \cdot \phi + 1 = 0.$$

It should be stressed that the above equations are valid in the presence of total finite deformations provided that $c \approx I$.

Observe that to within the present approximation equation (27) which defines $\dot{R}$ becomes

$$\dot{R} = \Omega.$$

Furthermore within the approximations introduced here we find from (45)

$$V^p = -\tfrac{1}{2}\phi \cdot \left[\frac{\partial Y}{\partial \varepsilon^e} \cdot V \right],$$

$$\Omega^p = \Omega$$

and plastic incompressibility takes the form

$$\operatorname{tr} \phi = 0.$$

Note that in the present case the inequality (41) associated with energy considerations becomes

$$\frac{\partial \Psi}{\partial \varepsilon^e} \cdot \phi \leqq 0 \quad \text{when} \quad Y(s, \varepsilon^e) = 0. \tag{48}$$

But now

$$\frac{\partial \Psi}{\partial \varepsilon^e} = \sigma \quad \text{and} \quad \phi = -2 \frac{V^p}{[(\partial Y / \partial \varepsilon^e) \cdot V]}$$

so that (48) yields

$$\sigma \cdot V^p \geqq 0,$$

which now has the meaning that the energy dissipated must be non-negative.

11. Further remarks on the tensorial nature of the state of stress free elements

We end this paper by considering briefly the question of experimental determination of the rank and the number of tensors $q_1, \ldots, q_m$ which represent the state of a stress free element. Since the q_i characterize the relevant and salient features of dislocation distribution within an element, one may be tempted to state that the information on the nature of these quantities must come from microscopic considerations. The purpose of this section is to point out that such information can also be obtained from pheno-

menological experiments.[1]) An important class of experimental information is contained in the form of the initial and subsequent yield surfaces.
Thus it is reasonable to expect that information on the nature of q_i could
be obtained from a study of the shape of yield surfaces.

Here we consider yield surfaces in the c-space with the coordinates c_{ij}.
Such a surface can, in principle, be obtained experimentally.

For a given value of s the equation

$$Y(s, c) = 0$$

determines a surface $\mathscr{S}$ in the c-space. Let us now apply to the points of
this surface constant orthogonal transformation Q which takes the point c
to the point $QcQ^T = P_Q c$.

This transformation creates another surface, say $\mathscr{S}_1$. In view of the
invariance requirement (16) this new surface is the yield surface associated
with the stress free state

$$s_1 = P_Q s.$$

Now we let Q take on all the values in $O^+(3)$ to obtain a family of yield
surfaces. We assert that a study of this family will shed light on the tensorial
nature of q_i.

Suppose that it is claimed, as it is often done in literature on plasticity,
that the state s is characterized by a single symmetric second rank tensor q:
Can we support or reject this claim by a study of the above family? Note
that when $s = q$ where q is a symmetric second rank tensor we have

$$s = P_Q s$$

for those Q's which represent 180° rotations about the principal directions
of q. For such a Q we have in view of the invariance requirement (16) and (49)

$$Y(P_Q s, P_Q c) = Y(s, P_Q c) = 0.$$

This means that for such a Q the surfaces $\mathscr{S}$ and $\mathscr{S}_1$ coincide. Thus the
above claim would be rejected, for instance, if the class of yield surfaces
discussed above do not have such property of coincidence.

---

[1]) For an extended discussion of this claim the reader is referred to [19] where the determination of the number of state variables from phenomenological experiments has also
been discussed.

References

[1] Green, A. E. and P. M. Naghdi, A general theory of an elastic-plastic continuum. *Arch. Ratl. Mech. Anal., 18* (1965) 251–281.

[2] Lee, E. H. and D. T. Liu, Finite strain elastic-plastic theory. *Proc. IUTAM Symp., Vienna (1966)*, Springer, Berlin 1968, 213–222.

[3] Lee, E. H., Elastic-plastic deformations at finite strains. *J. Appl. Mech., 36* (1969) 1–6.

[4] Perzyna, P. and W. Wojno, On the constitutive equations of elastic-viscoplastic materials at finite strain. *Arch. Mech. Stos., 18* (1966) 85–99.

[5] Willis, J. R., Elastic-plastic deformations at finite strains. *J. Mech. Phys. Solids, 17* (1969) 359–369.

[6] Owen, D. R., Thermodynamics of materials with elastic range. *Arch. Ratl. Mech. Anal., 31* (1968) 91–112.

[7] Mandel, J., Relations de comportement des milieux elastiques-plastiques et elastiques-viscoplastiques. Notion de rèpere directeur. *Foundations of Plasticity*, edited by A. Sawczuk, Warsaw (1972), Noordhoff Intl. Leyden 1973, 387–400.

[8] Onat, E. T., The notion of state and its implications in thermodynamics of inelastic solids. *Proc. IUTAM Symp., Vienna (1966)*, Springer, Berlin 1968, 292–314.

[9] Onat, E. T., Representation of Inelastic mechanical behavior by means of state variables. *Proc. IUTAM Symp., East Kilbride (1968)*, Springer, Berlin 1970, 213–225.

[10] Geary, J. A. and E. T. Onat, On the construction of state space for non-linear materials with memory. (In preparation).

[11] Noll, W., A mathematical theory of the mechanical behavior of continuous media. *Arch. Ratl. Mech. Anal.,* 2, (1958–59) 197–226.

[12] Onat, E. T., Description of mechanical behavior of inelastic solids. *Proc. 5th U.S. National Congress of Appl. Mech.* (Minneapolis, Minn.), ASME, New York 1966, 421–434.

[13] Gel'fand, I. M., R. A. Minlos, and Z. Ya. Shapiro, *Representation of the rotation and Lorenz groups and their applications.* Pergamon Press, Oxford 1963.

[14] Kröner, E., Dislocation: A new concept in the continuum theory of plasticity. *J. of Math. and Phys., 42* (1963) 27–37.

[15] Handelman, G. H., C. C. Lin, and W. Prager, On the mechanical behavior of metals in the strain-hardening range. *Quart. Appl. Math., 4* (1947) 397–407.

[16] Breuer, S. and E. T. Onat, On recoverable work in linear viscoelasticity. *Zeit. Angew. Math. Phys., 15* (1964) 12–21.

[17] Day, W. A., Reversibility, recoverable work and free energy in linear viscoelasticity. *Quart. J. Mech. Appl. Math., 23* (1970) 1–15.

[18] Farren, W. S. and G. I. Taylor, The heat developed during plastic extension of metals. *Proc. Royal Soc., 107* (1925) 422–451.

[19] Onat, E. T. and F. Fardshisheh, *Representation of creep of metals.* U.S. Atomic Energy Commission, Oak Ridge National Laboratory Report No. ORNL-4783 (1972).

DISCUSSION

F. Sidoroff [1]).   Am I right in saying that the first two terms $\dot{s} - T(\dot{R})s$ which appear in the equation giving the time variation of the state s define some kind of corotational time derivative, in the case when s is a symmetric second order tensor?

E. T. Onat.   That is the reason why I only said 'some kind of' corotational time derivative. Well, this ensures that this equation is properly frame independent.

---

[1]) University Paris VI, Paris, France.

Elastic-plastic theory at finite strain

E. H. Lee

Stanford University, Stanford, California, USA

P. Germain

University Paris VI, Paris, France

It is shown that in order to solve certain technological problems, it is necessary to develop elastic-plastic theory for which both components of strain can be finite. This poses a challenging problem because of the need to include nonlinear discontinuous response to loading history, that is to determine nonlinear discontinuous functionals. Various approaches to this challenge are considered and contrasted. The possible advantages of direct extension of classical approaches is expressed. Questions of the validity and applicability of suggested theories are considered from the standpoints of invariance, convenience of representation for specific materials and for introduction into stress and strain analysis evaluations. Thermodynamic aspects are also discussed.

1. Introduction

The topic of the analysis of elastic-plastic deformation at finite strain has already received considerable attention at this Symposium, with other papers in this area still to come. It would therefore be repetitious for us to attempt a survey of recent work – even if it were feasible in view of the time allotted and the limitation of our knowledge of the many developments being pursued. We will therefore try to assess philosophies of formulating theories, discuss some aspects that arise, and hopefully leave ample time for discussion, so that related aspects, based on approaches not adequately dealt with by us, can be introduced.

The introduction to this talk is addressed to non-specialists in finite-deformation theory, to explain why practitioners become involved in what may appear to be over-elaborate theory. One of us was quoted in a recent Applied Mechanics Reviews article [1][1]) as stating that plastic flow is an

---

[1]) Numbers in square brackets refer to the bibliography at the end of the paper.

extremely complicated process, and we here endorse this opinion. In analysing plastic flow, one might wish to include, for example, an involved, discontinuous, non-linear and anisotropic memory (including the Bauschinger effect and the stress cycling phenomena discussed at this Symposium by Valanis and by Fong and Simmons), thermal effects including thermodynamic irreversibility and strain rate influences. In practice, for practical applications, simple models have been devised which encompass the basic phenomena of elastic and plastic deformation but which neglect the more elaborate aspects just mentioned.

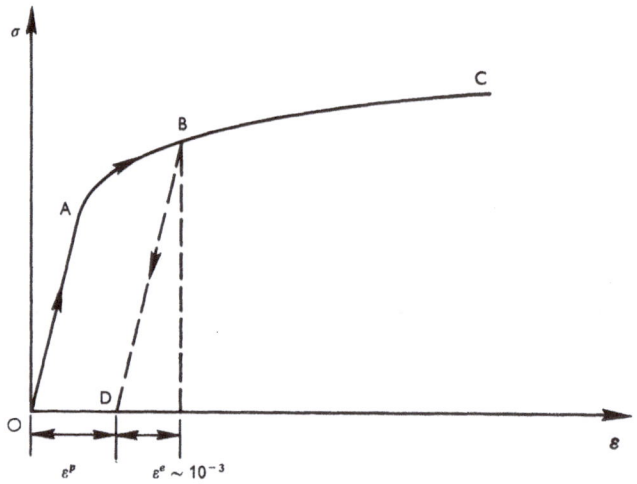

Fig. 1. Stress-strain curve in simple tension.

Consider a typical stress-strain curve in simple tension as shown in Fig. 1. Hooke's law applies until the elastic limit is reached at A. Plastic flow occurs along ABC. Unloading from B is elastic and parallel to the initial elastic line OA, so that the total strain at B can be considered to be composed of the sum of plastic and elastic components:

$$\varepsilon = \varepsilon^p + \varepsilon^e. \tag{1}$$

The elastic strain is given by the yield stress at B divided by the Young's modulus and is of the order 10^{-3}. For ductile metals, plastic flow can continue along BC extended until strains of the order unity are reached.

For simple tension $\varepsilon^e = \sigma/E$, where E is Youngs modulus. For combined stresses with possibly non-linear response, the elastic law can be written

118

formally:

$$\sigma_{ij} = f_{ij}(\varepsilon_{kl}^e) \tag{2}$$

a function relation between the stress and elastic deformation. Plasticity is governed by incremental or flow type functional relations, for example

$$\dot{\varepsilon}_{ij}^p = g_{ij}(\sigma_{kl}, \dot{\sigma}_{mn}) \tag{3}$$

where the superposed dot denotes the time derivative. Because of the different mathematical structures of the relations (2) and (3) the (stress – total strain) relation, needed for the analysis of stress and strain distribution problems, is cumbersome. However, if the total strains are of the order 10^{-3}, the plastic and elastic components will be of the same order, and so must both be included in spite of the difficulties of elastic-plastic theory. For this strain magnitude, infinitesimal strain theory is adequate, and for general loading (1) becomes:

$$\varepsilon_{ij} = \varepsilon_{ij}^p + \varepsilon_{ij}^e. \tag{4}$$

For metal forming problems, strains of the order unity arise, and the elastic strains are usually limited to the order 10^{-3}, because the stress is limited by the yield condition and the corresponding strains are determined by (2). Hydrostatic pressure is not thus limited by the onset of yield, but usually the inability of shear or deviator components to increase without limit because of yield, combined with the boundary conditions, does limit all stress magnitudes. Since the elastic strains are then negligible compared with the total strain, it is common to neglect them and so utilize a plastic-rigid model. Deformation is then entirely governed by a law of type (3), in which ε_{ij}^p can be replaced by ε_{ij}. Since only rate of strain appears in this relation, and not the strain itself, it is a fluid type law, and, in fact, in the terminology of non-linear continuum mechanics, a material in plastic flow can be specified as a simple fluid with a volume constraint. As in fluid mechanics, it is then common to define rate of strain as velocity strain:

$$\dot{\varepsilon}_{ij} = \dot{\varepsilon}_{ij}^p = \frac{1}{2}\left(\frac{\partial v_i}{\partial x_j} + \frac{\partial v_j}{\partial x_i}\right) \tag{5}$$

where $v_i(x, t)$ is the velocity field, t the time and $x \equiv (x_1, x_2, x_3)$ is the current position coordinates. Stress distribution problems are then studied in terms of velocity fields, and final deformations are obtained by integrating these over time. Thus, even though finite strains arise in metal forming

119

problems, the utilization of the kinematics of finite strain is not required. Not all problems associated with metal forming can be so treated; for example, the analysis of residual stresses generated by plastic flow is inextricably dependent on the elastic unloading strain. For such problems the complications of elastic-plastic theory, with its combined solid and fluid type response associated with (2) and (3), cannot be avoided.

There are situations of highly constrained flow commonly brought about by the symmetry of configuration and loading or the influence of inertia forces for rapid loading, in which large stresses arise in spite of the limiting effect of plastic yield. These must have a dominant hydrostatic component, and the resulting elastic strains given by (2) may fall outside the range of satisfactory infinitesimal strain analysis, and no longer be negligible so as to permit the use of plastic-rigid theory, even for finite plastic strains. Such a situation arises in the analysis of plane shock waves in metals generated by explosive or impact loading. In this case both elastic and plastic strain components are of the same order and can be finite [2, 3]. Elastic strains greater than 25% can readily arise in explosive loading [4]. In order to develop a satisfactory theoretical basis for the analysis of such situations, elastic-plastic theory is needed for which both elastic and plastic strain components are free to be of finite magnitude. Then non-linearity, in both its geometrical and material aspects, must be considered. The former concerns the non-linear kinematics of the geometry of finite strain, and the latter the non-linear material characteristics, both elastic and plastic, commonly associated with finite deformation. It is important to consider these influences in shock wave analysis, since the influence of non-linearity plays a major role in shock wave theory [5]. Although shock waves generate elastic and plastic strains of the same order, it can be that for high pressure loading (> 100 kilobars) the change in stress from pure hydrostatic pressure, caused by material strength, is small compared to the pressure magnitude. It may then be reasonable to neglect plasticity effects and use hydro-dynamic theory, the shape deformation of an element then occurring freely as in an inviscid fluid, without appreciable deviator stress required to generate it by elastic or plastic deformation. However, even at high pressures, plastic analysis may still be required adequately to analyse some aspects of wave propagation, such as wave front attenuation associated with unloading [6].

It was to develop a theory for the propagation of shock waves in metals that the approach to finite deformation elastic-plastic theory instituted at Stanford was proposed [2, 3, 7, 8, 9]. Although developed with a practical application in mind, the attempt to remove the restriction of infinitesimal elastic strains does call for a reconsideration of the basic structure of plasticity theory and poses interesting fundamental problems. Some of these aspects are considered in the present paper, as is also the comparison of this particular approach with parallel investigations of basic elastic-plastic theory.

2. Approaches to elastic-plastic theory

Since, as described in the previous section, plastic flow is an extremely complicated phenomenon involving a non-linear functional relationship between stress and deformation, it is virtually impossible to write down a detailed mathematical representation which is both general and definitive. Thus authors have tended to stress some particular aspect of the theory.

For example, the approach to elastic-plastic theory presented by A. E. Green and P. M. Naghdi [10, 11] provides a general theoretical structure within a rate-type functional representation. General function forms are utilized for constitutive functions, and the body is permitted to be an-isotropic in its initial state. Specialization to isotropic behavior is then carried out. The inclusion of plastic strain and a hardening parameter in the elastic relation and in the various plastic constitutive functions which appear, involves a generality which at the present time would be difficult to utilize through measurements of specific material properties, or to intro-duce into methods of solution of stress-analysis problems.

In contrast B. Budiansky [12] developed a finite-deformation elastic-plastic theory based on incremental strains for both elastic and plastic deformation, designed to simplify the solution of stress and deformation analysis problems. This theory has been applied to the analysis of necking in a tensile bar [12]. However, the incremental elastic law is not deducible from a stress-elastic strain relation and hence the principle of conservation of elastic energy may be violated in a loading and unloading cycle. It is claimed, however, that the error thus introduced is negligible, so that the ease of application achieved justifies the approximation.

A. C. Pipkin and R. S. Rivlin [13] considered constraints on a functional elastic-plastic law for finite strain which guarantees a rate independent response as commonly observed in quasi-static loading. The mathematical representation of the functional was not specified, but the conditions for

121

elastic unloading, and the use of a scalar deformation variable to ensure rate independence were presented.

In discussion of the development of continuum mechanics, Truesdell [14], and Truesdell and Noll [15] have stated that the classical theories of plasticity fail to meet basic requirements of continuum mechanics, and so they have not included this topic in reviews of the development of non-linear theory. The difficulty appears to be that as the laws are commonly written, they apply directly only for infinitesimal displacements. However, because plasticity is governed by incremental or fluid type relations and elastic strain components can often be considered small, proper interpretation can yield satisfactory laws for finite plastic strain. Essentially, this is so because in the classical theories, memory of the previous deformation history is embodied in the current shape and a scalar parameter which defines the isotropic work hardening which has taken place. In the case of ideal plasticity, the work hardening parameter does not arise, and the memory of the history of deformation is embodied in the current plastically deformed geometry. In developing elastic-plastic constitutive relations when both components of strain can be finite, it is clearly important to satisfy the basic requirements of continuum mechanics in an unequivocal manner in the formal statement of the relations. This has been achieved in the recent developments of finite deformation theory referred to in the bibliography.

The development of the classical theories of plasticity, including work hardening, has grown from an extensive experimental effort, in which the yield condition has usually been determined in terms of the Cauchy or 'true' stress in the plastically deformed specimen. In developing finite deformation theories, some advantage can therefore be gained by maintaining the use of Cauchy stress in the development of the theory, since our knowledge of, for example, the yield condition in terms of Cauchy stress is already established. If, from the point of view of the structure of the theory, some other stress measure seems more appropriate, how this will modify the established functional forms for the yield surface must be kept in mind.

In a similar way, it is known that effective uncoupling between the elastic and plastic laws arises if the unstressed (permanently plastically deformed) configuration is used as a reference state for deformation changes. The yield surface is insensitive to $-\sigma_{ii}$, the average hydrostatic pressure or equivalently to the elastic dilatational strain [16]. Also the elastic constants are known to be insensitive to prior plastic flow [17, 18, 19]; indeed it is common practice in design to assume a fixed value for the Young's modulus of a particular material irrespective of the method of fabrication of the component, which may have involved cold-forming. Such effective un-coupling arises when the current state (which commonly is close to the un-stressed state) is used as a reference state for defining strain or strain incre-ments, and Cauchy stress for the loading or unloading variable. Such uncoupling of the elastic and plastic laws can greatly reduce the complexity of the combined constitutive relations. Use of another reference configu-ration, such as the initial state, or other variables, can introduce coupling of the elastic and plastic laws, which may add appreciable complications both in establishing the laws by measurement and in utilizing them in the solution of stress and deformation analysis problems. Again, as for the yield function, the use of a particular mathematical structure of continuum theory may dictate against direct extension of classical theory in terms of the same variables, but possible benefits of mathematical structure will have to be weighed against the possibility of more involved material properties representations.

3. Comments on a particular form of theory

The theory of elastic-plastic deformation at finite strain developed at Stan-ford [2, 7, 8, 9] was proposed in order to introduce non-linear influences in a consistent manner for application to the analysis of shock-wave pro-pagation in metals. Earlier work in this area [20, 21], while possibly providing an adequate approximation, utilized Hooke's law incrementally with a pressure dependent bulk modulus for the elastic component of the strain. Such a mixture of linear and non-linear concepts seems to be unsatisfactory in principle and was replaced by classical non-linear thermo-elastic theory. The analytical structure was developed as an extension of classical elastic-plastic analysis, to take maximum advantage of known material charac-teristics as discussed in the previous section. Some idealizations of classical theory, such as isotropic work hardening, were retained, although the mathematical structure developed could be used with more general model characteristics. Some aspects of this theory have been subject to criticism, and it therefore seems worthwhile to comment here on questions raised. The theory developed was applied to the analysis of shock wave propagation in metals (see [5, 9] and also the paper on this topic in the present Sym-posium).

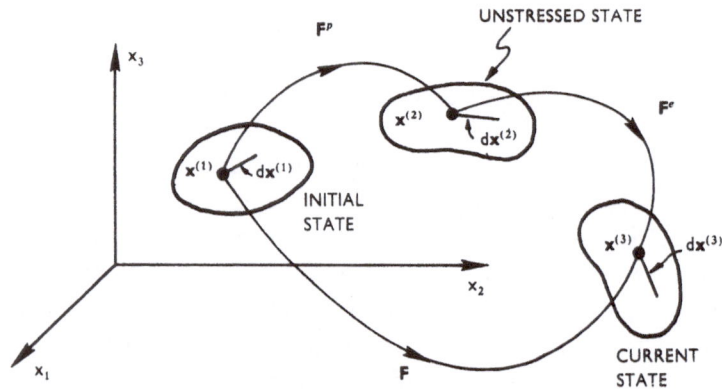

Fig. 2. Plastic and elastic deformation.

Plastic and elastic strains are defined through a thought experiment depicted in Fig. 2. With a fixed Cartesian coordinate system $x \equiv (x_1, x_2, x_3)$, the position of particles in the initially undeformed body is prescribed by $x^{(1)}$, where the bold type face represents a vector quantity. During elastic-plastic deformation, at time t the particle positions are given by:

$$x^{(3)} = x^{(3)}(x^{(1)}, t).\qquad(6)$$

The configuration $x^{(1)}$, for example, corresponds to the origin O of the stress strain curve in Fig. 1, and the deformed state $x^{(3)}$ to the point B. The kinematics of deformation at a point can be investigated in terms of the deformation gradient:

$$F = \frac{\partial x_i^{(3)}}{\partial x_j^{(1)}} = \frac{\partial x^{(3)}}{\partial x^{(1)}}\qquad(7)$$

which is a 3×3 matrix. If the body is now unstressed and the temperature reduced to the initial uniform temperature, we will consider that the body takes on the new configuration:

$$x^{(2)} = x^{(2)}(x^{(1)}, t)\qquad(8)$$

which corresponds to the point D in Fig. 1. In general, following plastic flow, removing surface tractions will leave a residual stress field in the body, and unstressing, as represented in Fig. 2, will require division of the body into infinitesimal elements to produce a non-continuous map (8). This circumstance introduces no complication into the study of elastic-plastic constitutive relations, since only the value of the deformation gradients at each point appears, the laws being of the 'simple material' type as defined by Noll. If for any particular point $x^{(1)}$, history $F(x^{(1)}, t)$ were produced in uniform strain throughout the body, unloading would produce un-stressing, with a continuous differentiable deformation field, and so the unstressed deformation gradient

$$F^p(x^{(1)}, t) = \frac{\partial x^{(2)}}{\partial x^{(1)}} \tag{9}$$

could be achieved. Strictly speaking a body force field would have to be imagined to equalize inertia forces in this contemplated configuration change. Whereas we speak of deformation gradients $F^{(p)}$ and

$$F^e = \frac{\partial x^{(3)}}{\partial x^{(2)}} \tag{10}$$

these are simply point matrix functions in the non-uniformly strained body, which give the local limit of the deformation gradient at each point as the constraint of the surrounding material is removed to achieve state $x^{(2)}$.

Since the stress is zero in the state $x^{(2)}$ and the temperature is equal to the initial reference temperature, the thermoelastic strains are zero, and the strain in this configuration is the plastic strain, which is that generated in the loading to $x^{(3)}$ if unloading from $x^{(3)}$ to $x^{(2)}$ is purely elastic. Thus F^p is the deformation gradient of plastic flow to the state $x^{(3)}$, and F^e the thermo-elastic deformation gradient from the current unstressed state $x^{(2)}$ to $x^{(3)}$.

Objections have been raised to the definitions of plastic and elastic deformations described above, because F^p and F^e are not strictly defined deformation gradients of continuously differentiable maps, but the presentation given above indicates that these conditions offer no essential limitation to the theory. In fact any arbitrary variation of $F(t)$ could, in principle, be achieved in homogeneous straining, all three deformation gradients F, F^p and F^e would then be unequivocally defined. The circumstance of the non-continuous configuration in the unstressed state has been observed by Sedov [22], who points out that convected coordinates become non-Euclidean in the configuration $x^{(2)}$. The matter was considered in a round table discussion at a meeting of the Society of Natural Philosophy where it was pointed out that these kinematical aspects were equivalent to the structure of the theory of materials with elastic range due to D. R. Owen [23, 24].

Using the unstressed configuration as a reference state for the repre-

125

sentation of strain, it is known that for many metals the elastic constants do not depend appreciably on prior plastic flow [17, 18, 19]. Hill [16] states that appreciable anisotropy induced by plastic flow does not arise for plastic strains less than about 30%. Bell [25] claims a marked effect of plastic flow on elastic constants, but his tests concerned soft annealed metals for which it is more difficult to delineate the plastic and elastic components of strain. Assuming no such influence, the thermo-elastic stress-strain relation is given by the classical one for purely elastic bodies [8]:

$$\sigma = \frac{2\rho_0}{\det(F^e)} F^e \frac{\partial \psi}{\partial C^e} F^{e^T} \tag{11}$$

where σ is the Cauchy or 'true' stress, ρ_0 the initial density, $\psi(C^e, T)$ is the Helmholtz free energy per unit mass, $C^e = F^{e^T} F^e$ is a measure of the elastic strain and T is the absolute temperature. In this case the function ψ replaces the usual elastic constants, and is assumed to be independent of plastic strain.

In contrast Green and Naghdi [11] use the relation (eqn. 3.6)(of [11]):

$$S = S(E', E'', \kappa, T) \tag{12}$$

where S is the symmetric Piola stress tensor, E' the elastic strain, E'' the plastic strain with

$$E' + E'' = E = (F^T F - I)/2 \tag{13}$$

the total Lagrange strain, and κ a strain hardening scalar parameter. Relation (12) is much more general than (11), and has been shown by Green and Naghdi to include (11). However, it contains a tensor and a scalar as additional variables, and the simplicity associated with plastic flow not modifying appreciably elastic properties is hidden. Even if the elastic properties of the material are unchanged by plastic flow, the plastic strain E'' would still appear in (12) in order to compensate for the fact that E' gives the elastic strain measured in a convected system of coordinates based on the initial undeformed configuration. Thus the influence of pure geometry change of the unstressed state, which is caused by plastic flow, is intertwined with the effect of plastic flow on physical properties. The elastic constitutive relation is commonly most conveniently expressed in terms of the unstressed reference state, which for an elastic body remains unchanged. With plastic flow, this reference state deforms continuously. These complexities introduce two types of difficulties. It would be awkward to establish the insensitivity of elastic properties to plastic flow in terms of (12), and application to stress analysis problems would lead to complexities not introduced in (11).

The formulation (12) is similar to the development in terms of convected coordinates presented by Sedov [22]. The use of convected coordinates yields the total strain tensor as the sum of elastic and plastic strain tensors, but since the metric of the convected coordinates in the unstressed reference state varies with plastic flow, so does the elastic law [27], even if the elastic characteristics of the material are unchanged. In this connection, the statement by Lee [8] concerning non-additivity of elastic and plastic strains at finite deformations is misleading, since it was made on the assumption, unstated, that plastic strain was expressed as permanent deformation from the undeformed state, and elastic strain as deformation from the plastically deformed unstressed reference state, in order to achieve effective uncoupling of elastic and plastic properties. Additivity can be achieved at finite strain if such uncoupling is not demanded, as illustrated by [11, 22].

For practical applications of the theory, it is suggested that the simplification in mathematical structure and in the determination of the physical constants and functions, due to the uncoupled formulation, carries considerable weight in the selection of the most convenient representation of constitutive relations. In the example under discussion, merely the reduction in the number of independent variables in the elastic relation may spell the difference between the feasibility or impracticality of applying the theory.

The question of the requirements of invariance of elastic-plastic theory under rigid body rotation is discussed in [26]. There arbitrary independent rotations are considered in the final deformed configuration and in the unstressed configuration. We suggest that this requirement is excessive, since the unstressed state, devised as a reference state to provide a simple elastic law, is merely a mathematical construct to simplify the formulation of the theory. In the course of plastic deformation, the body, in general, never attains this state, so that invariance under rigid body rotation need only be demanded of the deformed state $x^{(3)}$. Demanding additional conditions on the unstressed state $x^{(2)}$ has a certain philosophical connection with the criticism, mentioned earlier, that validity of the theory is compromised because the state $x^{(2)}$ is not a continuous map of $x^{(1)}$. Such difficulties are avoided if the state $x^{(2)}$ is simply considered as a way of defining variables F^p and F^e which permit a more concise mathematical structure for the constitutive relations. In fact the development in [8] utilized the simplification achievable by considering unloading without rotation, so that $F^{(e)}$ is a symmetric matrix. Clearly this form of the theory would not be invariant under independent rotations of the states $x^{(2)}$ and $x^{(3)}$, but this has no bearing on its validity.

Since thermo-mechanical coupling can play a significant role in problems of plastic deformation, it is necessary to include thermodynamics in the constitutive relations. It is known that the irreversibility of plastic flow does not fall within the scope of the current theory of irreversible thermodynamics, since, as pointed out by Bridgeman [28], a plastically deformed state cannot be reached by a reversible path, hence leading to difficulties in the definition of state variables. Green and Naghdi [10, 11] introduce the Helmholtz free energy and entropy as functions of the elastic and plastic strains, the hardening parameter and temperature, and impose the conditions of energy balance and the entropy production inequality, that is the first and second laws of thermodynamics. In contrast [7, 8] utilize experimental findings by Taylor, Farren and Quinney [29, 30] where it was shown that the temperature rise in specimen subject to elastic-plastic deformation was deducible from the thermoelastic relations, with dissipation of plastic work into heat equal to

$$\dot{Q} = \gamma \dot{W}^p \tag{14}$$

where W^p is the work expended in plastic flow, and γ a factor varying slowly between 0.9 and 1.0 as plastic flow proceeds. In the shock wave application [5], the associated temperature rise is generally small compared with the thermo-elastic temperature rise, so that a crude approximation to (14), say γ held constant at 0.9, provides an adequate approximation for the whole process.

The treatment of this question in [10, 11] is clearly more general, but correspondingly less specific. For application of the theory it will be necessary to select the free energy function:

$$\psi = \psi(E', E'', \kappa, T). \tag{15}$$

While forms of free energy function for thermoelastic solids are well established, the introduction of plasticity parameters calls for new developments. Since this includes irreversible processes, it will be necessary to investigate the nature of entropy production in them. Thus an assessment of the magnitude of the entropy inequality will be needed, and not merely the condition that it not be violated. The simple model (14) based on experimental findings does satisfy thermodynamic requirements for the formulation presented in [8], but more general specific laws seem not to be available at present for applications of the theory.

4. Discussion

From the assessment given already of the various models which have been suggested to represent plastic flow, it is clear that much progress is needed if the response of actual materials is to be predicted accurately for general loading histories. This is even true within the limitation of small strains. For example, the available theories prescribing the variation of the yield surface with plastic flow have in the main been confined to isotropic hardening and kinematic hardening, both of which are gross simplifications of the actual situation. Such characteristics can have a dominant influence in some problems, for example the propagation of plastic waves of combined stress [31], for which the current shape of the yield surface determines the wave profile, since the stress point may execute a broad traverse of that surface after initial plastic flow.

It seems clear that the non-linear history influence, requiring an applicable adequate non-linear functional relation between the variations of stress and deformation is likely to be too complicated to be deducible purely from the general mathematical structure, as are the laws of thermoelasticity and viscosity. These are function relations, and in the former, the laws can be generated from the consideration of reversible processes only. The entropy production inequality then becomes an equation, yielding a definitive function form for the constitutive relations, which can be determined by an experimental testing program. For plasticity the availability of mathematical representations of discontinuous non-linear functionals appears to rule out such an approach. It is necessary to study experimental test results to suggest forms of the laws and the appropriate variables with which to express them. Also information from the basic physical theory of plastic flow must be drawn on, again to indicate the structure of the theory and to suggest the most appropriate variables to utilize.

This symposium confirms that many are actively seeking solutions to these problems from many standpoints. We have exposed our prejudices on these questions. We will be glad if others will express theirs in the discussion.

References

[1] Johnson, W., *Appl. Mech. Reviews 24* (1971) 977.

[2] Lee, E. H., *Proc. 5th US National Congress of Appl. Mech.* (1966) 405.

[3] Lee, E. H. and D. T. Liu, *Proc. IUTAM Symp.* on Irreversible aspects of continuum mechanics, Eds. H. Parkus and L. I. Sedov, Springer-Verlag (1968) 213.

[4] Rice, M. H., R. G. McQueen and J. M. Walsh, *Solid state physics*, Vol. 6, Eds. F. Seitz and D. Turnbull, Academic Press, New York (1958) 1–63.

[5] Germain, P. and E. H. Lee, On shock waves in elastic-plastic solids, to appear *Jour. Mech. Phys. Sol.*

[6] Lee, E. H. and D. T. Liu, *Proc. IUTAM Symp.* on Stress waves in anelastic solids, Eds. H. Kolsky and W. Prager, Springer-Verlag, Berlin (1964) 239.

[7] Lee, E. H. and D. T. Liu, *Jour. Appl. Phys. 38* (1967) 19.

[8] Lee, E. H., *Jour. Appl. Mech. 36* (1969) 1.

[9] Lee, E. H. and T. Wierzbicki, *Jour. Appl. Mech. 34* (1967) 931.

[10] Green, A. E. and P. M. Naghdi, *Arc. Ratl. Mech. Anal. 18* (1965) 251.

[11] Green, A. E. and P. M. Naghdi, *Proc. IUTAM Symp.* on Irreversible aspects of continuum mechanics, Eds. H. Parkus and L. I. Sedov, Springer-Verlag (1968) 117.

[12] Chen, W. H., *Int. Jour. Sol. Struct. 7* (1971) 685.

[13] Pipkin, A. C. and R. S. Rivlin, *ZAMP 16* (1965) 313.

[14] Truesdell, C., *IUTAM Symp.* Second order effects in elast. etc., Eds. M. Reiner and D. Abir, Pergamon Press, Oxford (1964) 9.

[15] Truesdell, C. and W. Noll, *Handbuch der Physik*, Ed. S. Flügge, III/3 (1965) 47.

[16] Hill, R., *The mathematical theory of plasticity*, Clarendon Press, Oxford (1950) 24.

[17] Dalby, W. E., *Phil. Trans. Roy. Soc. London, A221* (1921) 117.

[18] Howard, J. V. and S. L. Smith, *Proc. Roy. Soc. London, A107* (1925) 113.

[19] Lee, E. H., *Inelastic behavior of solids*, Eds. M. F. Kanninen, W. F. Adler, A. R. Rosenfield, R. I. Jaffee, McGraw-Hill, New York (1970) 445.

[20] Morland, L. W., *Phil. Trans. Roy. Soc. London, A251* (1959) 341.

[21] Fowles, G. R., *Jour. Appl. Phys. 32* (1961) 1475.

[22] Sedov, L. I., *Introduction to the mechanics of a continuous medium*, Addison-Wesley, Reading (1965) 230 ff.

[23] Owen, D. R., *Arch. Ratl. Mech. Anal. 31* (1968) 91.

[24] Owen, D. R., *Arch. Ratl. Mech. Anal. 37* (1970) 85.

[25] Bell, J. F., *Springer tracts in natural philosophy*, Springer-Verlag, New York (1968).

[26] Green, A. E. and P. M. Naghdi, *Int. Jour. Eng. Sci. 9* (1971) 1219.

[27] Green, A. E. and W. Zerna, *Theoretical elasticity*, Clarendon Press, Oxford (1968).

[28] Bridgeman, P. W., *Rev. Mod. Phys. 22* (1950) 56.

[29] Farren, W. S. and G. I. Taylor, *Proc. Roy. Soc., London, A107* (1925) 422.

[30] Quinney, H. and G. I. Taylor, *Proc. Roy. Soc., London, A143* (1934) 307.

[31] Ting, T. C. T., *Jour. Appl. Mech. 36* (1969) 203.

DISCUSSION

S. Nemat-Nasser[1]). I wish to point out that the division of any strain measure in finite deformation problems, into various constituents, can be done quite arbitrarily, and will involve no question of principles. For example, one can write for the Lagrangian strain with rectangular Cartesian components E_{AB},

$$E_{AB} = E_{AB}^{(1)} + E_{AB}^{(2)}. \tag{*}$$

However, if the first term in the right-hand side is to have the property that the second Piola-Kirchhoff stress S_{AB} is to be given by

$$S_{AB} = \frac{\partial \psi}{\partial E_{AB}^{(1)}}, \tag{**}$$

where ψ is the free energy, then by necessity $E_{AB}^{(1)}$ must represent the elastic part of the strain, the strain which gives rise to the stress in the sense of Cauchy. This elastic strain must correspond to Professor Lee's elastic strain, otherwise (**) cannot be valid. Since the left-hand side of (*) is the total strain, $E_{AB} - E_{AB}^{(1)}$ may be called whatever one pleases, although it is defined in terms of both the elastic and plastic parts of strain, as given by the decomposition $F = F^e F^p$.

E. H. Lee. We agree that in principle any strain measure can be divided into constituents arbitrarily. If one constituent is selected to be the elastic strain, then we agree that it must 'correspond' with the elastic strain given in our paper, in the sense that for deformation in an elastic range (either for initial loading or after unloading following plastic flow) the rate of work expression leads to a constitutive relation of the form (**) of the discussion or (11) of the paper.

Since Green and Naghdi [10, 11] choose to prescribe $E_{AB}^{(2)}$ to be the plastic strain, and hence require $\dot{E}_{AB}^{(2)}$ to be zero in an elastic range, the change of variables from $E_{AB}^{(1)}$ and $E_{AB}^{(2)}$ to E_{AB} and $E_{AB}^{(2)}$ ($\psi(E_{AB}^{(1)}, E_{AB}^{(2)}, ...) = \hat{\psi}(E_{AB}, E_{AB}^{(2)}, ...)$) in their case generates the relations:

$$S_{AB} = \frac{\partial \psi}{\partial E_{AB}^{(1)}} = \frac{\partial \hat{\psi}}{\partial E_{AB}} \tag{***}$$

[1]) Northwestern University, Evanston, Illinois, U.S.A.

Thus, apart from an additive matrix which is a function of plastic strain only, i.e. which remains constant within an elastic range, the elastic strain in (*) and (**) which also satisfies Green and Naghdi's requirements must be equivalent to Lagrange strain based on the undisturbed state as a reference state. Since the reference configuration is then not in general an unstressed state, ψ will not take on its simplest form, for example, if the material is isotropic elastically (i.e. for loading from the unstressed state) the law will not be expressible in terms of the usual invariants. This follows because the reference state may be a stressed state, and may not even be in the elastic range. It is to avoid such complexities that we have advocated using the unstressed reference state $x^{(2)}$. Note that the $\hat{\psi}$ in the second of (***) will depend on prior plastic flow purely for geometrical reasons, just as in elasticity theory, for which the particular reference configuration chosen in order to define strain will affect the form of the free energy function.

F. A. McClintock[1]).   Might it not be possible to simplify the formulation by omitting the original state, which has been 'forgotten' by the material except for preferred orientation? This would require carrying along internal parameters, or history functions.

E. H. Lee.   Although it is, perhaps, somewhat hidden by the mathematical structure, the type of theory discussed at some length in the paper does not in fact specifically include the initial state. Because of the assumption of isotropic hardening, the memory in the material of previous deformation consists of a scalar memory of previous plastic flow embodied in the current yield stress magnitude, and the change in configuration caused by the plastic flow which is embodied in the current unstressed shape or 'permanent' deformation. By using

$$F^e = \frac{\partial x^{(3)}}{\partial x^{(2)}}$$

to express the elastic deformation in the isotropic elastic law (eqn. (11) of the paper) the initial configuration $x^{(1)}$ does not appear in this. Nor does it appear in the plasticity relation which is incremental in nature and is expressed in terms of the velocity gradient, $\partial v^{(2)}/\partial x^{(2)}$, in the current natural

[1]) Massachusetts Institute of Technology, Cambridge, Mass., USA.

unstressed state $x^{(2)}$ (eqns. (38) and (40) of [8]). However, if one starts with the initial configuration and needs to compute the total deformation:

$$F = \frac{\partial x^{(3)}}{\partial x^{(1)}}$$

or the 'permanent' plastic deformation

$$F^P = \frac{\partial x^{(2)}}{\partial x^{(1)}}$$

the initial configuration appears. Because of the uncoupling of the structure of the incremental elastic-plastic law from the initial configuration as mentioned above, if one has no interest in the deformation from the initial state, change to the current unstressed configuration at any time as a new reference state or to any other reference state will not change the structure of the constitutive relation. Thus we believe that, in a sense, Professor McClintock's suggestion is embodied in the theory. The only internal parameters required are the current yield stress magnitude, a scalar, and the current shape. Perhaps the possibility of making this circumstance explicit by writing the whole in terms of the stress and stress rate and the current velocity gradient

$$L = \frac{\partial v^{(3)}}{\partial x^{(3)}}$$

should be investigated. It seems likely, however, that not to use the unstressed reference state $x^{(2)}$ in the formulation might make the mathematical structure cumbersome.

The suggestion focuses the analysis on a fluid mechanics approach, which is certainly appropriate for an incremental plasticity theory which is a fluid type constitutive relation. Such a form may be most appropriate for certain applications.

Director vectors and constitutive equations for plastic and visco-plastic media

Jean Mandel

Ecole Polytechnique, Paris, France

The paper is concerned with the structure of the constitutive relations of an elastic-plastic or an elastic-viscoplastic medium. The term elastic-viscoplastic means that the viscosity does not intervene in the elastic domain whose frontier, in particular, is well defined at every stage of the deformation,

1. Director vectors

It is assumed that the actual thermodynamic state of an element of the material can be specified by prescribing the values of the observable variables (deformations, temperature) and of a certain number of hidden variables. The important fact is that, for anisotropic plastic bodies, the introduction of internal hidden variables requires also that of an external hidden variable, which is a variable of orientation, distinct from the gradient of transformation.

This is because the plastic deformation results from discontinuities of displacements and rotations between the oriented microelements the assembly of which constitutes a macroelement. Three director vectors can be attached to the centroid of a microelement. Because of the discontinuities these vectors do not undergo the same transformation as the material lines which join the centroids of the microelements. Thus the knowledge of the gradient of the global transformation is insufficient for the description of the behaviour. *The plastic medium is a medium with directors*, [1], analogous to a certain extent to a Cosserat medium. But unlike the Cosserat medium, the relative rotation of two contiguous microelements is made possible by the discontinuities and does not imply the existence of couple stresses (or at least it seems legitimate to neglect the couple stresses in the first approximation).

2. Director frames

How the system of directors contained in a macroelement can be represented? Starting from two of them a rectangular triad can be constructed whose orientation with respect to a fixed frame is defined, for instance, by Euler's angles. These angles are the external hidden variables. The components of the director vectors along the axes of the triad will then be the internal hidden variables. The triad in question will be called *a director frame*. There exists obviously a broad arbitrariness in the choice of the frame.

In the case of a monocrystal a natural director frame is a triad attached to the atomic lattice of the crystal. For a polycrystal we can take a triad attached to the lattice of any one of the constituting crystals. In some problems it is preferable (but by no means necessary) to take the triad whose orientation corresponds to a weighted mean of the orientations of the different atomic lattices. It is possible to define an optimal director frame having the following property: if there is at any time statistical isotropy (in the relaxed configurations defined hereafter) the rate of rotation of the triad is the same as that of the macroelement.

3. Description of the thermodynamic state of a plastic element

It is possible to adopt either the lagrangian or the eulerian description, but in fact, neither of them is well adapted to the purpose. The most satisfactory procedure consists of introducing a configuration called an intermediate or relaxed one. Let (0) be the initial configuration under zero stress and at the temperature T_0, (a) the actual configuration under the stress σ and at the temperature T. Assume that at the moment t the element is unloaded very rapidly and brought back to the temperature T_0: this unloading is elastic. A *relaxed instantaneous* configuration (κ) is thus obtained, which is specified only to within a rotation. Denoting by F the gradient of the total transformation $(0) \rightarrow (a)$, by P that of the plastic transformation $(0) \rightarrow (\kappa)$, by E that of the elastic transformation $(\kappa) \rightarrow (a)$, we then have:

$$F = EP. \tag{1}$$

In order to fix the choice of the configuration (κ), let us specify that it is such that the director triad adopted all the time preserves the same orientation with respect to the fixed axes; we shall call this configuration $(\hat{k})$ *isoclinal*. Then the variables defining the thermodynamic state are:

$\hat{\varDelta}$ Green's deformation tensor with respect to $\hat{k}$ corresponding to the transformation $(\hat{k}) \to (a)$; i.e. $1 + 2\hat{\varDelta} = \hat{E}^t \hat{E}$ (t: transposed),

T temperature,

$\hat{\alpha}$ internal hidden variables, scalar or tensorial.

The latter include scalar products of the director vectors and other variables such as densities of dislocations, residual stresses in the relaxed configuration, etc.

Instead of $\hat{\varDelta}$ we can take as a variable of state $\hat{\pi}$, the Kirchhoff stress tensor with respect to the configuration $(\hat{k})$, connected to Cauchy's stress tensor σ by:

$$\hat{\pi} = (\det E)E^{-1}\sigma(E^t)^{-1}. \tag{2}$$

Using these variables a function $\hat{u}$ of the thermodynamic state will be denoted $\hat{u}(\hat{\pi}, T, \hat{\alpha}^j)$.

Now let (κ) be another relaxed configuration, obtained from the preceding one $(\hat{k})$ through the rotation:

$$\xi = \beta(t)\hat{\xi} \qquad (\xi: \text{position vector}). \tag{3}$$

The tensors become:

$$\pi = \beta\hat{\pi}\beta^t, \quad \alpha = \beta\hat{\alpha}\beta^t.$$

So that the state is now defined by $\varDelta$ (or π), T, α^j and β, β being the external hidden variable of orientation. A function of the thermodynamic state becomes $u(\pi, T, \alpha^j, \beta)$; it is connected to $\hat{u}(\hat{\pi}, T, \hat{\alpha}^j)$ by:

$$u(\pi, T, \alpha^j, \beta) = \begin{cases} \hat{u}(\beta^t\pi\beta, T, \beta^t\alpha^j\beta) & \text{if } \hat{u} \text{ is a scalar,} \\[2mm] \beta\hat{u}(\beta^t\pi\beta, T, \beta^t\alpha^j\beta)\beta^t & \text{if } \hat{u} \text{ is a 2nd order tensor.} \end{cases} \tag{4}$$

It is to be noted that if we take $\beta = g(\alpha^j)$, the state is again defined by $\varDelta$, T, α^j: this corresponds to the substitution of one director frame by another.

4. Definition of the elastic frontier

In order to obtain a satisfactory definition, in the case of finite transformation, we have to use the stress-tensor π. The elastic domain is then represented by:

$$G(\pi, T, \alpha^j, \beta) < 0. \tag{5}$$

In isoclinal relaxed configurations β disappears, whence

$$\hat{G}(\hat{\pi}, T, \hat{\alpha}^j) < 0. \tag{6}$$

The function $\hat{G}$ depends on the director triad adopted.

It is interesting to find out what this condition becomes in the lagrangian description, i.e. when we use, as the reference configuration, the fixed configuration (0) in respect to which the Kirchhoff stress-tensor is π_0. A polar decomposition of $\hat{P}$ is: $\hat{P} = \hat{Q}L$, $(\hat{Q}^t\hat{Q} = 1, L^t = L)$ gives:

$$\hat{G}(\hat{Q}L\pi_0 L\hat{Q}^t(\det L)^{-1}, T, \hat{\alpha}^j) < 0, \tag{7}$$

where $L = (P^tP)^{\frac{1}{2}}$.

If *and only if* $\hat{G}$ is an isotropic function of $\hat{\pi}$, i.e. if there is isotropy in the relaxed configurations, $\hat{Q}$ disappears from (7) and conditions of the type proposed by Green and Naghdi [2] and also by Perzyna [3] are obtained. The incidence of $\hat{Q}$ in the case of anisotropy is due to the director vectors.

5. Constitutive relations for viscoplastic transformation

The rate of plastic transformation $\dot{\hat{P}}\hat{P}^{-1}$, in respect to the director frame, does not depend on the configuration (0) used in the definition of $\hat{P}$. Following the hypothesis introduced in Sec. 1, it depends on $\hat{\pi}, T, \hat{\alpha}^j$ and possibly on their rates (the dependence on the latter is linear, provided that $\hat{P}(t)$ as a functional of $\hat{\pi}(\tau), T(\tau), \tau \leq t$, possesses the derivatives in respect to the actual values $\hat{\pi}(t)$ and $T(t)$. In the case where instantaneous plastic deformations do not occur $\dot{\hat{P}}\hat{P}^{-1}$ can not depend on the rates, whence:

$$\dot{\hat{P}}\hat{P}^{-1} = \hat{A}(\hat{\pi}, T, \hat{\alpha}^j)H(\hat{G}) \qquad (H = \text{Heaviside's function}). \tag{8}$$

If an arbitrary relaxed configuration (κ) – not isoclinal – is used, this equation becomes:

$$\frac{DP}{Dt} P^{-1} = A(\pi, T, \alpha^j, \beta)H(G). \tag{9}$$

A is connected to $\hat{A}$ by the equation (4) and DP/Dt is the derivative of P, in the motion of the configuration (κ) *in respect to the director frame*, i.e.:

$$\frac{DP}{Dt} = \dot{P} - \omega_D P, \tag{10}$$

$\omega_D = \dot{\beta}\beta^t$ designates the rate of rotation of the director frame in respect to the axes adopted S, $\dot{P}$ is the derivative of P in the frame S.

We must also define the evolution of the internal parameters. We assume relations of the form:

$$\frac{D\alpha^k}{Dt} = h^k(\pi, T, \alpha^j, \beta). \tag{11}$$

It is to be noted that (8) determines the entire rate of transformation in respect to the director triad (9 components), i.e. not only the rate of deformation, but also the rate of rotation of the element with respect to the director triad – or that of the director triad with respect to the element. *It was necessary*, because the director triad plays role of a hidden variable (otherwise it serves to determine β).

6. The case of instantaneous plastic deformations

Such deformations occur when the point π, T attains a frontier of instantaneous plasticity;

$$f(\pi, T, \alpha^j, \beta) = 0 \tag{12}$$

with a positive rate $\dot{f}$, thus:

$$\lambda = \text{tr}\left(\frac{\partial f}{\partial \pi} \frac{D\pi}{Dt} + \frac{\partial f}{\partial T} \dot{T} + \frac{\partial f}{\partial \alpha^j} h^{tj}\right) > 0. \tag{13}$$

The corresponding rates of transformation being linear functions of $D\pi/Dt$, $\dot{T}$, vanishing with λ, we can put:

139

$$\frac{DP}{Dt}P^{-1} = A(\pi, T, \alpha^j, \beta) + \lambda B(\pi, T, \alpha^j, \beta), \tag{14}$$

$$\frac{D\alpha^k}{Dt} = h^k(\pi, T, \alpha^j, \beta) + \lambda l^k(\pi, T, \alpha^j, \beta). \tag{15}$$

When $A = 0$ and $h^k = 0$ these relations are reduced to the flow rules of classical plasticity, but completed (antisymmetrical part of (14)) and made precise (influence of β and of $\omega_D = \dot{\beta}\beta^{-1}$).

What becomes of the relation (14) in lagrangian coordinates (fixed configuration (0))? It is found that if, and *only if* $\hat{f}$, $\hat{A}$, $\hat{B}$ are isotropic functions of $\hat{\pi}$, this relation (with $L = (P^t P)^{\frac{1}{2}}$ takes the form:

$$\dot{L} = A'(\pi_0, L, T, \hat{\alpha}^j) + \lambda B'(\pi_0, L, T, \hat{\alpha}^j) \tag{16}$$

proposed by Green and Naghdi [2] in the case of perfect plasticity, and by Perzyna [3] in the case of viscoplasticity. The non validity of (16) in the anisotropic case is due to the incidence of the director vectors.

7. Composition of thermoelastic and plastic transformations

This can be deduced from the relation (1). Let us restrict ourselves to indicate that, when the thermoelastic deformations are infinitesimal, we find (with $\omega_D^1 = \dot{\beta}_1\beta_1^t$ and the velocity v)

$$\text{grad } v = \omega_D^1 + \mathscr{D}_{el}^1 + A(\sigma, T, \alpha_1^j, \beta_1) + \lambda B(\sigma, T, \alpha_1^j, \beta_1). \tag{17}$$

The index 1 refers to the particular relaxed configuration (κ_1) which is deduced from the actual one (under the stress) by pure deformation (without rotation). It is seen that *in this case the rates of elastic and of plastic deformations are added together*. The rate of elastic deformations $\mathscr{D}_{el}^1$ is obtained from the derivative $(D\sigma/Dt)_1$, of Cauchy's stress tensor in *the motion of (κ_1) in respect to the director frame chosen*. This makes precise a point which remained obscure in classical plasticity.

On the other hand the incidence of β_1 and ω_D^1 shows that the relations concerning only the rate of deformation are insufficient; *the antisymmetric part of* (17) *is necessary*. In the case where there is isotropy in the relaxed configurations, it simply gives $\omega_D^1 = \omega$, the rate of rotation of the macro-element, and we come back to the usual concepts.

References

[1] Mandel, J., Relations de comportement des milieux élastiques-plastiques et elastiques-viscoplastiques. Notion de repère directeur, Foundations of Plasticity (Warsaw 1972) 387–400. Noordhoff, Leyden 1973.

[2] Green, A. E. and P. M. Naghdi, A general theory of an elasti-plastic continuum, *Arch. Rat. Mech. Anal. 4* (1965) 251.

[3] Perzyna, P., Thermodynamics of theological materials with internal changes, *Journ. Méc. 10* (1971) 391.

DISCUSSION

E. Kröner[1]).   I would like to point out a close relationship between Mandel's directors and the dislocation density tensor $\alpha_{ij}(x)$ described in my lecture. It is one of the basic results of dislocation theory (Nye 1953) that the combination $k_{ij} \equiv \alpha_{ji} - \frac{1}{2}\alpha_{kk}\delta_{ij}$ describes a lattice rotation through an angle $d\theta_j$, say, if one proceeds from points x^i to the neighbouring point $x^i + dx^i$. This rotation is also the rotation of three lattice vectors, which can be called directors, of course. Vectors of this type have been used since 1953 by the workers in the continuum theory of dislocations. Therefore, a good physical reason does, in fact, exist for the introduction of directors into the macroscopic theory of plasticity and viscoplasticity.

These are the situations in which excess dislocations of one sign occur in comparatively large numbers. There are situations where this will not be admissible; for instance in plastic bending of crystals at elevated temperatures. Since the relative lattice rotations certainly lead to an increase of internal energy as compared to the undisturbed crystal lattice, one expects a specific response in the form of (special) couple stresses, as discussed by myself and others (e.g. E. Kröner, *Int. J. Engng. Sci. 1* (1963) 261 and *Mechanics of generalized continua*, E. Kröner ed., Springer-Verlag, Berlin-Heidelberg-New York 1968, p. 330). The couple stresses can be neglected only if the corresponding energy is small compared to the energy of the ordinary macroscopic (i.e. force) stresses.

[1]) University of Stuttgart, Stuttgart, GFR.

J. Mandel. Les remarques de E. Kröner montrent bien qu'il est nécessaire d'introduire des directeurs et qu'on peut négliger les couples de contrainte lorsque le gradient de déformation reste petit vis à vis de l'inverse de la dimension des grains. Ainsi les dislocations justifient notre modèle mathématique et en indiquent les limitations. Qu'on me permette toutefois d'ajouter que ce modèle est plus général que le modèle des dislocations puisqu'il s'applique à tous les milieux plastiques, et par exemple aux sols ou à des matériaux composites formés de grains noyés dans une matrice.

H. Lippmann[1]). A few years ago, I dealt with a Cosserat-like approach to rigid-plastic flow (H. Lippmann, Eine Cosserat-Theorie des plastischen Fließens, *Acta mech.* *4* (1969)), and applied it to such specific problems as torsion of cylindrical bars and problems of plane strain. At that time, two major difficulties arose.

Firstly directors depend on the crystalline orientation of the grains in metals. Thus, if the grain size in an isotropic material tends to zero, all solutions of the basic equations should in the limit become the corresponding solutions of classical plasticity. No general proof of this property was given. Simply counting the variables, however, leads to the conclusion that, either the yield surface had to be of second order (and must not depend on asymmetric shear stresses), or the stress point had to lie on an edge of the yield surface. In the meantime, Dr. Besdo of Brunswick found that for special problems, the above restrictions may be weakened.

Moreover, after having solved some problems I realized that under realistic assumptions, the differences with classical results were so small that they fall far beyond the range of experimental scattering. Therefore, I stopped to consider Cosserat plasticity.

My question is simply whether in developing your theory you have been led to similar difficulties or experiences.

J. Mandel. Il doit être entendu que dans la théorie que je présente on néglige les couples de contraintes. De la conception des Cosserat on ne retient que la notion de trièdre directeur. Ceci paraît légitime si le gradient de déformation reste petit vis à vis de l'inverse de la dimension des grains. Dans des conditions on ne rencontre pas la difficulté théorique signalée dans le premier point de votre discussion, puisqu'on se place directement

[1]) University of Karlsruhe, Karlsruhe, GFR.

dans le cas limite. Le second point de votre discussion confirme le bien-fondé de cette approximation.

La seule différence avec la théorie classique, pour les milieux élasto-plastiques, c'est l'introduction d'un trièdre directeur, qui permet de préciser les relations de comportement dans le cas d'anisotropie.

Internal variable description of plasticity

Piotr Perzyna

Institute of Fundamental Technological Research, Warsaw, Poland

Three trends in the description of the plastic effects based on internal state variables are discussed. A mathematical structure of a thermodynamic theory of a material with internal state variables is presented. It has been shown to what extent thermodynamics leads to the proper description of internal dissipation when thermomechanical coupling is taken into account. The discussion of mechanisms responsible for internal dissipation of a material is given. It has been shown that two mechanisms, namely the thermal activation process and the damping of dislocation by phonon viscosity are the most important for proper explanation of the strain rate and temperature sensitivity of a material. Review of different proposition of physical interpretation of internal state variables is presented. A thermodynamic theory of viscoplasticity valid for the entire range of strain rate and temperature changes is developed. Theoretical assumptions are compared with experimental data available.

1. Introduction

The objective of this paper is to discuss the role the internal state variables play in the description of plastic behavior.

The internal state variables (hidden parameters) are broadly used in physics, especially in quantum mechanics and in chemical physics. The question arises whether the internal state variables can be of use in the description of internal dissipation in plastic materials, and if so, waht are advantages and features of such a procedure.

Three approaches can be distinguished in the description of plastic effects using internal state variables.

The first one is the formal approach based on the ideal model of plastic flow e.g. Vakulenko [61, 62] and Mróz [39].

The second approach is characterized by physical justification of internal state variables. It is assumed that the internal state variables can be interpreted as phenomenological macroscopic parameters needed to de-

145

scribe internal dissipation of a material. This interpretation is based on a broad analysis of physical mechanisms of plastic flow and on microscopic and macroscopic results of experiments e.g. Kröner [25], Perzyna and Wojno [51], Zarka [67, 68], Kratochvil and Dillon [22, 23], Kestin and Rice [21], Rice [53], Teodosiu [60], Mandel [33], Valanis [66], Perzyna [46–49, 52], and Olszak and Perzyna [41].

The third approach is based on statistical physics where internal state variables are interpreted as certain average quantities as a consequence of the method of statistical mechanics e.g. Kröner [26].

In this review a particular attention is given to the discussion of methods used in the second approach which unifies both physical and phenomenological principles.

In the presentation of a mathematical structure for a thermodynamic theory of a material with internal state variables we show to what extent thermodynamics is usefull to describe properly the internal dissipation when thermo-mechanical coupling is taken into account. This presentation is valid for finite deformations.

In order to show rules of physical interpretation for the internal state variables introduced the analysis of dissipative mechanisms for both rheological and time-dependent plastic flows is given. Review of different propositions of physical interpretation of internal state variables is presented.

Passing to specific theories of time-dependent plasticity we develop a thermodynamic theory of viscoplasticity describing strain rate sensitivity and thermal influences in the entire range of deformation rate and temperature changes.

Theoretical assumptions are compared with experimental data available.

2. Mathematical structure of a thermodynamic theory of a material with internal state variables

Let us consider a body $\mathscr{B}$ with particles X and assume that this body can deform inelastically and conduct heat.

To describe the deformation and the distribution of temperature in a body $\mathscr{B}$ it is necessary to know the right Cauchy-Green deformation tensor $C(X, t)$, the temperature $\vartheta(X, t)$ and the temperature gradient $\nabla\vartheta(X, t)$ in the particle X.

Definition 1. A triple

$$\Lambda(X, t) = \{C(X, t), \vartheta(X, t), \nabla\vartheta(X, t)\} \tag{2.1}$$

we shall call the actual deformation-temperature configuration of a particle X.

Definition 2. A local thermodynamic process at a material point X of a body $\mathscr{B}$ in the motion χ is a collection of functions

$$\mathscr{P}_X = \{\Lambda(X, t), \pi(X, t)\} \tag{2.2}$$

given for every $t \in (t_p, t_k)$, which satisfies the thermodynamic postulate[1])

$$-\dot\psi + \frac{1}{2\rho}\,\mathrm{tr}(T\dot C) - \eta\vartheta - \frac{1}{\rho\vartheta}\,q\cdot\nabla\vartheta \geqq 0, \tag{2.3}$$

where $\Lambda(X, t)$ is the deformation temperature configuration of a particle X and $\pi(X, t)$ represents the variables

$$\pi(X, t) = \{\psi(X, t), \eta(X, t), T(X, t), q(X, t)\} \tag{2.4}$$

where $\psi(X, t)$ is the specific free energy per unit mass, $\eta(X, t)$ represents the specific entropy per unit mass, $T(X, t)$ denotes the Piola-Kirchhoff stress tensor and $q(X, t)$ is the heat flux vector per unit surface in the reference configuration. The dot denotes the material differentiation with respect to time t and ρ is the mass density in the reference configuration.

We have implicitly assumed that the body force $b(X, t)$ can be determined by Cauchy's first law of motion

$$\mathrm{Div}\,(FT) + \rho b = \rho\ddot x, \tag{2.5}$$

Cauchy's second law of motion gives an assertion that the Piola-Kirchhoff stress tensor T is symmetric

$$T = T^T, \tag{2.6}$$

and the heat supply per unit mass and unit time $r(X, t)$ can be uniquely determined by the local form of the first law of thermodynamics

$$\tfrac{1}{2}tr(T\dot C) - \mathrm{Div}\,q - \rho(\dot\psi + \vartheta\dot\eta + \dot\vartheta\eta) + \rho r = 0. \tag{2.7}$$

---

[1]) Cf. Coleman and Noll [8].

Definition 3. A thermo-mechanical state at time t of a material point X of a body $\mathscr{B}$ is a collection of values which take the functions $\mathscr{P}_X$ for this particular time $t \in (t_p, t_k)$.

Our main purpose is to present a set of rules for prediction. These rules allow one to predict a 'future state' from a given 'actual state' i.e., to describe a thermodynamic process. However, we should remember that our prediction concerns a description of thermo-mechanical phenomena for a particular material at a body $\mathscr{B}$. As we shall show the material for a body $\mathscr{B}$ will be defined by a set of constitutive relations such that for the description of a thermo-mechanical state at time t of a material point X of a body $\mathscr{B}$ we do not need all values of functions $\mathscr{P}_X$.

The next proposition is crucial for our concept of a thermodynamic description of dissipative materials.

Proposition 1. (*Description of a state*). A thermomechanical state of a particle X corresponding to time $t \in (t_p, t_k)$ is described by the deformation-temperature configuration $\Lambda(X, t)$ of a particle X at time t and by the method of preparation of this configuration.[1]

The next definition will explain what we mean by the method of preparation.

Definition 4. By the method of preparation of the deformation-temperature configuration of a particle X we mean a document giving detailed instructions for this preparation.

To determine actual thermo-mechanical states of a particle during an irreversible thermodynamic process it is not sufficient to have the actual deformation-temperature configuration of a particle X but we additionally need the method of preparation of this configuration. In other words the method of preparation should give the additional information required to uniquely define the state of a particle X of a body $\mathscr{B}$ during an irreversible thermodynamic process. We need a method of preparation of the deformation-temperature configuration of a particle X to describe the internal dissipation of a material.

Owing to the proposition 1 a thermo-mechanical state of a particle X at time t is described by the value taken by the function

$$g = \{\Lambda(X, t); \text{ Method of Preparation}\}. \tag{2.8}$$

[1]) Cf. Perzyna [46–48]. This concept of state is similar to that introduced, in another connection, by Bridgman [2] and Giles [14, 15].

To specify the material structure in a body $\mathscr{B}$ we will introduce

Postulate 1. (Constitutive assumption). The thermo-mechanical principle of determinism for a dissipative material is expressed by the functional relation

$$\pi(X, t) = \mathscr{R}(g), \tag{2.9}$$

where

$$\mathscr{R} = \{\Psi, N, T, Q\} \tag{2.10}$$

represents the constitutive functions (functionals) for free energy Ψ, entropy N, stress T and heat flux Q.

We have assumed that all the constitutive equations describing the physical properties of a material satisfy the principle of material frame-indifference as formulated by Noll (1958).

Obviously we can use different descriptions of a method of preparation of the deformation-temperature configuration of a particle X. The methods lead to different concepts regarding description of the internal dissipation.

We may state now the main problem of thermodynamics of materials (cf. Coleman and Noll, [8]): In an assigned class of processes $\mathscr{P}_X$ and within an assigned class of functions (functionals) $\mathscr{R} = \{\Psi, N, T, Q\}$ to determine those that satisfy the Clausius-Duhem inequality (2.3).

Definition 5. A local thermodynamic process described by $\mathscr{P}_X$ is said to be admissible in $\mathscr{B}$ if it is compatible with the constitutive assumption (2.9) at even particle X of $\mathscr{B}$.

Thus, it can be said that the main problem of thermodynamics of materials is to determine an admissible thermodynamic process.

Our objective here is to discuss a thermodynamic theory of rheological materials with internal structural changes implied by plastic deformation. We shall use internal parameters (hidden variables) to describe the internal dissipation.

In the case of a rheological material with internal changes there are two sources of internal dissipation. The first one is associated with rheological effects or viscous properties and the second is related to internal structural changes due to plastic deformations. To describe both effects simultaneously we shall introduce the following.

Proposition 2. The method of preparation of the deformation-temperature configuration of a particle X for a rheological material with internal changes can be determined by two groups of internal parameters $\alpha(X, t)$ and $\omega(X, t)$.[1]

The first group $\alpha(X, t)$ is meant to describe rheological effects and the second $\omega(X, t)$ is responsible for internal changes generated by plastic deformations.

Then we have

Postulate 2. A thermo-mechanical state of particle X at time t for a rheological material with internal changes is described by the value of function

$$g(X, t) = \{A(X, t); \alpha(X, t), \omega(X, t)\} \tag{2.11}$$

at $t \in (t_p, t_k)$, and by the initial-value problem for differential equations

$$\dot{\alpha}(X, t) = A(g(X, t)), \quad \alpha(X, t_0) = \alpha_0(X),$$
$$\dot{\omega}(X, t) = \Omega(g(X, t)), \quad \omega(X, t_0) = \omega_0(X), \tag{2.12}$$

where t_0 is also from (t_p, t_k) and $t_0 < t$.

The material structure in a body $\mathscr{B}$ is defined by the relations (2.9) and (2.10) with the function $g(X, t)$ given by (2.11) and the initial-value problem (2.12).

To be sure that we can have the collection of functions $\mathscr{P}_X$ for every time $t \in (t_p, t_k)$ which describes a local thermodynamic process at a material point X it is sufficient to assume that the set of integral equations

$$\alpha(X, t) = \alpha_0(X) + \int_{t_0}^{t} A(g(X, \xi))\mathrm{d}\xi,$$
$$\omega(X, t) = \omega_0(X) + \int_{t_0}^{t} \Omega(g(X, \xi))\mathrm{d}\xi, \tag{2.13}$$

has a unique solution for $t \in [t_0, t_k)$. This condition imposes certain restrictions on the functions A and Ω. Namely, the functions A and Ω should be Lipschitz continuous functions with respect to α and ω and continuous functions with respect to A.

Let us investigate the restrictions imposed on the constitutive equations by the thermodynamic postulate (2.3). To this end we assume that the function Ψ is piecewise continuously differentiable with respect to $g(X, t)$, i.e. jointly with respect to C, ϑ, $\nabla\vartheta$, α and ω.

---

[1] Cf. Perzyna [49].

Let us compute

$$\dot{\psi}(t) = tr[\partial_C \Psi \dot{C}] + \partial_\vartheta \Psi \dot{\vartheta} + \partial_{\nabla\vartheta} \Psi \cdot \overline{\nabla\vartheta} + \partial_\alpha \Psi \cdot \dot{\alpha} + \partial_\omega \Psi \cdot \dot{\omega}. \tag{2.14}$$

The thermodynamic inequality (2.3) becomes

$$\frac{1}{2\rho} tr[(T - 2\rho\partial_C \Psi)\dot{C}] - (\partial_\vartheta \Psi + \eta)\dot{\vartheta} - \partial_{\nabla\vartheta} \Psi \cdot \overline{\nabla\vartheta}$$

$$-\partial_\alpha \Psi \cdot \dot{\alpha} - \partial_\omega \Psi \cdot \dot{\omega} - \frac{1}{\rho\vartheta} q \cdot \nabla\vartheta \geqq 0. \tag{2.15}$$

Choosing arbitrary values of $\dot{C}$, $\dot{\vartheta}$ and $\overline{\nabla\vartheta}$ it is possible to specify an admissible local thermodynamic process at a material point X of a body $\mathscr{B}$. Hence the inequality (2.15) yields[1]

$$\partial_{\nabla\vartheta}\Psi = 0, \tag{2.16}$$

$$T(X, t) = 2\rho\partial_C \Psi(g^*(X, t)), \tag{2.17}$$

$$\eta(X, t) = -\partial_\vartheta \Psi(g^*(X, t)), \tag{2.18}$$

$$\partial_\alpha \Psi(g^*) \cdot \dot{\alpha} + \partial_\omega \Psi(g^*) \cdot \dot{\omega} + \frac{1}{\rho\omega} q \cdot \nabla\vartheta \leqq 0, \tag{2.19}$$

$$g^*(X, t) = \{C(X, t), \vartheta(X, t); \alpha(X, t), \omega(X, t)\}. \tag{2.20}$$

The internal dissipation of a rheological material with internal structural changes is determined by the function

$$\sigma(g(X, t)) = -\frac{1}{\vartheta}\{\partial_\alpha \Psi(g^*) \cdot A(g) + \partial_\omega \Psi(g^*) \cdot \Omega(g)\}. \tag{2.21}$$

Since there are two sources of internal dissipation in a rheological material with internal structural changes the function $\sigma(g(X, t))$ consists of two terms.

3. Analysis of mechanisms of internal dissipation

In order to give a proper physical interpretation of internal parameters introduced so far we should analyse mechanisms which govern the internal dissipation.

[1] Cf. Coleman and Gurtin [7] and Valanis [64].

Let us assume that the cause of rheological dissipation is the internal friction which may arise by means of a number of mechanisms depending on many sources. In specific situations and for particular materials we can determine the most probable mechanism of internal friction experimentally. In this case, we can specify the function A in Eq. (2.12$_2$). It is worth to note that in a linear approximation the theory of each of these mechanisms leads to the well known Boltzmann constitutive equations. To describe this linear case we would assume a linear form of equations (2.12$_2$).

The second kind of dissipation caused by internal changes of a material during plastic deformation can be explained by different mechanisms.

The basic result of the microscopic investigations is that the elementary process of plastic deformation is the motion of a line-shaped crystal defect called dislocation. It is generally agreed that the thermal obstacles impeding the motion of dislocations through a crystal are responsible for the dynamic aspects of plastic deformation.

Let us consider mild steel as an example. Rosenfield and Hahn [55] have shown that in the temperature-strain rate spectrum of plain carbon steel we can consider four regions which reflect different mechanisms of plastic deformation, Fig. 1. It is convenient to discuss the material behaviour in relation to each of the regions.

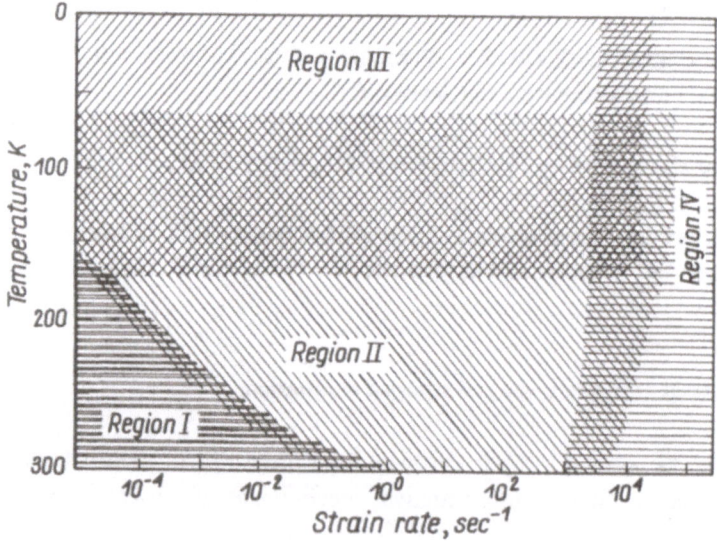

Fig. 1. Regions of the temperature strain-rate spectrum of low carbon steel that reflect different mechanisms of yielding. After Rosenfield and Hahn [55].

Region I is characterized by a yield stress relatively insensitive both to strain rate and temperature. The plastic flow is governed by non-thermal mechanisms. The dominant factor in Region I seems to be the long-range internal stress fields due to dislocations, precipitate particles, grain boundaries, etc. (cf. Campbell and Ferguson [4]).

The distinctive features of Region II lie not only in the fact that the yield stress is more markedly temperature and rate sensitive but also that the semilogarithmic rate sensitivity is independent of temperature, Campbell and Ferguson [4].

It is generally accepted that at low temperatures the plastic strain rate in c.p.h., f.c.c. and b.c.c. metals is controlled by the thermal activation of dislocation motion, the activation energy being a function of the applied stress and temperature.

Common thermal obstacles or mechanisms in pure metals are the Peierls-Nabarro stress, forest dislocations, the motion of jogs in screw dislocations, cross-slip of screw dislocations, and climb of edge dislocations (cf. Fig. 2).

The mechanisms of overcoming the dislocation forest which may appear in c.p.h., f.c.c. and b.c.c. crystals of metals in various temperature ranges has been developed theoretically by Seeger [59], Gibbs [12, 13], Schoeck [57],

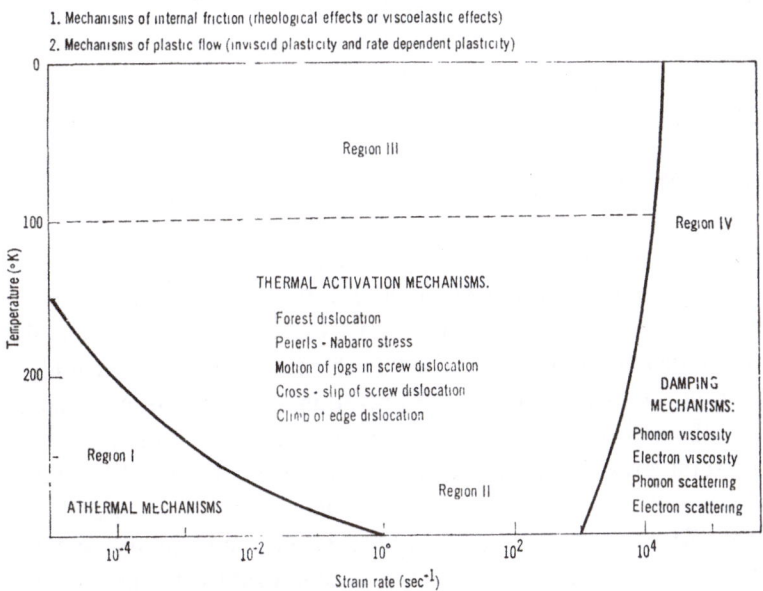

Fig. 2. Analysis of dissipative mechanisms.

and others cf. here the reviews by Conrad [9] and Evans and Rawlings [10].

When the plastic deformation P is controlled by a single thermally activated process we have

$$\dot{P} = v \exp(-U/k\vartheta),$$ (3.1)

where U is the energy that must be supplied by a thermal fluctuation for each successful activation, k denotes the Boltzmann's constant and v the frequency coefficient.

To fit the experimental data, the activation energy U is generally assumed as a nonlinear function of the excess stress over the yield point, i.e.

$$U = \varphi[a(T-Y)]$$ (3.2)

where a is a structural constant and Y denotes the non-thermal stress or the yield point of a material.

Thus, the rate of inelastic deformation is determined as follows

$$\dot{P} = v \exp\left\{-\frac{\varphi[a(T-Y)]}{k\vartheta}\right\}.$$ (3.3)

This implies the relation for stress

$$T = Y\left\{1 + \frac{1}{aY} \varphi^{-1}[k\vartheta \ln(v/\dot{P})]\right\}.$$ (3.4)

Region III is characterized by a lower rate and temperature dependence of the yield stress. Rosenfield and Hahn [55] have pointed out that this may be connected with the onset of twinning at the interface between Region II and Region III. The experimental results do not shed any light on the rate sensitivity in Region III.

Region IV encompasses very high strain rates, 10^3 to 10^6 sec^{-1}. The yield stress of mild steel and other metals in this region seems to become extremely strain-rate sensitive. Thermal activation is still operative in Region IV, but the increase in rate sensitivity in this region must be attributed to a reduction in strain rate due to the operation of an additional mechanism opposing the motion of dislocations. An alternative interpretation would be that the total stress is the sum of two parts, one determined by the short-range barriers and the other – by additional dissipative mechanisms (cf. Ferguson, Kumar and Dorn [11], Gilman [16]). A moving dislocation may dissipate energy on various mechanisms. Theoretical treatments of a number of these mechanisms have been reviewed by Nabarro [40], see Fig. 2,

Region IV. Campbell and Ferguson [4] have shown that the increase of the rate sensitivity in Region IV is due to viscous resistance to dislocation motion and the observed macroscopic viscosity is consistent with that expected from the theory of the damping of dislocation motion by phonon viscosity.

The phonon viscosity theory has been developed by Mason [34] and Mason and Rosenberg [35]. This theory predicts that the dislocation velocity is linearly proportional to the applied-resolved stress. Thus, for Region IV the dislocation drag coefficient is defined by the relation

$$F = Bv, \tag{3.5}$$

where F is the force acting on a unit length of the dislocation and v is the velocity of the dislocation. The plastic strain rate $\dot{P}$ is given by the relation

$$\dot{P} = \rho_M bv, \tag{3.6}$$

where ρ_M is the density of the mobile dislocations, b is their Burger's vector and v is their average velocity.

In the viscous damping region a linear dependence of stress on strain rate is observed and can be represented by the relation

$$T - T_B = \beta\dot{P}, \tag{3.7}$$

where T_B is attributed to the stress needed to overcome the forest dislocation barriers to the dislocation motion and is called the back stress, β is the slope of the stress vs. strain rate curve in Region IV.

The net force F on dislocations is given by the relationship

$$F = (T - T_B)b. \tag{3.8}$$

Combining equations (3.5) to (3.8) we obtain

$$B/\rho_M = b^2\beta. \tag{3.9}$$

The ratio B/ρ_M can be obtained from the slope of the linear portion of the stress vs. strain rate curve, and from this ratio the value of B can be calculated if a value of ρ_M is assumed (cf. here results for mild steel obtained by Campbell and Ferguson [4], for copper by Kumar and Kumble [28] and for aluminium by Ferguson, Kumar and Dorn [11], Kumar, Hauser and Dorn [27] and Gorman, Wood and Vreeland [17]).

REVIEW OF DIFFERENT PROPOSITIONS

1. Mechanisms of internal friction Biot 1954, Meixner 1954, Schapery 1964, Valanis 1967, 1968
2. Mechanisms of plastic flow

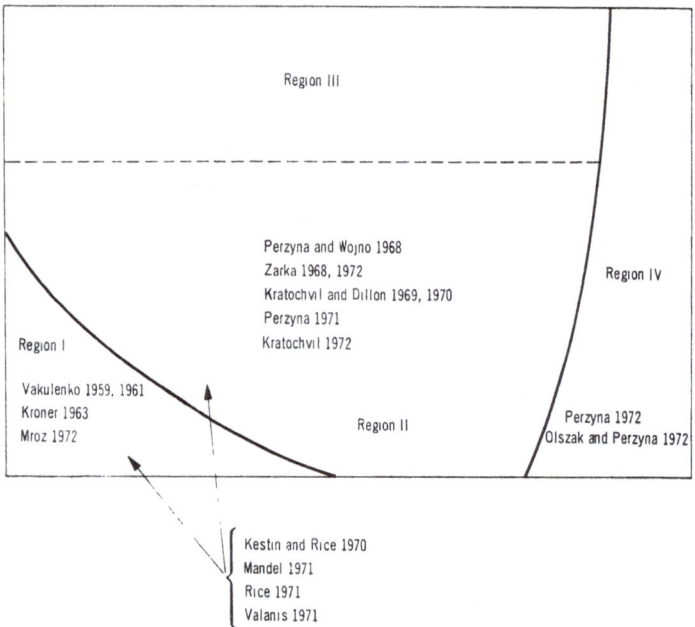

Fig. 3. Analysis of different propositions.

4. Review of earlier propositions

The first application of the internal state variable procedure to the description of dissipation for viscoelastic materials were given by Biot [1] and Meixner [36]. This concept has been further developed by Schapery [56] and Valanis [64, 65], see Fig. 3. By applying the technique of a linearized thermodynamic theory of irreversible processes Schapery [56] obtained a system of differential equations describing the behaviour of a viscoelastic material at small deformations. This result has been generalized to finite deformations by Valanis [64, 65], cf. also Perzyna [49].

The description of mechanisms of inviscid plastic flow (Region I) by means of internal state variables were given in the papers by Vakulenko [61, 62], Kröner [25], Teodosiu [60], and Mróz [39]. Kröner gave a physical interpretation of the internal state variables introduced which he based on the results of the theory of dislocations under the assumption that the dislocation arrangement tensors are responsible for work-hardening effects during plastic flow.

For rate dependent plastic flow (Region II) there are several propositions of interpretations of internal state variables, cf. Perzyna and Wojno [51], Zarka [67, 68], Kratochwil and Dillon [22, 23], Perzyna [46–49], Kratochvil [24], see Fig. 3. Most of these authors assumed that the thermally activated mechanism is responsible for both strain rate and temperature dependence of plastic flow.

Perzyna and Wojno [51] formulated a theory of rate sensitive plastic material within the framework of thermodynamics of materials with internal state variables. The deformation tensor and temperature were considered as thermodynamic state variables, while the components of the inelastic deformation tensor appear as internal state parameters. No connection between the deformation tensor and the inelastic deformation tensor was postulated.

A similar theory of rate-dependent plasticity was formulated independently by Zarka [67] and Kratochvíl and Dillon [22, 23]. Kratochvíl and Dillon assumed that quantities which are related to both motion and arrangement of dislocations in a material, play the role of internal state variables. This assumption was based on Kröner's idea.

The papers by Kestin and Rice [21], Mandel [33], Rice [54] and Valanis [66] are concerned with descriptions for both time-dependent and time-independent plasticity (Regions I and II).

Rice [54], Mandel [33] and Zarka [68] assumed that plastic behaviour of metals arises as a consequence of slip rearrangements of crystallographic planes through dislocation motion.

Perzyna [52] and Olszak and Perzyna [41] developed a thermodynamic theory of viscoplasticity describing both strain rate sensitivity and thermal influences for the entire range of strain rate and temperature changes (Regions II and IV). This conception will be discussed in detail in the next section.

5. Elastic-viscoplastic material

We intend to describe the behaviour of an elastic-viscoplastic material in the entire spectrum of strain rate and temperature. As we have show in Section 3, physical microscopic theories have suggested that two mechanisms, i.e. the thermal activation process (Region II) and the damping of dislocation motion by phonon viscosity (Region IV) are most important for explanation of the strain rate and temperature sensitivity of a material.

Let us postulate that

$$\alpha = \{\alpha^{(i)}\}, \quad i = 1, 2, ..., m; \quad A = \{A^{(i)}\}, \tag{5.1}$$

where $\alpha^{(i)}$, $i = 1, 2, ..., m$, are scalars and $A^{(i)}$, $i = 1, 2, ..., m$ are scalar functions, and

$$\omega = \{\omega_1, \kappa_1, \Gamma_1^{(j)}; \omega_2, \kappa_2, \Gamma_2^{(j)}\},$$

$$j = 1, 2, ..., n, \tag{5.2}$$

$$\Omega = \{\Omega_1, K_1, Z_1^{(j)}; \Omega_2, K_2, Z_2^{(j)}\},$$

where ω_1, ω_2, $\Gamma_1^{(j)}$, $\Gamma_2^{(j)}$, $j = 1, 2, ..., n$, are second order tensors and Ω_1, Ω_2, $Z_1^{(j)}$, $Z_2^{(j)}$, $j = 1, 2, ..., n$ are second order tensor functions, κ_1, κ_2 are scalars and K_1 and K_2 are scalar functions.

Additionally, we assume the relation for inelastic deformation tensor $P(X, t)$ as follows

$$\dot{P} = G(g(X, t)), \quad P(X, t_0) = P_0(X). \tag{5.3}$$

For our case the relation (5.3) takes the simple form

$$\dot{P} = \begin{cases} \dot{\omega}_1 & \text{in Region II,} \\ \lambda + \dot{\omega}_2 & \text{in Region IV,} \end{cases} \tag{5.4}$$

where λ is a constant value which can be determined from the condition of continuity for $\dot{P}$. The relation (5.4) furnishes a simple interpretation for the internal parameters ω_1 and ω_2. The quantities $\dot{\omega}_1$ and $\dot{\omega}_2$ are rates of inelastic deformation in the thermal activation mechanism and the mechanism of damping of dislocation motion by phonon viscosity, respectively.

We assume that the internal parameters $\{\omega_1, \kappa_1, \Gamma_1^{(j)}\}$ describe the thermal activation mechanism and $\{\omega_2, \kappa_2, \Gamma_2^{(j)}\}$ the mechanism of damping of dislocation motion by phonon viscosity. The scalars κ_1 and κ_2 represent the work hardening parameters and the symmetric second order tensors $\Gamma_1^{(j)}$, $\Gamma_2^{(j)}$, $j = 1, 2, ..., n$, are the dislocation arrangement tensors, respectively for these two mechanisms.

We assume that internal changes occur only during plastic deformations. It means that there are no changes of the work hardening parameters and the dislocation arrangement tensors when the rate of inelastic deformation tensor vanishes. This will be satisfied if we postulate

$$K_1 = \text{tr}[N_1(g)\dot\omega_1], \quad K_2 = \text{tr}[N_2(g)\dot\omega_2],$$
$$Z_1^{(j)} = S_1^{(j)}(g)[\dot\omega_1], \quad Z_2^{(j)} = S_2^{(j)}(g)[\dot\omega_2],$$

(5.5)

i.e. linear relations for K_1 and $Z_1^{(j)}$ with respect to the rate $\dot\omega_1$ and for K_2 and $Z_2^{(j)}$ with respect to the rate $\dot\omega_2$. This assumption implies that the second term in the internal dissipation (2.21) vanishes when there is no increment of rates respectively in Regions II and IV.

For the thermally activated Region II we assume that the rate $\dot\omega_1$ is a function of the excess of stress over the quasi-static yield condition.

A rheological material with internal changes before yielding has visco-elastic properties, hence the initial yield condition for such material, which will be called the quasi-static yield condition should depend on the internal parameters $\alpha^{(i)}$.

Thus, a quasi-static yield condition for an elastic-viscoplastic material can be defined as follows

$$\mathscr{F}_1(g) = \frac{f_1}{\kappa_1} - 1.$$

(5.6)

It is postulated that the following differential equation determines the internal state tensor ω_1 for an elastic-viscoplastic material[1])

$$\dot\omega_1 = \gamma_1(\vartheta)\left\langle\!\left\langle \Phi\left(\frac{f_1}{\kappa_1} - 1\right)\right\rangle\!\right\rangle M_1(g),$$

(5.7)

where $\gamma_1(\vartheta)$ is a temperature dependent viscosity coefficient of a material and M_1 is a second order symmetric tensor function. The dimensionless function $\Phi(\mathscr{F}_1)$ may be chosen to represent results of tests on the dynamic behaviour of materials and the proper choice of $\Phi(\mathscr{F}_1)$ permits to describe the influence of the rate of deformation and of the temperature on the yield limit of material in Region II. The symbol $\langle\!\langle\Phi(\mathscr{F}_1)\rangle\!\rangle$ is defined as follows

$$\langle\!\langle\ [\ ]\ \rangle\!\rangle = \begin{cases} 0 & \text{if } f_1 \leq \kappa_1 \quad \text{or} \quad f_2 > \kappa_2, \\ [\ ] & \text{if } f_1 > \kappa_1 \quad \text{and} \quad f_2 < \kappa_2. \end{cases}$$

(5.8)

---

[1]) General foundations of viscoplasticity were given by Hohenemser and Prager [20]. Further development of this idea is contained in the papers by Perzyna [43, 44, 45, 47], and Perzyna and Wojno [50, 51].

Based on phonon viscosity damping mechanism in **Region IV**, we assume the differential equation for ω_2 in the form

$$\dot{\omega}_2 = \gamma_2(\vartheta) \left\langle \frac{f_2}{\kappa_2} - 1 \right\rangle M_2(g), \tag{5.9}$$

where $\gamma_2(\vartheta)$ is a temperature drag coefficient of a material and the function, M_2 is a second order symmetric tensor function

$$\mathscr{F}_2(g) = \frac{f_2}{\kappa_2} - 1 \tag{5.10}$$

which defines the transition criterion from **Region II** to **Region IV**. The symbol $\langle f_2/\kappa_2 - 1 \rangle$ is defined as follows

$$\langle [\ ] \rangle = \begin{cases} 0 & \text{if } f_2 \leqq \kappa_2, \\ [\ ] & \text{if } f_2 > \kappa_2. \end{cases} \tag{5.11}$$

Thus, we assume that in **Region IV** the rate $\dot{\omega}_2$ is proportional to the excess of stress over the transition criterion.

Equation (5.7) yields the following dynamical yield condition for **Region II**

$$f_1 = \kappa_1 \left\{ 1 + \Phi^{-1} \left[-\frac{(\text{tr } \dot{P}^2)^{\frac{1}{2}}}{\gamma_1(\vartheta)} \, (\text{tr } M_1^2)^{-\frac{1}{2}} \right] \right\} \tag{5.12}$$

and the equation (5.9) gives the dynamical relation for **Region IV** as follows

$$f_2 = \kappa_2 \left\{ 1 + \frac{[\text{tr}\,(\dot{P} - \lambda)^2]^{\frac{1}{2}}}{\gamma_2(\vartheta)} \, (\text{tr } M_2^2)^{-\frac{1}{2}} \right\}. \tag{5.13}$$

A comparison of the theoretical dynamical yield criterion (5.12) with the physically justified relation (3.4) shows that the phenomenological yield criterion may be treated as a generalization to polycrystals in general stress state of the relation for stress (3.4).

Similar conclusion can be drawn from the comparison of the physically justified stress-strain rate relation (3.7) and the phenomenological dynamical relation (5.13) for **Region IV**.

From these comparisons we have a simple interpretation of the internal state parameters κ_1 and κ_2. The parameter κ_1 can be regarded as a simple generalization of the non-thermal stress Y and the parameter κ_2 as a generalization of the back stress T_B.

In the present theory the internal dissipation is determined by the relations

$$\sigma_{\text{II}} = -\frac{1}{\vartheta}\{[\sum_{i=1}^{m}\partial_{\alpha^{(i)}}\Psi A^{(i)}] + \text{tr}[(\partial_{\omega_1}\Psi + \partial_{\kappa_1}\Psi N_1 +$$

$$+ \sum_{j=1}^{n}\partial_{\Gamma_1^{(j)}}\Psi S_1^{(j)})\dot{P}]\}, \tag{5.14}$$

$$\sigma_{\text{IV}} = -\frac{1}{\vartheta}\{[\sum_{i=1}^{m}\partial_{\alpha^{(i)}}\Psi A^{(i)}] + \text{tr}[(\partial_{\omega_2}\Psi + \partial_{\kappa_2}\Psi N_2 +$$

$$+ \sum_{j=1}^{n}\partial_{\Gamma_2^{(j)}}\Psi S_2^{(j)})(\dot{P} - \lambda)]\}, \tag{5.15}$$

respectively for Region II and Region IV.

6. Comparisons with experimental results

Lindholm's [31] experimental results on deformation of aluminium under a wide range of deformation rate, temperature and load conditions have shown a good agreement with the constitutive assumptions for Region II. Lindholm presented experimental results on plastic flow of aluminium for pure compression, tension, and torsion as well as for combined states of stress. The rate of loading in his experiment was varied to produce strain rates within the range of 10^{-3} sec^{-1} to 10^3 sec^{-1}. Data regarding the temperature range from 300°K to 700°K were obtained for compression and tension. Thus, all Lindholm's results correspond to Region II. In Fig. 4 Lindholm's data of a large number of tests are plotted to show the relationship between the stress and the strain rate invariants at constant strain amplitude and temperature. In accordance with the linearized theory, Seeger [59], this should be a linear relationship on a semilogarithmic plot, cf. also Lindholm [31, 32].

Hauser, Simmons and Dorn [19] presented the strain rate dependence of the flow stress for polycrystalline aluminium for 295°K, 194°K and 77.4°K over strain rates from 2 to 12 × 10³ sec^{-1}. An example from these experimental results for 194°K is shown in Fig. 5. The results concern Regions II and IV for aluminium. In Region II on a semilogarithmic scale the relationship between strain rate and stress is linear whereas it is nonlinear one

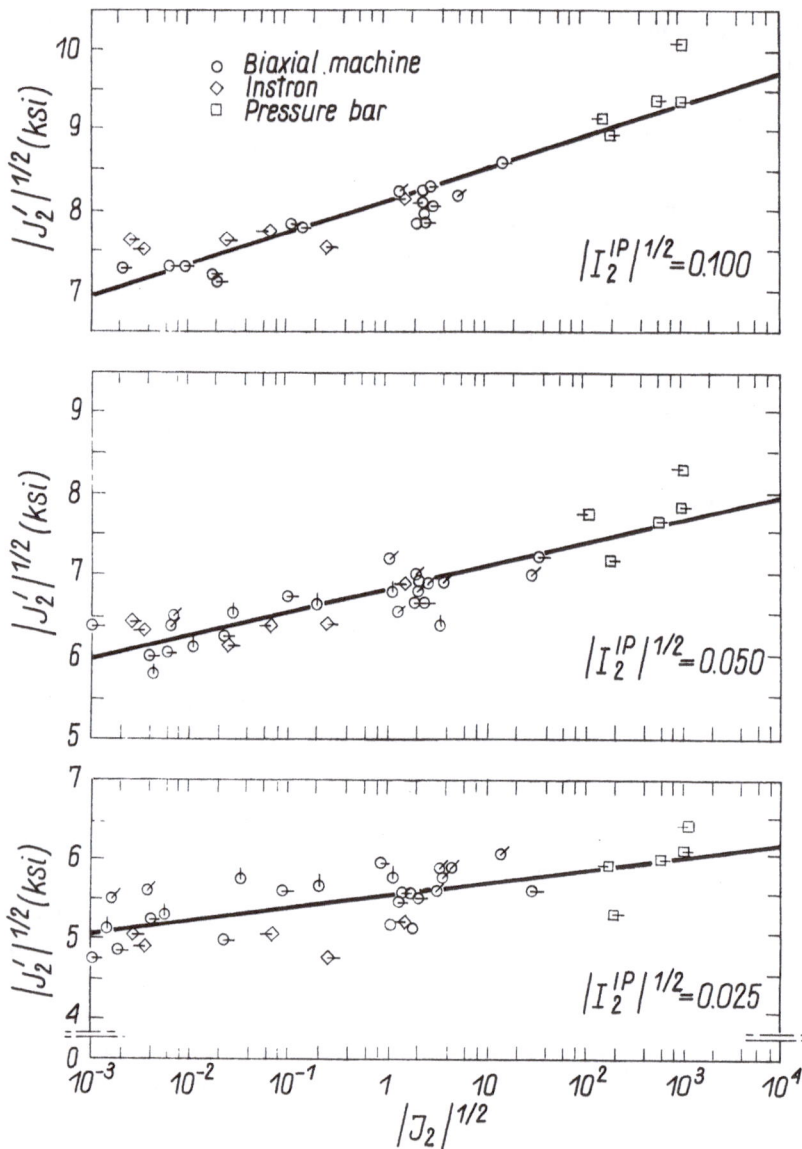

Fig. 4. Semilogarithmic stress-strain rate plots for aluminium in terms of the invariants at constant temperature $\vartheta = 294°K$ and strain. After Lindholm [31]; J_2' – second invariant of stress deviation, J_2 – second invariant of inelastic strain-rate deviation, $I_2'^P$ – second invariant of inelastic strain.

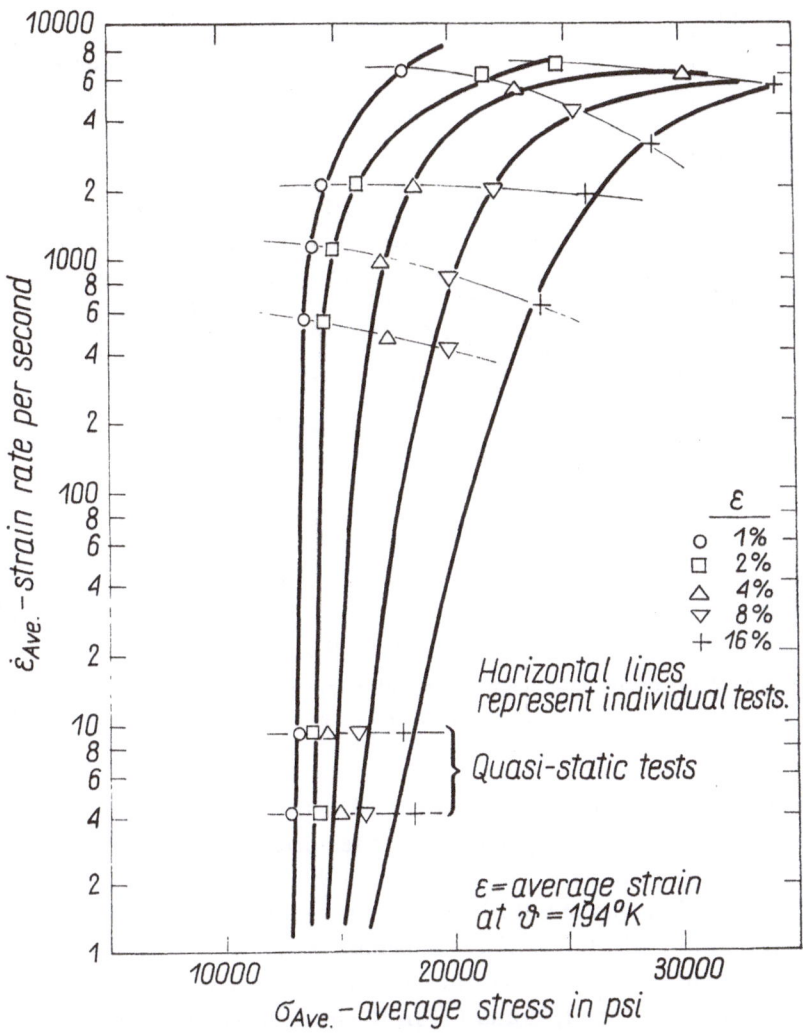

Fig. 5. Effect of stress and strain rate at constant strain and temperature $\vartheta = 194°K$. After Hauser, Simmons and Dorn [19].

in Region IV. Kumar, Hauser and Dorn [27] have shown that the same data of Hauser, Simmons and Dorn [19] give a nonlinear relationship between strain rate and stress in Region II and a linear one in Region IV if plotted on a linear scale. Thus, these experimental data show good agreement with the elastic-viscoplastic theory in the entire spectrum of strain rate changes, cf. also Campbell and Cooper [3].

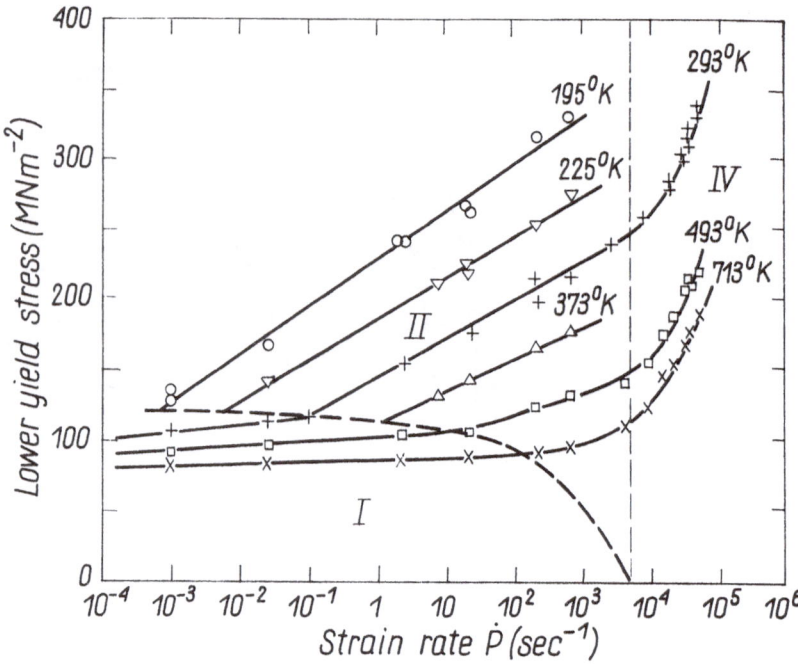

Fig. 6. Variation of lower yield stress with strain rate, at constant temperature. After Campbell and Ferguson [4].

Campbell and Ferguson [4] performed experiments in which the shear flow stress of mild steel was measured at temperature from 195°K to 713°K and the strain rates varied from 10^{-3} to 4×10^4 sec^{-1}. The flow stress at the lower yield is plotted in Fig. 6 as the shear stress against the logarithm of shear strain rate for various temperatures. These curves have been divided into three regions. Following Rosenfield and Hahn [55] these will be referred to as Regions I, II and IV. In Region I curves are characterized by a small, nearly constant slope, and the flow stress shows a little temperature dependence. The curves in Region II are also straight, but the rate and temperature dependences are considerably more pronounced. In Region IV the flow stress shows a further increase in the rate dependence, while the temperature dependence is unaffected.

The same data are replotted in Fig. 7 as the shear stress against the temperature at various constant strain rates.

The curves from Region IV are replotted in Fig. 8 on a linear scale.

The data have been interpreted by Campbell and Ferguson [4] in terms

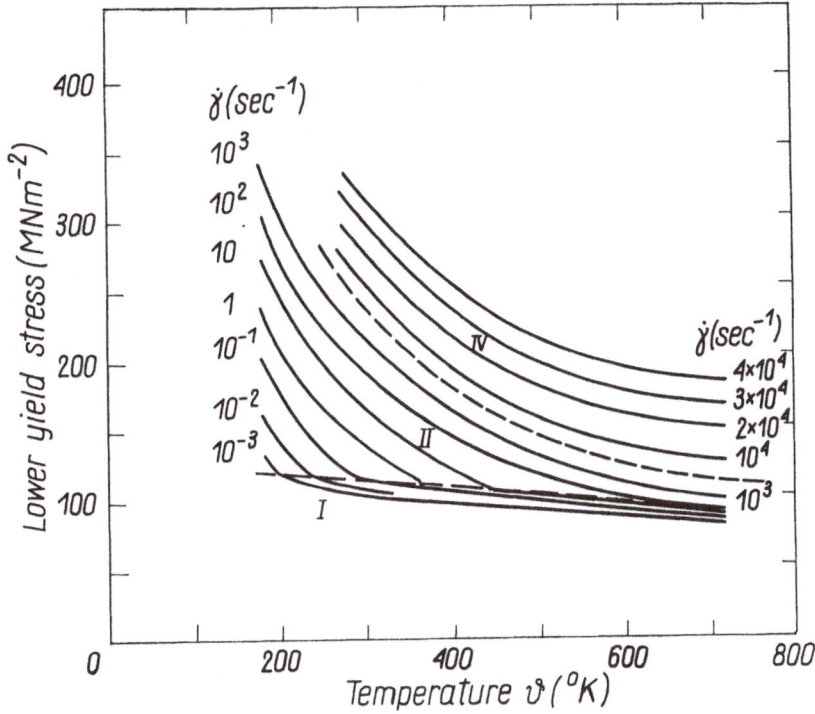

Fig. 7. Variation of lower yield stress with temperature, at constant strain rate. After Campbell and Ferguson [4].

of both the thermal activation rate theory (Region II) and the theory of damping of dislocation motion by phonon viscosity.

This comparison shows that the results in Region II are consistent with those expected from the theory based on thermally activated mechanism and in Region IV with the theory based on phonon viscosity damping mechanism.

Lindholm and Yeakley [32] investigated single crystal and polycrystalline specimens of high purity aluminium in compression at strain rates up to 500 sec^{-1} using the split Hopkinson pressure bar method. They obtained average stress-strain curves for the six orientations of a single crystal and similar curves for a polycrystalline material. The activation volume as a function of strain can be computed from the data obtained. Results for the single and the polycrystalline specimens of high purity aluminium are given in Fig. 9. An interesting feature of these curves is that the activation volume for the polycrystalline material falls within the bounds near the average of

165

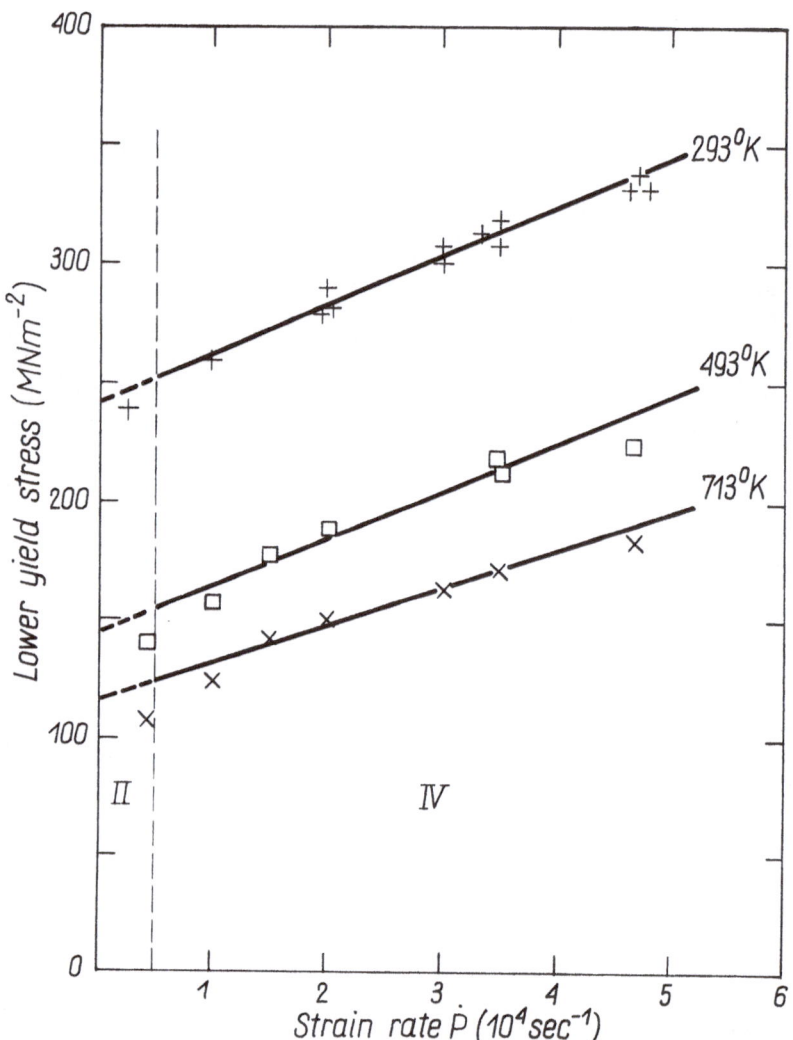

Fig. 8. Variation of lower yield stress with strain rate (Region IV). After Campbell and Ferguson [4].

the single crystal data. This implies that the same thermally activated mechanism controls the deformation in single and polycrystals as well and that the distributions of the activation barriers are essentially the same in both cases. This is in agreement with the results previously obtained by Mitra and Dorn [38] for aluminium at low temperature as well as by Mitra and Dorn [37] and Conrad [9] for iron and steel.

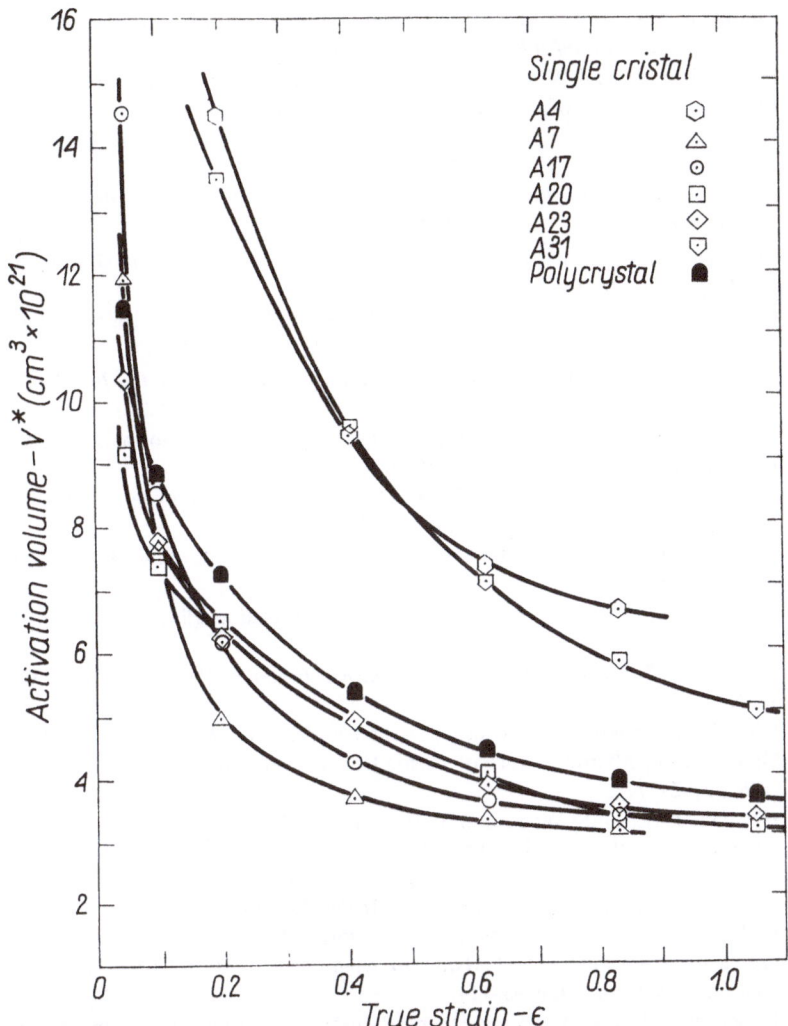

Fig. 9. Activation volumes for single and polycrystalline aluminium. After Lindholm and Yeakley [32].

References

[1] Biot, M. A., Theory of stress-strain relations in anisotropic viscoelasticity and relaxation phenomena. *J. Appl. Phys.*, *25* (1954) 1385–1391.
[2] Bridgman, P. W., The thermodynamics of plastic defromation and generalized entropy. *Rev. Mod. Phys.*, *22* (1950) 56–63.

[3] Campbell, J. D. and R. H. Cooper, Yield and flow of low-carbon steel at medium strain rates. *Proc. Conf. Phys. Basis Yield Fracture* (1967) 77–87.

[4] Campbell, J. D. and W. G. Ferguson, The temperature and strain-rate dependence of the shear strength of mild steel. *Phil. Mag., 81* (1970) 63–82.

[5] Coleman, B. D., Thermodynamics of materials with memory. *Arch. Ration. Mech. Anal., 17* (1964) 1–46.

[6] Coleman, B. D., On thermodynamics, strain impulses, and viscoelasticity. *Arch. Ration. Mech. Anal., 17* (1964) 230–254.

[7] Coleman, B. D. and M. E. Gurtin, Thermodynamics with internal state variables. *J. Chem. Phys., 47* (1967) 597–613.

[8] Coleman, B. D. and W. Noll, The thermodynamics of elastic materials with heat conduction and viscosity. *Arch. Ration. Mech. Anal., 13* (1963) 167–178.

[9] Conrad, H., Thermally activated deformation of metals. *J. Metls, 16* (1964) 582–588.

[10] Evans, A. G. and R. D. Rawlings, The thermally activated deformation of crystalline materials. *Phys. Stat. Sol., 34* (1969) 9–31.

[11] Ferguson, W. G., A. Kumar and J. E. Dorn, Dislocation damping in aluminium at high strain rate. *J. Appl. Phys., 38* (1967) 1863–1869.

[12] Gibbs, G. B., The thermodynamics of creep deformation. *Phys. Status Solidi, 5* (1964) 693–696.

[13] Gibbs, G. B., The activation parameters for dislocation glide. *Phil. Mag., 8, 16* (1967) 97–102.

[14] Giles, R., *Mathematical foundations of thermodynamics*. Pergamon Press, Oxford 1964.

[15] Giles, R., Foundations for quantum mechanics. *J. Math. Phys., 11* (1970) 2139–2160.

[16] Gilman, J. J., Microdynamics of plastic flow at constant stress. *J. Appl. Phys., 36* (1965) 2772–2777.

[17] Gorman, J. A., D. S. Wood and T. Vreeland, Jr. Mobility of dislocation in aluminium. *J. Appl. Phys., 40* (1969) 833–841.

[18] Gurtin, M. E. and A. C. Pipkin, A general theory of heat conduction with finite wave speeds. *Arch. Ration. Mech. Anal., 31* (1968) 113–126.

[19] Hauser, F. E., J. A. Simmons and J. E. Dorn, Strain rate effects in plastic wave propagation. In *Response of Metals to High Velocity Deformation*, Wiley (Interscience) New York, 1961, 93–114.

[20] Hohenemser, K. and W. Prager, Über die Ansätze der Mechanik isotroper Kontinua. *Z. Angew. Math. Mech., 12* (1932) 216–226.

[21] Kestin, J. and J. R. Rice, Paradoxes in the application of thermodynamics to strained solids. *A Critical Review of Thermodynamics*, Mono Book Corp., Baltimore, 1970, 275–298.

[22] Kratochvíl, J. and O. W. Dillon, Thermodynamics of elastic-plastic material as a theory with internal state variables. *J. Appl. Phys., 40* (1969) 3207–3218.

[23] Kratochvíl, J. and O. W. Dillon, Thermodynamics of crystalline elastic-visco-plastic materials. *J. Appl. Phys., 41* (1970) 1470–1479.

[24] Kratochvíl, J., Finite-strain theory of inelastic behaviour of crystalline solids. *Foundations of Plasticity*, (Warsaw 1972). Noordhoff International Publishing, Leyden 1973, 401–415.

[25] Kröner, E., Dislocation: A new concept in the continuum theory of plasticity. *J. Math. Phys.*, *42* (1962) 27–37.

[26] Kröner, E., Initial studies of plasticity theory based upon statistical mechanics. The Battelle Institute Colloquium on the Inelastic Behaviour of Solids, September 1969; in *Inelastic Behaviour of Solids*, McGraw Hill, 1930, pp. 137–148.

[27] Kumar, A., F. E. Hauser and J. E. Dorn, Viscous drag on dislocations in aluminium at high strain rates. *Act. Metallurgica*, *16* (1968) 1189–1197.

[28] Kumar, A. and R. G. Kumble, Viscous drag on dislocations at high strain rates in copper. *J. Appl. Phys.*, *40* (1969) 3475–3480.

[29] Lindholm, U. S., Some experiments with the split Hopkinson pressure bar. *J. Mech. Phys. Solids*, *12* (1964) 317–335.

[30] Lindholm, U. S., Dynamic deformation of metals. In *Behaviour of Materials under Dynamic Loading*. Am. Soc. Mech. Eng., New York, 1965, 42–61.

[31] Lindholm, U. S., Some experiments in dynamic plasticity under combined stress. *Symp. Mech. Behav. Mater. Dyn. Loads* (1967) 77–95. Springer, Berlin 1962.

[32] Lindholm, U. S. and L. M. Yeakley, Dynamic deformation of single and polycrystalline aluminium. *J. Mech. Phys. Solids*, *13* (1965) 41–53.

[33] Mandel, J., Plasticité classique et viscoplasticité. *Cours C.I.S.M.*, Udine, 1971.

[34] Mason, W. P., Phonon viscosity and its effect on acoustic wave attennation and dislocation motion. *J. Acoustical Soc. Amer.*, *32* (1960) 458–472.

[35] Mason, W. P. and A. Rosenberg, Phonon and electron drag coefficients in single-crystal aluminium. *Phys. Rev.*, *151* (1966) 434–441.

[36] Meixner, J., Thermodynamische Theorie der elastischen Relaxation. *Zeitschr. Naturforschung*, *9a* (1954) 654–665.

[37] Mitra, S. K. and J. E. Dorn, On the nature of strain hardening in face-centred cubic metals. *Trans. AIME 224* (1962) 1062–1071.

[38] Mitra, S. K. and J. E. Dorn, On the nature of strain hardening in polycrystalline aluminium and aluminium-magnesium alloys. *Trans. AIME 227* (1963) 1015–1024.

[39] Mróz, Z., A description of workhardening of metals with application to variable loading. *Foundations of Plasticity*, (Warsaw, 1972). Noordhoff International Publishing, Leyden 1973, 551–570.

[40] Nabarro, F. R. N., Theory of Crystal Dislocations. Oxford Univ. Press, London and New York, 1967.

[41] Olszak, W. and P. Perzyna, Physical theory of viscoplasticity for small deformations. *XIII Inter. Congress of Theoretical and Applied Mechanics*, Moscow, 1972.

[42] Onat, E. T., The notion of state and its implications in thermodynamics of inelastic solids. *IUTAM Symp. Irreversible Aspects Continuum Mech.*, (Vienna, 1966) 292–314, Springer, Wien 1968.

[43] Perzyna, P., The constitutive equations for rate sensitive plastic materials. *Quart. Appl. Math.*, *20* (1963) 321–332.

[44] Perzyna, P., The constitutive equations for work-hardening and rate sensitive plastic materials. *Proc. Vibr. Probl.*, *4* (1963) 281–290.

[45] Perzyna, P., Fundamental problems in viscoplasticity. *Advan. Appl. Mech.*, *9* (1966) 243–377.

[46] Perzyna, P., Thermodynamics of rheological materials with internal changes. *Jour. Méc.*, *10* (1971).

[47] Perzyna, P., Thermodynamic theory of viscoplasticity. *Advanced in Applied Mechanics* vol. 11 (1971) 313–354.

[48] Perzyna, P., Memory effects and internal changes of a material. *Int. J. Non-Linear Mechanics, 6* (1971).

[49] Perzyna, P., A gradient theory of rheological materials with internal changes. *Arch. Mechanics, 23* (1971) 845–850.

[50] Perzyna, P. and W. Wojno, On the constitutive equations of elastic/viscoplastic materials at finite strain. *Arch. Mech. Stosowanej, 18* (1966) 85–100.

[51] Perzyna, P. and W. Wojno, Thermodynamics of a rate sensitive plastic material. *Arch. Mech. Stosowanej, 20* (1968) 499–511.

[52] Perzyna, P., Physical theory of viscoplasticity, in print (1974).

[53] Rice, J. R., On the structure of stress-strain relation for time-dependent plastic deformation in metals. *J. Appl. Mech., 37* (1970) 728–737.

[54] Rice, J. R., Inelastic constitutive relations for solids: an internal variable theory and its application to metal plasticity. *J. Mech. Phys. Solids, 19* (1971) 433–455.

[55] Rosenfield, A. R. and G. T. Hahn, Numerical descriptions of the ambient low-temperature, and high-strain rate flow and fracture behaviour of plain carbon steel. *Trans. Am. Soc. Metals, 59* (1966) 962–980.

[56] Schapery, R. A., Application of thermodynamics to thermomechanical, fracture, and birefringent phenomena in viscoelastic media. *J. Appl. Phys., 35* (1964) 1451–1465.

[57] Schoeck, G., The activation energy of dislocation movement. *Phys. Status Solidi, 8* (1965) 499–507.

[58] Seeger, A., The temperature dependence of the critical shear stress and of work-hardening of metal crystals. *Phil. Mag., 7, 45* (1954) 771–773.

[59] Seeger, A., The generation of lattice defects by moving dislocations and its application to the temperature dependence of the flow-stress of f.c.c. crystals. *Phil. Mag., 7, 46* (1955) 1194–1217.

[60] Teodosiu, C., A dynamic theory of dislocations and its applications to the theory of the elastic-plastic continuum. *Fundamental Aspects of Dislocation Theory*, J. A. Simmons, R. de Wit and R. Bullough, Eds., Spec. Publ. (1970) 317, II, 837–876.

[61] Vakulenko, A. A., On stress-strain relations for inelastic bodies. *Dokl. Akad. Nauk SSSR, 118* (1958) 665–668 (in Russian).

[62] Vakulenko, A. A., Thermodynamical analysis of stress-strain relations in isotropic, elastic-plastic solids. *Doklady ANSSSR, 126* (1959) 136–139 (in Russian).

[63] Valanis, K. C., Thermodynamics of large viscoelastic deformations. *J. Math. Phys., 45* (1966) 197–212.

[64] Valanis, K. C., Entropy, fading memory and Onsager's relations. *J. Math. Phys. 46* (1967) 164–174.

[65] Valanis, K. C., Unified theory of thermodynamical behaviour of viscoelastic materials. *Symp. Mech. Behav. Mater. Dyn. Loads* (1967) 343–364. Springer, Berlin 1968.

[66] Valanis, K. C., A theory of viscoplasticity without a yield surface. *Arch. Mechanics, 23* (1971) 517–551.

[67] Zarka, J., *Thesis*, Paris, 1968.

[68] Zarka, J., Generalisation de la théorie du potentiel plastique multiple en viscoplasticité. *J. Mech. Phys. Solids, 20* (1972) 179–195.

DISCUSSION

A. Baltov[1]). Professor Perzyna has shown in his lecture a possibility of describing the visco-plastic deformation of bodies, using hidden parameters. The same approach can be extended to polymers in which a microfracture process takes place. The process can be regarded as a micromechanism of plastic and viscous deformation in polymers.

The following assumptions about the microfracture process can be made on the basis of results of one-dimensional experiments on fracture of specimens under constant tension:

i. The fracture process begins with rupture of the chemical bonds in the polymer molecule chains, when the energy of the thermal fluctuation of the atoms reaches the activation bound of the chemical reaction of degradation. In general a lower polymer is obtained in the chemical reaction and volatile gases evade. Atmospheric oxygen can also take part in the reaction.

ii. When the stress increases at a certain point of the polymer body, the activation energy of the fracture process decreases.

iii. When the temperature increases, the energy of thermal motion increases and the initiation of the fracture process is facilitated.

iv. Formation of a multitude of submicrocracks, located in the volume of the polymer body in the microfracture zone is accompanied by the initiation of the fracture process in the considered polymer. As soon as the external mechanical and thermal actions become sufficiently intense and a sufficient time has passed, the microcracks orientate themselves, join, and grow into magistral cracks, which brings to the total fracture of the polymer.

Microfracture is a complex process which comprises different interconnected processes: kinematic process, deformation process, process of change of structure, thermal process, thermal-diffusion process, diffusion, diffusion-chemical and chemical ones.

The macroscopic characteristics of the process are:

a) kinematical characteristics: the law of motion

$$x_i = \chi_i^\alpha(X_K^\alpha, t), \qquad (i = 1, 2, 3; K = 1, 2, 3; \alpha = \text{I, II, III, IV}),$$

$$\chi_i^\beta = \begin{cases} 0, & \text{at} \quad t_0 \leqq t \leqq t^*, \\ x_i, & \text{at} \quad t > t^* \end{cases}$$

[1]) Bulgarian Academy of Sciences, Sofia, Bulgaria.

171

where t^* is the time at which the microfracture process begins, $\chi_i^I = \chi_i^{III} = \chi_i$, $X_K^I = X_K^{III} = X_K$ for every $t > t^*$ and the function $X_K = X_K(x_i, t)$ exists; the velocity of the particles X^α at time t, $v_i^\alpha = (\partial\chi_i^\alpha/\partial t)|_{X_K = \text{const.}} = v_i^\alpha(x_i, t)$, $v_i^I = v_i^{III} = v_i$; the relative velocity of the particle X^α with respect to the solid phase particles $w_i^\alpha = v_i^\alpha - v_i$, $w_i^I = w_i^{III} = 0$.

b) deformational characteristics: the material deformation tensor $E_{KL} = \frac{1}{2}(x_{i,K}x_{i,L} - \delta_{KL})$; the material strain-rate tensor $\dot{E}_{KL}$; the second Piola-Kirchhoff stress tensor $\tilde{T}_{KL}$; the Cauchy stress tensor t_{ij}; $t_{ij,j} = 0$.

c) chemical characteristics: the extent of reaction ξ; the chemical reaction rate $\dot{\xi}$; the molecular mass M^α of the α-constituent; the stoichiometric coefficient v^α; $\sum_{\alpha=1}^{IV} M^\alpha v^\alpha = 0$.

d) diffusional and diffusion-chemical characteristics: the mass density of the α-th constituent $\rho^\alpha = \hat{\rho}^\alpha(x_i, t)$, $\hat{\rho}^\beta = 0$ for $t_0 \leq t \leq t^*$ and $\hat{\rho}^\beta \neq 0$ for $t > t^*$, (β = II, III, IV), $\rho_0^I = \rho_0(X_K)$ at $t = t_0$, $\rho = \sum_\alpha \rho^\alpha$, $\rho^* = \rho^I + \rho^{III}$; the chemical potential μ^α of the α-th constituent; the diffusion flux vector $I_i^\alpha = \rho^\alpha w_i^\alpha$, $\rho^* \dot{\bar{C}}^\alpha = \dot{\rho}^\alpha + \rho^\alpha v_{i,i} - M^\alpha v^\alpha \dot{\xi}$, $\dot{\bar{C}}^\alpha = -(1/\rho^*)(\rho^\alpha w_i^\alpha)_{,i}$.

e) thermo and thermal-diffusional characteristics: the heat flux vector q_i; the absolute temperature θ; the entropy per unit mass of the solid phase $\bar{\eta}$; the entropy η^α of the α-th constituent per unit mass of the α-constituent; the reduced heat flux vector $q_i^* = q_i + \theta \sum_\alpha \eta^\alpha I_i^\alpha$; the heat supply $\bar{q}$; the dissipation energy per unit time and per unit mass of the solid phase $(1/\theta)\bar{\sigma}$;

$$\dot{\bar{\eta}}\rho^* + \bar{\eta}\dot{\rho}^* = -(q_i^*\theta^{-1})_{,i} + \theta^{-1}(\bar{q} + \bar{\sigma})\rho^*, \quad \dot{\rho}^* = (M^I v^I + M^{III} v^{III})\dot{\xi}.$$

f) characteristics of internal structural changes: the material microdamage tensor E_{KL}^a which represents inelastic deformation.

g) energy characteristics: the internal energy $\bar{U}$ per unit mass of the solid phase; $\bar{U} = \hat{\bar{U}}(E_{KL}, \bar{\eta}, \bar{C}^\alpha, E_{KL}^a, \xi)$, E_{KL}^a and ξ being internal state variables; the free enthalpy $\bar{Z}$ per unit mass solid phase;

$$\bar{Z} = \bar{U} - \frac{1}{\rho_0}\tilde{T}_{KL}E_{KL} - \bar{\eta}\theta - \sum_\alpha \mu^\alpha \bar{C}^\alpha, \quad \bar{Z} = \hat{\bar{Z}}(\tilde{T}_{KL}, \theta, \mu^\alpha, E_{KL}^a, \xi).$$

The microfracture process begins when $\bar{Z} = \bar{Z}_0$, $\bar{Z}_0$ being an experimentally determined limit value. The process passes into its second phase – a process of crack propagation and of fracture as soon as $\max\{E_A^a\} = E_0$, where E_A^a, (A = 1, 2, 3) are the principal values of the tensor E_{KL}^a and E_0 is the limit value, experimentally determined.

The basic equations of the problem are: at

$$\bar{Z} > \bar{Z}_0, \quad E_{KL} = -\rho_0 \partial_{\tilde{T}_{KL}} \bar{Z}, \quad \bar{\eta} = -\partial_\theta \bar{Z}, \quad \bar{C}^\alpha = -\rho^* \partial_{\mu^\alpha} \bar{Z},$$

$$(\rho/\rho^*) X_{M,i} q_i = \Psi_M(\tilde{T}_{KL}, \theta, \theta_{,K}, \mu^\alpha, \mu^\alpha_{,K}, E^a_{KL}, \xi),$$

$$(\rho/\rho^*) X_{M,i} I^\alpha_i = F_M(\tilde{T}_{KL}, \theta, \theta_{,K}, \mu^\alpha, \mu^\alpha_{,K}, E^a_{KL}, \xi),$$

$$\dot{E}^a_{KL} = \langle \Phi(\mathscr{F}) \rangle \varphi_{KL}(\tilde{T}_{KL}, \theta, \mu^\alpha, E^a_{KL}, \xi), \quad \mathscr{F} = \bar{Z} - \bar{Z}_0,$$

$$\dot{\xi} = \langle \Phi(\mathscr{F}) \rangle f(\tilde{T}_{KL}, \theta, \mu^\alpha, E^a_{KL}, \xi), \quad \Psi = \bar{U} - \theta\bar{\eta},$$

$$\rho^* \partial_\xi \bar{Z} \cdot \dot{\xi} + \rho^* \partial_{E^a_{KL}} \bar{Z} \cdot \dot{E}^a_{KL} + \theta_{,i} q_i^* \theta^{-1} + \sum_\alpha \mu^\alpha_{,i} I^\alpha_i + \bar{\Psi}\rho^* \leqq 0,$$

where $\bar{Z}, \Psi_M, F_M, \varphi_{KL}, \Phi$ and f are to be determined experimentally.

F. Sidoroff[1]). I want to comment on the mathematical structure of the hidden variables formulation of plasticity. I shall use the very general framework which has been presented by Professor Perzyna and which is convenient to discuss general aspects of such theories. First, I shall slightly generalize Perzyna's formulation in order to include most, if not all, hidden variables formulations of plasticity.

The behaviour of a viscoplastic-plastic material is described by three sets of variables:

independant variables $\Lambda = (C, \theta, g)$;

dependant variables $\pi = (\psi, \eta, T, q)$;

internal variables, ω describing the internal state of the material. (For simplicity I do not distinguish between α and ω, between viscous effects and internal changes). According to the physical phenomena which we have in mind, these internal variables may be chosen in many different ways. For example, we can choose the variables which have been described by Perzyna, we can also choose $\omega = (P, \alpha_j)$ and then we get the theory which have been presented by Professor Mandel.

These three sets of variables are related by:

a constitutive equation which gives the present values of the dependant variables: $\pi = L(\Lambda, \omega)$;

a 'plastic flow law' which gives the time derivatives of the internal variables, $\dot{\omega}$. We can distinguish two types of behaviour, namely

i) Viscoplastic behaviour: $\dot{\omega} = \Omega(\Lambda, \omega)$.

[1]) University Paris VI, Paris, France.

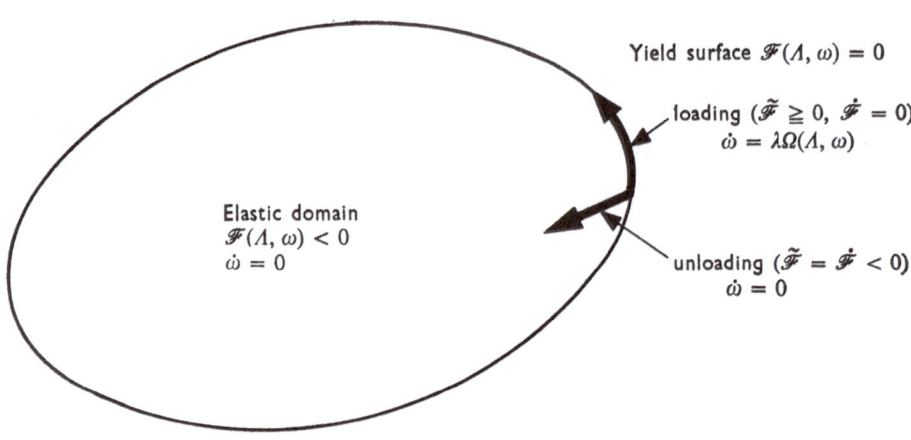

Fig. A. Plastic Behaviour.

ii) Plastic behaviour: we must always have $\mathscr{F}(\Lambda, \omega) \leqq 0$, and the plastic flow law is the following, (Fig. A):

  a) inside the elastic domain $(\mathscr{F}(\Lambda, \omega) < 0)$: $\dot{\omega} = 0$.

  b) on the yield surface $(\mathscr{F}(\Lambda, \omega) = 0)$

      $\dot{\omega} = 0$ during unloading (the state moves inwards the elastic domain)

      $\dot{\omega} = \lambda\Omega(\Lambda, \omega)$ during loading (the state remains on the yield surface, this condition determining the unknown scalar λ).

Of course we can combine both behaviours, like Professor Mandel does, but for the sake of simplicity, I shall only speak here about the plastic behaviour.

What we need now is a loading-unloading criterion in order to know whether the material undergoes loading or unloading. This is usually done by means of a loading parameter $\tilde{\mathscr{F}}$ which

a) constitutes a part of the total time derivative $\dot{\mathscr{F}}$

b) coincides with $\dot{\mathscr{F}}$ during unloading.

The loading-unloading criterion is then the following

    if $\tilde{\mathscr{F}} < 0$ the material undergoes unloading

    if $\mathscr{F} \geqq 0$ the material undergoes loading.

It has to be remembered moreover that the material must obey the principle of causality which asserts that the history $\Lambda(s)$ of Λ must uniquely determine $\omega(t)$ and $\pi(t)$. Clearly this will be fullfilled if and only if $\dot{\omega}$ can be calculated as a function of Λ, $\dot{\omega}$ and $\dot{\Lambda}$.

We turn now to the choice of the loading parameter $\tilde{\mathscr{F}}$. Obviously the quantity

$$\hat{\mathscr{F}} = \frac{\partial \mathscr{F}}{\partial \Lambda} \, \dot{\Lambda}$$

is an admissible loading parameter and the principle of causality is satisfied without further assumptions. For this reason $\hat{\mathscr{F}}$ will be called the 'natural loading parameter'. Many authors take the yield criterion in some special form $\mathscr{F}(\Lambda, \omega) = f(\zeta(\Lambda, \omega), \omega)$ and choose

$$\hat{f} = \frac{\partial f}{\partial \zeta} \, \dot{\zeta}$$

as the associated loading parameter.

With such a choice it can be shown that the principle of causality will be satisfied if and only if the coefficient

$$\mathscr{N}(\Lambda, \omega) = \left\{ 1 + \left[\frac{\partial f}{\partial \omega} \, \Omega \right]^{-1} \frac{\partial f}{\partial \zeta} \, \frac{\partial \zeta}{\partial \omega} \, \Omega \right\}^{-1}$$

is positive

$$\mathscr{N}(\Lambda, \omega) > 0.$$

Provided this inequality is satisfied, then it can be shown that the loading parameter $\hat{f}$ is equivalent to the natural one $\hat{\mathscr{F}}$.

Choosing properly the internal variables and giving special forms to the functions L, Ω and $\mathscr{F}$ (or L and Ω when dealing with viscoplasticity) we can obtain most of the hidden variables theories which have been presented at this meeting. (Since all loading parameters are equivalent we need not give a special one).

I will conclude in saying that when we use a loading parameter which is not the natural one $\hat{\mathscr{F}}$, then we must remember about the fundamental inequality $\mathscr{N}(\Lambda, \omega) > 0$.

P. Germain[1]). May I give my personal views on the various theories which have been discussed so far at this Symposium by Mandel, Onat, Lee, Perzyna and Sidoroff.

If we restrict ourselves first to *isotropic* elasticity, Green and Naghdi and Perzyna have proposed theories which are quite satisfactory. The theory of Lee appears as a special case as it was clearly shown by Sidoroff.

---

[1]) University Paris VI, Paris, France.

Although not as general as the previously quoted, Lee's theory on the other hand has the advantage of being more easily fitted with the physical property of invariance of elasticity with respect to plastic deformation.

The theories of Mandel and Onat go beyond the restriction of isotopic elasticity. They are more general and probably may be more realistic, but are also more complicated.

It is interesting to note that these two theories assume the multiplicative decomposition of the matrix gradient, first given by Lee. Mandel in particular has pointed out that the Green and Naghdi theory is not convenient if one wants to include anisotropic elasticity effects.

The microdynamics of plastic flow

J. J. Gilman

Allied Chemical Corporation
Morristown, New Jersey, USA

Advances in understanding the physical mechanisms of plastic flow and in the use of this understanding to interpret and to predict macroscopic plastic behavior are discussed. How macroscopic mechanical phenomena can be related to and predicted by microscopic dislocation physics is reviewed. Areas for further study are indicated.

1. Introduction

It has been a pleasure to have the opportunity for presenting this lecture since it has been some twelve years since I spoke to a similar group of people with a general interest in plasticity. The previous occasion was a Symposium on Plasticity that was held at Brown University. At that time the first direct measurements of the velocities of individual dislocations had recently been made. Dislocation multiplication had been studied experimentally to some extent; and a new interprrtation of the plastic instability at the upper yield point had been proposed.

Since that time many advances have occurred in understanding the physical mechanisms of plastic flow, and in the use of this understanding to interpret and predict macroscopic plastic behavior, [1].

There are many possible descriptions of macroscopic plastic flow, and we have heard a number of them during the course this week. An optimum one should be neither so simple that it covers only a few special cases, nor so complex that it is not useful for designing machines or gaining physical insight into plastic phenomena. Microscopic parameters must be included in the description because it is well-known that not only the external shape of a substance changes during plastic flow, but also the internal structure; and especially the internal distribution of dislocations. Time must also play a central role because it is well established experimentally that plastic

177

flow is a dissipative process. In the case of metals, about 95 percent of the plastic work gets converted immediately into heat, and only about 5 percent is stored in the material in the form of additional dislocations and other defects. Therefore, any time independent description is doomed to failure.

A description of plastic flow must also be statistical because the internal structure is exceedingly complex, as shown in the photographs that Professor Kröner presented earlier in the week. The most simple type of statistical theory is one in which only the average values of the parameters are considered, and fortunately this is adequate for most purposes.

From this point of view, no attempt is made to account for each dislocation as it traverses a specimen. Instead, just as the kinetic theory of gases deals with statistical averages that can be related to macroscopic properties, the microdynamical theory of plasticity smears all the dislocations in a specimen into a homogeneous distribution that is described by a strain-dependent density function, and a stress-dependent average velocity.

2. The strain-rate equation

The starting point of the description is a strain-rate equation that is then combined with various equilibrium and boundary conditions in order to describe the effects of creep, stress-strain, and plane elasto-plastic waves in materials. The purpose of the strain-rate equation is to connect stress, strain, and time in a way is determined by the behavior of dislocations, by their distributions, and by their interactions between the various dislocations. In order to understand the nature of the strain-rate equation it is best to start with the strain-rate acceleration equation:

$$\ddot{\varepsilon}_p = b(N\dot{v} + \dot{N}v). \tag{1}$$

The second derivative of the plastic strain-rate is given by the Burgers displacement, times the sum of two terms. The first consists of the average acceleration, $\dot{v}$ times the number that are accelerating, N. The second consists of the rate of change of the number of the dislocations that is present times their average velocity. The quantity, N can either be a flux-density or a loop-density. The time integral of the strain acceleration gives the plastic strain-rate. One simplification that occurs immediately is that the transient contributed to the strain-rate by the dislocation accelerations is very short because they have very small effective masses. Therefore, dislocations attain their steady-state average velocities very quickly and further transient behavior is entirely associated with multiplication and annihilation phenomena.

Fortunately, the rate of change of the dislocation density also has a relatively simple behavior, described by an equation of the form:

$$\dot{N} = \alpha v N + \beta e^{-\mu/\sigma}. \tag{2}$$

Under most conditions the second term is small so the rate of change is simply proportional to the instantaneous density; thus first order kinetics is obeyed. Also, as might be expected, the rate of change depends on the velocity v because the faster the dislocations move, the faster they can multiply. Alpha in (2) is the multiplication coefficient.

However, in some cases, especially under hard shock conditions, the stress, σ will initially rise to a very large value so that dislocations will either spontaneously nucleate or be heterogeneously nucleated at small precipitate particles and other structural defects. This behavior can be described by the second term which is small when $\sigma \ll \mu$ but increases rapidly when σ becomes comparable with μ. In this term β and μ are constants.

If the second term of Eq. (2) is neglected, and it is substituted into Eq. (1), and this is integrated with respect to time, the result is that the strain-rate grows exponentially with time. It is usually more convenient to work in terms of strain so the following form for the strain-rate equation is combined with the multiplication equation

$$\dot{\varepsilon} = bNv = \frac{bN}{\alpha}. \tag{3}$$

Integration yields:

$$N = N_0 + M\varepsilon,$$

where N_0 is the initial density and M is a coefficient that equals α/b. Thus the dislocation density equals the initial density plus a term that is proportional to the plastic strain. Accordingly, the strain-rate is proportional to the strain which leads to plastic instability if there is no strain-hardening:

$$\dot{\varepsilon} = bv(N_0 + M\varepsilon). \tag{4}$$

In more general form the plastic strain-rate on any particular glide system with index 'j' is given by the line integral of the Burgers displacement that

179

corresponds with the glide system '*j*', times the scalar product of the velocity (which is a function of stress) and the outer normal vector to the dislocation line:

$$\dot{\varepsilon}_j = \int_j b[V(\sigma) \cdot \boldsymbol{n}]dl. \tag{5}$$

In general the velocity is a function of position along the line. The integration is over the total length of dislocation line in the solid.

Another factor to be remembered about this equation is that in general *b* is a polar vector. Therefore it can make a difference whether one is looking in one direction or the other along the glide system. For example, one has to consider this possibility in the case of an iron crystal. On the other hand in the case of ionic crystals, one does not have to consider this because *b* is an axial vector.

To obtain the total plastic strain-rate it is necessary to sum Eq. (5) over all the glide systems. For a uniaxial case, the total plastic strain-rate consists of the sum of the strain-rate on each glide system times the direction cosines of the normal to the glide plane and the glide direction:

$$\dot{\varepsilon} = \sum_1^m \dot{\varepsilon}_j \cos \lambda_j \cos \phi_j \tag{6}$$

One can further generalize, of course, by writing the equation in its tensorial form.

3. The microscopic parameters

The average velocity of a dislocation line depends on a balance between the driving force on it that it is exerted by the applied shear stress, and various drag forces that are exerted by the matrix crystal structure and various imperfections within the structure. There is a very large literature at the present time that deals with experiments and theories on the nature of the drag forces. Since this literature is primarily of interest to physicists, I will not deal with it here. Instead I shall assume that a functional relationship between the average velocity and the stress is known from experiment or from the theory of drag forces and then I shall review some of the macroscopic consequences of the assumption. For a particular material the Burgers displacement is known from X-ray crystallography, and the velocity stress function can be determined by methods that were originated by W. G. Johnston and myself [2].

Then if one knows how the growth of the dislocation line length occurs, the strain rate equation is completely determined and various kinds of macroscopic responses as predicted by the microscopic parameters can be investigated. Various workers have applied this procedure to predict macroscopic creep curves, stress relaxation rates, stress-strain curves, stress-distance profiles, and one-dimensional strain wave behavior. Considerable success has been experienced in all cases. The most demanding application is the case of strain-waves and I shall concentrate on it, but I shall show a few results for the other cases.

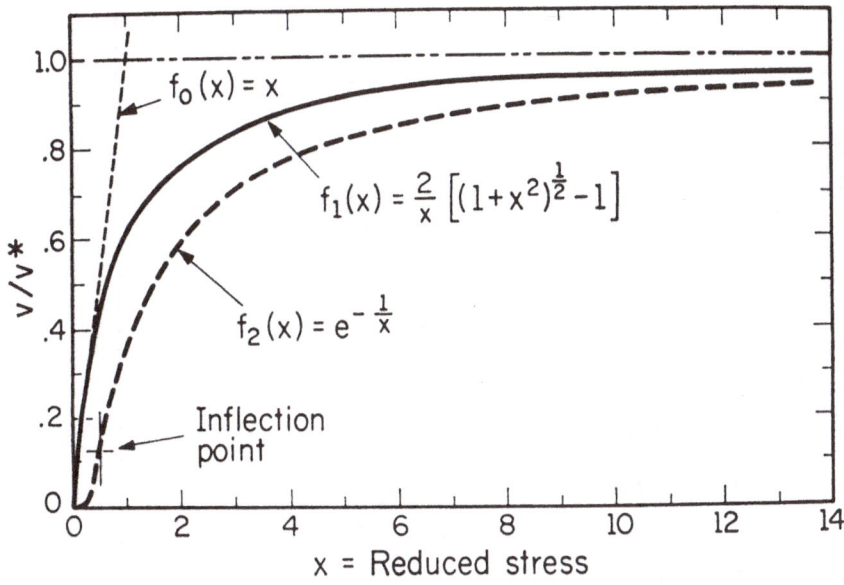

Fig. 1. Dislocation velocity-stress functions.

Before considering the macroscopic responses, I shall briefly review the velocity-stress curves and dislocation-density strain behavior. Fig. 1 summarizes the two qualitative possibilities. In this figure reduced velocity (i.e. velocity divided by the terminal velocity is plotted as a function of reduced stress (stress divided by a characteristic stress). The first curve, f_0 describes Newtonian viscous behavior at low stresses. This is observed for isolated individual dislocations in high-purity simple metals. The behavior at velo-

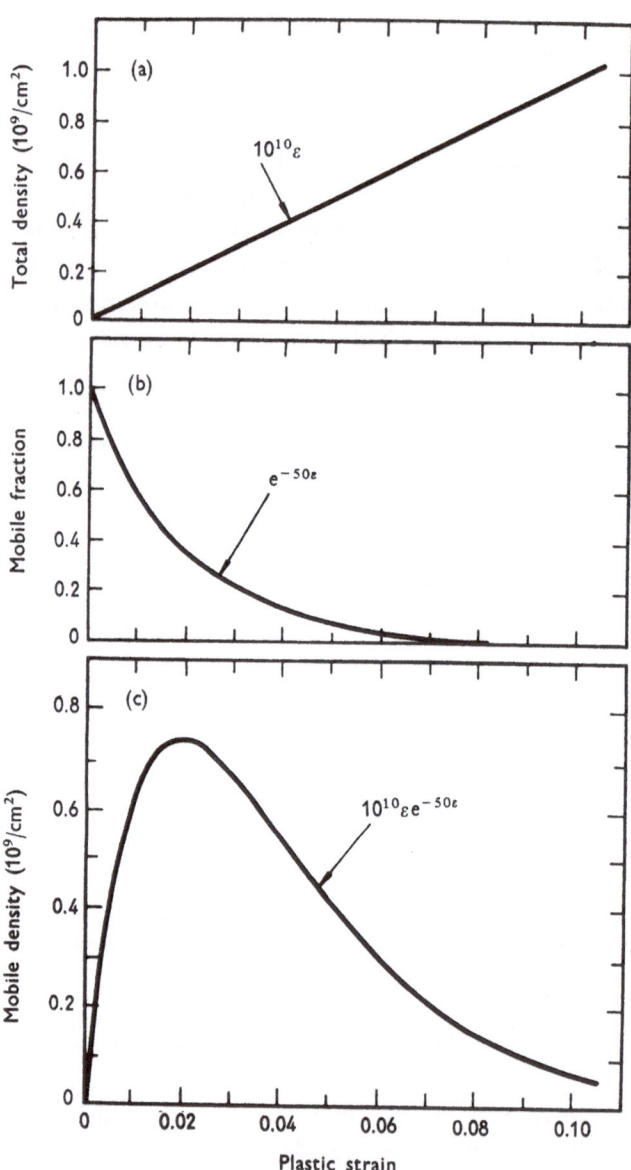

Fig. 2. Functions that describe mobile dislocation density as it depends on increasing strain.

cities near the terminal value is given by relativity theory; and the functional form, f_1 is shown in the figure. The curve, f_2 describes the behavior when rapid motion does not occur unless some minimum stress (yield stress) is exceeded. Then rapid motion begins and the exponential function with a reciprocal argument describes the behavior. For low stresses the velocity is very small, but when the stress reaches the order of the characteristic stress the velocity increases very rapidly. Later as the terminal velocity is approached the velocity stops changing rapidly.

Some dislocation density functions are outlined in Fig. 2. I have already mentioned that as a result of the first order kinetics for dislocation multiplication, the total density is proportional to the strain. But as soon as one gets many dislocations in a specimen they begin to interact with one another. Therefore, only a fraction of them are mobile, and this fraction decays exponentially. The net result is that the mobile density increases at first with plastic strain and then decays away. The former effect is associated with the initial yielding; especially with the instability just beyond the upper yield point. The latter effect is associated with strain hardening.

4. Macroscopic plastic responses

Using the functions described above containing constants measured from microscopic experiments, one can predict various macroscopic behaviors. The first example I will show is a creep curve (Fig. 3). The solid line is the measured creep curve for a LiF specimen and the points were calculated from the equations that I have given [3]. The only disposable parameter is the strain-hardening coefficient in this case. Similar results have been obtained for a variety of materials, including: stainless steel; high-temperature alloys; alkali halides; aluminum oxide; for soft metals like zinc and aluminum; etc. In all cases one gets quite good agreement between theory and experiment.

Next uniaxial stress-strain curves have been calculated (Fig. 4). In this case no attempt was made to predict the strain-hardening part of the curve [4]. The actual measured curve is the solid line and the dotted one was calculated. For the small strain region no disposable parameters were used. Thus the small strain behavior can be predicted quite successfully in terms of the microscopic parameters.

Another feature that comes out of the theory quite simply is the dependence

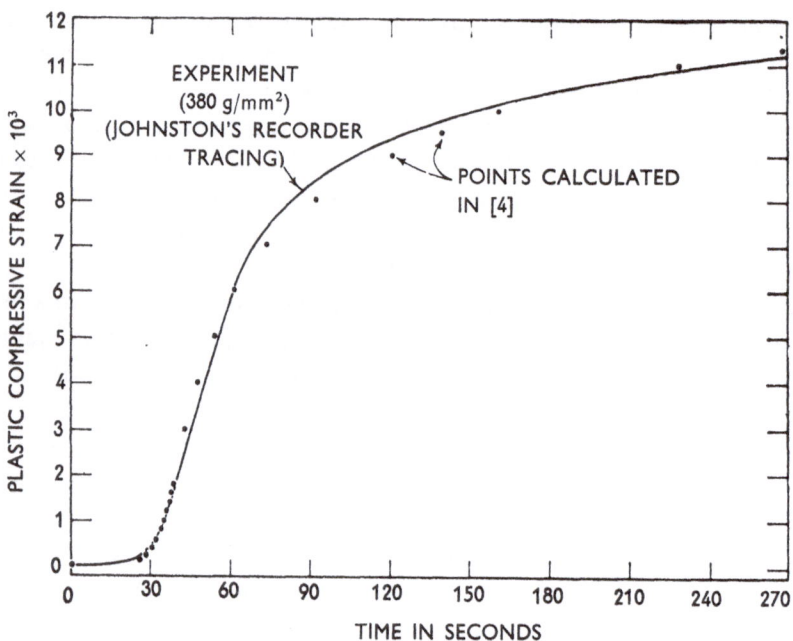

Fig. 3. Comparison of measured and calculated creep curves for an LiF crystal.

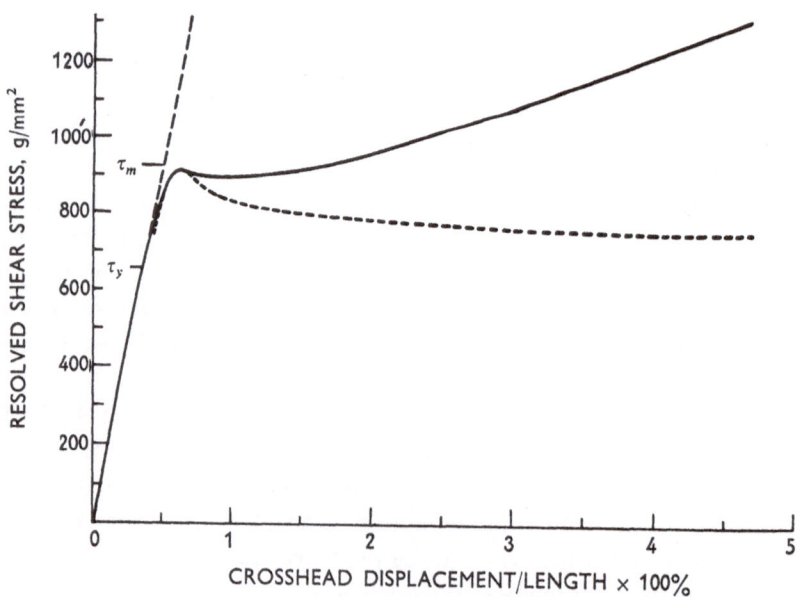

Fig. 4. Comparison of measured (solid) and calculated (dotted) stress-strain curves for an LiF crystal.

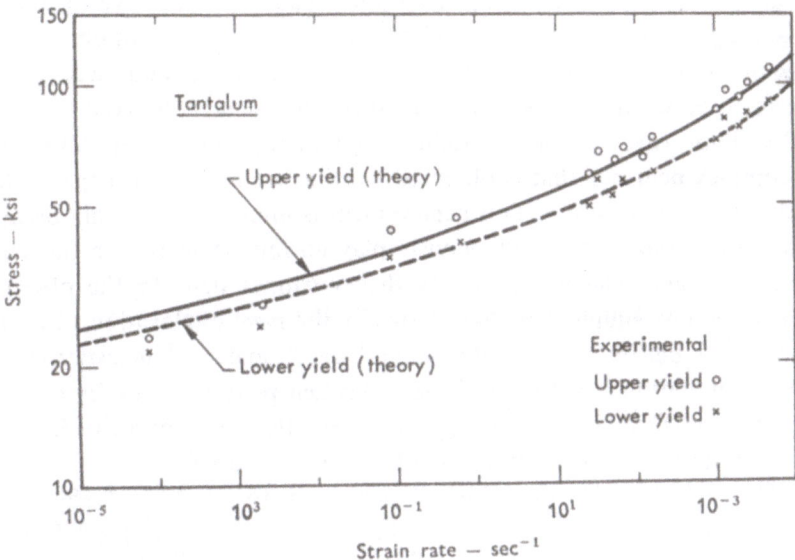

Fig. 5. The dependence of the yield point of tantalum on the strain-rate.

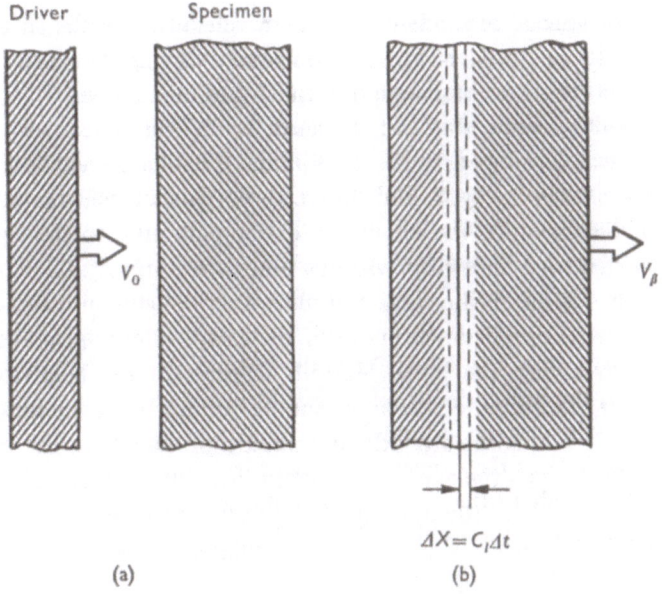

Fig. 6. Impact of a high velocity driver plate onto a specimen plate.

of the upper-yield-point on the strain-rate. This has been investigated by many people. The example I have chosen for Fig. 5 shows data obtained by Hoge and Gillis [5]. The data were obtained for tantalum and there is good agreement between what the theory predicts and what is observed.

Most impressive of all is the ability of the theory to interpret and predict the complex behavior that is observed when a strain-wave propagates thru an elastic-plastic medium. This achievement is impressive not only because stress, strain and time simultaneously play an important role in this case, but also because classical plasticity theory cannot describe the observed behavior in any simple way. Schematically the plastic relaxation of a one-dimensional plastic wave is shown in Figs. 6 and 7. The experimental situation is outlined in Fig. 6. A flat specimen plate is struck by a driver plate that moves toward it at a high velocity. After the impact elastic waves move out from interface and into the specimen and driver.

Relaxation of the elastic wave is illustrated in Fig. 7. A small element in the material prior to the presence of the wave is sketched at the left in the figure. The wave causes elastic compression in one-dimension when it reaches the element so that both the volume changes and the state of internal shear. Plastic relaxation relaxes the internal shear leaving only the volume change.

The time or spacial dependence of plastic relaxation of the shear part of elastic wave can be predicted by simultaneously solving the elastic equation of motion and the strain-rate equation that I outlined above.

One computing code that can be used to do this effectively is called HEMP and was developed by M. L. Wilkins. Results of calculations using this code are shown in Fig. 8. This figure shows particle velocity curves as a function of distance. To the left is the driver plate and to the right is the specimen. Curves are shown for various times after impact: 0.2, 0.4, ..., 1.8 microseconds. As the elastic wave continues to propagate into the specimen, additional relaxation occurs so more of the elastic energy gets absorbed as the wave moves thru the plate. Various features of the behavior can be checked. First, the initial disturbance should propagate at the longitudinal-wave velocity whereas the second front should propagate at the bulk-wave velocity. In both cases the expected behavior is confirmed. A second feature is that the magnitude of the upper yield point decays roughly exponentially. Finally, the details of the shapes of the calculated curves can be compared with those of measured curves.

Before showing some experimental data, I should like to make the point

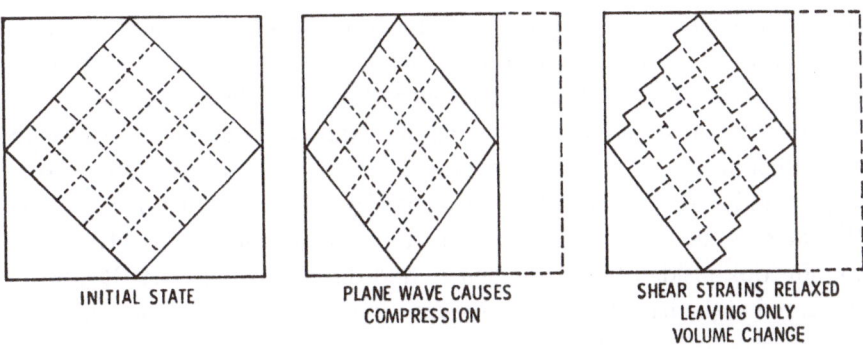

INITIAL STATE | PLANE WAVE CAUSES COMPRESSION | SHEAR STRAINS RELAXED LEAVING ONLY VOLUME CHANGE

Fig. 7. Schematic response of a volume element to a plane one-dimensional strain wave.

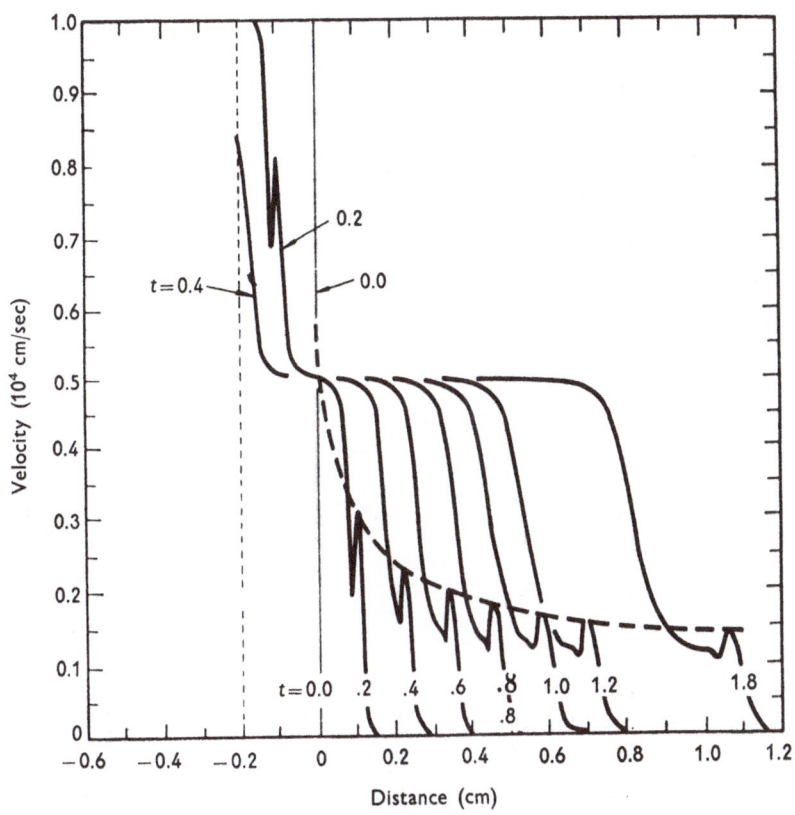

Fig. 8. Results of HEMP code calculations.

187

that from a physical point-of-view, there is no such thing as a plastic wave involved in this behavior. What exists is an elastic wave whose shear component is being absorbed as it propagates through a plastic material.

Experimental measurements of the attenuation of the upper-yield-point as a function of distance are shown in Fig. 9. These data were obtained by Taylor and Rice [6]. The upper-yield particle velocity, which is essentially the same as the stress, as a function of specimen thickness is given. One cannot look inside of the specimen to find the particle velocity. It must be measured at the back face when the wave emerges. Therefore, measurements

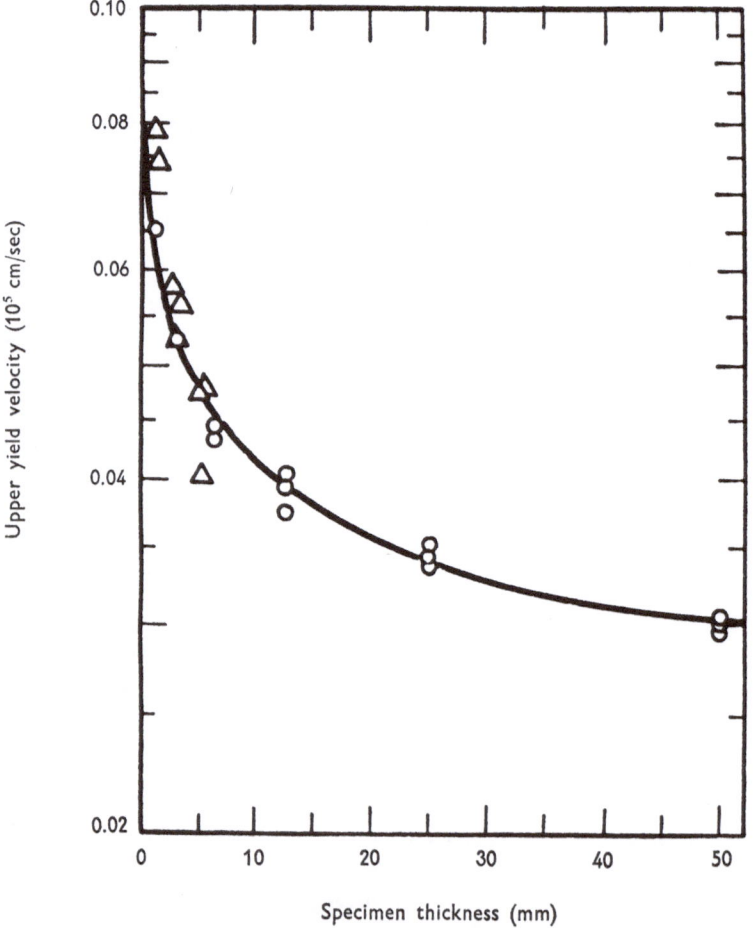

Fig. 9. Attenuation of the upper-yield-point particle-velocity in Armco iron.

are made on specimens having various thicknesses which is equivalent to looking inside at various distances from the impact surface. The solid line was calculated using reasonable values for the microscopic parameters. It compares well with the experimental points.

Assay et al. [7] have done quite detailed experiments on the propagation of plain strain-waves in lithium fluoride as shown in Fig. 10. The particle velocity (or stress) profiles for lithium fluoride specimens are shown here after various time periods. Again one can see the same general type of decay that as predicted by the theory. Also, the general shapes of these curves are those that would be predicted by the theory.

In some cases a complication occurs (especially the case of various iron-base alloys) because twinning can occur in addition to translation-gliding. Then the strain-rate equation must be written in terms of the twinning behavior as follows:

$$\dot{\varepsilon} = kN\dot{\Omega}.$$

This strain-rate equation is analogous to the one for gliding. The twinning shear, k is the analog of Burgers displacement, the twin concentration, N is the analog of dislocation concentration, and the rate of twin volume change, $\dot{\Omega}$ is the analog of the dislocation velocity. Equations of this form have been applied by Johnston and Rhode [8], to the case of dynamic yielding in iron at low temperatures where twinning tends to dominate translation-gliding. Some of their results are shown in Fig. 11.

Again stress-time curves are considered for plate impact experiments. The solid lines are the experimental results; the dotted and dashed lines are calculated curves, which assume that deformation occurs mostly by three-dimensional twin growth at rates determined by the characteristic time, T_3. The best agreement is obtained by assuming a characteristic time of 6μ sec.

5. Concluding remarks

I have reviewed how macroscopic mechanical phenomena of various kinds can be related to and predicted by microscopic dislocation physics. Much remains to be done. One of the areas that has not been studied enough is dislocation multiplication. How it occurs; what the modes are; how they depend on stress, time and other factors such as the defect structure of material: all these are topics that need more detailed study. This is

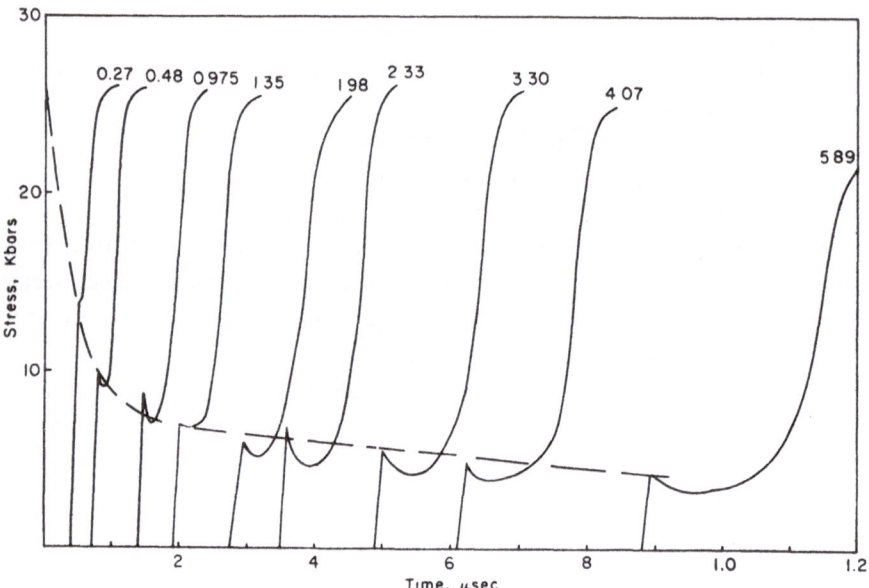

Fig. 10. Experimental stress-time profiles for plane waves in LiF crystals. Curves are shown for nine specimen thicknesses ranging from 0.27 to 5.89 mm.

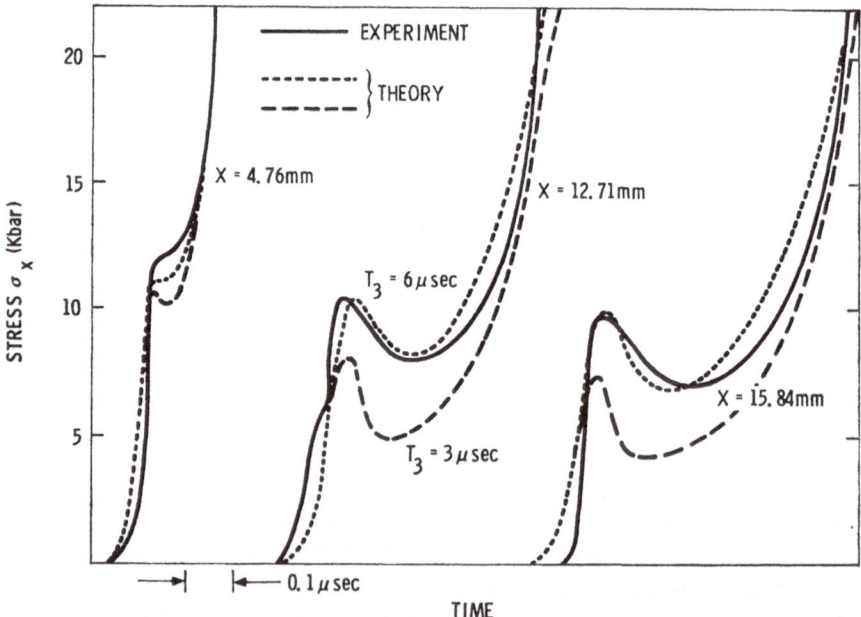

Fig. 11. Stress-time profiles for shock-loaded Ferrovac E-iron. Solid lines show experimental results for specimens of three different thicknesses (propagation times). Theoretical curves assume three-dimensional twin-growth at rates determined by the characteristic time, T_3.

necessary in order to predict behavior during very rapid changes in strain rates. Internal thermal effects also need to be taken into account, especially in high strain rate situations where significant amounts of adiabatic heating can occur.

Furthermore, methods for predicting dislocation mobilities in terms of chemical constitution are needed. These could eliminate the necessity for measuring drag stresses.

References

[1] Gilman, J. J., *Micromechanics of flow in solids.* McGraw-Hill, New York, 1969.

[2] Gilman, J. J. and W. G. Johnston, Dislocations in lithium fluoride crystals. Solid State Physics, *13* (1962) 147.

[3] Gilman, J. J., Microdynamics of plastic flow at constant stress. *J. Appl. Phys., 36* (1965) 2772.

[4] Johnston, W. G., Yield points and delay times in single crystals. *J. Appl. Phys., 33* (1962) 2716.

[5] Hoge, K. G. and P. P. Gillis, Rheological properties of annealed tantalum. *Metall. Trans., 2* (1971) 261.

[6] Taylor, J. W. and M. H. Rice, Elastic-plastic properties of iron. *J. Appl. Phys., 34* (1963) 364.

[7] Assay, J. R., G. R. Fowles, G. E. Durall, M. H. Miles and R. F. Tinder, Effects of point defects on elastic precursor delay in LiF. *J. Appl. Phys., 43* (1972) 2132.

[8] Johnson, J. N. and R. W. Rohde, Dynamic deformation twinning in shock-loaded iron. *Appl. Phys. 42* (1971) 4171.

Experimental plasticity. Some thoughts on its present status and possible future trends

A. Phillips

Yale University, New Haven, Connecticut, USA

The point of view of this paper is a phenomenological one. It discusses experimental work on the fundamental laws of plasticity. After a short discussion of plastic flow under uniaxial stress where the concept of the quasistatic stress-strain curve is emphasized, recent experimental work on yield surfaces is considered. In particular, the importance of the definition of yielding on the determination of the yield surface is discussed. A new hardening law discovered by the author is discussed. The topics of loading surfaces, normality, linearity, and corners are treated and the article ends with a brief consideration of creep behavior as well as strain-rate effects. The paper ends by emphasizing that the testing of the validity of any theory of plasticity must be based on experiments showing the validity of each fundamental assumption of the theory independently of the other assumptions of the theory.

1. Introduction

A general lecture on 'Experimental Plasticity' within the framework of a Symposium on the Foundations of Plasticity must be devoted to those parts of experimental plasticity which are vital for the foundations. We shall therefore restrict our considerations to a number of items which we consider today as of particular importance for the foundations of plasticity and we shall present them in a form which may facilitate future experimental research on this subject. The selection of the items to be considered depends of course on this writer's interests. Testing techniques, as well as detailed discussion of material behavior under uniaxial conditions, will not be considered. We shall take the point of view of the phenomenologist. We shall consider phenomena as they occur without regard to the microstructural reasons for which they occur. A major part of the lecture will be devoted to the determination of the yield conditions since this topic has been a major interest of the author for many years. As the subtitle of this lecture indicates we do not aim towards completeness. This is not a survey. We present only some thoughts on the present status and some possible future trends of experimental plasticity. Only a limited number of arbitrary selected representative papers will be cited.

As mentioned by Lin [1] plasticity can be divided into a mathamatical theory and a physical theory. The mathematical theory uses highly idealized assumptions in order to solve boundary value problems successfully. The physical theory, on the other hand, is preoccupied with the derivation of the constitutive relations of plasticity from physical considerations. Obviously this lecture will be closer to the physical theory than to the mathematical.

Our purpose in this lecture is not to mention experimentally derived empirical formulas which are useful to design, primarily because of their mathematical simplicity. Our purpose is rather to determine which experimentally derived empirical formulas represent real laws of behavior. In the literature the interest lies in simple expressions which will be useful to the solution of boundary value problems irrespectively of their fundamental validity, so long as they do not deviate too much from the empirical representation of real behavior. The drawback of the latter approach is that we are never sure whether the expression used can be employed under conditions very different from the ones it was experimentally derived and therefore it remains a very crude approximation of reality instead of an expression representing reality in a fundamental way.

2. Plastic flow under uniaxial stress

The total strain can be divided into an elastic portion and a plastic portion. The plastic portion is the only one of importance to plasticity. Equation

$$\varepsilon = \varepsilon^{el} + \varepsilon^{pl} \tag{1}$$

is valid only for small strains. For finite strains it is preferable to use the deformation gradients. Restricting ourselves to small strains, we use equation (1) where ε^{el} is proportional to the stress; thus we can differentiate the plastic strain from the elastic one by means of observing the deviation from proportionality. The proportional limit is the stress where we first observe a deviation from proportionality. Such observations are very difficult since they depend to some extent on the sensitivity of the instruments used and, of course, on the skill of the observer. Such observations fall within the province of *microplasticity*, see for example [2], which deals with the transition from the elastic to the plastic behavior in the region of small strains. The microplastic region is defined as the region where the plastic strain is equal to or greater than the elastic part of the strain. Information about microplastic phenomena is of great value for applications in which purely elastic behavior is needed (for example, vibrating reeds, inertia guidance systems, etc.). These phenomena are at the other end of the spectrum from gross plastic behavior such as seen in metal forming applications. In addition microplastic observations are of great importance to the solution of some problems of plastic stability, since plastic stability depends to a great extent on the laws of transition from elastic to plastic region, see Phillips [3], Hutchinson and Koiter [4], and Sewell [83].

It is important to know whether or not the deviation from proportionality is a genuine expression of the beginning of permanent deformation; therefore, it is necessary to follow the first small deviation from proportionality by a small unloading from which we judge whether there is permanent strain or not. In what follows, when we discuss the proportional limit we shall assume that it is in reality the limit of elasticity.

The plastic strain is difficult to measure in the above way with certainty. Hence, a proof strain method has been used which acknowledges a certain error in measurement. This proof strain method is useful to intermediate strains or large strains for which the error assumed diminishes in influence. On the other hand for small strains the influence of this error is quite significant. In order to avoid human errors, recently Phillips and his co-workers [5, 6] standardized the definition of deviation from proportionality by introducing a backward linear extrapolation. This extrapolation is different from the one used thirty years ago by Taylor [7], which is based on intermediate strains. The latter will be called Lode-extrapolation method since it was first suggested by Lode[1]) [8] and it should not be confused with the one used by Phillips and co-workers [5, 6]. The two extrapolation methods are shown in Fig. 1. Note that the author's backward extrapolation assumes a readability of $\frac{1}{2}\mu$ in/in minimum and an extrapolation of less than 3μ in/in.

Another reason why the observation of the proportional limit is difficult is because permanent deformation is truly time dependent deformation. Whereas elastic strain occurs instantaneously, (except when viscoelastic

---

[1]) See Paul [9] for this name.

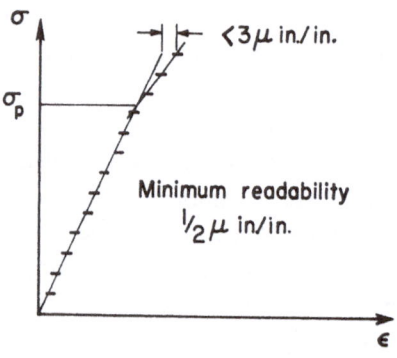

Author's extrapolation method

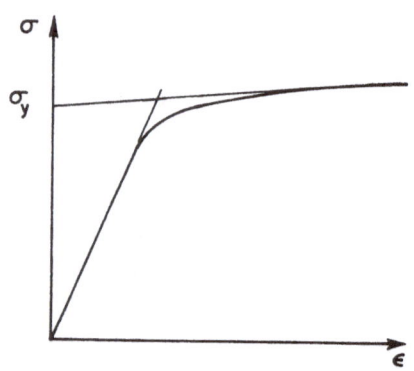

Lode extrapolation method

Fig. 1. The two backward linear extrapolation methods.

strain appears) permanent deformation needs time to develop fully for a given stress and temperature. An arbitrary division into plastic and creep strains does not represent physical reality although it might be convenient for solving boundary value problems.

In this lecture we shall take the position that the tensile stress-strain curve at a given temperature in a quasistatic case is a sequence of equilibrium positions due to successively larger values of stress. Each increment of stress is assumed to be applied after the total permanent strain due to the previous stress increment had the time to appear fully. Therefore the quasistatic stress-strain curve is in reality an incremental stress-strain curve where the infinitesimal increments are in the stress. To each increment of stress $\Delta\sigma$

at a given temperature corresponds an increment in elastic strain $\Delta\varepsilon^{el}$, and an increment in the plastic strain $\Delta\varepsilon^{pl}$ which needs a time t to appear.

In fact each increment of stress is applied while the total plastic strain due to the previous stress increments is still developing. Therefore, upon the application of an increment of stress not only that plastic strain appears which is due to that particular increment of stress but also plastic strain may have continued to appear, which is due to some of the previous stress increments. It is, of course, impossible to differentiate experimentally between the portions of appearing plastic strains due to different stress increments.

If the material is forced to extend faster than it needs to express the entire permanent strain, as for example, when the deformation is with a constant stress rate, the stress-strain curve appears elevated, as curve OK instead of curve OL in Fig. 2, since the permanent portion of strain AB does not have

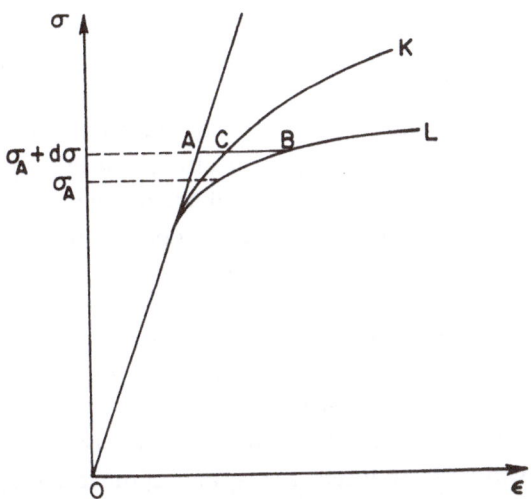

Fig. 2. The quasistatic stress-strain curve OL is the lowest stress-strain curve possible at a given temperature.

the time to develop fully when the stress increases from the level σ_A to a level $\sigma_A + d\sigma$. Thus, only the portion AC of the permanent strain appears. The quasistatic stress-strain curve OL is the lowest stress-strain curve possible at a given temperature.

Suppose that in Fig. 3 curve OA represents the equilibrium stress-strain curve as defined previously. A moderate decrease of the stress below the

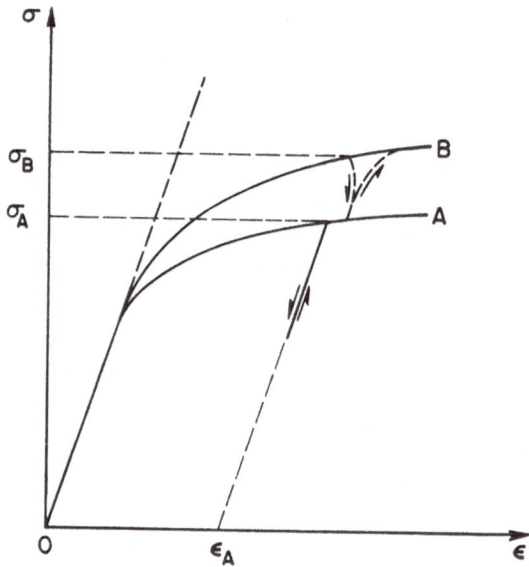

Fig. 3. Development of plastic strain between stress-strain curves OB and OA.

value σ_A at the permanent strain ε_A will produce only elastic strain. Subsequent increase of the stress to σ_A will also produce only elastic strains. Only after σ_A is exceeded plastic strain will again appear. On the other hand if OB is a stress-strain curve obtained under nonequilibrium conditions, for example, under constant finite stress rate (OB will always be higher than OA) both a slow decrease in stress from σ_B to σ_A as well as the slow increase again of the stress from σ_A to σ_B will produce some plastic strains.

It is important when decreasing the stress below σ_A not to go below the reverse proportional limit which may be situated above $\sigma = 0$ (as experimentally verified), Phillips et al. [5, 6]. Otherwise permanent strains of a negative value will be introduced, Fig. 4, and upon reloading σ_A will again be affected since it will correspond to a different value of the total permanent strain. Cyclic plastic straining will then occur. This occurrence of cyclic plastic straining due to loading below the reverse proportional limit (situated above $\sigma = 0$) is worth investigating as to its effect on yielding.

For the foundations an important question deals with the loading-unloading-reloading problem. Carefully made experiments are of importance in which the equilibrium stress-strain curve will be obtained by first loading at a constant stress rate then unloading at selected strains followed by very slow reloading. In this way the equilibrium stress-strain curve can be obtain-

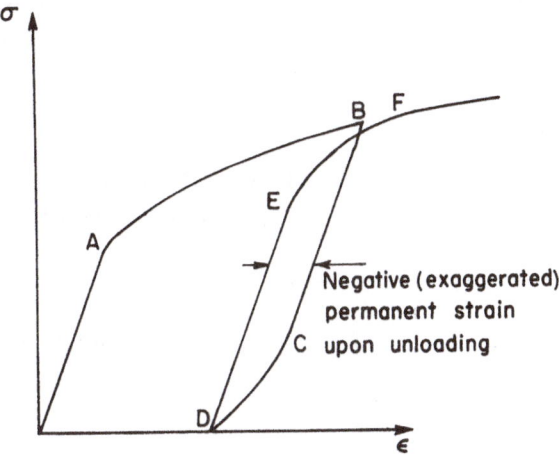

Fig. 4. Cyclic plastic straining.

ed, as the locus of all points were the first deviation from proportionality during reloading occurs, and the behavior of the material can be determined in the region between the equilibrium stress-strain curve and the curve of constant stress rate.

3. The initial yield surface

One of the fundamental assumptions of the theory of plasticity, at least as developed by Prager [10], Drucker [11] and Hill [12] is the existence of a yield surface in the six-dimensional stress space. This surface represents the locus of all points enclosing the region in the stress space in which any motion of the representative stress point produces only elastic strain. It is assumed that any excursion of the representative point outside the yield surface will produce permanent strains. Obviously a fundamental question to be answered is the establishment of the yield surface in stress space. We shall consider here the initial yield surface for an isotropic material before hardening occurs.

The concept of the yield surface is intrinsically connected with the definition of yielding and with the first appearance of plastic strain. To determine the yield surface it is necessary to define yielding as equivalent to the proportional limit, that is, to a first deviation from proportionality, if the yield surface is assumed to be the region in stress space in which for any motion of the stress point only elastic strains appear. Experiments by Phillips and his co-workers [5, 6, 13, 14] have shown that the yield surface for combined tension-torsion as well as combined tension-torsion-internal pressure for pure aluminum lies between a Mises and a Tresca surface, for the temperature range of 70°F to 305°F.

Fig. 5 gives the yield surfaces for pure aluminum as obtained in [5] for combined tension and torsion. They have been obtained for the temperatures

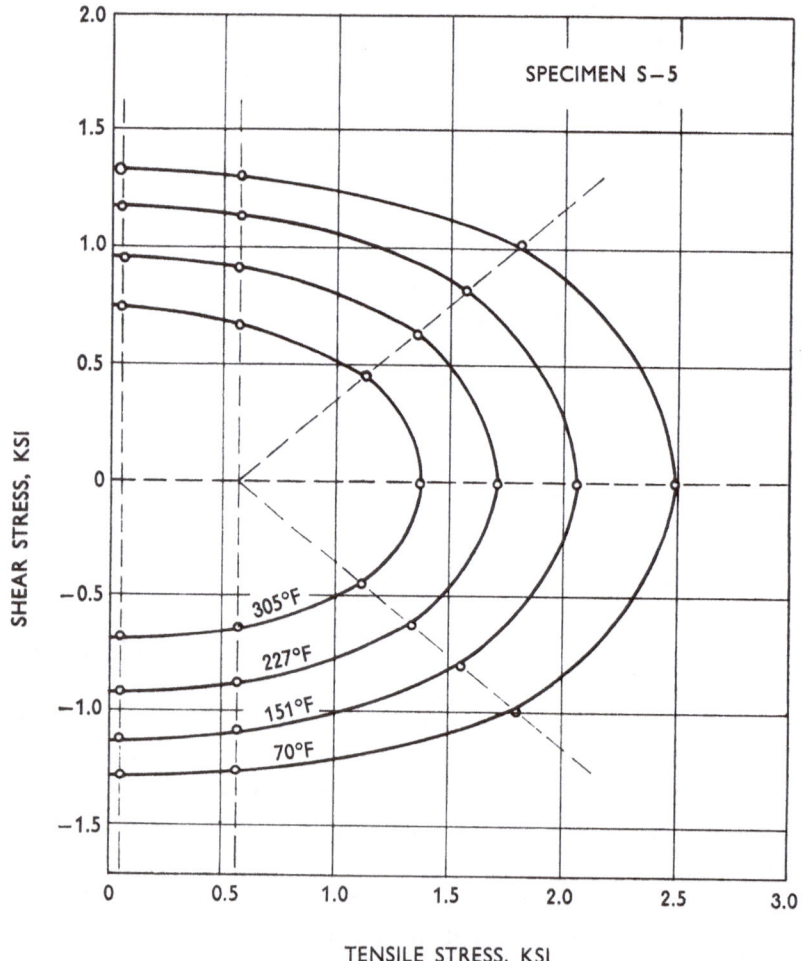

TENSILE STRESS, KSI
INITIAL YIELD SURFACE AT
ELEVATED TEMPERATURES

Fig. 5. Initial yield surfaces for pure aluminum for combined tension and torsion.

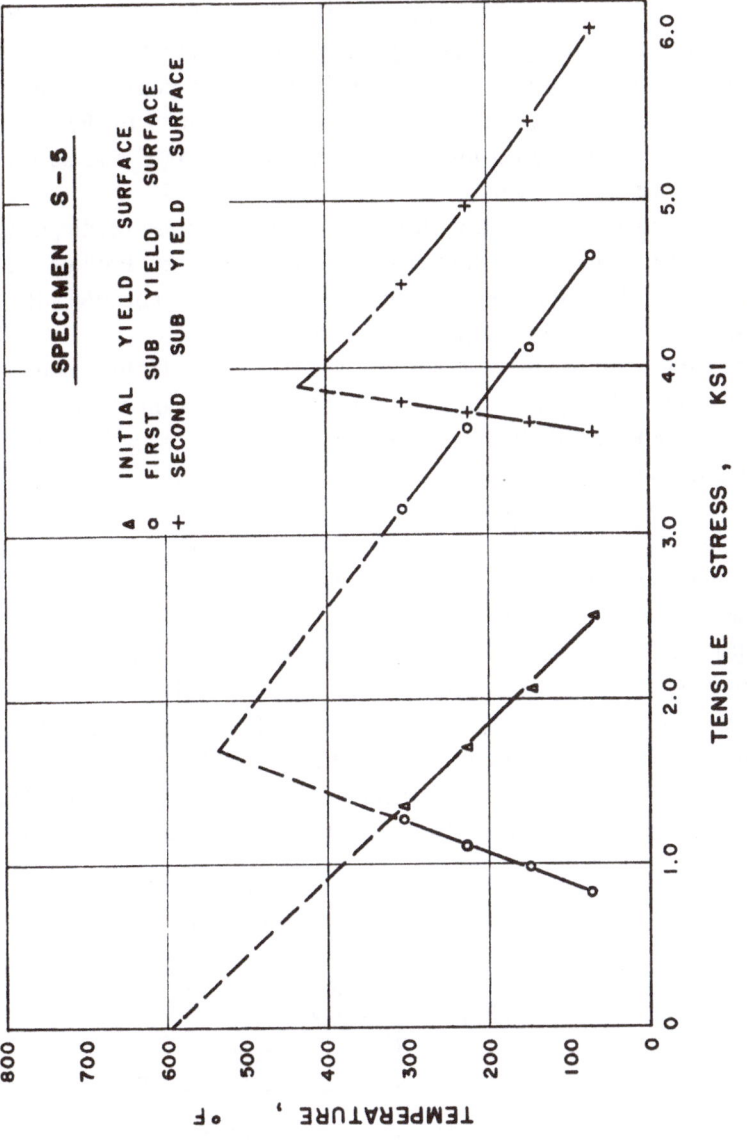

Fig. 6. Intersection of yield surface cone with the τ-T plane.

of 70°F, 151°F, 227°F, 305°F successively. The procedure for obtaining these curves was by using the same specimen for all data points and temperatures, making in the process small excursions into the plastic range at each point to be determined. All the data points on the 70°F isothermal plane were determined before raising the temperature to 151°F and so on until the data at the highest temperature were obtained.

The experimental curves are little different from curves predicted by the Mises theory. Other (still unpublished) experiments by Phillips and co-workers show that even the order in temperature does not affect the yield curves, at least when the temperature change is slow.

The virgin equilibrium yield surface is a truncated elliptical cone in the σ, τ, T space. The apex of this cone (extrapolated) is on the temperature axis at $T = 600°$ to 650°F and the basis at 70°F. The cone is truncated at 305°F since no data are available at higher temperatures. Fig. 6 shows the intersection between this cone and the $\tau - T$ plane. It is of importance to investigate effects of temperature above 305°F since it can be expected that an increase in temperature will produce a vanishing yield surface at a temperature much less than the annealing temperature of aluminum (650°F).

In [6] experiments are described with tension-torsion tests on pure aluminum in which for the determination of the yield surface thermal loading paths were used. Fig. 7 gives the type of loading paths used. The intersections of the thermal loading paths with the equilibrium yield surface are as predicted by the previously described isothermal tests. Thermal loading paths are of importance to obtain yield surfaces in the temperature-stress space.

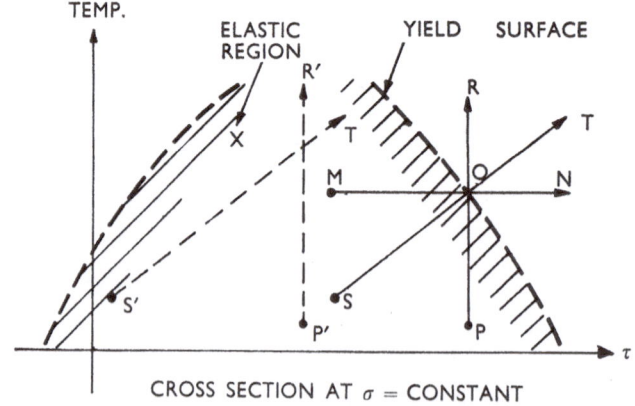

Fig. 7. Non-isothermal loading paths.

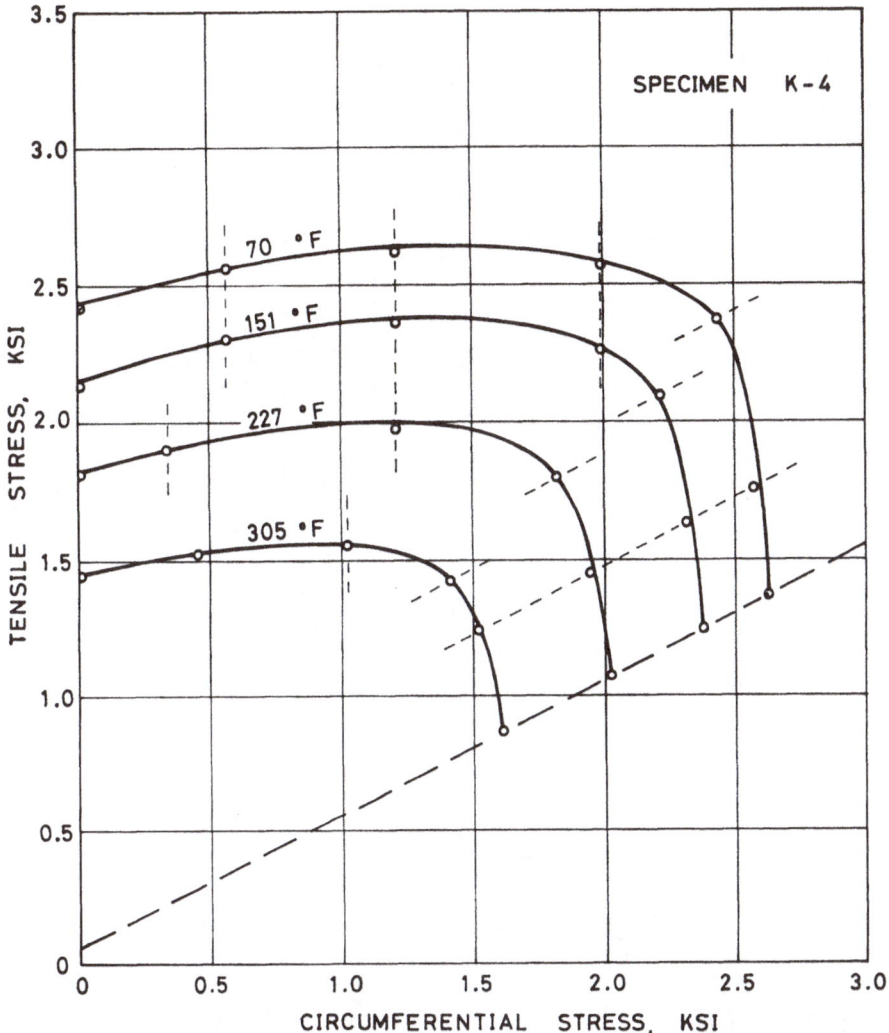

Fig. 8. Initial yield surface for pure aluminum for combined tension-tension.

In Figs. 8 and 9 we give the results of experiments by Phillips and Kasper [13] with the determination of yield surfaces in the tension-internal pressure space and in the tension-torsion-internal pressure space. We see again a remarkable validity of the Mises surface.

In the literature yield surfaces have also been obtained on the basis of either an offset in permanent strain (proof strain) or a Lode-backwards

203

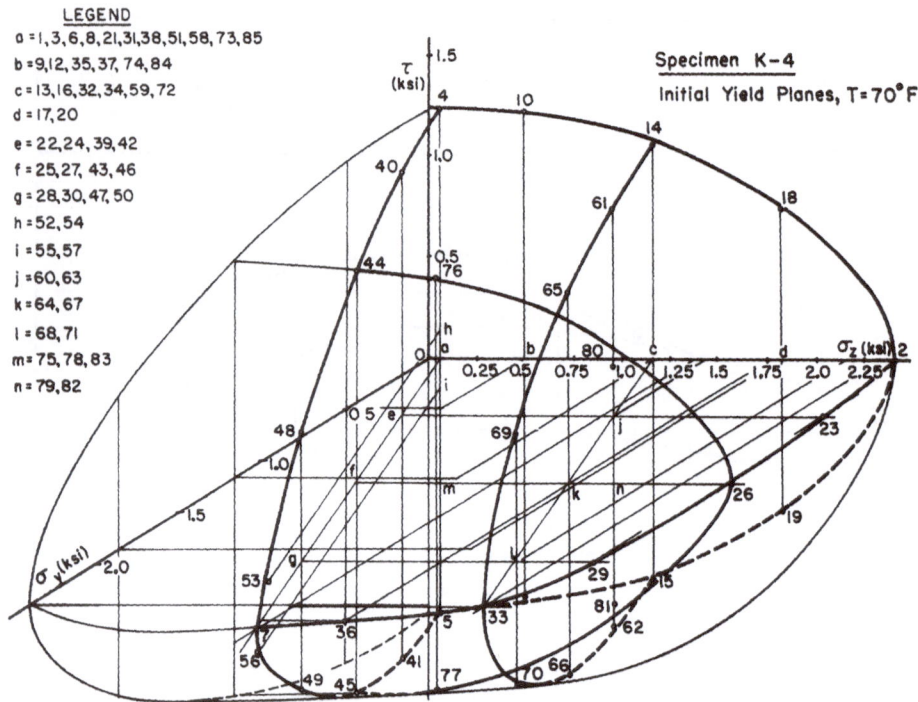

Fig. 9. Initial three-dimensional yield surface for pure aluminum.

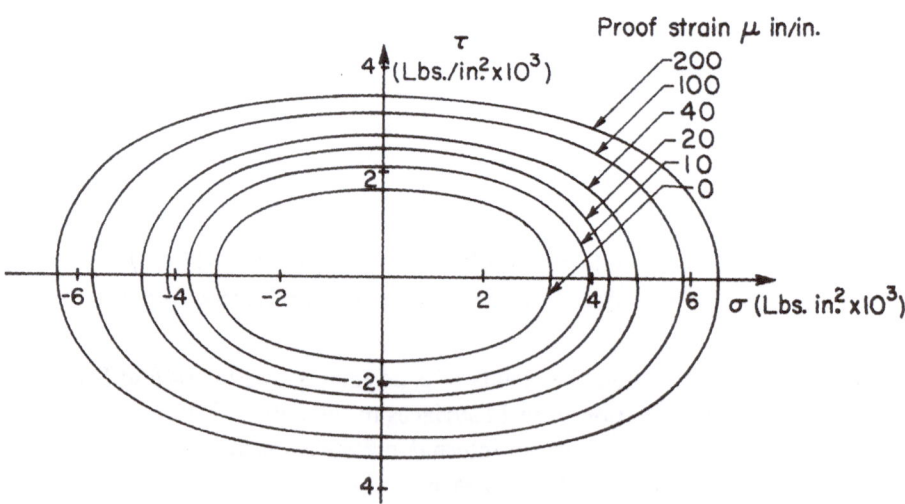

Fig. 10. Initial yield surfaces under combined tension and torsion under different values
of proof strain (from Williams and Svensson) [19].

extrapolation. Both of these definitions have serious drawbacks, one of which is that the yield surface obtained has no basic physical significance, and another one that for combined stresses they presuppose theories of equivalent strain and equivalent stress. In addition they include a region where a motion of the stress point will produce some plastic deformation. Fig. 10 shows results by Williams and Svensson [15] for the yield surfaces under combined tension and torsion under different values of proof strain. We see that there is a nearly proportional growth of the yield surface with an increasing 'equivalent' permanent set. The experimental results by Taylor and Quinney [7] can be seen in the same light. The yield surfaces determined by them are proportionally grown yield surfaces.

Whatever the investigator, the method used, and the material used the results show that the yield surface obtained is very similar to the Mises surface. This is true both at room temperature and at elevated temperatures.

4. Hardening

In the development of the yield surface prestressing is of paramount importance. The yield surface moves in stress space as prestressing develops. This is the phenomenon of hardening. Our definition of hardening is slightly different from the usual one. In the usual definition the yield surface is assumed to move with the plastic strain. Here it is assumed to move with

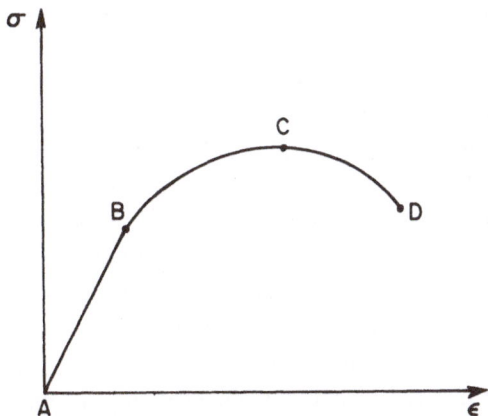

Fig. 11. Line BC corresponds to hardening. Line CD corresponds to hardening if prestraining definition accepted, to softening if prestressing definition accepted.

the prestressing. The difference between the two definitions can be made clearer by considering Fig. 11. In this figure the portion *BC* corresponds to hardening, while portion *CD* corresponds to softening, according to our definition, since in hardening the stress increases whereas in softening it decreases. In both cases the plastic strain increases so that according to the usual definition we would have had hardening always. While theories of hardening have been proposed in the past [16, 17, 18, 19, 20] they were introduced on purely intellectual grounds. Convincing experimental verification of them is lacking.

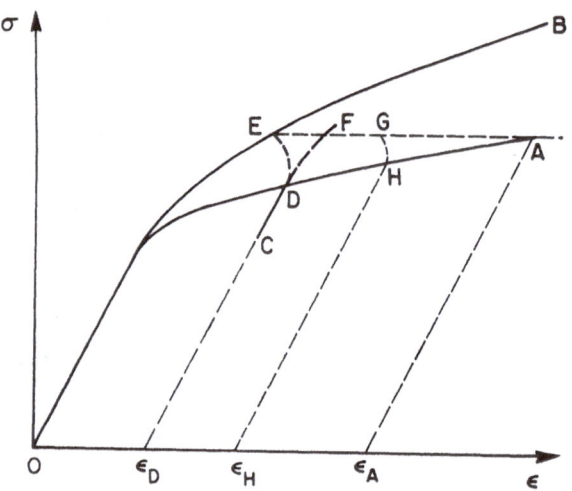

Fig. 12. Effect of the time-development of the plastic strain.

In the process of prestressing with a given stress rate the permanent strain corresponding to a given stress does not have sufficient time to develop. Suppose, for example, that in Fig. 12, line *OA* is the equilibrium stress-strain line and *OB* is a stress-strain line obtained with some finite stress rate. If while obtaining the curve *OB*, at the point *E* the stress remains constant, then the strain would continue to increase until the equilibrium curve *OA* would be reached at *A* with a final plastic strain ε_A. If, however, upon reaching *E* we decrease the stress to a point *C* below the equilibrium line crossing it at *D*, we freeze the amount of plastic strain accumulated to the amount ε_D which has been reached while crossing the equilibrium line at *D*. Reloading again we obtain line *CDF* where *CD* is the elastic reloading line coincident with the elastic unloading line while *DF* does not coincide with

ED. Thus the point *D* of the equilibrium line corresponds to the prestress point *E* and the plastic strain ε_D. If unloading would have been started at *E* but after some additional plastic strain has appeared, for example at *G*, and the equilibrium line had been crossed at *H* the plastic strain would have been frozen at *H*, with the plastic strain value ε_H which is different from ε_D. This is due to the time-development of plastic strain.

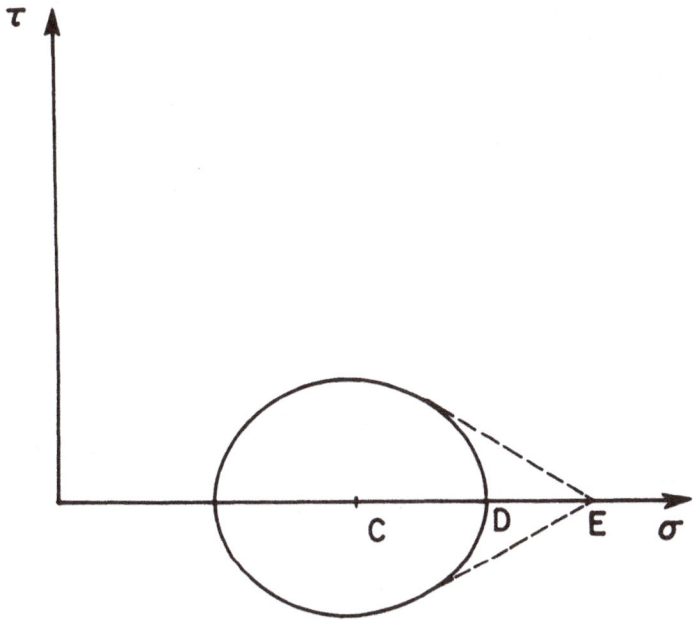

Fig. 13. The prestressing point lies outside the yield surface.

In combined stresses the axis σ of Fig. 12 is replaced by the stress space, Fig. 13, and we observe that the points *D* and *E* do not coincide. It follows that the prestressing point lies outside the yield surface which corresponds to that prestressing point. Experiments by Phillips and his co-workers [5, 6, 13, 14] have shown that for even a very slow rate of stressing as applied by dead load testing of the order of 50 lbs/sq. in. every 2 minutes, for commercially pure aluminum at room temperature, the equilibrium yield surface does not pass through the prestressing point. This is of course true also for higher temperatures, and as it has been shown [5] the yield surface does not change even if several days pass, while the state of stress is within the elastic region. On the other hand by waiting at σ_E before unloading the yield surface may change [13] because of the development of additional plastic strain and the crossing of the equilibrium yield surface at some other point than the point *D*. (Fig. 12). Thus, hardening although basically due to prestressing is also influenced by the amount of plastic strain developed in the process. This brings us to the concepts of yield and loading surfaces as developed by Phillips and co-workers [21, 22, 23]. The concept of the loading surface will be considered again in the next section.

Experimental work by Naghdi, Essenburg and Koff [24], as well as by Ivey [25], Fig. 14, has shown that there is a phenomenon called *absence of cross effect*. This means that if the prestressing is in torsion the yield surface

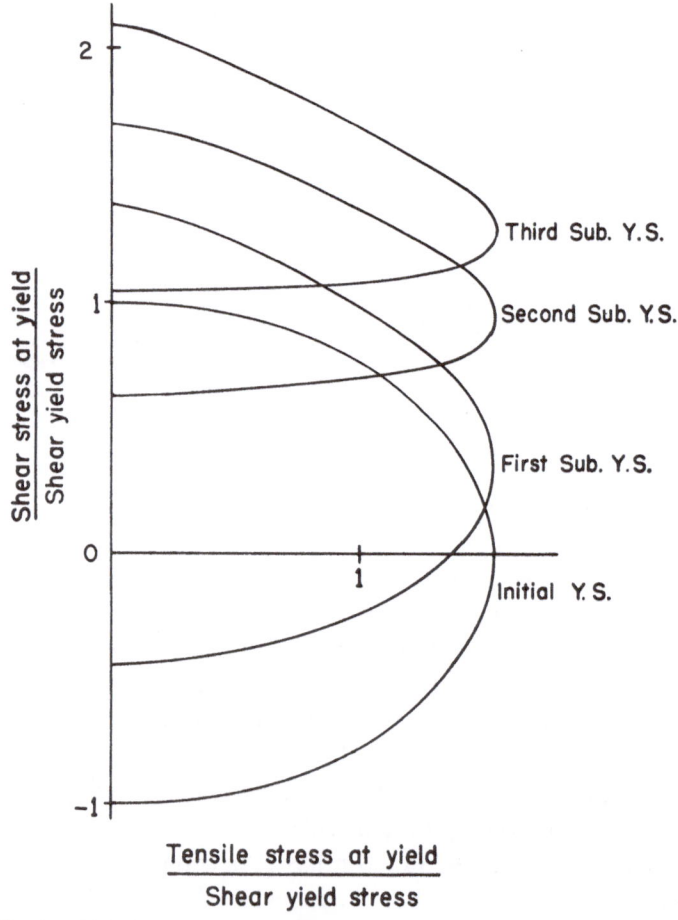

Fig. 14. The absence of cross effect as shown in Ivey's experiments [25].

does not change its diameter in the tension direction. This phenomenon has been observed in the past only in torsion.

Experimental work by Mair and Pugh [26] and others indicated that cross effect is existing. It follows that a very serious controversy existed until recently concerning the absence or presence of cross effect. In addition since the lack of cross effect appeared only for prestraining in torsion, the question arises of its possible existence in other directions or what happens when the

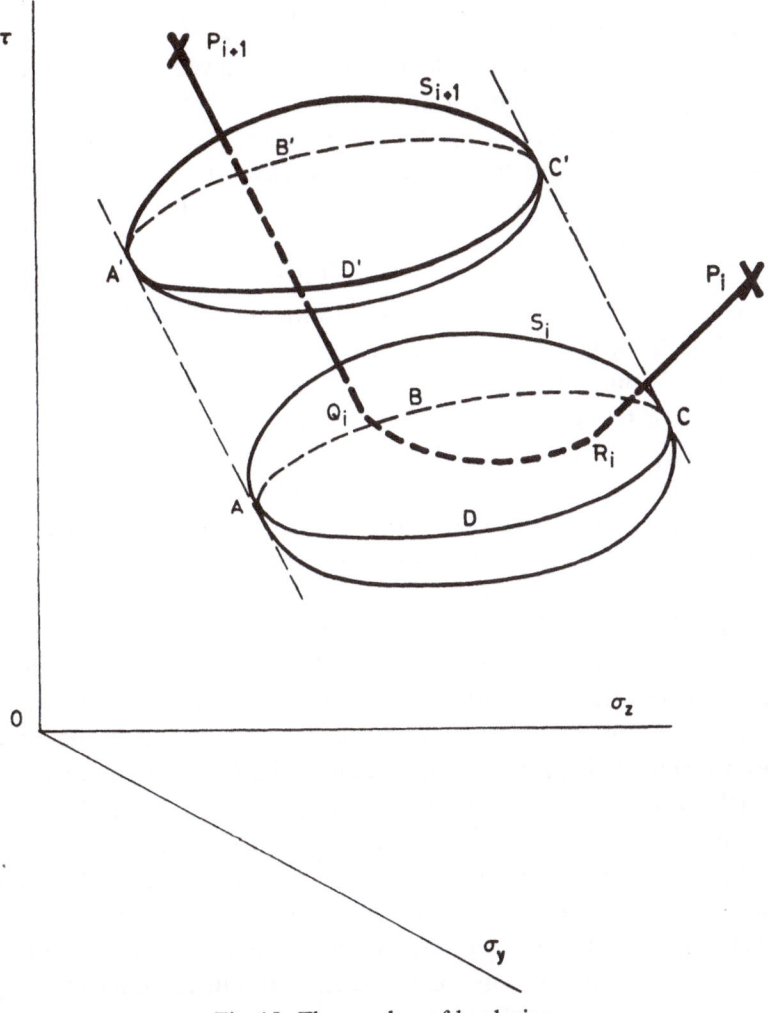

Fig. 15. The new law of hardening.

prestressing direction changes. Mair and Pugh [26] indicated that the disagreements concerning presence or absence of cross effect may be due to
the definition of yielding adopted by the different experimentalists.

Recent experiments by Phillips and co-workers [5, 6, 13, 14] have shown
that for the yield surface lack of cross effect is a universal law at every
tested temperature and every direction of prestressing even if the direction
of prestressing is changing provided the definition of yielding is that of
the proportional limit. Fig. 15 shows the law of hardening discovered by
Phillips et al. [14] which is explained below.

Suppose that the yield surface in the σ_y, σ_z, τ stress space at some constant
temperature and at some level of prestressing is given by S_i and that the
stress path which is responsible for the surface S_i terminates at the stress
point P_i. Note that the point P_i does not lie on the surface S_i but outside S_i.
We retrace the stress path leading to P_i backwards until we reach some
arbitrary position R_i inside S_i. Since any motion of the stress point inside
or on S_i will not alter S_i let us select a new arbitrary position Q_i inside or
on S_i not necessarily on the stress path leading to P_i. Suppose now that
additional prestressing is generated by arbitrary rectilinear motion of the
stress point from Q_i inside S_i to a position P_{i+1} outside S_i. The position
P_{i+1} may of course be the same as P_i. Then the new yield surface S_{i+1},
corresponding to the stress path terminating at P_{i+1}, is generated from
the surface S_i by a superposition of a rigid body translation in the direction
of prestressing Q_iP_{i+1} and of a deformation in the same direction Q_iP_{i+1}
independently of the direction of the normal to the yield surface at the
interaction between the yield surface and path Q_iP_{i+1}. The effect of the
deformation is that the width of the yield surface in the direction of prestressing will be decreasing when the motion is away from the origin, and
be increasing when the motion is towards the origin. Rigid body translation
is determined by the motion of the curve $ABCD$ to its new position $A'B'C'D'$.
The motion of this curve generates a cylinder with its axis in the direction
of prestressing. This cylinder is tangential to both the original and to the
new yield surface at the curves $ABCD$ and $A'B'C'D'$, respectively. This new
law of hardening is valid for pure aluminum for every temperature within
the range tested. Figs. 16, 17 and 18 show clearly our results in combined
tension and torsion from [5]. Fig. 19 shows schematically the results in
combined tension, torsion, and internal pressure, from [13]. Fig. 20 shows
in the tension-internal pressure plane the exact results as obtained in Fig. 19
for combined tension, torsion, and internal pressure.

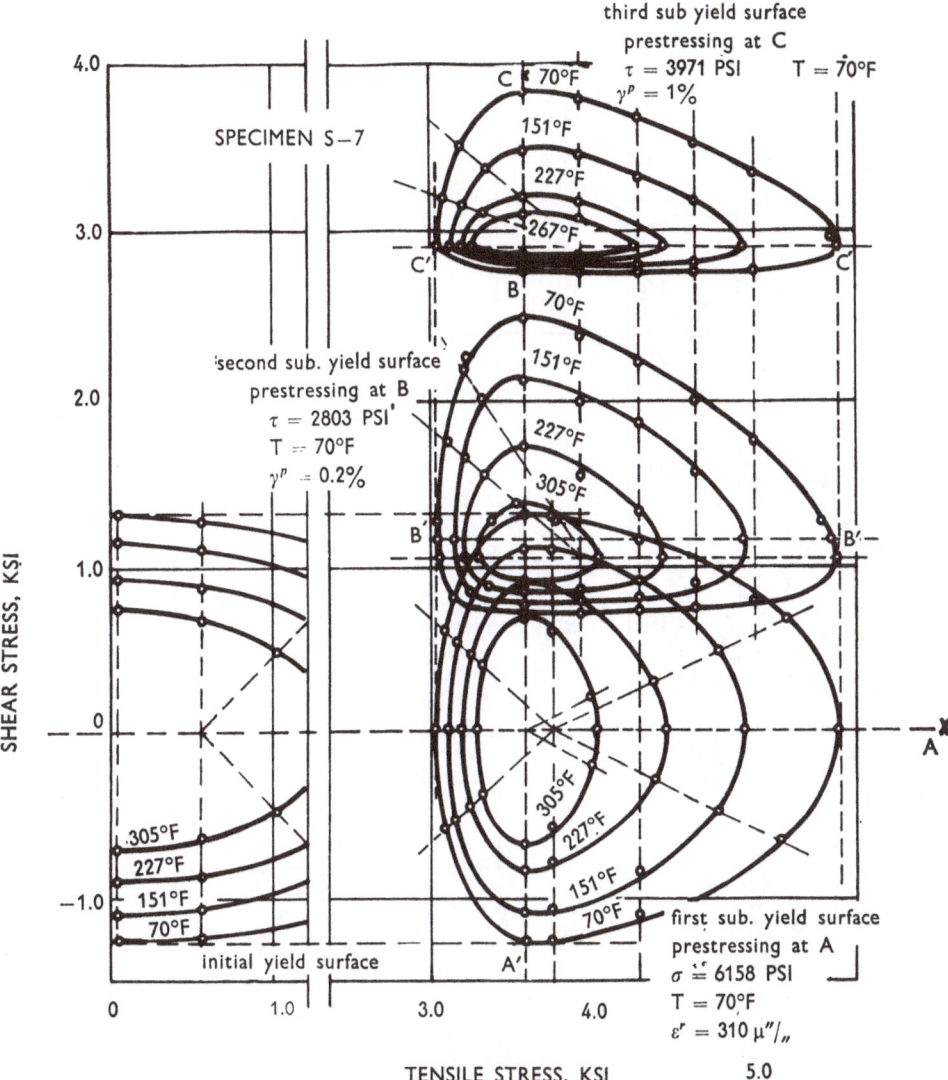

Fig. 16. Subsequent yield surfaces for pure aluminum for combined tension and torsion.

Fig. 16 shows that the maximum values of positive and negative shear stress occur at the same value of tensile stress for all temperatures. It is seen that the tensile stresses at which these maximum and minimum values of stress occur are independent of the temperature. This phenomenon, a variant of which is shown in Figs. 17 to 20 gives an additional basis to the

211

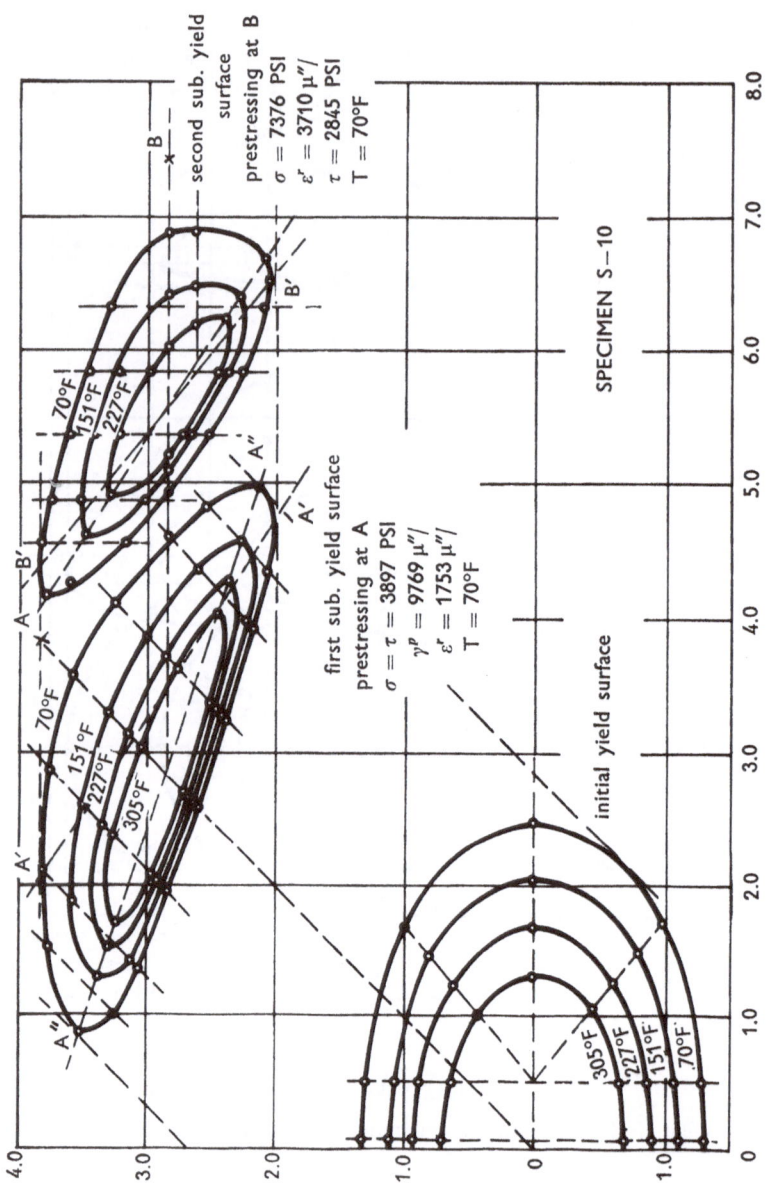

Fig. 17. Subsequent yield surfaces for pure aluminum for combined tension and torsion.

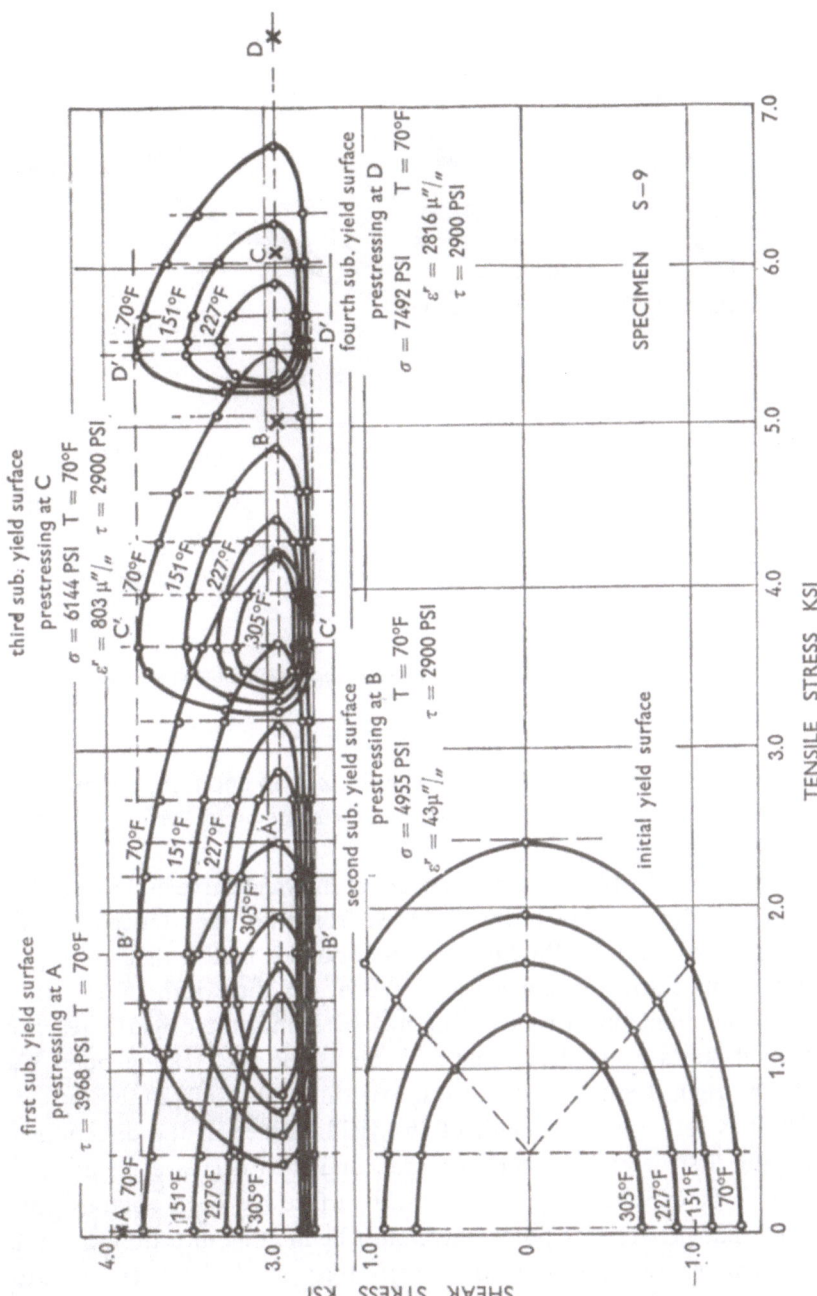

Fig. 18. Subsequent yield surfaces for pure aluminum for combined tension and torsion.

213

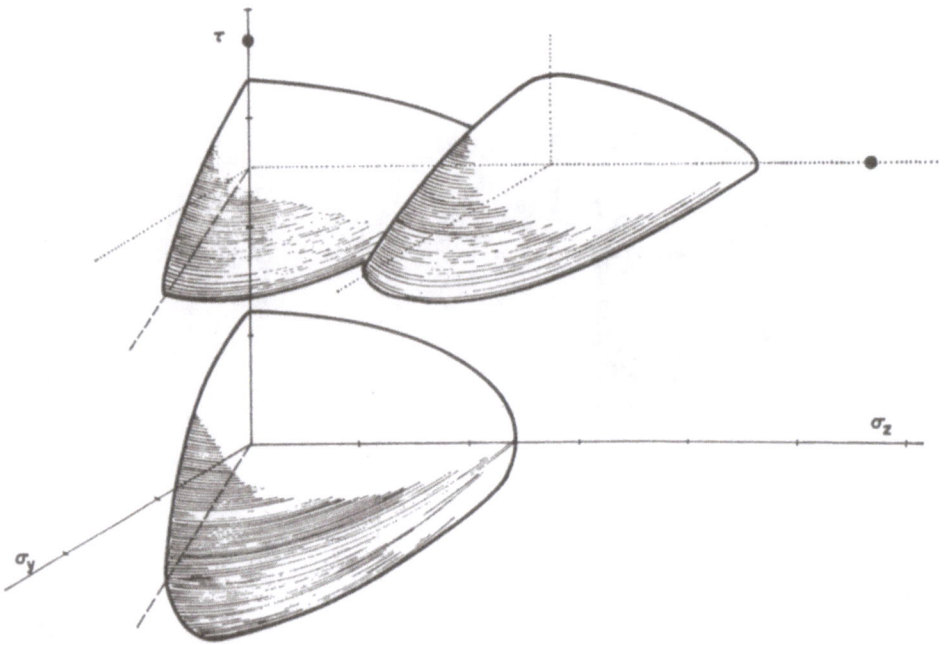

Fig. 19. A sketch of subsequent yield surfaces due to combined double tension and torsion.

analytical work given by Phillips et al. [14, 23, 27]. In addition, the previous figs. show that very soon after hardening starts the yield surface does not include the origin.

It is shown in Figs. 16 to 20 that the width of the yield curve in the direction of prestressing decreases the farther we move from the origin. Indeed in some still unpublished work by Phillips and co-workers it is seen that if the pre-stressing point starts moving towards the origin the width of the yield surface in the direction of prestressing increases until the origin is reached and then it starts again decreasing as soon as the prestressing point has passed the origin. Figs. 21 and 22 show experimental results in tension and torsion. It is seen that the width of the yield surface in the direction of prestressing first decreases, then increases, and then decreases again. It is therefore to be expected that when a yield surface has been obtained, due to a particular prestressing point, there must be limiting directions in stress space which separate the directions of prestressing of yield surfaces with decreasing width from the directions of increasing width yield surfaces.

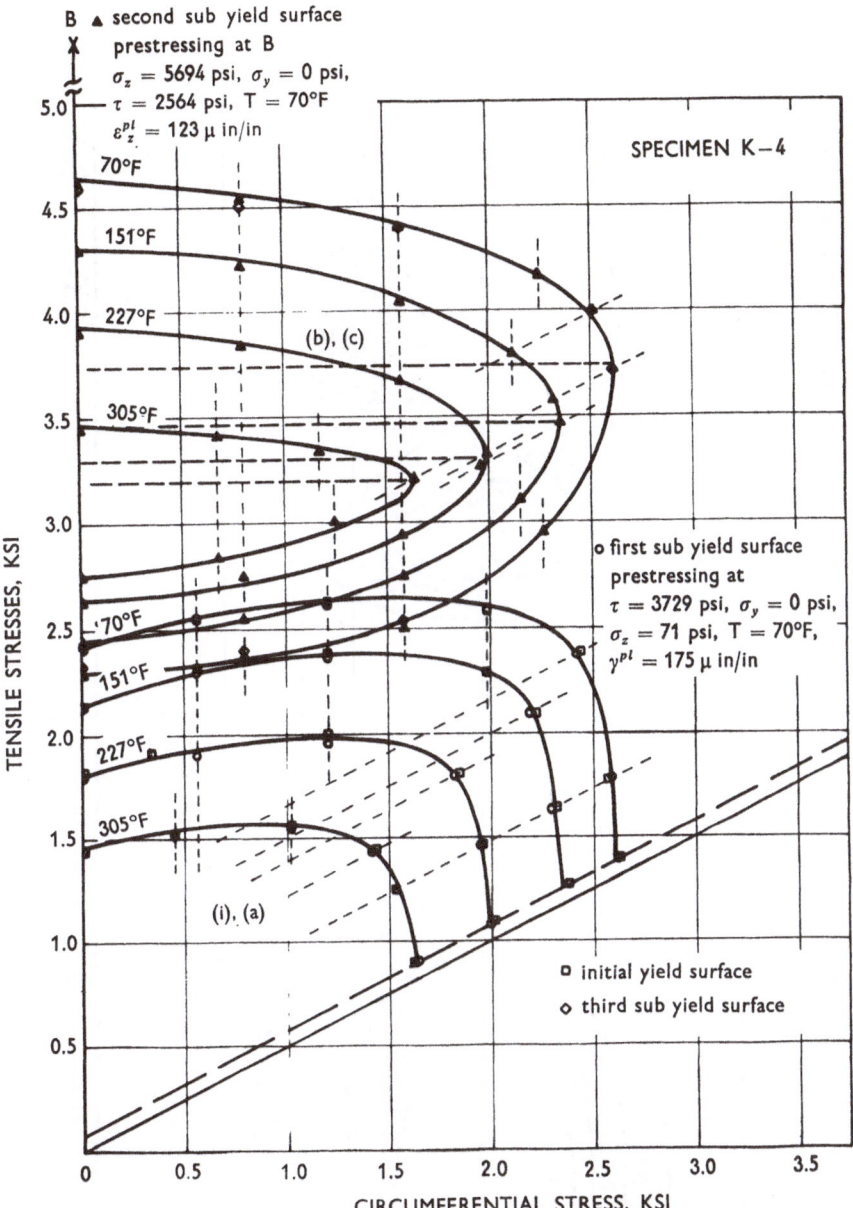

Fig. 20. Subsequent yield surfaces for pure aluminum for combined tension-tension.

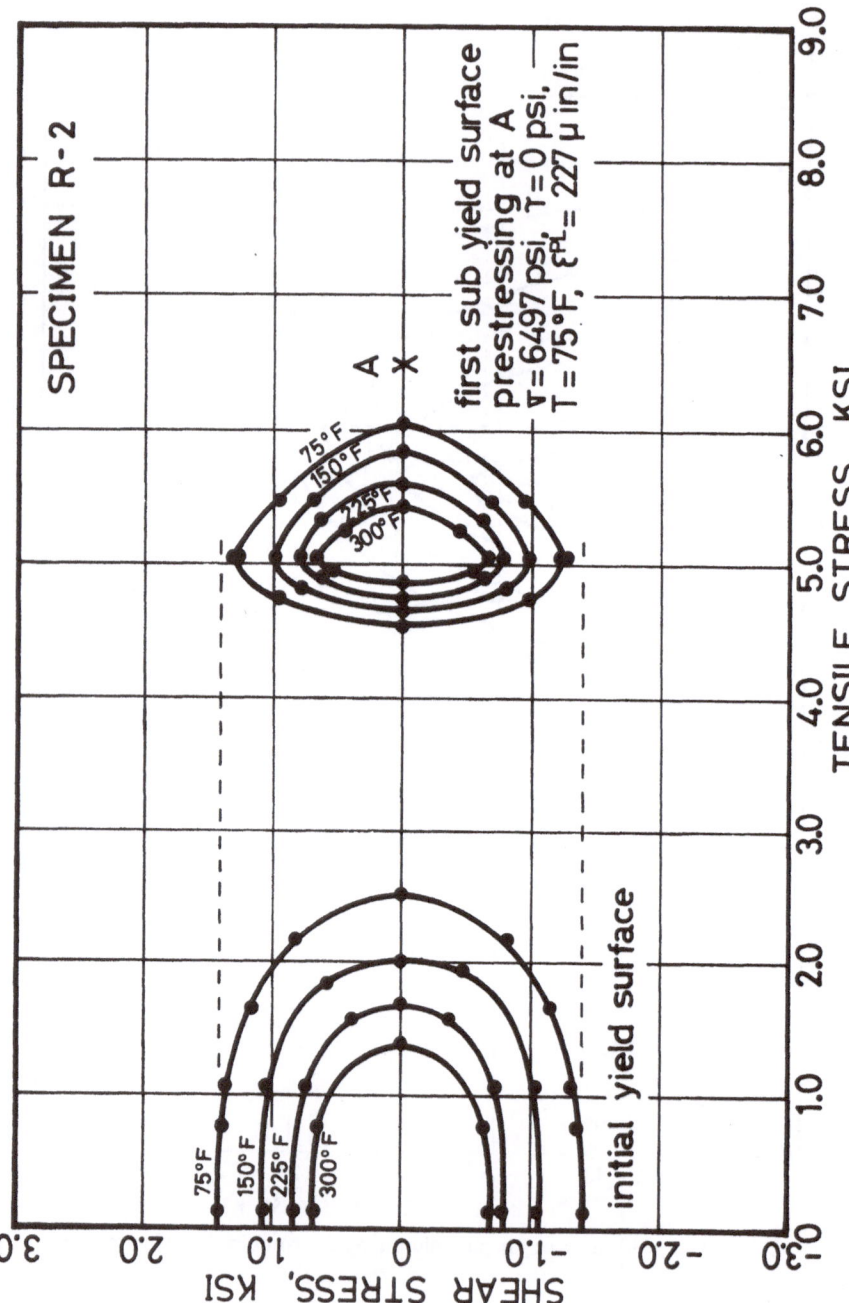

Fig. 21. Initial and subsequent yield surfaces for pure aluminum for combined tension and torsion.

Experimental plasticity

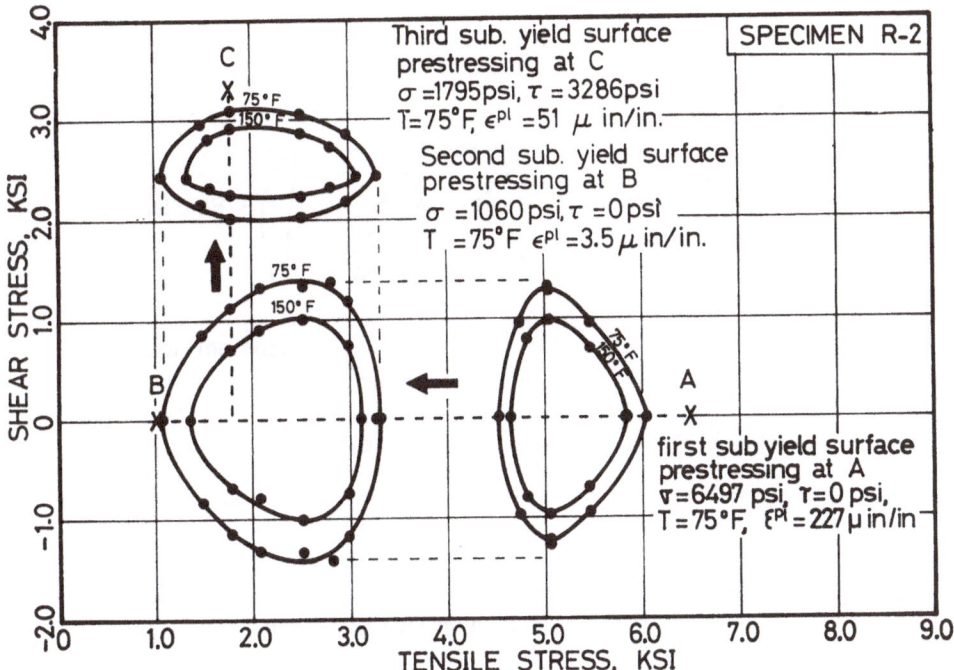

Fig. 22. Subsequent yield surfaces for pure aluminum for combined tension and torsion. Continuation of experiment in Fig. 21. Note that the width of the yield surface in the direction of prestressing first decreases then increases and then decreases again.

Fig. 23 illustrates this concept. The experimental determination of the corresponding limiting surface will be of great interest.

In Fig. 24 we see for prestressing in the τ-direction the intersections of the yield surfaces with the τ, T plane. With prestressing the width of the yield surface first decreases then upon reversal of prestressing direction it increases again until approximately the origin is reached; then for prestressing beyond the origin the width of the yield surface decreases again. We observe that for either direction of prestressing, the intersections change from straight lines to increasingly concave curves as prestressing increases. This phenomenon is shown in Fig. 6 also for prestressing in tension. This concavity may be of significance to thermodynamic considerations in plasticity.

Experiments [5, 6, 13, 14] are, to the author's knowledge, the only ones where the effect of constant or variable temperature on the yield surface has been investigated. It is important to investigate this effect, at levels of temperature which are outside the range investigated up to now.

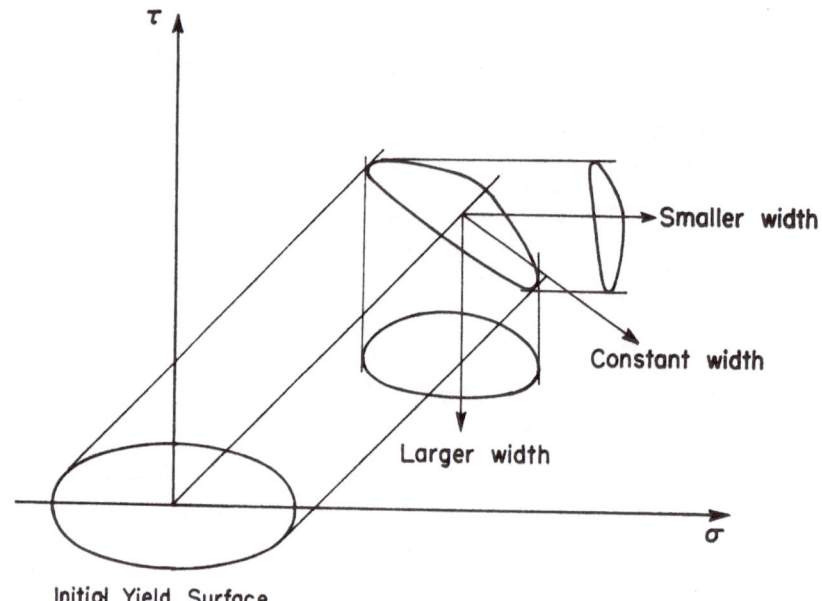

Initial Yield Surface

Fig. 23. The concept of the limiting direction.

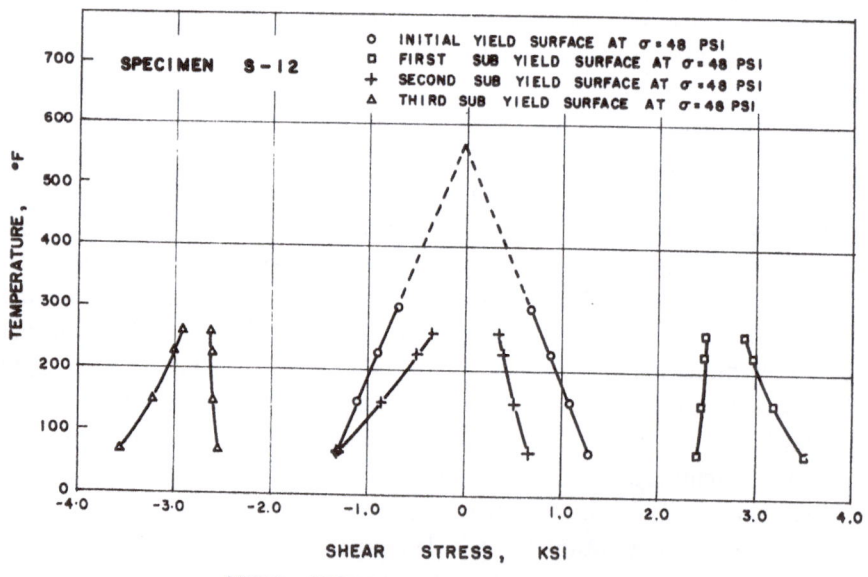

CROSS-SECTION OF YIELD SURFACES

Fig. 24. Concavity in the yield surface.

Since the definition of yielding is very important for the results obtained, as correctly predicted by Mair and Pugh [26], it is uncertain how the many experiments done in the past could be interpreted. This is also true for hydrostatic pressure effects, of or their absence, see for example, Bridgman [28], Hu et al. [29], Crossland [86] and Pugh [87].

Experiments by Miastkowski and Szczepiński [30, 35, 36], Williams and Svensson [15], as well as others [31, 32, 33, 34, 37, 38] determine yield surfaces based on definitions of several amounts of proof strain. These experiments may be of significance in the determination of a loading surface in a stress space, the question of linearity and the question of the flow rule. These questions will be considered in the next section.

A final word of caution. The experiments described in [5, 6, 13, 14] have been made with a specially designed very accurate dead-load testing machine and pressure equipment and with highly accurate and very carefully installed SR-4 gages. It has been the experience of the author and his colleagues that the results, such as the law of hardening presented in this section, depend to a large extent on whether all the elements of the experimental set-up are working properly.

5. Loading surfaces – normality – linearity – corners

In most investigations concerning a yield surface the definition of yielding is not that of the proportional limit but of the locus of points where an arbitrary amount of offset strain has been accumulated. Let us consider this concept a little closer. Consider loading in a given direction of stress. The yield point in that direction is assumed to be the stress where the offset strain is equal to some arbitrary value set beforehand (f.e. 0.5 %). If we obtain enough such points in different directions the locus of all these points will traditionally be called a yield surface. In reality it is an offset yield surface. The offset yield surface is not the limit of the elastic region but it is larger than the elastic region because it inclides a region where during motion of the stress points plastic strain up to say, 0.5 %, can be obtained. The offset yield surface depends on the paths used for obtaining it since the magnitude of offset strains will differ with the path used. In addition we must decide what we mean by a combined strain offset equivalent to a simple strain offset; in practice we must assume a theory of plasticity before the measurements will start and thus we introduce a certain bias towards the assumed theory.

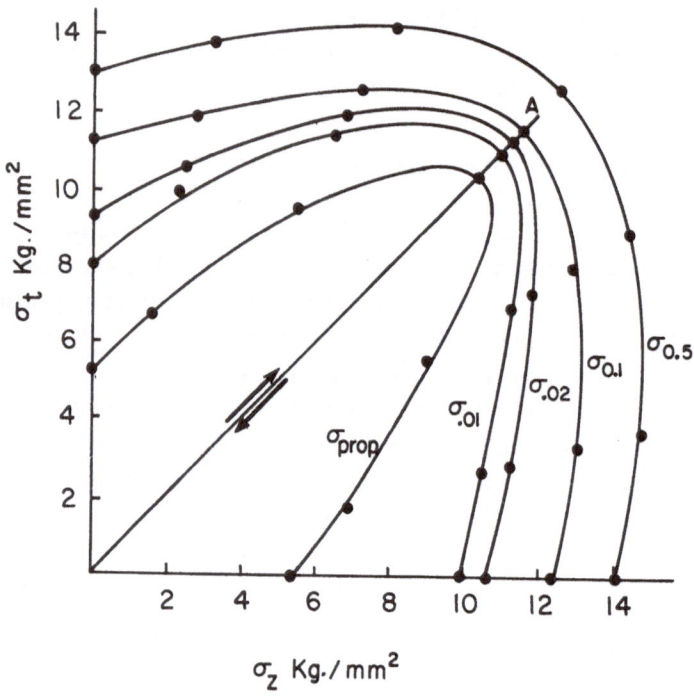

Fig. 25. A sequence of offset yield surfaces for prestressing to A (from Miastkowski and Szczepinski [30].

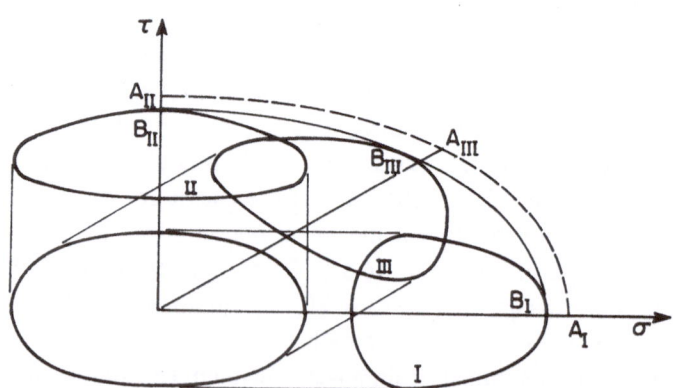

Fig. 26. The envelope of all proportional limit yield surfaces corresponding to a given offset yield surface.

Since usually the Mises theory is the assumed theory it is understandable why in such cases the Mises theory seems to be the one predicted by the experiments.

Fig. 25 from [30] gives a sequence of offset yield surfaces for prestressing to A and shows the influence of the amount of strain offset selected. As the offset value increases the surfaces resemble increasingly to the Mises surfaces.

To obtain the offset yield surfaces we cannot use one specimen only. We have to use one specimen for each point of each offset surface. The result is an inevitable influence from scattering of the results.

Consider the procedure of obtaining the offset yield surface and consider also the proportional limit yield surfaces defined previously, Fig. 26. We see that we have a certain conceptual relationship between these two yield surfaces. When loading in the σ-direction we obtain a proportional limit yield surface I for the loading point A_I. When loading in the τ-direction with another specimen we obtain for the loading point A_{II} (say an equivalent one to A_I by some offset theory) the proportional limit surface II. Finally when loading with a third specimen in an intermediate direction III to the loading (equivalent) point A_{III} we obtain the proportional limit surface III. The relationship between A_I, A_{II}, A_{III} is one of the same equivalent offset strain so that the surface passing through A_I, A_{II}, A_{III} is the offset yield surface. To this offset yield surface corresponds the envelope B_I, B_{II}, B_{III} of all the proportional limit yield surfaces corresponding to the offset yield surface by means of radial loading and of course to practically the same equivalent plastic strain. This envelope is probably normal to the strain rate vectors and is probably parallel to the offset yield surface which in [21] has been denoted as a loading surface since it passes through the loading point.

Experiments concerning the normality between the quasistatic yield surface and the strain-rate vectors are needed. Some experimental work concerning the normality between the offset yield surface and the strain rates is reported in [30, 31].

The question has been raised in the past concerning the existence of corners in the yield surface. As mentioned by the author [39] it is not feasible to show the existence of corners experimentally in a direct way although such corners may exist and may be shown indirectly [39, 40]. In fact experiments indicate some lack of linearity at the prestressing point and the existence of corners can be deduced. On the other hand other authors [41] indicate the existence of linearity and the absence of corners. The theoretical work of Lin [1] predicts corners and the slip theory [42] postulates a corner at the prestressing point. However, these theories are based on simplifying assumptions and the question may be raised whether the simplifying assumptions are not the ones which generate the corners. (See also the work of Hill [84] and Havner [85]). On the other hand the experimental work shown in Figs. 16 to 18 leads to the tentative conclusion that if corners exist they might also exist in positions different from those occupied by the prestressing points. Fig. 13 also shows that if with time the point *D* moves to reach the position *E* and at the same time the width of the yield surface normal to *DE* does not change we shall have ultimately a corner at *E*. Much depends on the growth of the yield surface with plastic strain.

Attempts to verify the theory of plasticity by experiments of a nature different than the previously considered have been made by several authors [43, 44, 45].

A very different theory of plasticity has been proposed by Ilyushin [46]. This theory is based on a representation in a strain space and on a postulate of isotropy and a principle of delay. Experimental work concerning the basic concepts of this theory as well as the theory as a whole has been reported by Lensky [47]. It will be of interest to compare the experimental findings of the author's group with the predictions of Ilyushin's theory particularly since the experiments of the author have been made with a stress machine while those of Ilyushin's group (to the author's knowledge) with a straining machine. For this comparison some theoretical work by Pipkin and Rivlin [48] may be of importance.

6. Strain rate and creep

We are here interested only in some experimental aspects of the influence of rate of straining on the constitutive equations. For two recent excellent surveys of this topic see Cristescu [49, 50].

The existence of a rate effect on the stress-strain curve of metals has been shown experimentally when stress-strain curves have been obtained at different strain rates. See for example Fig. 27 from [51] where the influence of strain rate is obvious. Stress-rate effect has been assumed existing from the beginning of this lecture by stating that plastic strain needs time to appear. The fact that the loading point is outside the equilibrium yield surface is also due to a stress rate effect.

An important criterion for the existence of a rate effect is the behavior

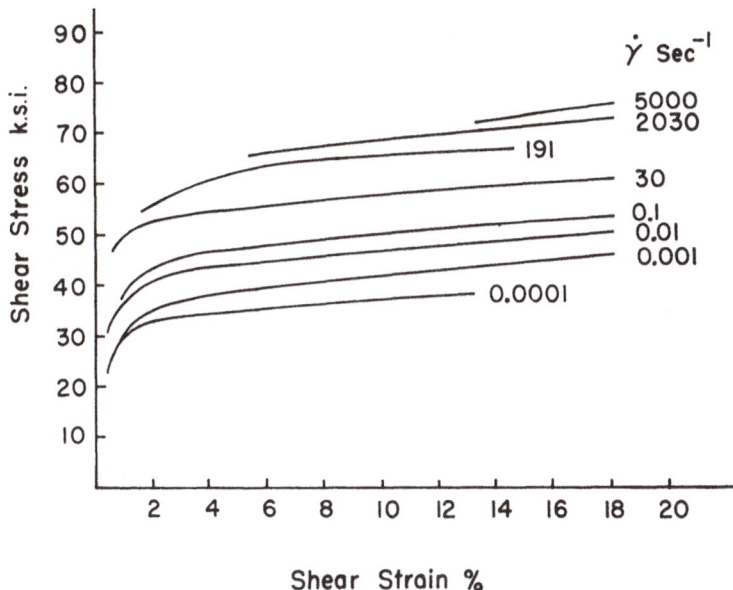

Fig. 27. The influence of strain-rate on the stress-strain curves (from Lawson and Nicholas [51].

of a prestressed bar when the load is abruptly increased by a very small amount either in a stress machine or by impact (stress wave propagation). The experiments show that the new stress strain curve has initially an elastic slope and that the wave propagates with elastic speed. These phenomena show that there is a rate effect. In the literature the statement has been made that for some metals the strain-rate effect is small. It is immaterial from the fundamental point of view how large this effect is. It is important that it exists although experimentation concerning this effect may preferably be concentrated on metals for which it is substantial.

When the additional load on a prestrained bar is of finite magnitude the stress-strain curve has the form of Fig. 28 from [52]. Such a behavior is compatible with the existence of a rate effect.

Extensive experimental work concerning rate-effects on metals is under way. A number of experimental techniques [53, 54, 55] have been developed concerning material behavior at high rates of strain. Each of the techniques available has its positive and negative sides [56, 57]. The influence of the strain rate history is obviously of importance and is currently under intensive investigation [58, 93].

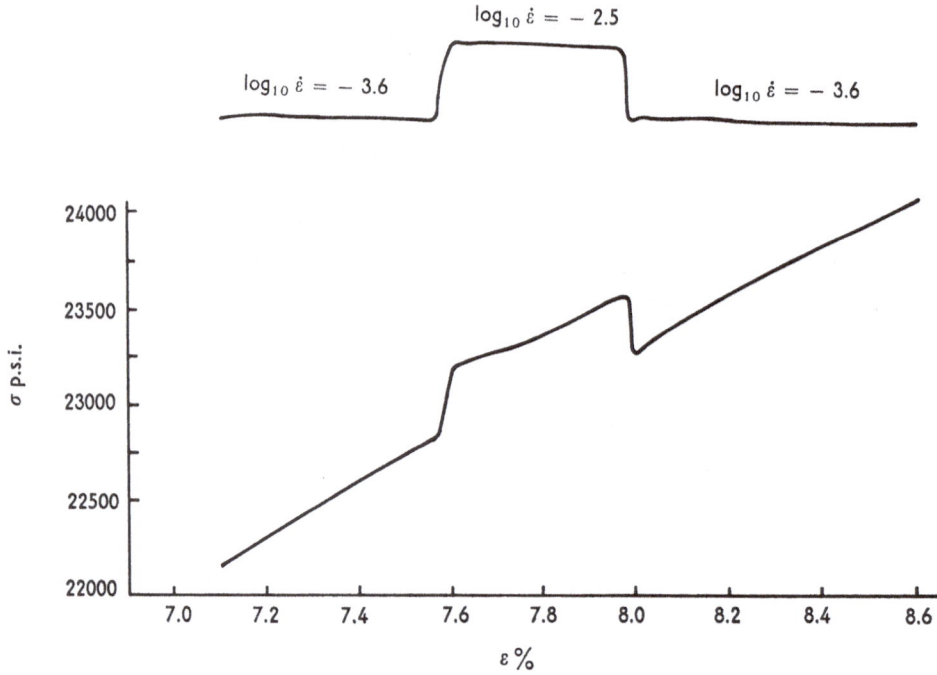

Fig. 28. Effects of rapid increase and decrease of finite magnitude loads on a stress-strain curve (from Lequear and Lubahn [52])

Distinguishing between the different theories by measuring out the wave profile characteristics is a nearly hopeless task, since as Ripperger [59] has shown such wave profile characteristics are not particularly sensitive as indicators of the form of the constitutive relationship. A better alternative may be to find by experimentation with a computer phenomena which are peculiarly due to particular theories or versions of theories and then experimentally try to observe if such phenomena really occur. In this way it will be possible to select between classes of theories. Experimentation by computer has been presented in the above mentioned paper by Ripperger [59] and in some old and forthcoming papers by the writer and his colleagues [60, 61, 62]. In particular [61] it has been shown theoretically that the constant strain plateau at the impact end of the bar predicted by the rate-independent theory is likewise predicted by the rate-dependent theory.

Another interesting question is whether and to what extent rate-effects are existing during unloading or reversed loading. As our yield surface experiments show there is a substantial Bauschinger effect; there is plastic deformation during unloading even while the stress is positive. It follows that there may exist rate effects during unloading while the stress is still of the same sign as during loading [50]. For experiments concerning reversed loading see [63].

The behavior of materials during combined stresses under dynamic loading is also of current interest. Experimental work on combined stress wave propagation has been done by Lindholm [64, 92] who has designed a tension-torsion machine for dynamic loading and by Lipkin and Clifton [65]. Additional experimental work in two dimensions is necessary, coupled with theoretical work based on *real* proportional limit yield surfaces generated as indicated in the previous sections.

The definition of the dynamic yield surface for strain-dependent materials is an interesting question. Let us consider again the quasistatic stress-strain curve and a stress-strain curve obtained under a constant stress rate different from zero. The second curve will be elevated with respect to the quasistatic stress-strain curve. For the same plastic strain there will be two values of stress, the quasistatic one and the dynamic one. The dynamic stress is one point in the stress space which is part of the dynamic yield surface corresponding to a given equivalent stress rate and a given value of equivalent permanent strain. The corresponding definition of the dynamic yield surface is a generalization of the offset yield surface to dynamic conditions and consequently it suffers from the same drawbacks as the offset yield surface. In particular the definitions of the combined equivalent stress rate and of the combined equivalent permanent strain give preferential treatment to particular theories. Of interest is also the relationship between the different yield curves when the stress rate is abruptly changing either by an increase or by a decrease in the stress rate. These questions are of importance to the theories of viscoplasticity; see for example Perzyna [66, 67] and Phillips and Wu [68].

An extreme case of stress rate effect is its absence which is the case of creep. We shall not discuss here the large literature on experiments on creep in uniaxial cases, since Rabotnov [69], Odqvist [70], Finnie [71], as well as Boresi and Sidebottom [72], give excellent surveys of these cases. We shall only point out that one of the fundamentally important problems in uniaxial creep of metals is the case when the stress and/or the temperature changes. Fundamental experiments with variable stress and temperature are needed by means of which the basic assumptions will either be proven or disproven. One of the questions to be considered is for example the amount of stress or temperature reduction necessary to terminate creep temporarily or permanently and the correlation of these findings with relaxation experiments [73, 74]. It is quite evident that from a fundamental point of view both the strain-hardening and time-hardening theories of creep are much too simple to represent the real behavior of metals. A beginning towards a more general theory has been given by Rabotnov [75] although his theory is a straight generalization of the strain-hardening theory of creep. Experimentation is needed to guide us towards a correct theory.

For creep in combined stress a large number of experiments have been made, particularly by Johnson [76], Nikishara et al. [77], Kennedy et al. [78], Findley [79], and others. These experiments generally try to prove or disprove the assumption that the strain rate vector is always normal to the Mises of Tresca surface and try to compare the total creep strains experimentally obtained with those predicted by means of an available isotropic theory of creep. Available creep theories are not yet satisfactory from a fundamental point of view: they will only be of an empirical nature until their basic assumptions will have been proven correct by experiment under a great variety of conditions.

In view of the anisotropy generated by plastic flow, as indicated in the previous sections, it should be expected that such anisotropic behavior occurs in creep also [80]. Therefore, the previous tests should be re-evaluated and new carefully formulated experiments should be performed to see whether the existing disagreements between theory and experiment are not due to the development of anisotropy during creep. The already mentioned experiments by Findley's group [79] give a hint that this may indeed be the case.

Related to the topics of this section is also the phenomenon of superplasticity which is of great technological interest and is discussed in a review by Underwood [88]. Experimental research on the constitutive relations applicable to this phenomenon, in the spirit described in this paper, is needed.

The work of Dillon [89, 90, 91] on the dynamic behavior of aluminum and particularly on serrated stress-strain curves, and on the heat generated during plastic deformation should also be mentioned because of its importance for future research.

7. Conclusions

As final comments I may add that quite drastically different theories of plasticity have been introduced recently, and in the past. The recent ones are based on the concept of hypoelasticity [81] and on the concept of an intrinsic time [82]. In order to verify their validity it is necessary to devise experiments in which their fundamental assumptions are tested. The testing of the validity of any theories on the basis of the total produced strain in a limited number of paths is not sufficient since many alternative theories of several types may produce approximately the same amounts of strain for those limited paths. The same criticism can be leveled at early experiments in plasticity. The experimentation must be directed towards verifying the structure of the theory instead of being of a semi-empirical nature.

The substantial number of excellent papers on experimental plasticity presented in this symposium shows that there is a very active interest for the subject of the foundations of plasticity from the experimentalist's point of view. It is the hope of the author that this lecture may contribute to keep this active interest alive.

Acknowledgement. This research has been supported in part by the National Science Foundation of the U.S.A. Government.

Bibliography

[1] Lin, T. H., Physical theory of plasticity. *Advances in Applied Mechanics*, vol. 11, (C. S. Yih, ed.), 256–312. Academic Press., New York, 1971.

[2] Brown, N., Observations of microplasticity. In *Advances in Material Research*, vol. 2, (C. J. McMahon, Jr., ed.) 45–73. Interscience, New York, 1968.

[3] Phillips, A., Some thoughts on the solution of a plastic buckling paradox. *Journal of AIAA*. July 1972, 951–953.

[4] Hutchinson, J. W. and W. T. Koiter, Postbuckling theory, *Applied Mechanics Reviews*, 23 (1970) 1353–1366.

[5] Phillips, A. and J. L. Tang, The effect of loading path on the yield surface at elevated temperatures. *Int. J. Solids Structures*, 8 (1972) 463–474.

[6] Phillips, A., K. Liu, and W. J. Justusson, An experimental investigation of yield surfaces at elevated temperatures. *Acta Mech.*, 14 (1972), 119–146.

[7] Taylor, G. I. and H. Quinney, The plastic distortion of metals. *Phil. Transactions Royal Society*, London, Ser. A (230), (1931) 323–362.

[8] Lode, W., *Z. Physik*, 36 (1926) 913–939.

[9] Paul, B., Macroscopic criteria for plastic flow and brittle fracture. In *Fracture*, vol. II (edited by Liebowitz), 313–496. Academic Pres, New York, 1968.

[10] Prager, W., Recent developments in the mathematical theory of plasticity. *Journal of Applied Physics, 20* (1949) 235–241.

[11] Drucker, D. C., A more fundamental approach to plastic stress-strain relations. *Proceedings 1st USA Nat. Congress of Applied Mechanics*, 487–491. ASME, New York, 1951.

[12] Hill, R., *Plasticity*, Oxford University Press Oxford, 1950.

[13] Phillips, A. and R. Kasper, On the foundations of thermoplasticity – an experimental investigation. *J. Appl. Mech.* (in press) (1973).

[14] Phillips, A., Yield surfaces of pure aluminum at elevated temperatures. *IUTAM Symposium in Thermoinelasticity* (B. Boley, editor), 1968, 241–258, Springer, Berlin, 1970.

[15] Williams, J. F. and N. L. Svensson, Effect of torsional prestrain on the yield locus of 1100-F aluminum, *Journal of Strain Analysis, 6*, (1971) 263–272.

[16] Prager, W., A new method of analyzing stress and strains in workhardening plastic solids. *J. Appl. Mech., 23* (1956) 493–496.

[17] Hodge, P. G., Jr., Discussion of Ref. 16. *J. Appl. Mech. 24* (1957) 482.

[18] Ziegler, H., A modification of Prager's hardening rule, *9. Appl. Math., 17* (1959) 55–65.

[19] Baltov, A. and A. Sawczuk, A rule of anisotropic hardening. *Acta Mech., 1* (1965) 81–92.

[20] Ishlinskii, I. U., General theory of plascticity with linear strain hardening, *Ukr. Mat. Zh., 6* (1954) 314–324.

[21] Phillips, A. and R. L. Sierakowski, On the concept of the yield surface. *Acta Mech., 1* (1965) 29.

[22] Justusson, J. W. and A. Phillips, Stability and convexity in plasticity. *Acta Mech., 2* (1966) 251.

[23] Eisenberg, M. A. and A. Phillips, A theory of plasticity with non-coincident yield and loading surfaces. *Acta Mech., 11* (1971) 247–260.

[24] Naghdi, P. M., F. Essenburg, and W. Koff, An experimental study of initial and subsequent yield surfaces in plasticity. *J. Appl. Mech., 25* (1958) 201–209.

[25] Ivey, H. J., Plastic stress-strain relations and yield surfaces for aluminum alloys. *J. Mech. Eng. Sci., 3* (1961) 15–31.

[26] Mair, W. M. and H. Ll D. Pugh, Effect of prestrain on yield surfaces in copper. *J. Mech. Eng. Sci., 6* (1964) 150–163.

[27] Phillips, A. and M. Eisenberg, Observations on certain inequality conditions in plasticity. *Int. J. Nonlinear Mechanics, 1* (1966) 247–256.

[28] Bridgman, P. W., The compressibility of thirty metals as a function of pressure and temperature. *Proc. Amer. Acad. Arts Sci., 58* (1923) 166.

[29] Hu, L. W., J. Markovitz and T. A. Bartush, A triaxial-stress experiment on yield condition in plasticity. *Exper. Mechanics.*, January (1966).

[30] Miastkowski, J. and W. Szczepinski, An experimental study of yield surfaces of Prestrained brass. *Int. J. Solids Structures, 1* (1965) 189–194.

[31] Bertsch, P. K. and W. H. Findley, An experimental study of subsequent yield surfaces,

corners, normality, Bauschinger and allied effects. *Proc. 4th U.S. Nat. Congress Appl. Mech. 893–907. ASME*, New York, 1962.

[32] Dudderar, T. D. and J. Duffy, Neutron irradiation and the yield surfaces in copper. *J. Appl. Mech., 84* (1967) 200–206.

[33] Talypov, G. B., Investigation of influence of preliminary plastic strain and natural aging on behavior of low carbon steel. In *Investigation on Elasticity and Plasticity*, (in Russian) Izdatel'stvo Leningradskovo Universiteto, Leningrad 1963.

[34] Shiratori, E. and K. Ikegami, Experimental study of the subsequent yield surface by usung cross-shaped specimens. *J. Mech. Phys. Solids, 16* (1968) 373–394.

[35] Szczepiński, W. and J. Miastkowski, An experimental study of the effect of the prestraining history on the yield surfaces of an aluminum alloy. *J. Mech. Solids, 16* (1968) 153–162.

[36] Szczepiński, W., On the effect of plastic deformation on yield condition. *Arch. Mech. Stos., 15* (1963) 275–296.

[37] Miastkowski, J., Analysis of the memory effect of plastically prestrained material. *Arch. Mech. Stos., 20* (1968) 261–276.

[38] Szczepiński, W. and K. Turski, On the lines of influence of plastic deformation. *Arch. Mech. Stos., 21* (1969) 49–57.

[39] Phillips, A., Pointed vertices in plasticity. In *Plasticity, Proceedings 2nd Symposium Naval Structural Mechanics*, 202–214. Pergamon Press, Oxford, 1960.

[40] Phillips, A. and G. A. Gray, experimental investigation of corners in the yield surface. *Journal Basic Engineering, 83* (1961) 275–289.

[41] Paul, B., W. Chen, and L. C. Lee, An experimental study of plastic flow under step-wise increments of tension and torsion. In *Proceedings of the 4th U.S. National Congress of Applied Mechanics*, vol. 2, 1031–1038, ASME, New York, 1964.

[42] Batdorf, S. B. and B. Budiansky, A mathematical theory of plasticity based on the concept of slip. *NACA TN* 1871 (1949).

[43] Theocaris, P. S. and C. R. Hazell, Experimental investigation of subsequent yield surfaces using the moire method. *J. Mech. Phys. Solids, 13* (1965) 281–294.

[44] Budiansky, B., N. F. Dow, R. W. Peters, and R. P. Shepherd, Experimental studies of polyaxial stress-strain laws and plasticity. *Proc. 1st U.S. Nat. Congress, Appl. Mech.* 503–512, *ASME*, New York (1952).

[45] Phillips, A. and L. Kaechele, Combined stress tests in plasticity. *J. Appl. Mech., 23* (1956) 43–48.

[46] Ilyushin, A. A., *Prikl. Mat. Mekh., 18* (1954) 641–666.

[47] Lensky, V. S., Analysis of plastic behavior of metals under complex loading. In *Plasticity, 2nd Symposium Naval Structural Mechanics* 259–278, Pengamon Press, Oxford, 1960.

[48] Pipkin, A. C. and R. S. Rivlin, Mechanics of rate-independent materials. *Brown University Technical Report*, No. 95, Office of Naval Research Contract NONR 562 (10), November 1964.

[49] Cristescu, N., *Dynamic plasticity*. North-Holland, Amsterdam, 1967.

[50] Cristescu, N., Dynamic plasticity. *Appl. Mech. Reviews, 21* (1968) 659–668.

[51] Lawson, J. E. and T. Nicholas, The dynamic mechanical behavior of titanium in shear. *J. Mech. Phys. Solids, 20* (1972) 65–76.

[52] Lequear, H. A. and J. D. Lubahn, Some transient effects during creep and tensile test of the aluminum alloy 61 ST. *J. Metals* (1956) 497.

[53] Kolsky, H., An investigation of the mechanical properties of materials at very high rates of loading. *Proc. Phys. Soc.*, London *62* B (1949) 676–700.

[54] Bell, J. F., *The physics of large deformation of crystalline solids.* Springer, Berlin, 1968.

[55] Duffy, J., J. D. Campbell, and R. N. Hawley, 'On the use of a torsional split hopkinson bar to study rate effects in 1100–0 aluminum. *J. Appl. Mech.*, *38* (1971) 83–91.

[56] Jahsman, W. E., Reexamining of the Kolsky technique for measuring dynamic material behavior. *J. Appl. Mech.*, *38* (1971) 75–82.

[57] Nicholas, T., An analysis of the split Hopkinson bar technique for strain-rate-dependent material behavior. *J. Appl. Mech.* Paper No. 72-APM-26.

[58] Nicholas, Th. and J. N. Whitemire, The effects of strain rate and strain rate history of the mechanical properties of several metals. *AFML*-TR-70-218 (1971).

[59] Ripperger, E. A. and H. Watson, Jr., The relationship between the constitutive equation and one-dimensional wave propagation. In *Mechanical Behavior of Materials under Dynamic Loads* (Lindholm, ed.), 294–313. Springer, Berlin, 1968.

[60] Wood, E. R. and A. Phillips, On the theory of plastic wave propagation in a bar. *J. Mech. Phys. Solids*, *15* (1967) 241–254.

[61] Phillips, A., E. R. Wood, M. Zabinski, P. Zannis, On the theory of plastic wave propagation in a bar-unloading waves. *Intern. Journal Nonlinear Mechanics 8*, (1972) 1–16.

[62] Phillips, A., M. Zabinski, Spherical wave propagation in a viscoplastic medium. *Ingenieur-Archiv 41*, (1972) 367–376.

[63] Nevill, G. E., Jr. and C. D. Myers, Strain rate effects during reversed loading. *J. Mech. Phys. Solids*, *16* (1968) 187–194.

[64] Lindholm, P. S. and L. M. Yeakley, A dynamic biaxial testing machine. *Exp. Mech.*, *7* (1967) 1–7.

[65] Lipkin, J. and R. J. Clifton, An experimental study of combined longitudinal and torsional plastic waves in a thin-walled tube. *Proc. 12th Intern. Congress Appl. Mech.*, (Stanford 1968), Springer, Berlin, 1970. 292–304.

[66] Perzyna, P., The constitutive equations for rate-sensitive plastic materials. *Quart. Appl. Math.*, *20* (1969) 321.

[67] Perzyna, P., Fundamental problems in viscoplasticity. In *Advances in Applied Mechanics*, *9* (1966) 243–387.

[68] Phillips, A. and H. C. Wu, A theory of viscoplasticity. *Intern. Journal of Solids and Structures 9*, (1972) 15–30.

[69] Rabotnov, Iu, *Creep problems in structural members.* North-Holland, Amsterdam, 1969.

[70] Odqvist, F. G. J. and J. Hult, Kriechen der mehellischen Werkstoffe. Springer, Berlin, 1962.

[71] Finnie, I. and W. R. Heller, *Creep of engineering materials.* McGraw Hill, New York, 1959.

[72] Boresi, A. P. and O. M. Sidebottom, Creep of metals under multiaxial states of stress. *Nuclear Engineering and Design*, *18* (1972) 415–456.

[73] Onat, E. T. and T. T. Wang, The effect of incremental loading on creep behavior of

metals. *Proc. IUTAM Colloquium on Creep in Structures*, 125–136, Springer, Berlin, 1962.

[74] Rosen, A., Delay time during creep of aluminum. *Journal Inst. Metals, 99* (1971) 111–114.

[75] Rabotnov, Iu, On the equations of state of creep. *Proc. Joint Intern. Conf. on Creep.* (1963) 2–117 to 2–122.

[76] Johnson, A. E., J. Henderson and B. Khan, *Complex stress creep, relaxation and fracture of metallic alloys*, Edinburgh, H. M. Stationery Office, Department of Scientific and Industrial Research, National Engineering Laboratory (1962).

[77] Nikishara, T., K. Tanaka and Shima, Creep and tension and torsion. *Proc. 5th Japan Nat. Congress. Appl. Mech.*, 1955, Tokyo (1956).

[78] Kennedy, C. R., W. O. Harris and D. A. Douglas, Multiaxial creep studies on inconel at 1500°F. *ASME*, paper No. 58-A-193 (1958).

[79] Blass, J. J. and W. N. Findley, Short-time, biaxial creep of an aluminum alloy with abrupt changes of temperature and state of stress. *J. Appl. Mech., 38* (1971) 489–501.

[80] Berman, I. and D. H. Pai, A theory of anisotropic steady-state creep. *Int. J. Mech. Sci. 8* (1966) 341–352.

[81] Tokuoka, T., Yield conditions and flow rules derived from hypoelasticity. *Arch. Rat. Mech. Anal., 42* (1971) 239.

[82] Valanis, K. C., A theory of viscoplasticity without a yield surface. *Arch. Mech. Stos., 23* (1971) 517–551.

[83] Sewell, M. J., A general theory of elastic and inelastic plate failure. *J. Mech. Phys. Solids*, 12, pp. 279–297, (1964).

[84] Hill, R. J., *Surveys in mechanics* (G. I. Taylor 70th Anniversary Volume), 7–31, Cambridge University Press, Cambridge, 1956.

[85] Havner, K. S., A discrete model for the prediction of subsequent yield surfaces in polycrystalline plasticity. *Int. J. Solids Structures, 7* (1971) 719.

[86] Crossland, B., The effect of fluid pressure on the shear properties of metals. *Proc. Inst. Mech. Engrs.*, London, *168* (1954) 935.

[87] Pugh, H. L. D. and D. Green, The effect of hydrostatic pressure on the plastic flow fracture of metals. *Proc. Inst. Mech. Engrs.*, 1956.

[88] Underwood, E. E., A review of superplasticity. *Journal of Metals, 14* (1962) 914.

[89] Dillon, O. W. Jr., Waves in bars of mechanically unstable materials. *J. Appl. Mech., 33* (1966) 267.

[90] Dillon, O. W. Jr., The heat generated during torsional oscillations of copper tubes. *Int. J. Solids Structures, 2* (1966) 181.

[91] Dillon, O. W. Jr., Plastic deformation waves and heat generated near the yield point of annealed aluminum. In *Mechanical Behavior of Materials under Dynamic Loads* (Lindholm, ed.), 21–59, Springer, Berlin, 1968.

[92] Lindholm, U. S., Some experiments in dynamic plasticity under combined stress. In *Mechanical Behavior of Materials Under Dynamic Loads* (Lindholm, ed.), 77–95, Springer, Berlin, 1968.

[93] Klepaczko, J., Discussion of the incremental wave effects in a bar. *Arch. Mech. Stos., 24* (1972) 187.

DISCUSSION

J. Zarka[1]).   I should like to put some questions to Professor Phillips.

1. How long did you wait in order to find the elastic limit? What about the relaxation and creep?

2. Some of your field curves in the $(\sigma_{xx}, \sigma_{x\theta})$ plane were closed but only when σ_{xx} is positive; the Bauschinger effect was then so important that the elastic limit in the opposite direction was positive.

I have to say that Dr. Bui gave data for the whole plane ($\sigma_{xx} > 0$ and $\sigma_{xx} < 0$) in 1964 and 1966. One of his results has been shown in our communication at this symposium.

3. When the temperature was changed, did you change also your sample?

A. Phillips.   Dr. Zarka's questions 1 and 3 may be answered by references [5] and [6] where a very detailed explanation of the procedure used in the experimentation is given. The same sample was used for all temperatures and prestress levels; for example, for the case explained in Figure 16 all sixteen yield curves were obtained with the same specimen. Since the machine used was a dead-load one there is no relaxation. No difficulty was encountered because of creep.

I am pleased to know that Dr. Bui did obtain data for the entire stress plane as early as 1964. Unfortunately I do not have access to these data. The yield surface on the compression side of the stress plane should be exactly symmetrical to those on the tensile side. Figure 3 of the communication by Dr. Zarka, Bui, and Zaoui presented in this symposium shows an initial yield surface which is symmetrical to the shear axis, while the prestrained yield surfaces are similar to such as obtained by a number of investigators, for example, Williams and Svensson (*Journal of Strain Analysis 5* (1970) 128) who investigated both the tensile and compressive parts of the stress plane. A comparison of those results with Figure 3 in Dr. Zarka's communication gives the strong impression that the data in the communication have been obtained with a definition of the yield limit which is different from the one used in the present lecture. It is of course already mentioned in my lecture that the yield limit definition affects the yield surface to a considerable extent.

[1]) Ecole Polytechnique, Paris, France.

G. Gudehus[1]). It is interesting to compare the results reported in this lecture with measurements made with samples of dry sand under cuboidal deformations (to be published in the series 'Veröffentlichungen des Instituts für Bodenmechanik und Felsmechanik der Universität Karlsruhe').

Letting aside the problem of precisely finding yield points in experimental stress-strain-curves (which is the same for all elastic-plastic materials), yield surfaces for dry sand are cones in the space of principal stresses. For comparison with metals one has to consider section curves of these cones with a unit deviator plane. Similar to the behaviour of metals we have observed: the section curves are shifted and deformed by deformation histories, indicating an increasing degree of anisotropy. This anisotropy can be swept out by a reversal of the deformation path. The envelope of all these yield curves is an isotropic curve, however.

As the microscopic mechanism of plastic deformations in sand is completely different from those in metals, this striking similarity is a challenge for theoreticians: it should be possible to cast this typical anisotropic hardening behaviour of all plastic materials into one constitutive equation.

A. Phillips. The author is intrigued with the similarity between behavior of metals and behavior of sand mentioned by Professor Gudehus. To assess this similarity it is of importance, however, to know what is the real meaning of a yield surface in dry sand. Is it the boundary of the region of purely elastic deformation? If so a comparison between yield surfaces in metals and in dry sand is well taken. If, on the other hand, it is not, then we are considering two different concepts which have the same name and a comparison may not have much of a meaning.

---

[1]) University of Karlsruhe, Karlsruhe, GFR.

Mathematical methods in plasticity theory

H. G. Hopkins

University of Manchester Institute of Science and Technology
Manchester, England

A discussion is given of the application of mathematical methods to the study of stress and strain in plastically-deformed solids, especially metals. First, the purpose and value of mathematical methods is considered. Secondly, attention is given to the mathematical formulation of problems in continuum mechanics, that is, as systems of partial differential equations and finite equations and associated initial-value or boundary conditions for dependent (physical) variables as functions of independent (space and time) variables, and to the classification of the mathematical structure of these systems in terms of the classical Cauchy problem and characteristics theory. For systems expressed as first-order equations in any finite number of dependent variables that are functions of just two independent variables (one space and the other either space or time) their mathematical structure is best treated by means of algebraic eigensystem theory and hence, if applicable, by the matrix presentation of characteristics theory, which is simple, elegant and unifying. Next, this approach is illustrated by reference to a variety of problems involving either quasi-static or dynamic deformations in solids, attention being drawn in passing to such ideas as Riemann invariants, boundary-layers, simple waves, fans, similarity and shocks. Lastly, the interplay between the physical properties of solids and the mathematical structure of their representative equations is touched upon, and also some areas are indicated where further research is needed.

1. Introduction

The present purpose is to discuss briefly some aspects of certain mathematical methods that play an important role in the theoretical study of stress and strain in plastically-deformed solids, especially metals.

It is appropriate to begin with some remarks that concern the teaching of plasticity theory. New mathematical methods and concepts and ideas or re-discoveries and developments of older ones are most likely to emerge in research, rather than in teaching, and it is only subsequently that they become incorporated in lecture courses. It is, naturally, a question of the greatest importance to know just what are the basic mathematical methods

235

needed in plasticity theory and also the manner in which they are best organized for teaching. So far as the latter aspect is concerned, there are the usual two alternatives, namely, a separate treatment rather aside from applications or a unified treatment embracing both theory and applications. Each method of treatment has its own advantages and disadvantages, but certainly for engineering scientists the second is undoubtedly the one normally likely to be preferred. It may be noted that Hill [1] gives in a series of appendices details of mathematical methods whereas Prager and Hodge [2] review or quote mathematical methods in the main text. In the last decade or so, there has been a spate of books published on the mechanics of continuous media and some of these (for example, Malvern [3]) seem to recognize the desirability of reviewing or including in chapters and/or appendices particular mathematical methods, so that these are right to hand and presented in the manner needed for the reader. However, this said, consideration will now be given to mathematical methods in plasticity theory *per se*.

It is to be noted at the outset that the basic mathematical formulation of the physical problems of plasticity theory is given in terms of relations between tensors of various kinds and that these relations are normally not linear, for either geometrical or physical reasons or both. Thus, reverting briefly for the moment to considerations of the teaching of plasticity theory, the mathematical methods that are needed are immediately different both in character and in complexity from those which most students may be expected to have normally been exposed to in their earlier mathematics courses. There results, accordingly, the considerable challenge of the effective teaching of advanced mathematical methods of a nature hitherto unfamiliar to students.

This is not the time to argue for the place of mathematical methods, their purpose and value, in scientific and engineering subjects, like plasticity theory. This question has been discussed *in extensio* on numerous occasions, but its essentials are only too easily lost in philosophical or semantic detail. It is important, of course, although not easy, to attempt to steer a middle course that seeks to avoid the mistake of either undervaluing (most likely for engineers) or overvaluing (most likely for mathematicians) mathematical methods, and also important to recognize that plasticity theory is still generally in formative stages of development (although certain domains within it, for example, limit analysis, are achieving a more-or-less definitive form), so that the existing stock of mathematical methods is likely to be enlarged as time passes. It is strange that, despite the fact that plasticity is now scarcely a young science, there has been no comprehensive discussion of mathematical methods in plasticity theory. The performance of this task would certainly be extremely valuable but it would require altogether far more space (and detailed mathematical argument) than there is available in the present paper. It is not proposed to set out here a catalogue of mathematical methods – the most important ones undoubtedly would include those associated with tensors, matrices, partial differential equations, and extremum and variational principles – but rather to give attention primarily to those particular mathematical methods associated with algebraic eigensystem theory and characteristics theory in two dimensions. This choice is made here, since, in the present writer's view, characteristics theory, if applicable, does have a central importance because it furnishes the proper means of classification of the basic governing equations and also provides methods for their integration, analytically and numerically. Furthermore, this theory has its simplest, most elegant and unifying form when presented in terms of matrix theory. Characteristics theory is illustrated here by reference to a variety of problems involving either quasi-static or dynamic deformations in solids, attention being drawn in passing to such ideas as Riemann invariants, boundary-layers, simple waves, fans, similarity and shocks. Lastly, the interplay between the physical properties of solids and the mathematical structure of their representative equations is touched upon, and also some areas are indicated where further research is needed.

The notation adopted in this paper is standard and it is therefore for the most part not explicitly defined.

Part I

Classification of systems of partial differential equations in two independent variables

2. Algebraic eigensystem theory and characteristics theory

In problems of continuum mechanics, the totality of governing equations invariably comprises the following: (i) equations of conservation of mass, momentum and (total) energy (applicable to all continua whatever be their particular nature), (ii) constitutive equations (depending upon the particular nature of the continuum), and (iii) initial-value or boundary conditions (depending upon the particular problem in hand). Equations (i) and (ii) constitute the field equations. Normally, equations (i) apply throughout a region of the space being considered and are written as partial differential equations, that is, on the assumption that the field variables are continuous and differentiable, but otherwise they are written in finite form as jump or discontinuity equations; it is simply that the allowed mathematical expression of the physical conservation laws depends upon the smoothness of behaviour of the field variables. Here, we shall not be especially concerned with discontinuity conditions (but see, for example, Prager [4] and Hill [5]). The field equations comprising the conservation and constitutive equations furnish a set of partial differential equations for the dependent (physical) variables as functions of the independent (space and time) variables. This set may be written in various equivalent ways; for example, the constitutive equations if in finite (algebraic) form may be used to eliminate some of the field variables or they may be satisfied identically through the introduction of new field variables, or the conservation equations may be simplified through the introduction of potential-type or generating functions (cf. the Airy stress function as used to satisfy the two-dimensional plane-strain equilibrium equations). The question of the nature of the mathematical structure of these equations is of crucial importance. The viewpoint is properly taken that this structure is dependent upon the conditions under which the Cauchy problem may be formulated and solved (see, for example, Hadamard [6]). The Cauchy problem is simplest when the number of independent variables is restricted to just two (either both in space or one in space and the other in time). However, even then, the normal treatment of the problem tends to be complicated when either a large number of low-order equations or a small number of high-order equations is involved. In this paper, for reasons of simplicity, the restriction is made to the case of just *two* independent variables; also, it is assumed without restriction that the set of field equations comprises solely *first-order* partial differential equations. The mathematical structure of this set is then determined by the method of characteristics theory as presented by Courant and Hilbert [7] but developed in terms of algebraic eigensystem theory (cf. Hopkins [8] and Band [9], and see Wilkinson [10]). Such a presentation has the advantages of simplicity, elegance and unification.

2.1 *Single first-order equation*

It is instructive first to consider the very simple case of the Cauchy problem for a single first-order partial differential equation in a single dependent variable $u(x, y)$ that is a function of two independent variables x, y (either both 'space-like' or one 'space-like' and the other 'time-like'). The general, quasi-linear, inhomogeneous equation to be considered is of the form

$$P(x, y, u)u_x + Q(x, y, u)u_y + R(x, y, u) = 0, \tag{2.1}$$

where P, Q, R are given functions and subscripts denote partial differentiation with respect to the indicated variables. The Cauchy problem for (2.1) is posed as follows. Let $u(x, y) = U(s)$ on Γ, where Γ is a curve with equations $x = x(s)$, $y = y(s)$ (s being a parametric coordinate, such as arc length) in the (x, y)-plane, i.e.

$$u(x, y) = u(x(s), y(s)) = U(s) \text{ on } \Gamma \tag{2.2}$$

is the given (continuous and differentiable) boundary data for u. The Cauchy problem for u in some non-zero neighbourhood of Γ is said to be solvable if from the equation (2.1) and the data (2.2) it is possible to determine the first-order and all the higher-order derivatives of u with respect to x, y on Γ. In this case, the coefficients in the Taylor power-series expansion for $u(x, y)$ about an arbitrary point of Γ are known, and then within its radius of convergence, this Taylor series will generate a solution of (2.1) that satisfies (2.2). The differential expression formed from (2.2) is

$$u_x dx + u_y dy = U'(s)ds \text{ on } \Gamma, \tag{2.3}$$

where $U'(s) \equiv dU(s)/ds$. Then, (2.1) and (2.3) determine a unique solution for u_x, u_y on Γ if, and only if,

$$\begin{vmatrix} P & Q \\ dx & dy \end{vmatrix} \neq 0, \tag{2.4}$$

i.e.

$$P dy - Q dx \neq 0. \tag{2.5}$$

If (2.5) holds, then it is straightforward to show (assuming conditions of differentiability on P, Q, R, U) that all higher-order derivatives of u are found from solving equations formed by differentiation of (2.1) and (2.2); conversely, if (2.5) does not hold, then the first-order derivatives (and *a fortiori* the higher-order ones as well) are not found from the differential equation (2.1) and the prescribed data (2.2) (and the impasse is not avoided by the provision of higher-order boundary data). Thus, the condition (2.5) is a necessary one for the Cauchy problem to be solvable. The integrals of

$$Pdy - Qdx = 0 \qquad (2.6)$$

are called characteristics, and in general through every point of the plane (singular points, where $P = Q = 0$, excepted) there will pass just one characteristic, so that the characteristics form a one-parameter family. Also, since in general P and Q depend upon u, the characteristics are only definite when a definite solution u is specified. Thus, in this case, the characteristics may be said to 'float' with the solution; and otherwise they are 'fixed' and independent of the solution u (the equation (2.1) is then called 'semi-linear' rather than 'quasi-linear'). However, (2.1) always has a real family of characteristics associated with it, and if Γ coincides with a member of this family then the Cauchy problem is not solvable. Now let Γ be a characteristic C and let $x = x(s)$, $y = y(s)$ on Γ. Then,

$$0 = Pdy - Qdx = (Py_s - Qx_s)ds \text{ on } C$$

and hence (2.1) may be written as

$$0 = x_s(Pu_x + Qu_y + R) = P(x_s u_x + y_s u_y) + Rx_s,$$

$$0 = y_s(Pu_x + Qu_y + R) = Q(x_s u_x + y_s u_y) + Ry_s,$$

(which are equivalent relations if $(x_s, y_s) \neq 0$), i.e.

$$Pu_s + Rx_s = 0, \quad Qu_s + Ry_s = 0 \text{ on } C, \qquad (2.7)$$

which are clearly seen to be equivalent in view of (2.6). Equations (2.7) show the special feature that differentiation *only with respect to s* (and not with respect to n, where n is a coordinate measured normal to C) is involved, and hence an immediate formal integration is possible:

$$\int (Pdu + Rdx) = \text{const.}, \quad \int (Qdu + Rdy) = \text{const. on } C. \qquad (2.8)$$

Thus, on C, the differential equation (2.1) takes the form (2.7), which has the integral (2.8), and since neither (2.7) nor (2.2) provides information on u_n, it becomes quite obvious just why the Cauchy problem fails to be solvable; also, the data (2.2) must be consistent with (2.7). In fact, the differential equation (2.7) may then apparently permit discontinuities in u normal to C, but in this case (2.1) is not strictly valid on C itself and some ancillary condition is required to link the two solutions applying on either side of C. Let α be a suitable parametric coordinate for the characteristics C, i.e. on any particular C, $\alpha = $ const. Then, (2.8) may be written as

$$\int (Pdu + Rdx) = R_1(\alpha), \quad \int (Qdu + Rdy) = R_2(\alpha) \text{ on } \alpha = \text{const.} \quad (2.9)$$

If now the Cauchy data (2.2) are given on a non-characteristic curve, then (2.9) determines the solution for u throughout some region. The quantity $R_1(\alpha)$ (or $R_2(\alpha)$) is called a Riemann invariant, and it is constant on any one characteristic. It may be noted that (2.9) may have to be integrated numerically (certainly, when P, Q, R involve u) but since (2.7) may be approximated by the difference relations

$$P\Delta u + R\Delta x = 0, \quad Q\Delta u + R\Delta y = 0, \quad (2.10)$$

Δu is calculable for given Δx (or Δy), and so knowing $u(x, y)$, $u(x + \Delta x, y + \Delta y) = u + \Delta u$ is found, and then $dy/dx = Q/P$ at $(x + \Delta x, y + \Delta y)$ is also found. In practice, the numerical procedure is often iterative in form. Hence, the solution is developed progressively away from the non-characteristic curve on which the basic data are given. Of course, this process must terminate either when adjacent characteristics meet at a point or when characteristics form an envelope (a limit line). In particular, if adjacent characteristics run together, then in general $R_{1,2}(\alpha)$ become many-valued at the point of intersection; the derivatives u_x, u_y may then no longer be finite; and the equation (2.1) then fails to have a meaning and it must be replaced by some other condition.

As a simple example, consider the equation

$$uu_x + u_y = 0. \quad (2.11)$$

This is equivalent to

$$u = \text{const. on } dx/dy = u, \quad (2.12)$$

and if

$$u(x, y)|_{y=0} = f(x) \quad (2.13)$$

then it follows that

$$u = f(\xi) \text{ on } x = f(\xi)y + \xi \quad (2.14)$$

241

or, eliminating the parameter ξ,

$$u = f(x - uy),\tag{2.15}$$

which apparently is the solution of (2.11) that satisfies (2.13). Equation (2.15) is an implicit equation for u, and it clearly satisfies (2.13). On substitution of (2.15) into (2.11) it is found that

$$(1 + yf')(uu_x + u_y) = 0,\tag{2.16}$$

and hence (2.11) is satisfied if

$$1 + yf' \neq 0.\tag{2.17}$$

But, if (2.17) is not satisfied, then (2.11) does not follow, and it is easy to show that

$$1 + yf' = 0\tag{2.18}$$

corresponds to a situation in which two adjacent characteristics run together and the derivatives u_x, u_y become infinite so that (2.11) ceases to have meaning. In particular, if $f(x) = \sin x$ ($0 \leq x \leq \pi$) then the graphs of the functions $u^*(x) = u(x, y)|_{y = \text{const.}}$ for y increasing from 0 show a progressive flattening for $0 \leq x < \frac{1}{2}\pi$ and steepening for $\frac{1}{2}\pi < x \leq \pi$, and in fact $u^*(x) = u(x, 1)$ shows a vertical tangent at $x = \pi$. Beyond this stage the formulation of the problem would require modification, if a single-valued solution is to obtain. This type of breakdown of the solution of a nonlinear partial differential equation is quite common.

The above suggests that the type of problem as represented by (2.1) must under certain conditions be re-formulated. From a purely mathematical viewpoint, one possible procedure is as follows. If (2.1) is valid throughout a region $\mathscr{R}$ in the (x, y)-plane, then

$$\iint_{\mathscr{R}^*} (Pu_x + Qu_y + R)\mathrm{d}x\mathrm{d}y = 0,\tag{2.19}$$

where $\mathscr{R}^*$ is contained in $\mathscr{R}$. Now,

$$Pu_x + Qu_y + R = (Pu)_x + (Qu)_y - (P_x + Q_y)u + R,$$

and if $P_x + Q_y = 0$, $R = 0$ then (2.19) may be transformed by means of Gauss's divergence theorem into

$$\int_{\partial\mathscr{R}^*} (lPu + mQu)\mathrm{d}s = 0,\tag{2.20}$$

where l, m are the direction cosines of the outwards-drawn normal to $\partial \mathcal{R}^*$ (the boundary of $\mathcal{R}^*$) and s is arc length measured along $\partial \mathcal{R}^*$. Now, if (2.20) is valid for arbitrary $\partial \mathcal{R}^*$ and if u is continuous and differentiable, then reversal of the argument shows that (2.1) is valid. However, (2.20) is valid when u is not continuous and differentiable, and hence applies when (2.1) does not apply. In effect, the strong integral condition (2.20) is a generalization of the weak differential condition (2.1). Next suppose that u is discontinuous across a curve Γ. Then, from the usual argument taking $\partial \mathcal{R}^*$ to be a small quasi-rectilinear contour $\delta s \times \delta n$ ($1 \gg \delta s \gg \delta n$) enclosing part of Γ and then letting first $\delta n \to 0$ and second $\delta s \to 0$, it follows since $l = \mathrm{d}y/\mathrm{d}s$, $m = -\mathrm{d}x/\mathrm{d}s$ that

$$\mathrm{d}y[Pu] - \mathrm{d}x[Qu] = 0, \tag{2.21}$$

and this discontinuity condition must apply across Γ. If the conditions $P_x + Q_y = 0$ (a divergence condition), $R = 0$ do not apply then the procedure must be modified. Consider

$$T(Pu_x + Qu_y + R) = (TPu)_x + (TQu)_y - \{(TP)_x + (TQ)_y\}u + R,$$

where T is an arbitrary function. If T can be chosen so that $(TP)_x + (TQ)_y = 0$, then (2.20) is replaced by

$$\int_{\partial \mathcal{R}^*} (lTPu + mTQu)\mathrm{d}s + \iint_{\mathcal{R}^*} R\mathrm{d}x\mathrm{d}y = 0. \tag{2.22}$$

If R is finite then the earlier limiting process leads now to

$$\mathrm{d}y[TPu] - \mathrm{d}x[TQu] = 0 \tag{2.23}$$

and this new discontinuity condition must apply.

In regard to (2.11), this may be written in the form

$$(\tfrac{1}{2}u^2)_x + u_y = 0 \tag{2.24}$$

and accordingly the corresponding discontinuity condition may be written

$$\mathrm{d}y[\tfrac{1}{2}u^2] - \mathrm{d}x[u] = 0. \tag{2.25}$$

It may be noted that these generalization procedures are not necessarily unique (e.g. (2.11) can also be written $(\tfrac{1}{3}u^3)_x + (\tfrac{1}{2}u^2)_y = 0$ and so forth). However, the present interest lies of course normally with partial differential equations that arise from the application of some definite physical law, in which case the possible ambiguity may not arise.

Finally, it should be noted that the essence of characteristics theory is that the original *partial* differential equation (2.1) is replaced by an *ordinary* differential equation (2.7) referred to the characteristics (2.6) (which, however, for a true quasi-linear equation are initially unknown). In detail,

$$Pu_x + Qu_y + R = 0, \quad \text{where} \quad u = u(x, y), \tag{2.26}$$

is replaced by

$$Pu_s + Rx_s = 0 \quad \text{or} \quad Qu_s + Ry_s = 0 \quad \text{on} \quad Py_s = Qx_s \ (\alpha = \text{const.}), \tag{2.27}$$

where now

$$u = u(\alpha, s), \quad x = x(\alpha, s), \quad y = y(\alpha, s),$$

so that u, x, y are now dependent variables and α, s are independent variables. Furthermore, the characteristics formulation leads straightforwardly to a corresponding difference approximation and also to the necessary modification to cater for discontinuities.

2.2 *Systems of first-order equations*

The general system of n first-order, quasi-linear, inhomogeneous partial differential equations in n dependent variables $u_i(x, y)$ $(i = 1, 2, ..., n)$ that are functions of two independent variables x, y may be written as

$$L_i \equiv a_{ik}(x, y, u_j)u_{k,x} + b_{ik}(x, y, u_j)u_{k,y} + c_i(x, y, u_j) = 0$$

$$(i = 1, 2, ..., n), \tag{2.28}$$

where the a_{ik}, b_{ik}, c_i are given functions of their arguments $(x, y, u_1, u_2, ..., u_n)$ and the usual conventions for subscripts in regard to summation and partial differentiation apply. The representation (2.28) is a scalar one, and now the following equivalent matrix representation will be adopted:

$$\boldsymbol{L} \equiv \boldsymbol{A}(x, y, \boldsymbol{U})\boldsymbol{U}_x + \boldsymbol{B}(x, y, \boldsymbol{U})\boldsymbol{U}_y + \boldsymbol{C}(x, y, \boldsymbol{U}) = \boldsymbol{0}, \tag{2.29}$$

where A, B (both $n \times n$) and C $(n \times 1)$ are known matrix functions of x, y, U and U $(n \times 1)$ is the column vector of dependent variables. The reader is referred generally to Wilkinson [10] for the details of the matrix algebra needed subsequently in this section. At the outset it should be noted that, in general, in either (2.28) or (2.29) the differentiation of the u_i or U is not in a *single* direction $dx/dy = a_{ik}/b_{ik}$ at any point of the (x, y)-plane. Remembering the remarks made at the end of Section (2.1) the objective now is to reduce (2.29) to an equivalent system of equations in each of which U (that is, the u_i) is differentiated in some *single* direction (which may, however, vary over the plane and with the solution). *As will be seen the essential question is whether or not such a transformation exists and, if it does, then whether it is real or not.*

The initial problem is then set as the determination from $L = 0$ (equation (2.29)) scalar differential equations in each of which U is differentiated in some single direction. It will be assumed that B (or A) is non-singular, i.e. $\det(B) \neq 0$, and then

$$M \equiv B^{-1}(L-C) = B^{-1}AU_x + U_y = DU_x + U_y, \tag{2.30}$$

where

$$D = B^{-1}A.$$

For the moment, it will be more convenient to work with M rather than directly with L. Now let F be an arbitrary column vector (n components), and let the superscript T denote the transpose of a matrix. Then,

$$N = F^T M = M^T F \ (= N^T) \tag{2.31}$$

is a single entity (a scalar), viz. a linear combination of the components of M, say $F_k M_k$. Now, (2.31) may be written as

$$N = F^T(DU_x + U_y) = (U_x^T D^T + U_y^T)F$$

or

$$N = (U_x^T D^T + U_y^T I)F, \tag{2.32}$$

where I is the identity matrix ($n \times n$). Equation (2.32) defines a single (scalar) partial differential expression. Suppose now, that in N, U (that is, all the u_i) is differentiated in a *single* direction $x_s : y_s$ along a curve $x = x(s)$, $y = y(s)$ drawn in the (x, y)-plane, where s is a convenient parametric coordinate. Then, N (assuming for the moment, that F can be appropriately chosen to satisfy the requirement) is necessarily of the form

$$N = U_s^T G, \tag{2.33}$$

where G is a column vector (n components) as yet undetermined, i.e.

$$N = (U_x^T x_s + U_y^T y_s)G. \tag{2.34}$$

If now (2.32) and (2.34) are to be identical, then

$$D^T F = x_s G, \quad IF = y_s G, \tag{2.35}$$

and hence eliminating G it follows that

$$(D^T - (x_s/y_s)I)F = 0 \qquad (y_s \neq 0) \tag{2.36}$$

or, equivalently,

$$F^T(D - (x_s/y_s)I) = 0 \qquad (y_s \neq 0). \tag{2.37}$$

Thus, for non-trivial F (i.e. non-zero F), x_s/y_s must be an eigenvalue of $D = B^{-1}A$ (and also of D^T) and F is the corresponding *right-hand* eigenvector of D^T (F^T is the left-hand eigenvector of D for the same eigenvalue). Let λ_i ($i = 1, 2, ..., n$) be the eigenvalues of D, and then the λ_i are the roots of

$$\det(D - \lambda I) = 0. \tag{2.38}$$

The simplest case, to which attention is confined in this section, is that when the λ_i form a *distinct real* set of quantities, in which case the same is true of the corresponding right-hand (or left-hand) eigenvectors. It now remains to determine an appropriate form for N and hence for $F^T B^{-1} L$ along the directions $\mathrm{d}x/\mathrm{d}y = (\mathrm{d}x/\mathrm{d}y)_i = \lambda_i$, called characteristic directions. Now, (2.37) is written as

$$(D^T - \lambda_i I)F_i = 0 \qquad (i = 1, 2, ..., n), \tag{2.39}$$

where λ_i and F_i ($i = 1, 2, ..., n$) are associated eigenvalues and eigenvectors and $\lambda_i = x_s/y_s$ is the corresponding characteristic direction. The eigenvalues λ_i may be assembled into a diagonal matrix $\Lambda = \mathrm{diag}(\lambda_i)$ (if Λ has components λ_{ij} then $\lambda_{ij} = \lambda_j \delta_{ij}$ (not summed) where δ_{ij} is the Kronecker delta). The eigenvectors F_i may be respectively assembled into the columns of a matrix E (so that $E = [F_1, F_2, ..., F_n]$), and then the system of vector equations (2.39) may be written compactly as the single matrix equation

$$D^T E = E\Lambda, \tag{2.40}$$

and hence

$$D^T = E\Lambda E^{-1}, \tag{2.41}$$

since E is non-singular. Transformations of the type HAH^{-1} (or $H^{-1}AH$) of a matrix A, where H is non-singular, are called similarity transformations and A and HAH^{-1} are said to be similar. What has been shown here is that there is a similarity transformation which reduces D^T to diagonal form Λ; the matrix of the transformation has its columns equal to the right-hand eigenvectors of D^T; and the diagonal components of Λ are the eigenvalues of D (the eigenvalues and eigenvectors are to be ordered similarly). Now, from (2.30),

$$E_i^T M = E_i^T(DU_x + U_y), \text{ where now } E_i \equiv F_i,$$

and, hence from (2.39),

$$E_i^T M dy_i = E_i^T(\lambda_i U_x + U_y)dy_i = E_i^T(U_x dx_i + U_y dy_i)$$

$$= E_i^T dU_i \text{ on } dx/dy = (dx/dy)_i = \lambda_i \qquad (2.42)$$

(i not summed). Since $M = B^{-1}(L-C)$ it follows that

$$0 = E_i^T B^{-1} L dy_i = E_i^T(dU_i + B^{-1}C dy_i). \qquad (2.43)$$

Now define

$$\mathcal{D}_i = x_s \frac{\partial}{\partial x} + y_s \frac{\partial}{\partial y} = y_s\left(\lambda_i \frac{\partial}{\partial x} + \frac{\partial}{\partial y}\right) \quad (i = 1, 2, ..., n), \qquad (2.44)$$

and then (2.43) is written

$$E_i^T(\mathcal{D}_i U + B^{-1}C y_{i,s}) = 0 \text{ on } x_s/y_s = \lambda_i. \qquad (2.45)$$

Equations (2.45) are the equations that hold along the characteristics. Thus, the objective of transforming the original *partial* differential equations (2.29) into an equivalent system of *ordinary* differential equations referred to characteristic directions has been achieved. In the present case, when the eigenvalues are assumed to be real and distinct the system (2.29) is called *totally hyperbolic*, and then the relations (2.45) may be written approximately as difference relations along the characteristics, which may then be used as a basis for numerical integration along the characteristics (which, if necessary, are themselves found in the course of this numerical integration). Also, exactly as in Section (2.1), if the characteristics run together (excluding the limit line case), so that (2.29) fail, then it is possible to obtain the corresponding discontinuity relations.

Part II
Quasi-static deformations in solids

3. Perfect plasticity

In the theory of quasi-static deformations of rigid/perfect-plastic solids the mathematical structure of the field equations varies with the geometry of flow and with the type of yield criterion and (associated or non-associated) flow rule. Thus, in plane strain, the stress and velocity equations are hyperbolic for all possible yield criteria, and the same result is true also in torsion of prismatic bars; but, for example, in plane stress and axially-symmetric stress/strain, the field equations are not uniformly hyperbolic and changes in their structure (say, to or from hyperbolic type) occur with change in the location of the representative stress point on the yield surface, and this applies also in bending of thin plates (there the governing equations involve moments and curvature-rates). The special cases of plane strain and torsion are altogether exceptional in the uniformity and relative simplicity of the structure of their field equations, and in this regard they stand in marked contrast to most other cases. This, in fact, is the fundamental reason why, in general, most progress in the solution of problems of rigid/perfectly-plastic flow has been made for situations of plane strain and torsion (although the situation of axially-symmetric stress/strain, when the Haar/von Kármán criterion applies, is similar to that of plane strain). It seems therefore to be particularly important to be able to determine easily the mathematical structure of the field equations and especially to identify situations that are either totally hyperbolic or, at least, have some partially hyperbolic features.

In the case of torsion, the governing equations are

$$\tau_{xz,x} + \tau_{yz,y} = 0,$$
$$\tau_{xz}^2 + \tau_{yz}^2 = k^2, \tag{3.1}$$
$$w_{,x}/w_{,y} = \tau_{xz}/\tau_{yz},$$

being respectively equilibrium, yield, and flow equations. Now, (3.1_2) may be satisfied identically by writing

$$\tau_{xz} = -k \sin \psi, \quad \tau_{yz} = k \cos \psi, \tag{3.2}$$

and then $(3.1_{1,3})$ are written as

$$\cos \psi\, \psi_{,x} + \sin \psi\, \psi_{,y} = 0,$$

$$\cos \psi\, w_{,x} + \sin \psi\, w_{,y} = 0. \tag{3.3}$$

The system (3.3) (a statically determinate one) is very simple indeed, being virtually a case of that considered in Section (2.1). It may also be written in matrix form

$$L \equiv A(U_x + V_y) = 0,$$

where

$$A = \begin{bmatrix} \cos \psi & \sin \psi & 0 & 0 \\ 0 & 0 & \cos \psi & \sin \psi \end{bmatrix}, \tag{3.4}$$

$$U_x^T = [\psi_{,x}\ \ 0\ \ w_{,x}\ \ 0], \qquad V_y^T = [0\ \ \psi_{,y}\ \ 0\ \ w_{,y}].$$

But, clearly, directly from (3.3) and the earlier analysis of Section (2.1), equations (3.3) are both hyperbolic each with the same families of characteristics given by

$$x_s/y_s = \cos \psi/\sin \psi = -\tau_{yz}/\tau_{xz}, \tag{3.5}$$

and it is straightforward to complete the analysis.

In the case of plane strain the governing equations are

$$\sigma_{x,x} + \tau_{xy,y} = 0, \quad \tau_{xy,x} + \sigma_{y,y} = 0,$$

$$\tfrac{1}{4}(\sigma_x - \sigma_y)^2 + \tau_{xy}^2 = k^2, \tag{3.6}$$

$$u_{,x} + v_{,y} = 0, \quad 2\tau_{xy}/(\sigma_x - \sigma_y) = (u_{,y} + v_{,x})/(u_{,x} - v_{,y}),$$

being respectively equations of equilibrium, yield, and flow (plastic incompressibility and the Saint-Venant condition of coincidence of principal directions of stress and plastic strain-rate). These equations are most easily handled when slightly re-formulated by means of the introduction of new stress variables p and ϕ defined by

$$\sigma_x = -p - k \sin(2\phi), \quad \sigma_y = -p + k \sin(2\phi), \quad \tau_{xy} = k \cos(2\phi), \tag{3.7}$$

so that the yield condition (3.6_3) is identically satisfied, and then the remaining equations (3.6) are written as

$$-p_{,x} - 2k\cos(2\phi)\phi_{,x} - 2k\sin(2\phi)\phi_{,y} = 0,$$

$$-2k\sin(2\phi)\phi_{,x} - p_{,y} + 2k\cos(2\phi)\phi_{,y} = 0, \qquad (3.8)$$

$$u_{,x} + v_{,y} = 0, \quad -\cot(2\phi) = (u_{,y} + v_{,x})/(u_{,x} - v_{,y}).$$

Clearly, the equations (3.8) are 'statically determinate', and they have the following matrix representation:

$$AU_x + BU_y = 0, \quad CV_x + DV_y = 0, \qquad (3.9)$$

where

$$A = \begin{bmatrix} -1 & -\cos(2\phi) \\ 0 & -\sin(2\phi) \end{bmatrix}, \quad B = \begin{bmatrix} 0 & -\sin(2\phi) \\ -1 & \cos(2\phi) \end{bmatrix}, \quad U = \begin{bmatrix} p \\ 2k\phi \end{bmatrix},$$

$$C = \begin{bmatrix} 1 & 0 \\ \cos(2\phi) & \sin(2\phi) \end{bmatrix}, \quad D = \begin{bmatrix} 0 & 1 \\ \sin(2\phi) & -\cos(2\phi) \end{bmatrix}, \quad V = \begin{bmatrix} u \\ v \end{bmatrix},$$

or, alternatively, these two separate matrix equations may be written together in the form

$$FW_x + GW_y = 0, \qquad (3.10)$$

where

$$F = \begin{bmatrix} A & 0 \\ 0 & C \end{bmatrix}, \quad G = \begin{bmatrix} B & 0 \\ 0 & D \end{bmatrix}, \quad W = \begin{bmatrix} U \\ V \end{bmatrix}.$$

Note that the components of U have been chosen deliberately here to have the same dimensions (the components of V automatically have the same dimensions); and that the components of W *should* have been chosen to have the same dimensions, but that the point is met if it is supposed that all dependent variables are non-dimensionalized (it is not, however, necessary here to non-dimensionalize the independent variables). The matrices A, B, C, D, U, V now play the role of sub-matrices of F, G, W. The mathematical structure of the field equations (3.8) or (3.9) or (3.10) therefore depends upon the eigenvalues of $G^{-1}F$ or, equivalently, of $B^{-1}A$ and $C^{-1}D$. Now, it is straightforward to show that

$$B^{-1} = \begin{bmatrix} -\cot(2\phi) & -1 \\ -\cosec(2\phi) & 0 \end{bmatrix}, \quad D^{-1} = \begin{bmatrix} \cot(2\phi) & \cosec(2\phi) \\ 1 & 0 \end{bmatrix}, \qquad (3.11)$$

and hence that

$$B^{-1}A = \begin{bmatrix} \cot(2\phi) & \operatorname{cosec}(2\phi) \\ \operatorname{cosec}(2\phi) & \cot(2\phi) \end{bmatrix}, \quad D^{-1}C = \begin{bmatrix} 2\cot(2\phi) & 1 \\ 1 & 0 \end{bmatrix}. \quad (3.12)$$

It follows that the eigenvalues λ of both $B^{-1}A$ and $D^{-1}C$ satisfy

$$\lambda^2 - 2\cot(2\phi)\lambda - 1 = 0, \tag{3.13}$$

viz.

$$\lambda = \lambda_{1,2} = \cot\phi, -\tan\phi = -\tan(\phi + \tfrac{1}{2}\pi), -\tan\phi \tag{3.14}$$

in accord with the well-known results. The matrices (3.12) are therefore similar. It is straightforward now to determine the eigenvectors of $B^{-1}A$ and $D^{-1}C$ and thence to find the characteristic relations. The right-hand eigenvectors of $(B^{-1}A)^T$ and $(D^{-1}C)^T$ are given by the columns of the matrices

$$E^{(1)} = \frac{1}{\sqrt{2}}\begin{bmatrix} 1 & 1 \\ 1 & -1 \end{bmatrix}, \quad E^{(2)} = \begin{bmatrix} \cos\phi & \sin\phi \\ \sin\phi & -\cos\phi \end{bmatrix}, \tag{3.15}$$

these eigenvectors having been appropriately normalized (furthermore, they are already non-dimensional). The relations (2.41) are now easily verified. The characteristic relations are obtained from (2.45) and the results are

$$\mathscr{D}(p \pm 2k\phi) \text{ on } x_s/y_s = \cot\phi, -\tan\phi \tag{3.16}$$

respectively (which are, of course the well-known Hencky relations) and

$$\cos\phi\mathscr{D}u + \sin\phi\mathscr{D}v = 0, \quad \sin\phi\mathscr{D}u - \cos\phi\mathscr{D}v = 0 \text{ on}$$

$$x_s/y_s = \cot\phi, -\tan\phi \quad (3.17)$$

respectively; but, since

$$V = \begin{bmatrix} \cos\phi & \sin\phi \\ \sin\phi & -\cos\phi \end{bmatrix}\bar{V}, \tag{3.18}$$

where $\bar{V} = (\bar{u}, \bar{v})$ referred to the characteristic directions, it follows that (3.17) is equivalent to

$$\mathscr{D}\bar{u} - \bar{v}\mathscr{D}\phi = 0, \quad \mathscr{D}\bar{v} + \bar{u}\mathscr{D}\phi = 0 \text{ on } x_s/y_s = \cos\phi, -\tan\phi \tag{3.19}$$

respectively (which are, of course, the well-known Geiringer relations).

251

The matrix $(G^{-1}F)^T$ therefore has the eigenvalue spectrum

$$\lambda_1 = \cot \phi, \; \lambda_2 = -\tan \phi, \; \lambda_3 = \lambda_1 = \cot \phi, \; \lambda_4 = \lambda_2 = -\tan \phi \quad (3.20)$$

with corresponding right-hand eigenvectors as the columns of the matrix

$$E = \begin{bmatrix} E^{(1)} & 0 \\ 0 & E^{(2)} \end{bmatrix}, \quad (3.21)$$

where the submatrices of E are given by (3.15). The characteristic polynomial of $G^{-1}F$ is

$$(\cot \phi - \lambda)^2 (-\tan \phi - \lambda)^2. \quad (3.22)$$

Thus, the field equations for plane strain therefore show some multiplicity (or degeneracy) from the point of eigensystem theory: the distinct eigenvalues are merely two (and not four) in number (equations (3.20)) and any linear (non-dimensional) combination of the eigenvectors formed from respective columns of the two matrices $E^{(1)}$, $E^{(2)}$ (equations (3.15)) are eigenvectors. There are two (and not four) distinct families of characteristics and any linear combination of the respective Hencky relations (3.16) and Geiringer relations (3.18) are characteristic relations. However, despite the degeneracy, the characteristic formulation of the partial differential system (3.6) as the ordinary differential system (3.14), (3.16) and (3.18) is a complete one, in the sense that direct analytical (or numerical) procedures for the solution of specific problems may be based upon the latter system.

The field equations for other situations show a mathematical structure that, in general, is more complicated than that described above for torsion and plane strain. Thus, for example, for plane stress and axially-symmetric stress/strain, and for bending and extension of plates and shells, the mathematical structure of the governing equations varies according to the yield condition and flow rule and the location of stress points on the yield surface. The complexity and variation of the results is seen for axially-symmetric stress/strain from Cox, Eason and Hopkins [11] and for bending of plates from Hopkins [12], but the point to be noted here is that their analyses of the field equations may now be re-cast in the present matrix form to considerable advantage.

4. Work-hardening plasticity

In Section 3, the mathematical structure, from the point of view of eigen-system theory, of the field equations for various situations of rigid/perfectly-plastic deformation has been considered. In this section, consideration is given now to the type of modification that occurs for rigid/work-hardening deformation. Although it might perhaps be thought that when the yield strength is dependent upon plastic strain (or plastic work) the characteristic directions should change from being *real* to being *complex*, this happens *not* to be the case in general. The particular case of plane strain is of especial interest. The field equations (see, for example, Pearce (13)) are

$$\sigma_{x,x} + \tau_{xy,y} = 0, \quad \tau_{xy,x} + \sigma_{y,y} = 0,$$
$$\tfrac{1}{4}(\sigma_x - \sigma_y)^2 + \tau_{xy}^2 = k^2, \tag{4.1_{1,2,3}}$$

where

$$k = F\left(\int dW^p\right),$$

and

$$u_{,x} + v_{,y} = 0, \quad 2\tau_{xy}/(\sigma_x - \sigma_y) = (u_{,y} + v_{,x})/(u_{,x} - v_{,y}), \tag{4.1_{4,5}}$$

that is, equations (3.6) now modified to allow for the dependence of k upon the plastic work $\int dW^p$ for each material element. As before, p and ϕ may be introduced as defined by

$$\sigma_x = -p - k\sin(2\phi), \quad \sigma_y = -p + k\sin(2\phi), \quad \tau_{xy} = k\cos(2\phi), \tag{4.2}$$

so that the (work-hardening) yield condition (4.1_3) is satisfied. Straight-forward albeit somewhat lengthy analysis shows (see Pearce [13]) that the field equations (4.1) may be written in the form

$$p_{,\alpha} + 2k\phi_{,\alpha} - k_{,\beta} = 0, \quad -p_{,\beta} + 2k\phi_{,\beta} + k_{,\alpha} = 0, $$
$$hq^{-1}q_{,\gamma} = \sin(2\psi)k_{,\gamma}, \tag{4.3_{1,2,3}}$$

where

$$q^2 = u^2 + v^2 = \bar{u}^2 + \bar{v}^2$$

(see (3.18)) and $h = \frac{2}{3}H'$, H' being the slope of the equivalent stress vs. plastic strain curve, and

$$u_{,\alpha} - v\phi_{,\alpha} = 0, \quad v_{,\beta} + u\phi_{,\beta} = 0, \tag{4.3$_{4,5}$}$$

where subscripts α, β refer to space differentiation along the directions $x_s/y_s = \cot\phi$, $-\tan\phi$ and the subscript γ refers to space differentiation along the direction of *particle flow* (this direction being inclined at an angle ψ to the α-direction). If $k = $ const. in (4.3) then manifestly the earlier results obtained in Section 3 for the characteristic directions and the characteristic relations (the Hencky and Geiringer relations) are recovered. Clearly, the Hencky relations are *destroyed* whereas the Geiringer relations are *preserved* in the presence of work-hardening. Also, from (4.3), it is apparent that the α-, β-, γ-directions are characteristic directions with corresponding characteristic relations as the Geiringer relations (unchanged, since these devolve from relations independent of yield) and the work-hardening relation

$$hq^{-1}\mathscr{D}_\gamma q - \sin(2\psi)\mathscr{D}_\gamma k = 0 \tag{4.4}$$

in the obvious notation. The remaining equations (4.3$_{1,2}$) however *cannot be uncoupled* to form a hyperbolic system at all. In more detail, analysis shows that the system (4.3) has merely *three (not five)* distinct families of characteristics, viz. the α (twice repeated)-, β (twice repeated)- and γ-directions, and it cannot be simplified beyond a system comprising *three* ordinary differential equations and *two* (coupled) partial differential equations. The result that *real* characteristic directions still obtain when there is work-hardening is at first sight surprising (but it is rather not surprising at all when the physical origin of (4.3$_{3,4,5}$) is remembered), but it is nonetheless somewhat peculiar that there are no *imaginary* characteristic directions at all. The results (4.3) are essentially first due to R. Hill (unpublished work, 1949; see Pearce [13]). Rather the same type of result applies for the situation of axially-symmetric stress/strain (under the assumption of the Haar/von Kármán criterion) (see Devenpeck and Weinstein [14]). However, especially for situations of rigid/work-hardening deformation, since the mathematical structure of the field equations is likely to be complex and perhaps unusual, it is best to develop the analysis in terms of the present matrix theory. This must be a subject for future study, but it appears to be likely that this will involve the reduction of matrices $D = B^{-1}A$ occurring in (2.29) to Jordan canonical form (rather than to purely diagonal matrices) by means of similarity transformations (see Wilkinson [10]).

It should be noted that the type of mathematical structure possessed by (4.1) makes it more difficult to obtain solutions to problems (such as, for example, of steady-state extrusion through a die) and although perturbation methods (writing, say, $\bar{\sigma} = k(1+\alpha\bar{\varepsilon})$ where α, $0 < \alpha \ll 1$, is a small parameter) have been employed (see Pearce [13]), these methods are not too successful or reliable and rather what is required is a boundary-layer type of approach (cf. Everstine and Pipkin [15]).

Part III
Dynamic deformations in solids

5. Plastic waves

The most elementary situation of the propagation of one-dimensional elastic-plastic waves is that associated with a semi-infinite bar, and subject to certain approximations the governing equations are

$$\sigma_{,\xi} = \rho_0 v_{,t}, \quad \varepsilon_{,t} = v_{,\xi}, \quad \sigma = f(\varepsilon) \tag{5.1}$$

(assuming strain-rate *independent* mechanical behaviour), being respectively the equation of motion, the strain/velocity compatibility condition, and the stress-strain relation (see, for example, Hopkins [8]). If the stress σ is eliminated from (5.1), then the resulting system of equations is

$$\varepsilon_{,t} - v_{,\xi} = 0, \quad \varepsilon_{,\xi} - c^{-2} v_{,t} = 0, \tag{5.2}$$

where

$$c(\varepsilon) = (f'(\varepsilon)/\rho_0)^{\frac{1}{2}},$$

$c(\varepsilon)$ of course being of the nature of a velocity. In matrix form, (5.2) is written as

$$L \equiv A U_\xi + B U_t = 0, \tag{5.3}$$

where

$$A = \begin{bmatrix} 0 & -1 \\ 1 & 0 \end{bmatrix}, \quad B = \begin{bmatrix} 1 & 0 \\ 0 & -c^{-2} \end{bmatrix}, \quad U = \begin{bmatrix} \varepsilon \\ v \end{bmatrix}.$$

For convenience, it is supposed that all dependent and independent variables have been suitably non-dimensionalized. Now, it is very easy to show that

$$B^{-1} = \begin{bmatrix} 1 & 0 \\ 0 & -c^2 \end{bmatrix}, \quad B^{-1}A = \begin{bmatrix} 0 & -1 \\ -c^2 & 0 \end{bmatrix}, \tag{5.4}$$

and the eigenvalues of $B^{-1}A$ are obviously

$$\lambda_1 = c, \quad \lambda_2 = -c \tag{5.5}$$

(and since, in general, there is work-hardening, $f' \neq 0$, it follows that these eigenvalues are *distinct* as well as *real*). The corresponding right-hand eigenvectors of $(B^{-1}A)^T$ are respectively given by the columns of the matrix

$$E = (1+c^2)^{-\frac{1}{2}} \begin{bmatrix} c & c \\ -1 & 1 \end{bmatrix}, \tag{5.6}$$

these eigenvectors having been suitably normalized. The relation (2.41), i.e. $(B^{-1}A)^T = E\Lambda E^{-1}$, is now easily verified. The characteristic relations are obtained from (2.45) and are

$$c\mathscr{D}\varepsilon \mp \mathscr{D}v = 0 \text{ on } x_s/t_s = \pm c. \tag{5.7}$$

The results (5.5) and (5.7) are of course well-known (see, for example, Hopkins [8]). The analysis therefore converts the original *partial* differential equations (5.2) into an *ordinary* system of differential equations (5.7), this latter system being referred to the characteristic directions which in general (the non-elastic case) 'float' with the solution. Particular problems are easily solved, either analytically or numerically, depending upon circumstances. Ideas of simple waves, centred and non-centred fans of characteristics, and similarity all find application in the solution of particular problems (see Hopkins [8]).

In the case of strain-rate *dependent* mechanical behaviour, as treated, for example, by L. E. Malvern (see Hopkins [8]), the governing system of equations is now

$$\varepsilon_{,t} - v_{,\xi} = 0, \quad \rho_0 v_{,t} - \sigma_{,\xi} = 0, \quad E\varepsilon_{,t} - \sigma_{,t} - g(\sigma, \varepsilon) = 0 \tag{5.8}$$

(where the static stress-strain relation $g(\sigma, \varepsilon) = 0$ is known), being respectively the strain/velocity compatibility relation, the equation of motion, and the constitutive equation. In matrix form, (5.8) is written as

$$L \equiv A U_\xi + B U_t + C = 0, \tag{5.9}$$

where

$$A = \begin{bmatrix} 0 & -1 & 0 \\ 0 & 0 & -1 \\ 0 & 0 & 0 \end{bmatrix}, \quad B = \begin{bmatrix} 1 & 0 & 0 \\ 0 & \rho_0 & 0 \\ E & 0 & -1 \end{bmatrix}, \quad C = \begin{bmatrix} 0 \\ 0 \\ -g \end{bmatrix}, \quad U = \begin{bmatrix} \varepsilon \\ v \\ \sigma \end{bmatrix}.$$

Here, it is assumed, of course, that all dependent and independent variables have been non-dimensionalized. It is easy to show that

$$B^{-1} = \begin{bmatrix} 1 & 0 & 0 \\ 0 & \rho_0^{-1} & 0 \\ E & 0 & -1 \end{bmatrix}, \quad B^{-1}A = \begin{bmatrix} 0 & -1 & 0 \\ 0 & 0 & -\rho_0^{-1} \\ 0 & -E & 0 \end{bmatrix}, \tag{5.10}$$

and the eigenvalues of $B^{-1}A$ are

$$\lambda_1 = 0, \ \lambda_2 = c_0, \ \lambda_3 = -c_0, \ \text{where} \ c_0 = (E/\rho_0)^{\frac{1}{2}} \tag{5.11}$$

(c_0 is recognized as the bar elastic wave velocity), and these eigenvalues are, of course, *distinct* as well as *real*. The corresponding right-hand eigenvectors of $(B^{-1}A)^T$ are respectively given by the columns of the matrix

$$E = \begin{bmatrix} 1 & 0 & 0 \\ 0 & 1 & 1 \\ -E^{-1} & -(\rho_0 c_0)^{-1} & (\rho_0 c_0)^{-1} \end{bmatrix}, \tag{5.12}$$

the normalization factors having been omitted for simplicity (these factors are $(1 + E^{-2})^{\frac{1}{2}}$, $(1 + (\rho_0 c_0)^{-2})^{\frac{1}{2}}$). The relation (2.41), i.e. $(B^{-1}A)^T = E\Lambda E^{-1}$, is now easily verified. The characteristic relations are obtained from (2.45) and are

$$E \mathcal{D}\varepsilon - \mathcal{D}\sigma - g t_s = 0 \ \text{on} \ \xi_s/t_s = 0,$$

$$\mathcal{D}\sigma - \rho_0 c_0 \mathcal{D}v + g t_s = 0 \ \text{on} \ \xi_s/t_s = c_0, \tag{5.13}$$

$$\mathcal{D}\sigma + \rho_0 c_0 \mathcal{D}v + g t_s = 0 \ \text{on} \ \xi_s/t_s = -c_0.$$

The results (5.11) and (5.13) are, of course, well known.

It is possible to consider systems of greater complexity. This has been done for elastic-plastic waves in thin-walled circular cylindrical tubes under combined torsion and compression by Clifton [16] for strain-rate independent mechanical behaviour and by Hopkins (unpublished work) for strain-rate dependent mechanical behaviour. The same viewpoint has been further developed by Ting ([17] and also unpublished work). Also, Horie [18] has considered elastic-plastic waves in bars with due account of certain thermal effects. Such situations, of course, result in larger systems of field equations, or larger matrix equations, and the present type of matrix analysis is then really indispensable, as all these workers recognize. However, generally for dynamic plasticity situations, the present matrix methods are not as yet greatly in use or recognized to be valuable.

6. Shock waves

Under certain circumstances, the characteristics may run together, and then the present *differential* field equations fail and must be replaced locally by *finite* discontinuity conditions. An indication of the analysis required has been given in Section 2, and this will not be further considered here (but see Hopkins [8]).

7. Conclusion

This paper has drawn attention primarily to the possibilities and the merits for the extended use of algebraic eigensystem theory for determining the mathematical structure of either quasi-static or dynamic field equations in plasticity theory when there are just *two* independent variables (space-space or space-time). Such an approach seems to be virtually indispensable for situations involving large systems of equations, and it also leads straightforwardly to the characteristic directions and relations when the equations show hyperbolic features. Also, in work-hardening theory, the equations are likely to show peculiar features (perhaps best described as incompletely hyperbolic), and it is suggested that the development of boundary-layer methods rather than perturbation methods would be useful in order to help with the solution of specific problems. The inter-play between the assumed plasticity properties of solids as represented in the constitutive equations and the mathematical structure of the field equations is very clear, being associated with sub-matrices.

References

[1] Hill, R., *The mathematical theory of plasticity*. Clarendon Press, Oxford 1950.

[2] Prager, W. and P. G. Hodge, Jr., *Theory of perfectly-plastic solids*. Wiley, New York 1951.

[3] Malvern, L. E., *Introduction to the mechanics of a continuous medium*. Prentice-Hall, Englewood Cliffs, N.J. 1969.

[4] Prager, W., Discontinuous fields of plastic stress and flow. *Proceedings Second U.S. Nat. Congr. Appl. Mechs.* (edited by P. M. Naghdi), Amer. Soc. Mech. Engrs., New York 1955, p. 21.

[5] Hill, R., Discontinuity relations in mechanics of solids. *Progress in Solid Mechanics*, Vol. II (edited by I. N. Sneddon and R. Hill), Ch. VI. North-Holland, Amsterdam 1961.

[6] Hadamard, J., Lectures on Cauchy's problem in linear partial differential equations. Dover, New York 1952.

[7] Courant, R. and D. Hilbert, *Methods of mathematical physics*. Vol. II: *Partial differential equations* (by R. Courant), Ch. V. Interscience, New York 1962.

[8] Hopkins, H. G., The method of characteristics and its application to the theory of stress waves in solids. *Engineering Plasticity* (edited by J. Heyman and F. A. Leckie), Cambridge University Press 1968, p. 277.

[9] Anderson, G. D. and W. Band, Compressible fluid flow and the method of characteristics. *Amer. J. Phys.*, *30* (1962) 831.

[10] Wilkinson, J. H., *The algebraic eigenvalue problem*, Ch. 1. Clarendon Press, Oxford 1965.

[11] Cox, A. D., G. Eason and H. G. Hopkins, Axially symmetric plastic deformations in soils. *Phil. Trans. Roy. Soc.*, London *A 254*, (1962) 1.

[12] Hopkins, H. G., On the plastic theory of plates. *Proc. Roy. Soc.*, London *A 241*, (1957) 153.

[13] Pearce, J. H. B., *Theory of plane strain for a rigid/work-hardening material*. M. Sc. Thesis. University of Manchester 1970.

[14] Devenpeck, M. L. and A. S. Weinstein, Experimental investigation of workhardening effects in wedge flattening with relation to nonhardening theory. *J. Mech. Phys. Solids*, *18* (1970) 213.

[15] Everstine, G. C. and A. C. Pipkin, *Boundary layers in fiber-reinforced materials*. Division of Applied Mathematics, Brown University Tech. Rep. 1971.

[16] Clifton, R. J., An analysis of combined longitudinal and torsional plastic waves in a thin-walled tube. *Proceedings Fifth U.S. Nat. Congr. Appl. Mechs.* (edited by L. E. Goodman), Amer. Soc. Mech. Engrs., New York 1966, p. 465.

[17] Ting, T. C. T., A unified theory on elastic-plastic wave propagation of combined stress. *Foundations of Plasticity*, (edited by A. Sawczuk), Vol. 1, Noordhoff, Leyden 1973, p. 301.

[18] Horie, Y., *Longitudinal motion of bars by impact with temperature effects*. North Carolina State University (Raleigh, N.C.) Tech. Rep. 69–11, December 1969.

259

DISCUSSION

S. Nemat-Nasser[1]). Professor Hopkins gave us a differential equation account of conservation laws in plasticity. A complementary approach might be based on an integral formulation of conservation laws. The integral formulation would admit solutions from a wider class of functions, since it only assumes integrability of the involved quantities. Moreover, it includes the differential equations, as well as various jump conditions, characterizing the corresponding conservation laws. A significant feature of such a formulation then is that one can easily transfer these laws into their variational counterparts, and use numerical methods directly. In this connection I would like to suggest that general variational methods with several continuous or discontinuous independent fields can be effectively used.

H. G. Hopkins. Professor Nemat-Nasser makes a valid and interesting observation that the analysis that I presented is restricted to differential formulations rather than integral ones so that supplementary analysis is required to discuss shock discontinuities (as distinct from weak discontinuities) in physical variables. In a more general treatment I agree that it is essential to start with an integral formulation of conservation laws which would have the advantages that he states, but space limitations precluded this being done in the present paper. However, I thank Professor Nemat-Nasser for drawing attention to this question.

[1]) Northwestern University, Evanston, Illinois, USA.

Computer solutions of plasticity problems[1])

Philip G. Hodge, Jr.

University of Minnesota, Minneapolis USA

In recent years, finite-element methods have proved to be a very useful tool in obtaining solutions to problems of practical importance for both elastic and inelastic structures. Because the need for solutions is so great, and because the results predicted by finite-element solutions are reasonable, primary interest has been in the establishment of efficient computer programs for the solution of large scale problems. As a result, some theoretical considerations which may be important in extended applications of the methods have tended to be overlooked. The present paper reviews the current state of the art, focuses on some of these questions and recent work that has been done towards obtaining answers, and suggests some desirable directions for future research.

1. Introduction

The theory of plasticity has been a subject of interest to engineers and mathematicians for over a century.[2]) The finite element method of approximating continuum problems has developed primarily along with the high-speed digital computer,[3]) although, in a sense, the 'relaxation method' proposed by Southwell [5] may be thought of as falling in this category. However, it is only within the past dozen years [6] that the two ingredients have been combined. Despite its short history, the solution of plasticity problems by finite element methods has proved so amenable to implementation and has apparently answered such a real engineering need, that developments have been extremely rapid and 'all-purpose' computer programs are available [7, 8] which can purportedly solve any specific member of a wide class of problems.

---

[1]) This investigation was sponsored by the United States Office of Naval Research.

[2]) See, for example, Tresca's paper published in 1868 [1]. However, as Prager has pointed out [2], some of the basic ideas can be attributed to Galileo [3].

[3]) For a survey article with references to earlier work, see the 1966 paper by Argyris and Patton [4].

261

Despite, or perhaps because of, this rapid development in the implementation of methods, certain theoretical questions of interest have been bypassed or dealt with in an ad-hoc fashion; it is the purpose of the present paper to call attention to some of these questions.

Before dealing with these problems, we shall briefly review the continuum theory of plasticity in Sec. 2, and give a general description of the finite element method in Sec. 3. Then, in Sec. 4, we will present an alternative view of a finite element model as a direct structural model of the continuum.

As will be shown in Sec. 2, a general plasticity problem involves alternately solving for the field rates at a specific time and then integrating to obtain the field variables at a later time. In Secs. 5 and 6 we will discuss some of the questions that may arise when each of those two steps are carried out for a finite element model.

Section 7 considers a finite-difference model. Certain theoretical disadvantages are pointed out, and an alternative formulation which starts with a finite-element model is suggested.

Finally, in Sec. 8, we suggest some possible directions for future research in this area.

2. Theory of plasticity

In any continuum model of reality, we seek to determine the displacement field[1] u_i, the strain field ε_{ij}, and the stress field σ_{ij}; these are all to be found within a given domain D with boundary B which represents the physical body in the real world. In general, they are to be found as functions of time, possibly with D changing with time.

We restrict our consideration here to the quasi-static problem in which all changes with time take place so slowly that inertia effects may be ignored. Then, independently of the material of the body, the stresses and forces must satisfy the static requirements

$$\sigma_{ji,j} + F_i = 0 \qquad \text{in } D \tag{1a}$$

$$\sigma_{ji} n_j = T_i \qquad \text{on } B \tag{1b}$$

Here F_i is the prescribed body force in the interior, and at each point on the

---

[1] Subscripts have the range 1, 2, 3 for a three-dimensional model and the range 1, 2 for a two-dimensional model. Repeated subscripts are to be summed. A preceding comma denotes differentiation.

boundary each vector component of T_i is either prescribed or defined by (1b).

The strains and displacements must be related by

$$\varepsilon_{ij} = \tfrac{1}{2}(u_{i,j} + u_{j,i}) \qquad \text{in } D \tag{2a}$$

$$u_i = U_i \qquad \text{on } B \tag{2b}$$

where U_i are either prescribed boundary displacements or are defined by (2b). Equations (2) are also independent of the material.

In addition to satisfying (1) and (2), the stresses and strains must be related by appropriate constitutive equations. For a plastic material, these are expressed in terms of strain rates by

$$\dot{\varepsilon}_{ij} = \dot{\varepsilon}_{ij}^e + \dot{\varepsilon}_{ij}^p \tag{3a}$$

where the elastic strain rate satisfies Hooke's law

$$\dot{\varepsilon}_{ij}^e = C_{ijkl}\dot{\sigma}_{kl} \tag{3b}$$

At any given state with a specified history, it is assumed that there exists a yield function $f(\sigma_{ij})$ which depends explicitly on the state of stress, and a yield value f_0. Both f and f_0 may depend upon the prior history of stress and strain. At points where f is differentiable[1]) the plastic strain rate is given by

$$\dot{\varepsilon}_{ij}^p = \lambda \,\partial f / \partial \sigma_{ij} \tag{3c}$$

where

$$\text{if } (f < f_0 \text{ or } \dot{f} < 0) \text{ then } \lambda = 0 \tag{3d}$$

Further, the inequalities

$$f \leqq f_0 \tag{3e}$$

$$\lambda \geqq 0 \tag{3f}$$

must be satisfied. Finally, when $\lambda \neq 0$, a perfectly-plastic material must satisfy

$$\dot{f} = 0 \tag{3g}$$

and a strain-hardening material

$$\lambda = H\dot{f} \tag{3h}$$

---

[1]) As shown by Koiter [9], this concept can be generalized.

where H is a given function of stress and history. For an elastic-plastic material the constants C_{ijkl} in (3b) are fully symmetric and positive definite; for a rigid-plastic material $C_{ijkl} = 0$.

Viewed as a whole, Eqs. (1), (2), (3) are a non-linear set of partial differential equations in two or three space variables and time. For any given hypothesis as to which material points follow the branch in (3d), the remaining equality equations generally determine a unique solution, but only for the correct hypothesis will the inequalities (3e) and (3f) be everywhere satisfied.

Since the time-dependence occurs only in (3) and is linear and homogeneous, it is usual to regard it separately as follows. At any instant of time the complete state of stress, strain, and displacement is assumed known, the linear equations (1) and (2) are differentiated with respect to time, and the complete set of equations are solved for stress rates, strain rates, and velocities. If this can be done, we are left with a set of ordinary differential equations with time as the only variable. In all but the simplest of examples (see, for instance, [10–16]) regardless of what method is used to obtain the rates, the time integration must be done numerically.

Now, even viewed as a rate problem it is necessary to solve nonlinear equations with two or three independent variables, up to 10 dependent variables, and with a free boundary whose location is determined by the satisfaction of certain inequalities. Further, the continuity conditions across this boundary may be quite weak, depending upon the specific plasticity model, and in some cases the solution may not be fully unique.[1]) Obviously, there is little hope that complete solutions to practical problems can be obtained analytically.

However, despite the complexity of the problem certain general theoretical results have been obtained. In particular, various minimum and bounding principles are available.[2]) Many of these principles constitute generalizations of the well known theorems of minimum and complementary energy for elastic materials; some of these will be mentioned further in Secs. 4 and 5.

There is a particular branch of plasticity known as limit analysis in which a great deal of work has been done. In the limit analysis problem $F_i = 0$ in (1a) and at each point of the boundary either $U_i = 0$ in (2b) or (1b)

---

[1]) See, for example, Hill's paper in 1958 [17], where further references are also given.
[2]) See, for example [18] or, more recently [19] where references may be found.

can be written in the form

$$\sigma_{ji} n_j = \zeta T_i^0 \tag{4}$$

Here T_i^0 is given and does not depend on time, and ζ is an unknown constant. A value ζ^0 is defined as that number (proved to be unique) for which Eqs. (1–4) possess a solution other than $\dot{\varepsilon}_{ij} \equiv 0$ for a rigid-perfectly-plastic material. It has been shown that if ζ in (4) is thought of as slowly increasing from zero, then ζ^0 is the exact value at which a rigid-strain-hardening material will begin to strain, and it is the maximum value for which an elastic-perfectly plastic material can support a solution of (1)–(4), at least when geometry changes are neglected. Further, in several examples [16, 20, 21] it has been shown that ζ^0 is a very good approximation to the load intensity at which a general plastic material will undergo strains appreciably larger than those in a corresponding perfectly-elastic structure.

Not only is a complete solution of the limit analysis problem relatively simple since the very simple rigid-perfectly-plastic model is used, but by means of the theorems of limit analysis, the number ζ^0 may be bounded from above and below without solving all of Eqs. (1)–(4); without, in fact, necessarily finding any of the unknown field quantities σ_{ij}, ε_{ij}, or u_i. In particular, ζ^0 will be determined exactly if coincident bounds can be found by this method.

3. Finite-element methods

For definiteness, we shall first discuss one of the simplest finite-element models. We consider a two-dimensional problem, approximate the real domain by a polygon, divide that polygon into any number of triangular elements, and assume that the displacement field in any element is linear. Since there are two displacement components, the most general linear displacement field in a triangle is fully defined by six constants, and these constants may be chosen as the two displacement components of each of the three nodes of the triangle. This choice will automatically provide for continuity of displacements at the nodes over the set of triangles. Further, since triangles have straight sides and displacements are linear, this continuity of nodal displacements guarantees continuity of displacements along each triangle side, and hence throughout the field.

Linear displacements give rise to constant strains. Therefore, the state of strain in a given element can be represented symbolically by

$$\varepsilon = Au \tag{5}$$

Here u is a six-vector whose components are the displacements (u, v) at each of the nodes, and A is a 3×6 matrix whose elements depend only upon the geometry of the element; it may be computed once and for all. The strain ε is a three-vector whose components may be taken as $(\varepsilon_x, \varepsilon_y, \gamma_{xy})$ or as the extensional strains along each side of the triangle [7, 22]; the form of A will, of course, depend upon how the vector ε is chosen.

Turning to statics, let us assume that the state of stress is constant in each triangle. We represent it by the three-vector σ whose components are either $(\sigma_x, \sigma_y, \tau)$ or the tensile stresses along the three triangle sides; the definitions of ε and σ must correspond so that the internal work is $\varepsilon \cdot \sigma$ in either case.

We assume further that the triangles are connected only at their nodes so that the sides are stress free. Also, any external loads, either surface forces or body forces in the original continuum are replaced by concentrated loads at the nodes. Let (X_α^k, Y_α^k) denote the components of nodal force at node α which would be necessary to equilibrate α for triangle k. These forces are easily determined by considering a free body containing only one node of the triangle, so that we can write

$$F = B\sigma \tag{6}$$

where B depends only on the geometry of the triangle, and F is the six-vector of nodal forces (X_α^k, Y_α^k) for the three nodes of triangle k. Finally, the nodal forces for each triangle must be in over-all equilibrium at each node, so that

$$\Sigma X = T_x \qquad \Sigma Y = T_y \tag{7}$$

where T_x and T_y are the replacements for external loads (possibly zero).

For an elastic material, stress and strain are linearly related:

$$\varepsilon = C\sigma \tag{8}$$

Combining Eqs. (5), (6), and (8) we obtain

$$F = K^e u \tag{9}$$

where the elastic 'stiffness matrix'

$$K^e = BC^{-1}A \tag{10}$$

depends only upon the geometry of the triangles and the elastic constants. Therefore, for a triangle of any given dimensions and orientation, K^e may be computed once and for all, independently of any particular total configuration or loading condition. Finally, of course, substitution of (9) in (7) leads to $2N$ linear equations for the $2N$ nodal displacements, N being the total number of nodes.

The procedure described above can be easily generalized to three dimensions, [23] and analogous schemes can be established for structures such as plates and shells. Also, more general displacement fields may be considered. For example, if u and v are quadratic, they are defined by six constants each. If 'nodes' are assigned not only to the vertices but to the midpoints of the boundaries, the total displacement field will be continuous and the analysis continued in the same fashion. Another possible generalization is to consider rectangles or other polygons rather than triangles. This may be done *ab initio* or by considering these larger polygons as collections of triangles.

Conceptually, the combination of the ideas of this and the preceding section is obvious. Equation (8) is replaced by matrix equivalents of Eqs. (3) and Eqs. (5–7) are written in rate form. As a result (9) will be replaced by

$$\dot{F} = K^p \dot{u} \tag{11}$$

where K^p is no longer a constant but may depend upon the current state of stress and the history of the element; if $\lambda = 0$, then $K^p = K^e$. Then, if σ and u are known for all $0 \leq t \leq t_0$, K^p will be known at t_0, (11) can be substituted into the rate form of (7) and solved for $\dot{u}$, whence $\dot{\varepsilon}$ and $\dot{\sigma}$ are easily found. Finally, all quantities at $t_0 + \Delta t$ are evaluated by

$$\phi(t_0 + t) = \phi(t_0) + \Delta t \phi(t_0) \tag{12}$$

or any more sophisticated time integration technique, and the whole process repeated.

As mentioned in the Introduction this technique has been refined and exploited, and useful solutions to practical problems are being obtained every day.

However, there are certain unresolved (in some cases unasked) questions which form the basis of this paper, and which we will now proceed to elaborate on.

4. The finite-element model

The development in Sec. 3 was essentially historical. The subject of finite elements was presented in terms of a number of ad-hoc assumptions. No attempt was made to assess the merits of the individual assumptions, and the resulting model was accepted as a whole because it yielded valuable results.

However, the same results can be obtained by a quite different approach in which the finite-element model is viewed in the same light as any other structural theory such as beams or plates. Let us briefly recall how the simple Euler-Bernoulli beam theory may be developed by using the Principal of Virtual Work. One starts with the kinematic assumption that plane sections remain plane, etc. Assuming small slopes we then find the displacement field to be

$$u_x = -zw' \qquad u_y = 0 \qquad u_z = w(x) \tag{13}$$

whence the internal work reduces to

$$W_{\text{int}} = \int_a^b -w'' \left(\int_A z\sigma_x \mathrm{d}A \right) \mathrm{d}x \tag{14}$$

The form of (14) naturally suggests the definition of the bending moment or generalized stress

$$M = \int_A z\sigma_x \mathrm{d}A \tag{15}$$

Substitution of (15) in (14), two integrations by parts, and use of an appropriate expression for the external work, leads to

$$W_{\text{ext}} - W_{\text{int}} = \int_a^b (M'' + p)w\mathrm{d}x + \left[(Mw' - M'w) \right]_a^b \tag{16}$$

Since this expression should vanish for all choices of w, we are led to the equilibrium equation $(M'' + p) = 0$ and also to boundary conditions on moment and shear.

Following [24], we use the same approach to the finite-element model. After dividing up the domain into triangles, we make the single kinematic assumption that the displacement in each triangle is linear. Then the strain field in triangle k is the constant tensor ε_{ij}^k whence the total internal work is

$$W_{\text{int}} = \sum \varepsilon_{ij}^k \int_{T^k} \sigma_{ij} \mathrm{d}A \tag{17}$$

The form of (17) suggests the use of generalized stresses

$$\sigma_{ij}^k = (1/A^k) \int_{T^k} \sigma_{ij} \mathrm{d}A \tag{18}$$

Note that whereas in Sec. 3 we required an additional *assumption* that σ_{ij} was constant, here we have *defined* the set of constants σ_{ij}^k as the average of the generally variable real stress over the triangle. Similarly, when we consider the external work, we do not make further assumptions as to how a given external force system should be approximately replaced by concentrated loads at the nodes, but rather we are naturally led to definitions of generalized forces in terms of the given continuum force distributions. Finally, as shown in [24], the requirement that the Principle of Virtual Work hold for all sets of nodal displacements leads to the same equilibrium equations as were previously obtained by substituting (6) in (7).

To complete the analysis, the continuum constitutive equations can be substituted into (18) to yield appropriate constitutive relations between the sets of finite-element variables ε_{ij}^k and σ_{ij}^k [24, 25]. In Sec. 3, there was obviously no inconsistency between constant stress and constant strain for any conceivable material, but for a more sophisticated model there might be a real problem here. Even in the simple beam example, the strain is linear across the height, but the stress is not when the yield-limit has been reached over part of the cross section.

In addition to its practical value in deriving equations, the virtual-work approach has an important theoretical virtue. In almost all continuum models, the Principle of Virtual Work is used in a vital step in the proof of all uniqueness theorems and minimum principles. Therefore, when we use it to derive our equilibrium equations and constitutive relations, we know *a priori* that the resulting finite-element model will have the same uniqueness and minimum properties.

The importance of uniqueness in any model to be used numerically can scarcely be over-estimated. If the real problem truly has more than one solution, convergence of any iterative scheme will require very careful consideration. On the other hand, if the posed problem has several solutions of which only one is physically significant, one must be assured that the numerical method converges towards the 'right' solution.

The value of minimum principles for a finite-element model will be dealt with in more detail in Sec. 5. Here we shall discuss them from a somewhat general viewpoint.

A minimum principle considers a class of admissible functions f^* which satisfy some, but not all, of the defining conditions of a problem, and shows that a certain functional $\phi\{f^*\}$ is a minimum for that function f which satisfies the remaining conditions. As used in continuum mechanics, a minimum principle can find the 'best' approximate solution among some class f_n^* which depends upon a finite number of parameters a_n by determining a_n so as to minimize ϕ. However, since the complete solution f may not be a member of f_n^*, the result is still only an approximation.

However, for the finite-element model, the entire class of admissible functions contains only a finite number of parameters. Therefore, a minimum principle can be used not only for finding an approximate solution, but as an alternative technique for finding the complete solution.

Not only does the finite element model have minimum principles of its own, but it can be interpreted in terms of a minimum principle for the continuum. For example, for an elastic material the principle of minimum potential energy is concerned with the class of kinematically admissible fields. This class consists of all displacements u_i and strains ε_{ij} which satisfy strain-displacement relations, displacement boundary conditions, and certain continuity requirements. All of these requirements are satisfied by the displacement and strain fields defined in Sec. 3 for any set of nodal displacements. Therefore, the complete solution of the finite element model is not only the 'best' set of nodal displacements, but it provides a value of the potential energy Π which is not less than the value for the complete continuum solution.

5. The rate problem

The finite-element method of solution for plastic materials which was briefly presented in Sec. 3 has two main parts: the solution for rates at a generic instant of time t_0, and the integration with respect to time to obtain the field quantities at $t_0 + \Delta t$. In the present section we consider in detail some of the problems associated with finding the rates.

At the generic instant of time t_0 and for each triangle k, the yield function value f^k and yield value f_0^k are known, but the rate of change of f must be determined from the rate solution by

$$\dot{f}^k = (\partial f^k/\partial \sigma_{ij})\dot{\sigma}_{ij}^k \qquad (19)$$

If $f^k < f_0^k$, then triangle k satisfies the condition in (3d) so that we may set $\lambda^k = 0$ and use $\boldsymbol{K}^p = \boldsymbol{K}^e$ in Eq. (11). Such a triangle will be said to be in state E. Obviously, at any time t_0 it is known which triangles are E and which are non-E.

However, for the non-E triangles where $f^k = f_0^k$ there are two possibilities. On the one hand, $\dot{f}^k$ may be negative in which case Eq. (3d) must be satisfied, and the triangle will be said to be in state U. The alternative is state L for which λ^k is positive and Eq. (3g) or (3h) must be enforced. Thus, we need to known which of the non-E triangles are in U and which are in L before we can list the equations to be solved, but we need to solve the problem and find λ^k and $\dot{f}^k$ before we can correctly allocate the triangles to U or L.

There are various possible approaches to this impasse. The most simple-minded one is to assume that all triangles which are not in state E are in state L, and to proceed accordingly. This approach is certainly undesirable theoretically, but it works in many situations of practical importance. If no triangles ever are in U, the solution is obviously correct, and even if a few triangles should be in U, the effect of the error may be negligible on gross quantities of interest. Obviously it can only be used with any success when the external loading is increased approximately proportionally with time, but many important problems have this feature.

An obvious disadvantage to this method is that it is not self-checking. If used inappropriately, it will still present a 'solution' and there is no way of telling how good that solution is.

As a first refinement, the same assumption is made, the rates are found, and then $\dot{f}^k$ is computed from (19). If $\lambda^k \geqq 0$ for all non-E triangles, the assumption of L was correct and the solution is accepted. If $\lambda^k < 0$ for some triangles, those triangles are transferred to U and a revised solution for the rates is found.

In most reported solutions this revised solution is accepted. Again, in many practical situations it is either rigorously valid, or so close to valid as to give valuable results. However, if it is not checked, there is no warning that the solution may be drastically wrong.

The obvious remedy is to repeat the process of checking and refining. For each triangle in U, f^k is computed and, if it is positive, the triangle is transferred to L: each L triangle for which $\lambda^k < 0$ is changed to U. Then new revised rates are computed and checked. The process continues until no triangles are in an incorrect state.

The problem here is, that to the author's knowledge, the convergence of this process has never been investigated. It is at least conceivable that there would exist loading systems for which the process would eventually enter a closed loop which did not contain the solution. Since no counter-example has ever been presented, the question is certainly open. In fact, there is the additional disturbing possibility that convergence theoretically occurs but is so delicate that round-off error could destroy it.

A quite different approach to the problem of unloading, is to use a minimum principle directly to find the solution. Thus, we seek to minimize some functional $\phi\{f^*\}$ where the admissible class f^* of rate functions is subject to both equality and inequality constraints. The resulting mathematical problem is known as a nonlinear programming problem. A powerful technique, SUMT [26, 27] is available for quite general problems in this category, and has already been used for several plasticity problems [28, 29, 30].

In the particular case of an elastic-perfectly-plastic material, there exists a static minimum principle for the rate problem which is mathematically analogous to the theorem of minimum complementary energy in elasticity [18]. When this theorem is used, the functional ϕ to be minimized depends quadratically upon the unknown stress rates, and both the equality and inequality constraints are linear [31]. This specific form is known as a Quadratic Programming problem, and a specific mathematical procedure for solving it is available [32]. This approach to elastic-plastic problems has been successfully used with finite element models in torsion [33] and in plate-bending [34].

6. The integration problem

After the rates have been found, there remains the problem of determining the field quantities at some later time. Certain difficulties may arise here which are best illustrated in terms of a perfectly plastic material. Such a material is characterized by the fact that both the yield function f and the yield value f^0 are constant in history. Therefore, it follows from (3g) and (19), that if $\lambda^k > 0$ for triangle k, then

$$f^k = (\partial f^k / \partial \sigma_{ij}^k) \dot{\sigma}_{ij}^k = 0 \tag{20}$$

An obvious solution of (20) is $\dot{\sigma}_{ij}^k = 0$, and there are problems[1]) where that is, in fact, the relevant solution. However, in general $\dot{\sigma}_{ij}^k \neq 0$ so that the finding of $\sigma_{ij}^k(t_0 + \Delta t)$ involves an approximate time integration.

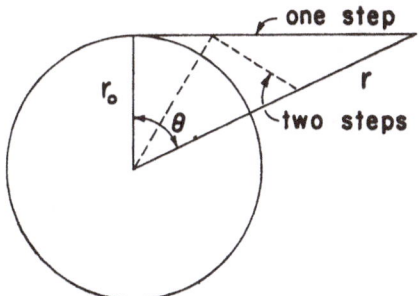

Fig. 1. Error of tangent motion.

The yield function f must be convex [37]. If it is strictly convex, then in the real continuous time solution of the finite-element model, the stress-point moves along the surface $f = f^0$, but if Eq. (12) is used to find $\sigma_{ij}^k(t_0 + \Delta t)$, the resulting motion is along a tangent to the surface. Now, if the total change in σ_{ij}^k is not very great, the error introduced by (12) will not be significant. For example, we consider as a problem in plane geometry in two dimensions the motion of a point on a circle, Fig. 1. The relative change in radius for a single tangent motion through an angle θ is

$$E = (r - r_0)/r_0 = \sec\theta - 1 \tag{21}$$

Thus, if the total change in θ is only 30°, $E = 0.155$.

Further, if we take n steps along the current tangent to achieve a total change θ, the relative growth in the radius is

$$E = (\sec\theta/n)^n - 1 \tag{22}$$

Table 1 shows the value of E as a function of the number of steps; with eight steps of about 4° each the error has been reduced to less than 2%.

[1]) For example, torsion of simply-connected bars [35, Chap. 3]. However, for multiply-connected bars the situation is more complicated [33, 36].

On the other hand, if the stress point moves entirely around the circle (i.e., through twelve times the angle), 100 steps of 3.6° each produces a relative error of over 20% and it is necessary to take 1000 steps to reduce it to 2%.

TABLE I

Relative error in radius

n	1	8	10	30	100	1000
$\theta = 30°$	.15	.020	.014	.0046	.0014	.00014
$\theta = 360°$	∞	15	7.3	.94	.22	.020

Therefore, if a very large change in stress direction is involved in the plastic range, an enormous number of time steps must be used to maintain reasonable accuracy.

Various schemes are possible to alleviate this problem. An obvious one is to use an improved formula for time integration. For example, at time t_0 the problem is solved as before, tentative values are computed at $t_0 + \Delta t$, and then improved values at $t_0 + \Delta t$ are obtained from

$$\phi(t_0 + \Delta t) = \phi(t_0) + (\tfrac{1}{2})\Delta t \ [\dot{\phi}(t_0) + \dot{\phi}(t_0 + \Delta t)] \tag{23}$$

However, although Eq. (23) will certainly yield better results than (12), it can be shown [31] that the problem is still qualitatively the same, namely, that every motion will take the stress point further outside of the yield curve, so that the errors will be commulative rather than random.

A different approach is to somehow force the stress point to return to the yield curve. The single point shown in Fig. 1 is misleadingly simple in that it should obviously be brought back along a radius at each step so that it starts each step from the correct curve. However, we recall that Fig. 1 represents merely one triangle of the model, and that if we change the values of that σ^k_{ij} only, we will violate the equilibrium relations with the stresses in other triangles. For the particular case of a framed structure, Morris and Fenves [38] have developed a scheme for drawing stress points back to the curve while still maintaining equilibrium, but it is not clear if their method can be generalized to other structures.

A closely related idea is to not allow the stress point to move outside of the yield curve in the first place. To the author's knowledge, this technique has not been implemented, although the theory is simple. One needs only to replace the constraint (20) which requires tangential motion, with one which requires the stress point to remain inside the yield surface. Thus, using (12) to compute $\sigma_{ij}^k(t_0 + \Delta t)$, we require the rates to satisfy

$$f^k[\sigma_{ij}^k(t_0) + \Delta t \, \dot{\sigma}_{ij}^k(t_0)] \leqq f_0 \tag{24}$$

instead of (20).

In general (24) represents a non-linear constraint on the rates, so that the solution of the rate problem may be more difficult. Also, it requires some decision on Δt before the rate problem is solved, whereas in other methods Δt needs to be chosen only after the rates are known. This latter drawback is not major, since we will not violate the yield condition if we use a larger Δt in (24) than the one finally decided upon.

Still another technique is to replace the strictly convex yield surface by a piecewise linear one [39]. Then, so long as Δt is kept small enough that no triangle moves from one face to another of the yield polygon, motion along the tangent is identical with motion along the surface, so that no error at all is introduced. Indeed, during a time interval in which no triangle changes state between E, U, or L or changes from one face to another in L, all the rate equations are linear in time, so that (12) is now a precise formula rather than an approximation.

An obvious drawback to this approach is that a large number of linear equations must be used to define the yield curve, rather than a single non-linear one. However, this drawback is not as severe as it might seem. For a polygonal yield surface, let the collection of equations of the faces be written $f_\alpha = 1$ where each f_α is linear. Then (3c) must be replaced by

$$\dot{\varepsilon}_{ij}^p = \Sigma_\alpha \dot{\lambda}_\alpha \; \partial f_\alpha / \partial \sigma_{ij} \tag{25}$$

where each f_α, $\dot{f}_\alpha$, and $\dot{\lambda}_\alpha$ is subject to the restrictions of (3d–g). At any given time, most of the f_α will be less than 1. Therefore, even though the total description of f might require, say, 20 faces, the only ones entering the equations will be the one or two for which $f_\alpha = 1$.

7. Finite difference methods

One of the big advantages of finite element methods is that they are readily applied to highly irregular boundaries. However, many significant problems have regular boundaries and simple boundary conditions; for such problems, finite difference methods are simple to formulate, and they are frequently found to give better answers for a given amount of computing [40, 41]. They are, however, subject to many theoretical disadvantages.

A primary objection is that the Principle of Virtual Work is not necessarily valid. Indeed, it has been shown [42] that for the simplest finite-difference model in which derivatives are approximated by lowest ordered centered differences, the principle holds in two-dimensions only for a rectangular domain. Although a slight modification has been introduced to extend the principle to any domain with only vertical and horizontal side segments [42], the method does not appear to be easily extended to more general boundaries or to formulations which involve more than first-order derivatives.

Another questionable feature of a finite-difference model for a plasticity problem is the role of discontinuities. It is well known that exact solutions to simple continuum problems for rigid-plastic and elastic-perfectly-plastic problems sometimes exhibit strong discontinuities in stress or velocity components. For example, the tangential stress along the diagonals of a square cylinder in fully-plastic torsion is discontinuous [35, Chap. 3] and most solutions of rigid-perfectly-plastic bodies in plane strain involve discontinuities in the tangential velocity [35, Chap. 6]. Such discontinuities are most disturbing to a finite-difference approach, since the usual derivation of the finite difference equations begins with expansion in a Taylor's series. We note that such discontinuities cause no particular difficulty in finite element methods since discontinuities of stress generally occur across the triangle boundaries as approximations to continuously varying stress fields. Therefore, if the continuum field has a discontinuity, it should merely show up as a larger discontinuity across the nearest element boundary.

Still another drawback is that even simple problems may have boundaries with corners or points of transition between traction and displacement conditions, and it is not clear what boundary conditions nor even how many conditions should be imposed at such points. Finally, if mixed derivatives are present, there is an ambiguity of formulas, in that expressions for the derivatives at a grid point can be expressed in terms of values of its eight neighbors in more than one way, each to the same accuracy h^2.

As a consequence of these various undesirable mathematical features, a well-posed continuum problem in elasticity may have two finite-difference solutions for displacements which cannot be resolved by a rigid-body motion [43].

Finite elements offer a possible technique for constructing finite difference models which are devoid of these difficulties. To this end, a finite element model is constructed with reference to the grid points of the finite difference model. Figure 2 shows three simple models with triangular elements. Figs. 2a and 2b, where the only nodes are grid points may be interpreted immediately in terms of the grid point values; hence they constitute a finite difference model directly. If symmetric formulas are desired, an average of those obtained from 2a and 2b may be used.

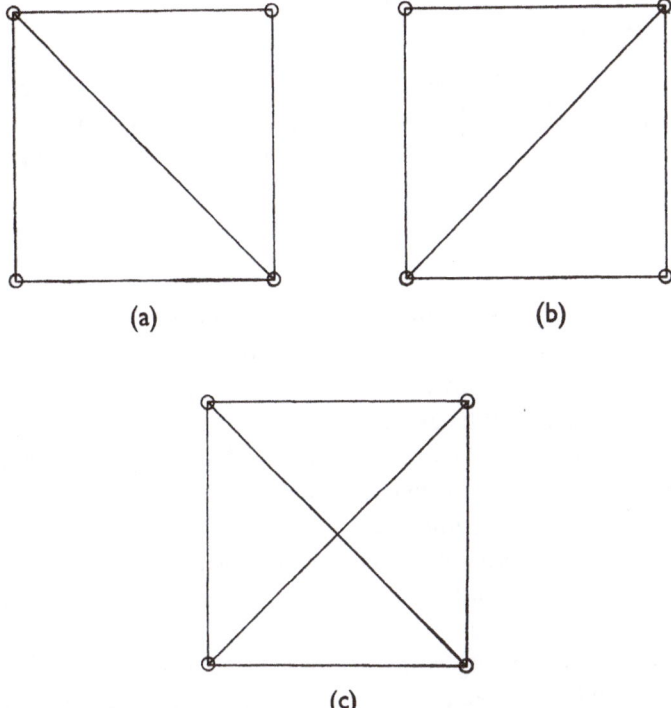

(a) (b)

(c)

Fig. 2. Finite element models for finite difference methods.

In Fig. 2c, an additional node has been introduced. However, the two equilibrium equations at this extra node may be used to explicitly solve for the nodal displacements there for any given constitutive equation, and hence all equations can again be reduced to grid form.

The rectangular mesh can be used directly as a finite element by allowing a displacement field of the form

$$u_i = a_i + b_i x + c_i y + d_i xy \tag{26}$$

in rectangle i. The four constants are determined by the nodal displacements at the nodes (grid points). Since u_i is linear along each line $x = $ const. or $y = $ const., this process produces a continuous displacement field. It turns out that the stress vector σ should now be a five-vector with three components given by (18) and the remaining two by

$$m_i^k = \frac{1}{A^k} \int_{T^k} (x_2 \sigma_{i1} + x_1 \sigma_{i2}) dA \tag{27}$$

Similarly, there are additional strain components which represent finite difference equivalents of $u_{,xy}^k$ and $v_{,xy}^k$. It is not surprising that (27) is not a proper tensor formula, since (26) is written for a specific coordinate system.

8. Future directions

It is obvious that the use of finite-element models on large computers is a practical method for obtaining useable answers to complex real problems which involve plastic behavior. It follows that existing programs will be used and that efforts will be expended to make these programs faster, more accurate, easier to use, and more widely applicable. These efforts will involve the creation of new finite-element models and of improved methods of solution for the resulting finite sets of equations. Finally, these resulting programs will be used to solve ever-more problems of ever-wider scope.

As new models are constructed, it seems important to bear in mind the Principle of Virtual Work. In the first place, it provides an automatic technique for constructing a model based on any appealing kinematic assumption. Natural definitions are encountered for generalized stresses and loads, and equilibrium and constitutive equations are easily derived.

A natural question that arises is whether a conjugate approach can be used in which static and kinematic roles are reversed. In such a model, we begin with basic static assumptions and use the Principle of Virtual Work to define generalized strains and obtain compatibility equations. Such an approach would usefully complement the kinematic approach in that both lower and upper bounds on the continuum solution could be obtained from the complementary minimum principles available for most materials.

However, static assumptions are by no means as easy to formulate as kinematic ones. A displacement field is subject only to weak continuity requirements so that it is trivial to construct one of any desired order. However, a static field is subject to internal equilibrium requirements and it is not simple to incorporate the resulting constraints. Some work in this direction has been done [44], but there is need for much more.

A problem of considerable computational interest is an analysis of the relative advantages of a more complex basic element as opposed to more elements. For an elastic material, it appears generally preferable to use more complex elements. We can see this quite simply if we examine a one-dimensional problem of curve fitting. Over a given length I, three values can be used to define either a parabola or a broken straight line. The first corresponds to a single quadratic kinematic field for I, the second to dividing I into two pieces with a linear field in each. The mathematical computation and storage will be about the same for each. Since elastic solutions are generally very smooth, it is obvious that the quadratic formulation will be more accurate for a given effort.

However, the conclusion for a plastic material is not obvious. The exact plastic solution may exhibit discontinuities on the elastic-plastic boundary, so that the broken line would actually give a better fit. Further, the linear displacement field gives rise to a constant stress field in each element which is either all elastic or all plastic. A more complex displacement field might lead to a partially plastic element which would tremendously complicate the use of the options in Eqs. (3).

Mention has already been made of the possible problems arising from the options of loading or unloading which are available to each element at yield. Since many useful production programs use an iteration method for apportioning elements between U and L, it would be highly desirable to know more about its convergence properties.

The integration problem is obviously not in a satisfactory state. Although the piecewise linear approach has great theoretical appeal, it may not be really practical for large problems for which great accuracy is required. A great disadvantage is that if there are many elements the allowable time steps may become exceedingly small. Not only should work be done on finding a better scheme for problems where the error is important, but a criterion should be established for deciding when the error resulting from the simple use of (12) or (23) is small enough to be ignored.

A final area which the author considers important is to continue the

279

reexamination of finite-difference methods through finite-element models. Prager [45] has used this approach for an elastic problem, but little work has been done for an arbitrary material.

References

[1] Tresca, H., Mémoire sur l'écoulement des corps solides, *Mém. pres. par div. sav. 18* (1868) 733–799.

[2] Prager, W., Théorie générale des états limits d'équilibre, *J. Math. Pure Appl. 34* (1955) 395–406.

[3] Galilei, G., *Discorsi e Dimonstrazioni Matematiche*, Leiden, 1638, p. 157.

[4] Argyris, J. H. and P. C. Patton, Computer oriented research in a university milieu, *Appl. Mech. Rev. 19* (1966) 1029–1039.

[5] Southwell, R. W., *Relaxation methods in engineering science*, Oxford Univ. Press, 1940.

[6] Padlog, J., R. D. Huff, and G. F. Holloway, Unelastic behavior of structures subjected to cyclic, thermal, and mechanical stressing conditions, *WADD Tech. Rep.* 60–271, Bell Aircraft Corp., Buffalo, New York, 1960.

[7] Argyris, J. H., *Continua and discontinua, Matrix methods in structural mechanics*, Wright-Patterson Air Force Base, Ohio, 1965, pp. 11–190.

[8] Marcal, P. V., Finite-element analysis of combined problems of non-linear material and geometric behavior, *Computational Approaches in Applied Mechanics*, ASME, New York, 1969, pp. 133–149.

[9] Koiter, W. T., Stress-strain relations, uniqueness, and variational theorems for elastic-plastic materials with a singular yield surface, *Q. Appl. Math.*, *11* (1953) 350–354.

[10] Hodge, P. G., Jr., On the plastic strains in slabs with cutouts, *J. Appl. Mech. 20* (1953) 183–188.

[11] Hodge, P. G., Jr., The effect of strain hardening on an annular slab, *J. Appl. Mech. 20* (1953) 530–536.

[12] Hodge, P. G., Jr., Displacements in an elastic-plastic shell, *J. Appl. Mech. 23* (1956) 73–79.

[13] Hodge, P. G., Jr. and F. Romano, Deformations of an elastic-plastic cylindrical shell with linear strain hardening, *J. Mech. Phys. Solids, 4* (1956) 145–161.

[14] Hodge, P. G., Jr. and R. Sankaranarayanan, On finite expansion of a hole in a thin infinite plate, *Q. Appl. Math. 16* (1958) 74–80.

[15] Perrone, N. and P. G. Hodge, Jr., Strain hardening solutions with generalized kinematic models, *Proc. Third U.S. Nat. Congr.* (Providence, 1958) 641–648.

[16] Hodge, P. G., Jr. and M. Balaban, Elastic plastic analysis of a rotating cylinder, *Int. J. Mech. Sci. 4* (1962) 465–476.

[17] Hill, R., A general theory of uniqueness and stability in elastic-plastic solids, *J. Mech. Phys. Solids 6* (1958) 236–244.

[18] Goodier, J. N. and P. G. Hodge, Jr., *Elasticity-Plasticity*, John Wiley and Sons, Inc., New York, 1958.

[19] Hodge, P. G., Jr., Numerical applications of minimum principles in plasticity, *Engineering Plasticity*, J. Heyman and F. A. Leckie, eds., Cambridge Univ. Press, Cambridge, 1968, pp. 237–256.

[20] Hodge, P. G., Jr., The practical significance of limit analysis, *J. Aerospace Sci. 25* (1958) 724–726.

[21] Hodge, P. G., Jr., Boundary value problems in plasticity, *Plasticity*, E. H. Lee and P. S. Symonds, eds., Pergammon Press, Inc., New York, 1960, pp. 297–337.

[22] Argyris, J. H. and D. W. Scharpf, Some general considerations on the natural mode technique, *Aero J. RAS 73* (1969) 218–226.

[23] Argyris, J. H., *Recent advances in matrix methods of structural analysis*, Macmillan Co., New York, 1964.

[24] Hodge, P. G., Jr., A consistent finite element model for the two-dimensional continuum, *Ing. Archiv. 39* (1970) 375–382.

[25] McMahon, A. and P. G. Hodge, Jr., A simple finite element model for elastic-plastic plate bending, Appendix A, *DOMIIT Rep.* 1–46, Illinois Institute of Tech. 1971.

[26] Fiacco, A. V. and F. P. McCormick, Programming under nonlinear constraints by unconstrained minimization, *Research Analysis Corporation*, McLean, Va. RAC-TP-96, 1963.

[27] Fiacco, A. V. and F. P. McCormick, Extensions of SUMT for nonlinear programming, *Management Sci. 12* (1966) 816–828.

[28] Hodge, P. G., Jr., Yield-point load determination by non-linear programming, *Proc. 11th Int. Congr. Appl. Mech.* (Munich 1964), 554–561, J. Springer, (1966).

[29] Hodge, P. G., Jr., Elastic-plastic torsion as a problem in nonlinear programming, *Int. J. Solids. Struct. 3* (1967) 989–999.

[30] Hodge, P. G., Jr. and T. Belytschko, Numerical method for the limit analysis of plates, *J. Appl. Mech. 35* (1968) 396–402.

[31] Hodge, P. G., Jr., T. Belytschko, and C. T. Herakovich, Quadratic programming and plasticity, *Computational Approaches in Applied Mechanics*, American Society of Mechanical Engineers, pp. 73–84, 1969.

[32] Cutler, L., *Product form quadratic programming code*, SHARE General Program Library, Program RS QRF4, 1965.

[33] Herakovich, C. T. and P. G. Hodge Jr., Elastic/plastic torsion of multiply connected cylinders by quadratic programming, *Int. J. Mech. Sci. 11* (1969) 53–62.

[34] Belytschko, T. and M. Velebit, *Finite element method for elastic plastic plates*, *J. Eng. Mech. Div. Proc. ASCE 98* EM1 (1972) 227–242.

[35] Prager, W. and P. G. Hodge Jr., *Theory of perfectly plastic solids*, John Wiley and Sons, Inc., New York, 1951; Dover Publications, Inc., New York, 1968.

[36] Hodge, P. G., Jr., On the sand-hill soap-film analogy for elastic-plastic torsion, *Progress in Applied Mechanics*, Prager Anniversary Volume, The Macmillan Co. (1963) 361–369.

[37] Drucker, D. C., some implications of work hardening and ideal plasticity, *Q. Appl. Math. 7* (1950) 411–418.

[38] Morris, G. A. and S. J. Fenves, Elastic-plastic analysis of frameworks, *J. Struct. Div. ASCE 96*, ST5 (1970) 931–946.

[39] Garg, V. K., S. C. Anand and P. G. Hodge, Jr., Elastic-plastic analysis of a wheel rolling on a rigid trock, AEM Report HI-7, Univ. of Minnesota, 1973.

281

[40] Bushnell, D. and B. O. Almroth, *Finite-difference energy method for nonlinear shell analysis*, Lockheed Palo Alto Research Laboratory, Palo Alto, California, 1970.

[41] Hodge, P. G., Jr., C. T. Herakovich, and R. B. Stout, On numerical comparisons in elastic-plastic torsion, *J. Appl. Mech. 35* (1968) 454–459.

[42] Hodge, P. G., Jr., A consistent finite difference model for the two-dimensional continuum, *Continuum mechanics and related problems of analysis*, L. I. Sedov, ed. Nauka Publ. House, Moscow 1972, pp. 605–614.

[43] Tanrikulu, M. and W. Prager, *Consistent finite difference equations for thin elastic disks of variable thickness*, IRPA Rep. 66–90 Univ. of Calif., San Diego, 1966.

[44] Belytschko, T. and P. G. Hodge Jr., Plane stress limit analysis by finite elements, *J. Eng. Mech. Div. ASCE 96* EM6 (1970) 931–944.

[45] Prager, W., Variational principles of linear elastostatics for discontinuous displacements, strains, and stresses, Recent Progress in Applied Mechanics; the Folke Odqvist Volume, Stockholm, 1969, pp. 463–474.

DISCUSSION

J. Zarka[1]).   I should like to make some remarks about the very interesting lecture of Professor Hodge.

1. Choice of Δt. When we want to integrate a differential equation $\dot{x} = f(x)$, we have to know the nature of the function f (for example does it satisfy the Lipschitz condition). In elastoplasticity we have the same kind of problem, but of course the functional f is rather complicated to study. Generally, empirically we take a not too big step Δt in order to have reasonably small strain increments. Often, we forget the influence of the choice of this Δt. Did you think about this problem?

That as you showed is very important, mostly for ideally plastic material because, for any $\Delta t \neq 0$, we go away from the yield criterion.

2. Theorem of convergence. We can prove that for ideally plastic or work hardening material an iterative method in elastoplasticity is convergent. In this method the field of the velocities of displacement is the unknown field and the initial stresses are introduced. There is convergence, of course, only when the limit load has not been reached, [1].

3. Viscoplastic regularisation. For a perfectly plastic material it has been proved by mathematicians that we can use a viscoplastic regularisation in order to find the field of the stress rate tensor. This regularisation is useful

[1]) Ecole Polytechnique, Paris, France.

in numerical calculus. From experience we can say that it is faster than your method which is based on quadratic programming.

References

[1] Nguyen, Q. S. et J. Zarka, Quelques méthodes de résolution numérique en elastoplasticité classique et en elastoviscoplasticité, Séminaire Plasticité et Viscoplasticité 1972, Ecole Polytechnique, Paris, France.

P. G. Hodge, Jr. The author wishes to thank Professor Zarka for his comments and suggestions.

With regard to integration with respect to time, the problem will be essentially as well-behaved mathematically as the time-dependence of the prescribed loads. In any physically significant problem with known loads the loads will be piecewise as well-behaved as one likes; say continuously differentiable. Provided we require that each Δt interval lies within only one such piece, there is no need to introduce such sophistications as Lipschitz conditions.

The convergence referred to is convergence of a specific process; namely the assigning of state L or U to each element based on the predictions of the previous incorrect assignments. The fact that other iterative methods of solution converge does not automatically answer the question as to whether this one does.

The idea of using a visco-plastic model with small but non-zero viscosity in place of the perfectly plastic model which corresponds to the limiting case of zero viscosity is certainly an intriguing one, and the author is grateful to Professor Zarka for mentioning his work on the subject.

O. C. Zienkiewicz[1]). Professor Hodge made the remark that most programs concerned with plasticity not checking on the positiveness of the value of the plasticity proportionality factor.

This certainly is done in the suite of plasticity programs which we use in Swansea and at all times a distinction is made between plastic or elastic regime or deformations.

However, it should be admitted that one of the main difficulties associated with numerical solution of plasticity is that of keeping the stresses exactly

[1]) University of Wales, Swansea, UK.

on the yield surface. To overcome this, recently a complete re-formulation of programs has taken place and this has invoked the theory of visco-plasticity. Problems are now being dealt with as creeping materials with the rate of creep proportional to the distance of the yield surface. This, while solving a new class of problems of realistic physical value, at the same time provides a means of obtaining pure plasticity solutions by introducing artificial viscosity. Here the stresses relax towards the plastic yield values at a decreasing rate finally reaching this with a reasonable degree of approximation. The program is of an initial strain type well known for its development in the solution of creep problems and is easier to achieve convergence with it than with standard plastic operations.

We feel that the introduction of visco-plasticity is one of the major achievements of the past year and opens up very many new doors.

G. Maier[1]) and O. De Donato[1]). Some points contained in the brilliant lecture of Professor Hodge might be supplemented, in the discussers' opinion, by the remarks which follow.

a) A formal analogy, [1] [2], exists between a class of elastoplastic flow laws (including those considered in Hodge's lecture) and a broad class of piecewise-linear, holonomic stress-strain relations. The description of both classes of constitutive laws may be centered on the algebraic structure called linear complementarity problem (LCP). E.g., let Eq. (25) be associated to the LCP in $\dot{\phi}_\alpha$, λ_β:

$$\dot{\phi}_\alpha \equiv \frac{\partial f_\alpha}{\partial \sigma_{ij}} \dot{\sigma}_{ij} - \sum_{1\,\beta}^{y} \frac{1}{H_{\alpha\beta}} \lambda_\beta, \; \dot{\phi}_\alpha \leqq 0, \; \lambda_\alpha \geqq 0, \; \dot{\phi}_\alpha \lambda_\alpha = 0, \qquad (25\text{-}a)$$

α and β being indices running over the set of y yielding modes which can be activated in the flow process at t_0. The fairly general 'corner' flow law thus defined reduces to that described by Eqs. (3-c-h) for $y = f$. When a finite element model is considered, both the rate problem with the former flow laws (even in the presence of significant geometry changes) and the analysis problem in finite terms with the latter constitutive laws, can be cast in LCP forms in the variables $\dot{u}$, λ (or u, λ) and in the variables λ (or λ) alone. Under a suitable condition (of overall 'stability'), the LCP forms lead, via Kuhn-Tucker theorem and duality, to pairs of quadratic programming (QP) formulations of the system analysis problems, [1] [3] [4] [5].

[1]) Technical University (Politecnico) Milan, Italy.

Thus the above analogy carries over to these classes of problems and is used below, inasmuch it permits to transfer theoretical results and computational experience from one class to the other.

b) An iterative procedure for solving LCP's by checking and refining subsequent 'incomplete' solutions were proposed first by Hildreth [6], have been used in both the above classes of mechanical problems, [4] [7], and their convergence has been proved, [4] [6], on the basis of the equivalent QP problem. Since the process outlined in Sect. 5 of the lecture bears a resemblance to Hildreth's, discussers feel that its convergence can be proved in a similar way.

As for the computational aspects, in the discussers' experience, the QP formulation in the λ (or $\dot{\lambda}$) variables alone, combined with a conjugate gradient algorithm, seems particularly efficient, [8], although also algorithms with finite termination have provided satisfactory results, [7] [9].

c) The piecewise linear approach to the integration problem (Sect. 6) exhibits, as emphasized in Sect. 8, the disadvantage that the time steps may become exceedingly small if there are many elements. This difficulty does not arise in the 'multistage' procedure, [7] [8], outlined below.

Let the loading history after the onset of yielding at $\bar{t}$ be *a priori* subdivided in stages $\Delta t_1 \ldots \Delta t_n$, such that the loads vary proportionally in each stage. The analysis along Δt_1 is performed by assuming holonomic piecewise linear laws, i.e. by solving a single LCP or QP problem. The additional response to the subsequent stage Δt_2 is evaluated in the same way, but after having adjusted the constitutive laws so that the plastic multipliers λ cannot decrease below the values attained at $\bar{t} + \Delta t_1$. The same is done for Δt_3, etc. Thus the irreversibility is fully allowed for from stage to stage, and errors of physical significance may arise only if the sequence loading-unloading of a yield mode occurs within a single stage Δt_i. This is an unlikely accurrence, particularly if Δt_i is kept small.

However even these errors may be detected if an LCP algorithm is used which incorporates 'monotonicity' checks, such that devised in [10]. The practical value of this method is currently under investigation.

References

[1] Maier, G., A quadratic programming approach for certain classes of non linear structural problems, *Meccanica*, *3*, N. 2 (1968).

[2] Maier, G., A matrix structural theory of piecewise linear elastoplasticity with interacting yield planes, *Meccanica*, *5*, N. 7 (1970).

[3] Maier, G., Incremental elastoplastic analysis in the presence of large displacements and physical instabilizing effects, *Int. J. Solids Structures*, *7* (1971) p. 345.

[4] Ceradini, G., Sul calcolo delle strutture elastoplastiche, *Costruzioni Metalliche*, *N. 3–4* (1965).

[5] Maier, G., Quadratic programming and theory of elastic perfectly plastic structures, *Meccanica*, *3*, *N. 4* (1968).

[6] Künzi, M. P. and W. Krelle, *Non linear programming*, Blaisdell, Waltham, Mass., 1966.

[7] De Donato, O. and G. Maier, Mathematical programming methods for the inelastic analysis of reinforced concrete frames allowing for limited rotation capacity, *Int. J. Num. Meth. Eng.*, *4* (May 1972).

[8] De Donato, O. and A. Franchi, A modified gradient method for finite element elastoplastic analysis by quadratic programming, *Technical Report N. 6, ISTC* (July 1972), to appear in *Comp. Meth. Appl. Mech. Eng.* (1973).

[9] Corradi, L., O. De Donato and G. Maier, Inelastic analysis of reinforced concrete frames allowing for second order effects, *Int. Symp. on Inelasticity and Nonlinearity in Reinforced Concrete Structures*, Univ. of Waterloo, Ontario (July 1972).

[10] Cottle, R. W., Monotone solutions of the parametric linear complementary problem, *Mathematical Programming*, *2*, *N. 4* (1972).

P. G. Hodge, Jr.   The author also wishes to thank Professors Zienkiewicz, Maier, and De Donato for their further comments which arrived somewhat later.

It is significant that both Professor Zarka and Professor Zienkiewicz should have suggested the idea of artificial viscosity. It appears that this technique may be a most practical one for eliminating the problem discussed in Sec. 6.

Any value which the paper may have has certainly been enhanced by the work and references countributed by Professors Maier and De Donato.

In conclusion, the author would like to point out that positive contributions to the field of his lecture have been submitted from three different countries. This fact certainly supports the validity of international conferences such as the one where these discussions took place.

Rate-type constitutive equations in dynamic plasticity

N. Cristescu

*University of Bucharest, Romania
and Center for Dynamic Plasticity, University of Florida, Gainesville, USA*

The main objective of the paper is to show in what way the informations furnished by experiments can be used in order to determine a constitutive equation. It was demonstrated that quasi-linear rate-type constitutive equations can successfully describe the main phenomena observed by experimentalists in one-dimensional dynamic plasticity tests. As a consequence of the analysis it was shown what kinds of experiments are necessary and in what priority these have to be performed in order to be useful for the determination of a constitutive equation.

1. Introduction

In the present paper it will be discussed in what way the information furnished by experimental data may be used in order to formulate a constitutive equation in dynamic plasticity.

Two types of such experiments will be discussed: the longitudinal impact of two identical bars and the combined stress experiments on thin-walled tubes.

For the first type of tests a great number of experimental results are already available in the literature. All the experimental data used in the present paper are due to J. F. Bell. These are: the strain-time curves at various cross-sections along the bar obtained with the diffraction grating technique (for some cross-sections of the bar these curves were obtained with electric resistance strain gauges), the variation in time of the surface angle (angle between the normal to the lateral surface of the specimen and the initial position of the same normal) at various cross-sections, piezocrystal measurements of the variation of stress in time at the impacted end of the specimen, measurements with an optical method of variation in time of the displacement at various cross-sections of the bar, the time of contact between the two bars measured optically and finally the final velocity of the specimen after rebound from the hitter. Therefore we have used seven sets of distinct experimental data (seven distinct experimental devices) obtained for one material – aluminium – in experiments of longitudinal impact of two identical bars.

Experimental data for combined stress tests reported to date are still scarce. Disparate results have been reported for various materials. The amount of data reported for a single material is far of being so complete as those obtained for bars. Sometimes only strain-time measurements obtained with electric resistance strain gauges have been reported, while the validity of such measurements for high plastic strains is questioned by many experimentalists. Even particle velocity measurements not accompanied by other kinds of data, are of little use when a constitutive equation is to be formulated since it is known from the analysis made for bars that from all unknown functions involved, the particle velocity is the less sensitive to quite significant changes in the constitutive equation. That is why for data of combined stress experiments only a short qualitative discussion of a rather theoretical character will be presented here.

2. Experimental data

The main experimental observations on which the following discussion will be based is the variation in time of the strain at various cross-sections along the bar, as obtained with the diffraction grating technique. A typical curve of this sort for copper is shown in Fig. 1 (Bell [2]). The full line is a theoretical curve plotted with a finite form of a constitutive equation. The dotted line represents the variation in time of the surface angle. Several details have to be observed. The $\varepsilon - t$ curve possesses an initial rise where strain is generally much higher than the strain at the yield point. At a certain point somewhere in the middle portion, the $\varepsilon - t$ curve changes the curvature (the magnitude of the strain corresponding to this point will be denoted by $\bar{\varepsilon}$) and it is at this point that the surface angle α reaches its maximum. This is the 'main' point of inflection, since the second one may occur very early on the $\varepsilon - t$ curve.

Curves similar to the one shown in Fig. 1 can be obtained with electric resistance gauges, but quantitatively such measurements are less accurate than those obtained with diffraction grating technique. However, for a qualitative discussion and cross checking such type of experimental data has been considered as well.

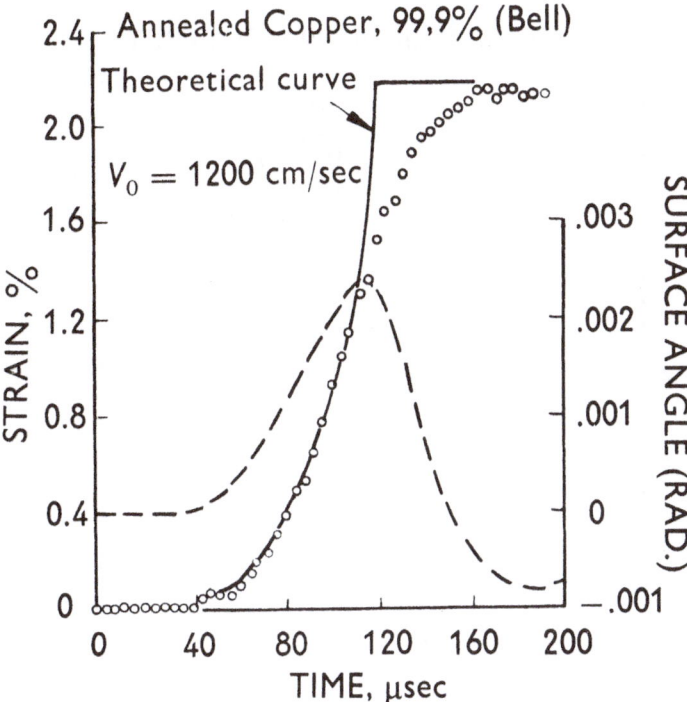

Fig. 1. Typical curves showing variation in time of the strain and surface angle at a certain cross-section of a copper bar (Bell) (3 D means 3 diameters from the impacted end, etc.).

The only place where the stress can be measured is at the end of the bar where the impact occured. This can be done either with a piezo-crystal placed between the two impacting bars or by measuring the strain in a hard transmitter bar (an elastic bar through which the loading pulse is transmitted). Two such typical curves are shown in Fig. 2 and Fig. 3 whereas Fig. 2 shows a piezo-crystal measurement as obtained by Filbey. It is of great importance for the following discussion to observe on this curve the presence of a quite long 'stress plateau' where stress is nearly constant. In the first microseconds following the impact a huge 'peak stress' is developed which decays very fast to the value of the plateau. An increase of the stress above the plateau as obtained in a hard transmitter bar is shown in Fig. 3. This time the stress maintains its higher value for a longer time than the stress of the plateau, but its magnitude is much smaller than that of the 'peak'. Such phenomenon is called 'overstress'. Both phenomena may be present in a certain experiment, while sometimes only one of them has been

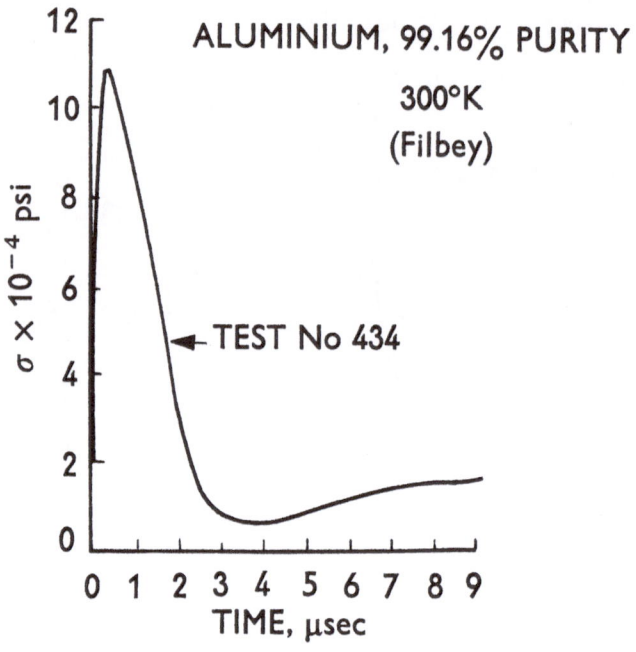

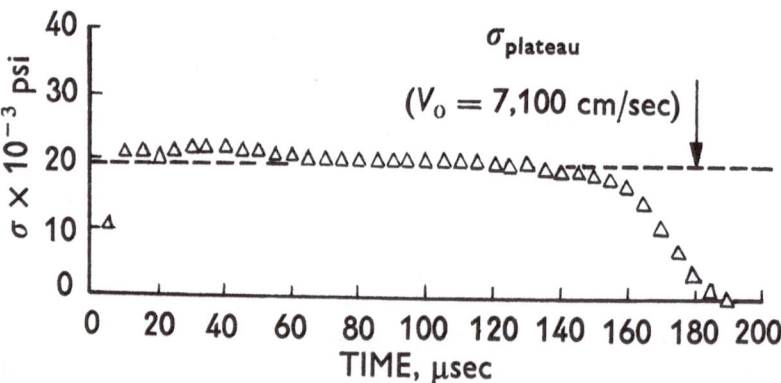

Fig. 2. The variation in time of the stress at the impacted end of the bar (Filbey).

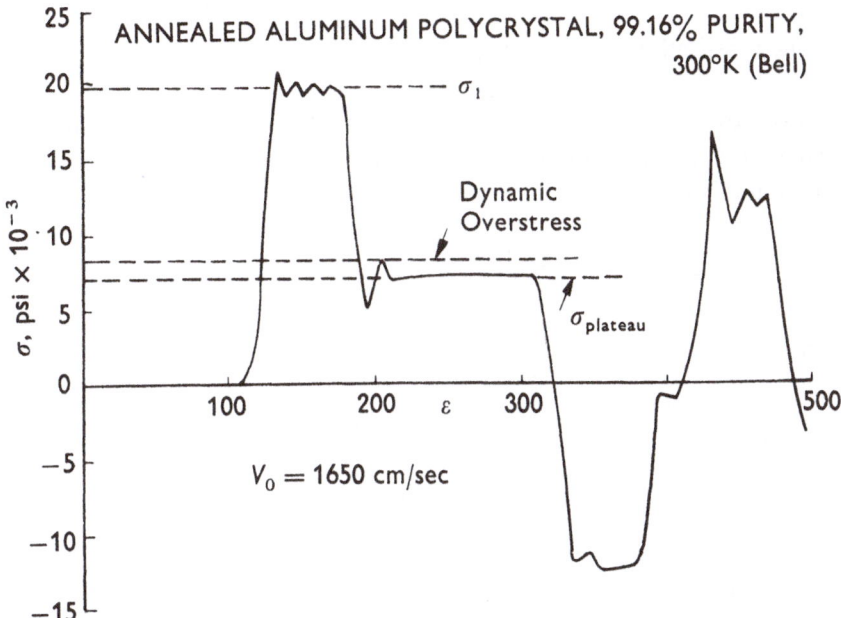

Fig. 3. Variation of stress as measured in a hard transmitter bar (Bell).

found. Generally the stress measurements at the impacted end furnish the stress history for the whole duration of the experiment. The main features of the stress-time curves at the impacted end are: the peak stress, the over-stress, the stress plateau and finally the last decreasing part of the curve which is connected to the unloading process. The moment when stress becomes zero at the impacted end is the 'time of contact' between the two bars. All the previously mentioned characteristics are of great significance for establishing an appropriate constitutive equation for the material under consideration.

The variation in time of the strain at various cross sections along the bar and that of the stress at the impacted end are the main experimental data which have been used to find a constitutive equation. The other experimental data mentioned in Sec. 1 have been used only for cross checking. All have been considered since each set of experimental data has been obtained with a different experimental device. Generally seven sets of experimental data obtained with seven types of experimental devices have been used. All these data correspond to the material aluminium and have been obtained from the experiment of impact with the same velocity of two identical bars.

291

3. The constitutive equations

In order to describe the phenomena presented in the previous section, for a single loading-unloading cycle, one may choose a rate-type constitutive equation of the form

$$\dot{\varepsilon} = \frac{1}{E} \dot{\sigma} + \Phi(\sigma, \varepsilon)\dot{\sigma} + \Psi(\sigma, \varepsilon) \tag{3.1}$$

with

$$\Psi(\sigma, \varepsilon) = \begin{cases} \dfrac{k(\varepsilon)}{E} \left[\sigma - f(\varepsilon)\right] & \text{if } \sigma > f(\varepsilon) \quad \text{and} \quad \varepsilon \geqq \varepsilon_Y, \\ 0 & \text{if } \sigma \leqq f(\varepsilon) \quad \text{or} \quad \varepsilon < \varepsilon_Y, \end{cases} \tag{3.2}$$

where

$$\sigma = f(\varepsilon)$$

is the equation of the relaxation boundary (see Cristescu [6, 7], for additional details). The existence of such a relaxation boundary was indicated by the presence of both stress and strain plateaus near the impacted end of the bar for a long period of time. For $f(\varepsilon)$ the following expression was used

$$f(\varepsilon) = \begin{cases} \sigma_Y & \text{if } \varepsilon \leqq \varepsilon_Y, \\ \sigma_Y + \dfrac{\beta}{2} \varepsilon_z^{-\frac{1}{2}}(\varepsilon - \varepsilon_Y) & \text{if } \varepsilon_Y \leqq \varepsilon \leqq \varepsilon_z, \\ \beta \varepsilon^{\frac{1}{2}} & \text{if } \varepsilon_z \leqq \varepsilon, \end{cases} \tag{3.3}$$

where

$$\varepsilon_z = \left(\frac{\beta \varepsilon_Y}{\sigma_Y - \sqrt{\sigma_Y^2 - \varepsilon_Y \beta^2}}\right)^2, \tag{3.4}$$

and β is a material constant.

The coefficient k can be considered to be a constant as a first approximation. However, varying values of k considered smaller for small strains are improving the solution significantly. The following law of variation for k has been used

$$k(\varepsilon) = \begin{cases} k_1 & \text{for} \quad \varepsilon \leqq \varepsilon_1, \\ \dfrac{k_1-k_2}{\varepsilon_1-\varepsilon_2}(\varepsilon-\varepsilon_2)+k_2 & \text{for} \quad \varepsilon_1 \leqq \varepsilon \leqq \varepsilon_2, \\ k_2 & \text{for} \quad \varepsilon_2 \leqq \varepsilon, \end{cases} \tag{3.5}$$

where k_1, k_2, ε_1 and ε_2 are constants.

The function $\Phi(\sigma, \varepsilon)$ describing instantaneous properties is defined in the domain $\varepsilon > \sigma/E$; for $\sigma < f(\varepsilon)$ we have $\Phi = 0$ while $\Phi \neq 0$ possibly only for $\sigma > f(\varepsilon)$. The following two forms for this function have been used

$$\Phi(\varepsilon) = \chi \left\{ \frac{3[\varepsilon-\varepsilon_Y-\varepsilon^*+(a/3E)^{\frac{3}{2}}]^{\frac{4}{3}}}{a} - \frac{1}{E} \right\} \tag{3.6}$$

and

$$\Phi(\sigma, \varepsilon) = \chi \left\{ \frac{3[\varepsilon-(\sigma/E)-\underset{\smile}{\varepsilon}+(a/3E)^{\frac{3}{2}}]^{\frac{4}{3}}}{a} - \frac{1}{E} \right\}. \tag{3.7}$$

In a first approximation one can take a as a constant. An improved solution, however, is obtained for

$$a = m+n\sqrt{\varepsilon} \tag{3.8}$$

with m and n constants. ε^* is defined by

$$\varepsilon^* = \frac{\sigma^*}{E} = h-\lambda\sqrt{|\varepsilon-f^{-1}[(\sigma_Y/E)+h]|}, \tag{3.9}$$

with h and λ as constants. By (3.9) a threshold stress σ^* is introduced so that in a narrow strip above the relaxation boundary we still have $\Phi = 0$.

In the expression (3.7) $\underset{\smile}{\varepsilon}$ is a constant (generally a very small one). The parameter $\chi(\sigma_m(x))$ depends on the maximum stress reached in each particular cross section of the bar and is defined as .

$$\chi = \begin{cases} 1 & \text{if} \quad \sigma = \sigma_m(x), \\ 0 & \text{if} \quad \sigma \leqq f(\varepsilon)+\sigma^* \quad \text{or} \quad \sigma < \sigma_m(x). \end{cases} \tag{3.10}$$

When choosing the expression for the function Φ, it was assumed that the slope of the instantaneous response of the material

$$\frac{d\sigma}{d\varepsilon} = \frac{1}{\Phi+(1/E)} \tag{3.11}$$

293

is smaller than E but greater than the slope of the relaxation boundary corresponding to the same strain (see Cristescu [6]).

Though the two expressions for Φ, namely (3.6) and (3.7) are furnishing generally quite close results, the expression (3.7) is qualitatively better for describing the peak stress.

4. Formulation of the problem and numerical integration

In order to obtain some numerical solutions for comparison with the experimental data, to the constitutive equation (3.1) the equation of motion

$$\frac{\partial v}{\partial t} = -\frac{1}{\rho}\frac{\partial \sigma}{\partial x} \tag{4.1}$$

and the compatibility equation

$$-\frac{\partial v}{\partial x} = \frac{\partial \varepsilon_E}{\partial t} + \frac{\partial \varepsilon_P}{\partial t} \tag{4.2}$$

have been added. Here the subscripts E and P means 'elastic' and 'plastic' (in a generalized sense) respectively and

$$\varepsilon = -\frac{\partial u}{\partial x}. \tag{4.3}$$

The system (3.1), (4.1), (4.2) is integrated using a method described elsewhere (Cristescu [4, 6]). In fact one uses the method of characteristics in which in each point of the characteristic plane the 'elastic' solution is considered to be the first approximation of the solution in that particular point (see Cristescu [4]).

Since the symmetric impact of two identical bars is considered the computations can be carried out for one of the bars only. For this bar, the 'specimen', the initial conditions are

$$\left.\begin{array}{c} t = 0 \\ 0 \leqq x \leqq l \end{array}\right\} : \sigma = \varepsilon = v = 0, \tag{4.4}$$

and the boundary conditions at the free end are

$$\left.\begin{array}{c} x = l \\ t > 0 \end{array}\right\} : \sigma = 0. \tag{4.5}$$

The velocity of impact V, is known at the impacted end and is the velocity of the hitter. It was assumed that this velocity is transmitted to the specimen in a linear way in a very short period of time, while the particle velocity of the specimen at the impacted interface equals that of the hitter after reaching its maximum. Thus

$$x = 0 \begin{cases} 0 \leqq t \leqq t_m: & v = \dfrac{t}{t_m}\, v_{max}, \\[2mm] t_m \leqq t < T_c: & v_H = v_S, \\[2mm] T_c \leqq t: & \sigma_S = 0. \end{cases} \tag{4.6}$$

Here t_m has been chosen as 0.5 μsec, $v_{max} = V/2$, while T_c is a computed quantity called 'time of contact' between the two bars. Thus the last condition (4.6) means that after rebound the stress at the impacted end is zero.

Due to the fact that just after impact the variation of functions near the impact end is very fast, the spacing between the characteristic lines in the basic grid was chosen in the interval $0 \leqq t \leqq 2$ μsec to be $\Delta t = \frac{1}{200}$ μsec. For 2 μsec $< t < 10$ μsec this spacing was gradually increased up to $\Delta t = = \frac{1}{4}$ μsec which was the spacing used throughout the whole remaining integration field (for $t > 10$ μsec).

5. Numerical examples and comparison with experimental data

Three examples will be presented and compared with experimental data. The formulas and constants used are listed in Table I. In Table II the re-

TABLE I

No. of example	Formulas used	Constants used
1	(3.1)(3.2)(3.3)(3.5) (3.6)(3.9)(3.10)	$\sigma_Y^0 = 1500.0$ psi; $\varepsilon_Y^0 = 0.0001471$; $\varepsilon_z = 0.002575$; $\alpha = 2$; $\beta = 5.6 \times 10^4$ psi; $k_1 = 100$; $k_2 = 1000$; $\varepsilon_1 = 0.0005$; $\varepsilon_2 = 0.004$; $m = 3.25 \times 10^4$ psi; $n = 9 \times 10^4$ psi; $h = 0.00003$; $\lambda = -0.00019$
2	(3.1)(3.2)(3.3)(3.5) (3.7)(3.9)(3.10)	same as 1, but $h = 0$; $\lambda = 0$; and $\overset{\vee}{\varepsilon} = 0$.
3	same as 2	same as 2 but $\overset{\vee}{\varepsilon} = 0.0004$

TABLE II

No. of example	σ_{max} (psi) peak	plateau	ε_{max} $x = 0$	plateau	T_c (μsec)	v_f (in/sec)	$\bar{\varepsilon}$ $x = 1$ D	Penetration of first unloading region $x\hat{U}$
1	10390	8394	0.02359	0.02248	306.5	884.2	0.0108	4 D
2	13475	8364	0.02403	0.02229	303.0	886.5	0.0107	4.25 D
3	30478	8328	0.02360	0.02210	301.0	882.4	0.0101	4.25 D
experiment	43000	8300	~0.0275	0.0219	310.8	891	0.0135	4 ~ 5 D

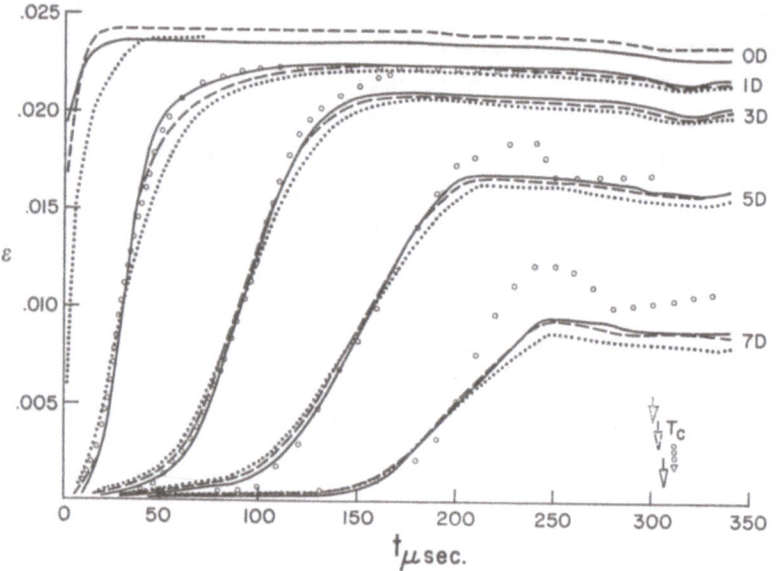

Fig. 4. Variation of strain at various cross-sections along the bar as furnished by various computed examples (Example 1 – full line; Example 2 – broken line; Example 3 – dotted line) and experiment.

sults obtained by numerical integration are given and the corresponding experimental values of the same quantities.

As already mentioned the greatest weight has been attributed to the graphs showing the variation of strain in time at various cross-sections along the bar, as obtained with the diffraction grating technique (see Fig. 4).

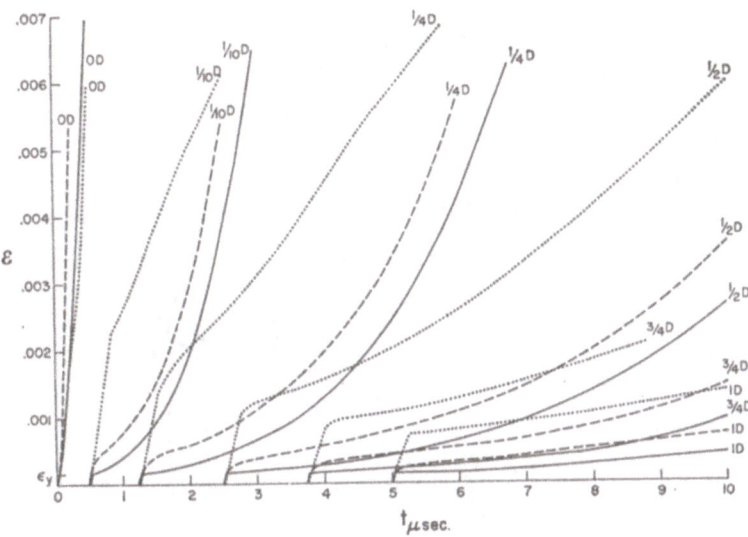

Fig. 5. Begining of the strain-time curves at cross-sections close to the impacted end (Example 1 – full line; Example 2 – broken line; Example 3 – dotted ine).

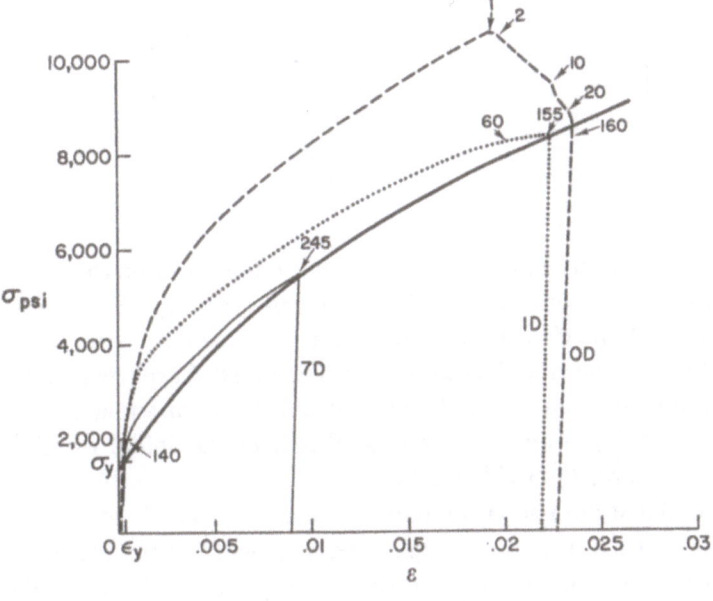

Fig. 6a.

Fig. 6. The stress-strain curves at various cross-sections; the solid line is the relaxation boundary. a – Example 1; b – Example 2; c – Example 3.

297

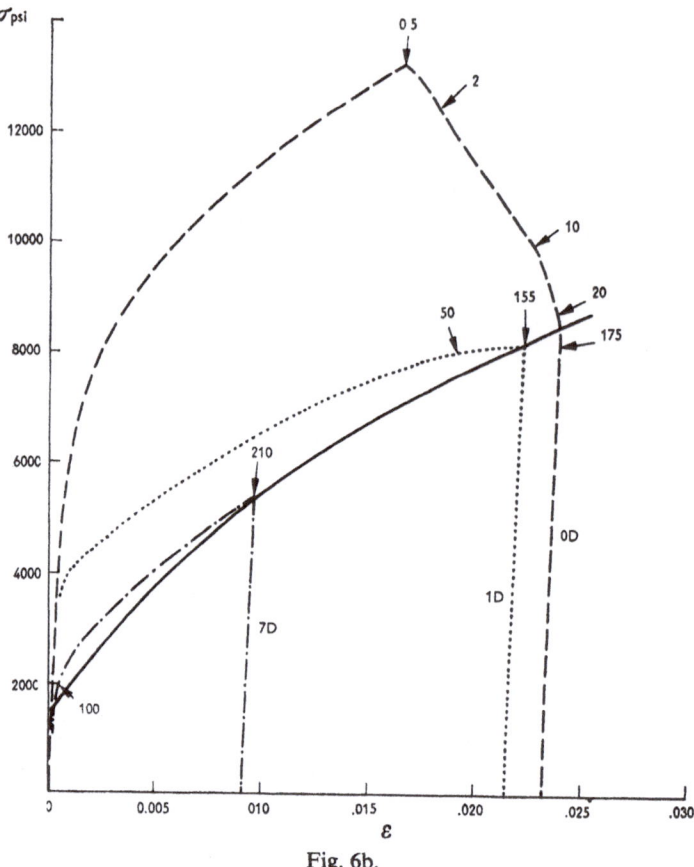

Fig. 6b.

The initial portion of these curves were not represented on this figure. Since the incipient portions of these curves are of great significance mainly for the cross-sections close to the impacted end of the bar, these were given in Fig. 5 (no experimental data available). The initial sharp rise of these curves raises the strain well above the yield strain. It is important to note, however, that during propagation the amplitude of this initial rise is decaying very fast towards the yield strain.

The initial strain rise at the impacted end corresponds to a nearly linear rise of the stress, before the peak-stress is reached. Thus the decay of the initial strain rise (or 'elastic precursor') takes place somehow simultaneously with the decay of the peak stress during propagation. This last decay can be seen on Fig. 6a to c. Here the solid line represents the relaxation boundary

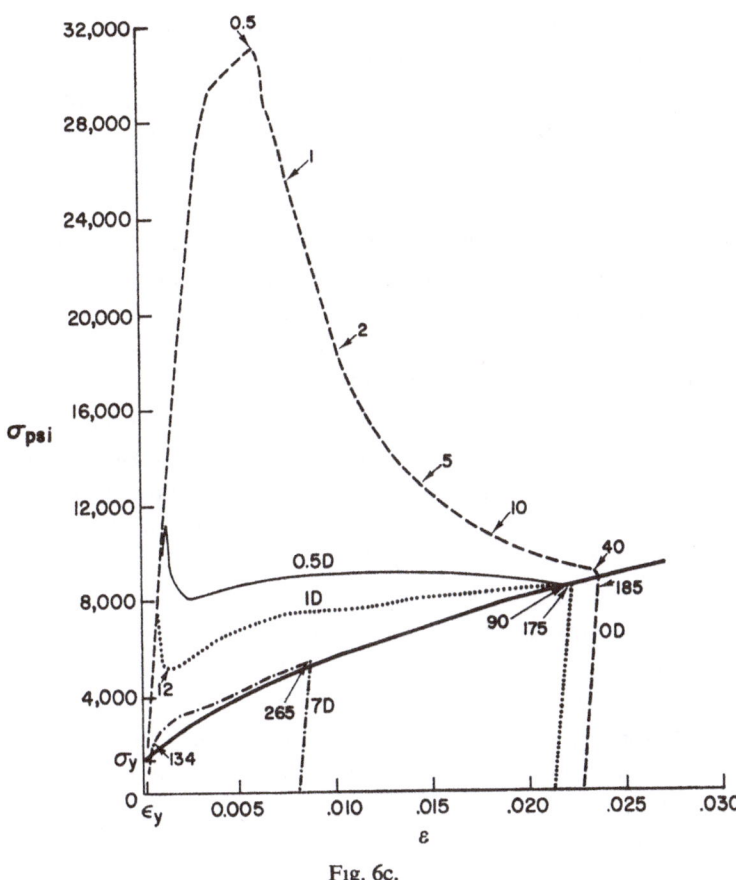

Fig. 6c.

(Eq. (3.3)) while the numbers written along various stress-strain curves indicate how many microseconds elapsed from the moment of impact between the two bars. For instance in Fig. 6c at the impacted end (0 D) the peak is reached in 0.5 μsec; after 40 microseconds the stress-strain point is very close to the relaxation boundary while it is only at 185 microseconds that this point is already significantly under the relaxtion boundary. In other words, the stress-strain states on the relaxation boundary (and on imediate vecinity) are quasistable states.

From Fig. 4 (solid lines) and Fig. 6a it is apparent that the expression (3.6) for Φ is poorly describing the peak-stress phenomenon, though otherwise it is furnishing a reasonable good solution, practically coinciding with the one obtained when the expression (3.7) for Φ is used. Therefore, in order to describe both peak stress and the initial rise of the strain-time curves one has to choose an expression function for Φ which would be either zero or negligible along the line $\sigma = E\varepsilon$, and possibly also in the neighbourhood of it. Thus, in the neighbourhood of this line the instantaneous response of the material would be 'nearly' elastic. A dependence of Φ on ε_p satisfies this requirement, while $\underset{\sim}{\varepsilon}$ introduces a strip along $\sigma = E\varepsilon$ where $\Phi = 0$.

One can therefore conclude that for those expressions for Φ possessing the previously mentioned property, the magnitude of the decaying peak stress and the initial strain rise can be governed by the rapidity of the loading at the end of the bar (i.e. t_m) and by $\underset{\sim}{\varepsilon}$. If the computed peak stress is still too small when compared with the experimental value (if only t_m and $\underset{\sim}{\varepsilon}$ are manipulated) then one may consider the introduction of a concave upwards instantaneous stress-strain curve with a slope possibly higher than the elastic one (Suliciu [10]); this will significantly increase the velocity of propagation of the waves, mainly in the portion of the bar close to the impacted end, if such an effect is desired.

The explicit expression for the function Φ is determined mainly from the arrival times of various magnitudes of strain at various cross sections of the bar (see Fig. 4). For instance at 1 D the computed arrival times for smaller strains ($\varepsilon < \bar{\varepsilon}$) as furnished by example 3 are too early when compared with the experimental data, but are too late for higher strains ($\varepsilon > \bar{\varepsilon}$). $\bar{\varepsilon}$ is the strain at the point of inflection and is given in Table II. Both effects are due to the fact that generally Φ is a bit too small. Therefore this solution can still be improved by increasing Φ, by changing for instance the coefficient a, or by decreasing $\underset{\sim}{\varepsilon}$ (example 2). A converse situation occurs for example 1, where Φ is slightly too big.

A somehow similar effect on the strain-time curve can be obtained by changing the parameter k entering in the expression of Ψ. In other words by increasing k or by increasing Φ one will produce more or less the same effect on the shape of the strain-time curves. However, if k is increased too much, both the peak stress and the initial sharp rise of the strain-time curves will be drastically diminished. This was one of the reasons why for small strains one had to choose a small value of k (see (3.5)). A second reason which imposes a small value of k for small strains is the behaviour near the yield point as described by the strain-time curves near the free end of the specimen (where strains are relatively small, i.e. surpassing a few times only the yield strain). Figs. 7 and 8 for instance are showing the variation in time of the strain at eight and nine diameters respectively, from the im-

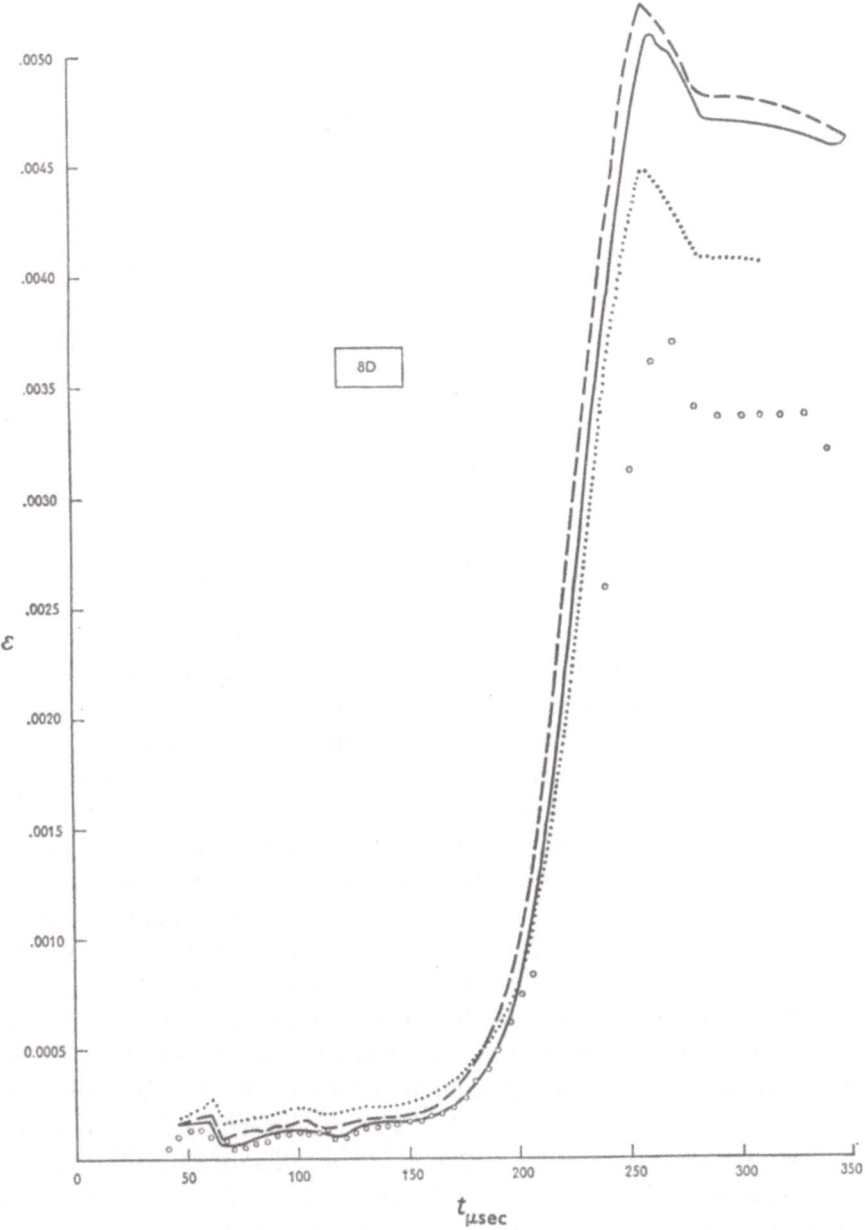

Fig. 7. Variation of strain at 8 D from the impacted end (Example 1 – full line; Example 2 – broken line; Example 3 – dotted line; experiments – circles).

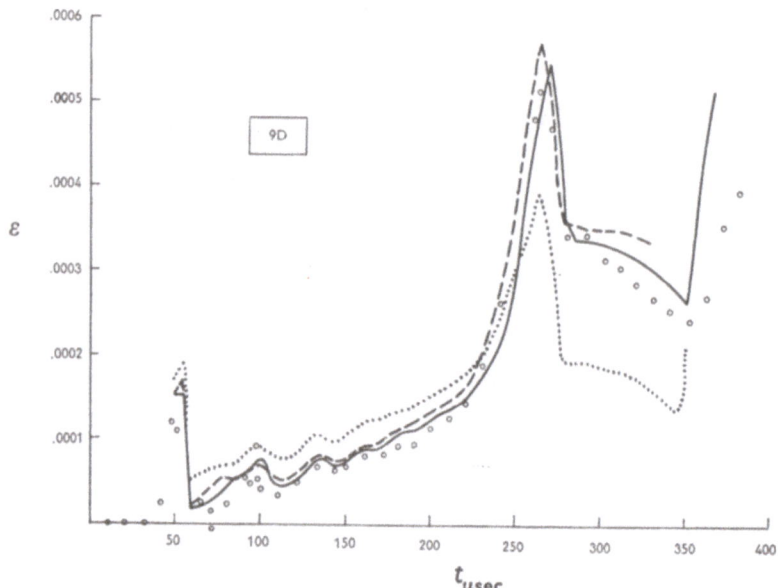

Fig. 8. Strain-time curves at 9 D (Example 1 – full line; Example 2 – broken line;
Example 3 – dotted line; experiments – circles).

pacted end. These figures show that the computed solutions are in excellent agreement with the experimental data. This is greatly due to a correctly chosen value for k. For higher k the maximum would be higher, raising portions of the curve too early, etc. (see Cristescu [7]). Finally a third argument in favour of a small value of k for small strains was the variation in time of the displacement at various cross-sections along the bar. Fig. 9 gives such experimentally obtained curves compared with the numerical solution of example 2. The general agreement seems reasonable. It is important to observe that the agreement for the cross-sections close to the free end is better if for smaller ε the value of k is smaller (see Cristescu [7]).

From all parameters and coefficient functions entering the constitutive equation, the most important and the first to be determined from experiments is the equation of the relaxation boundary. The starting point is the experimental observation that in the first few diameters near the impacted end, the strain is constant for a relatively long period of time after reaching its maximum (see Fig. 4) before unloading occurs, and that to this strain plateau corresponds a stress-plateau (see Fig. 10 for the variation of stress at the impacted end of the specimen) which again is quite long in time. It was shown elsewhere (see Suliciu et al. [11]) that quasilinear constitutive equa-

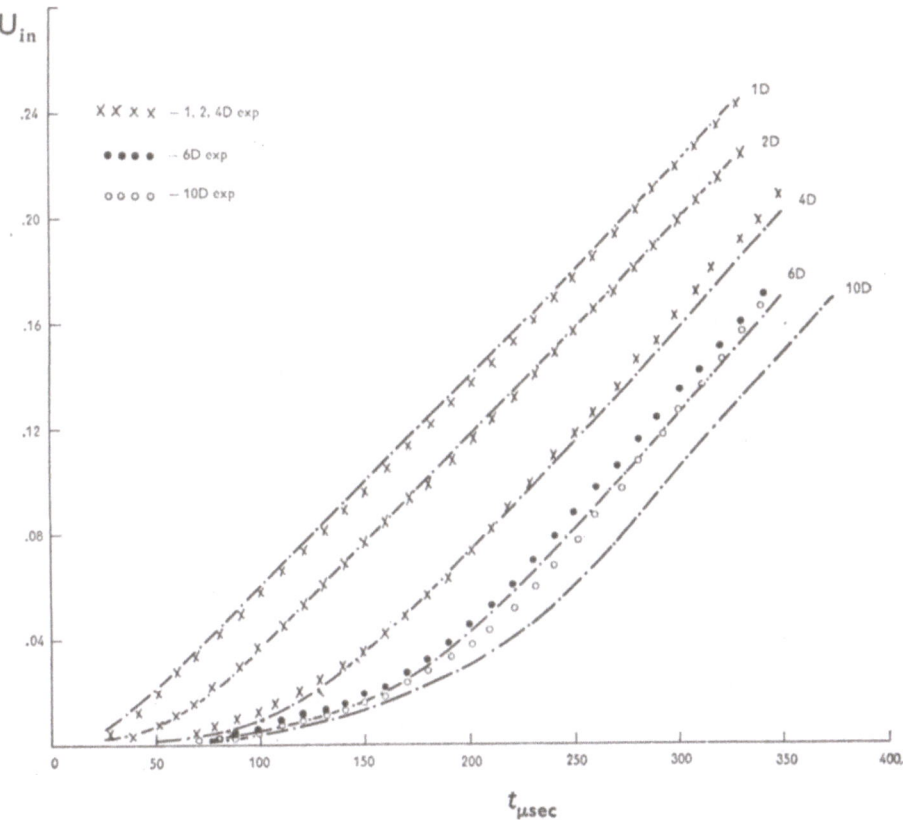

Fig. 9. Displacement-time curves at various cross-sections along the bar (Example 2 and experiments).

tion of the form (3.1) may describe even absolute plateaus, i.e. domains may exist in the characteristic plane where all three functions σ, ε and v are constant. The stress-strain states corresponding to these plateaus (for σ and ε) are by definition points on the relaxation boundary. Thus by making experiments with various velocities of impact and therefore by determining the stress-strain states corresponding to the stress and strain plateaus, one can determine the relaxation boundary for a certain considered material. Note that in Fig. 10 the raising portions of the curves shown have not been represented in order to simplify the figure.

The experiments are showing (Bell [1], Lee [9]) that the maximum strain is constant for a certain portion along the bar (generally several diameters near the impacted end) but close to the impacted end (one half diameter or so)

303

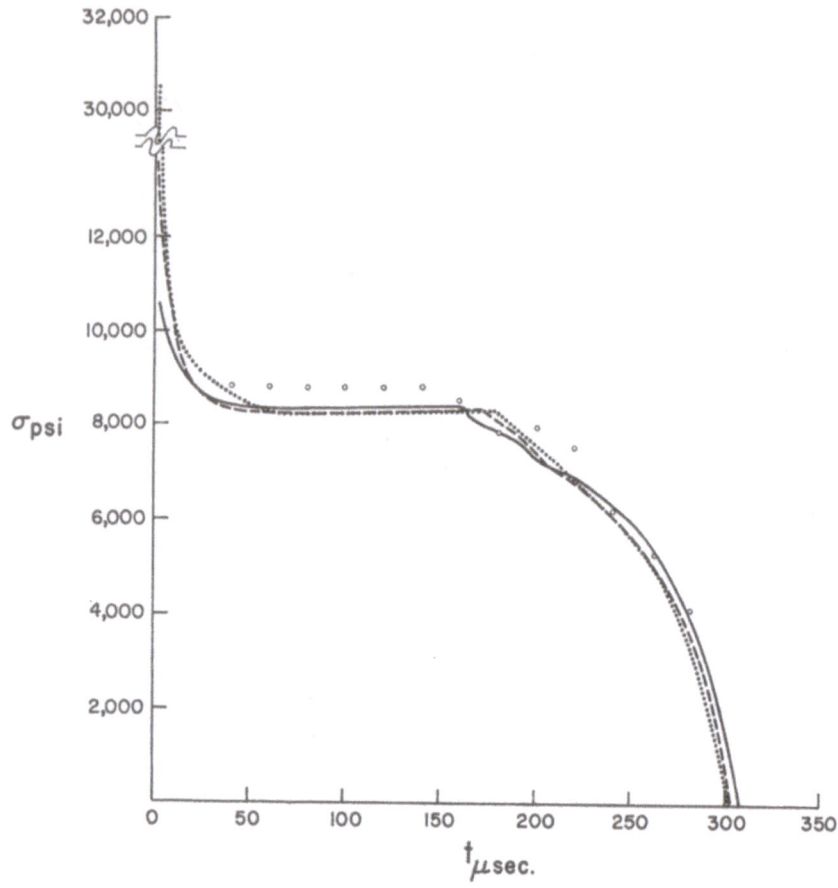

Fig. 10. Variation of stress in time at the impacted end of the bar (Example 1 – full line; Example 2 – broken line; Example 3 – dotted line).

the strain is somehow higher (Table II). This is also obtained by computation with the model considered (see Fig. 11 corresponding to example 2), and it is again due to the presence of the peak stress near the impacted end. In Fig. 11 the experimental points showing higher strains near $x = 0$ have not been represented since there were not available to the author.

Another quantity determined experimentally quite accurately is $\bar{\varepsilon}$, i.e. the strain magnitude at the point of inflection (generally the second one) of the strain-time curves. As it is well known this magnitude of the strain corresponds to the maximum of the surface angle α. It was shown in another

304

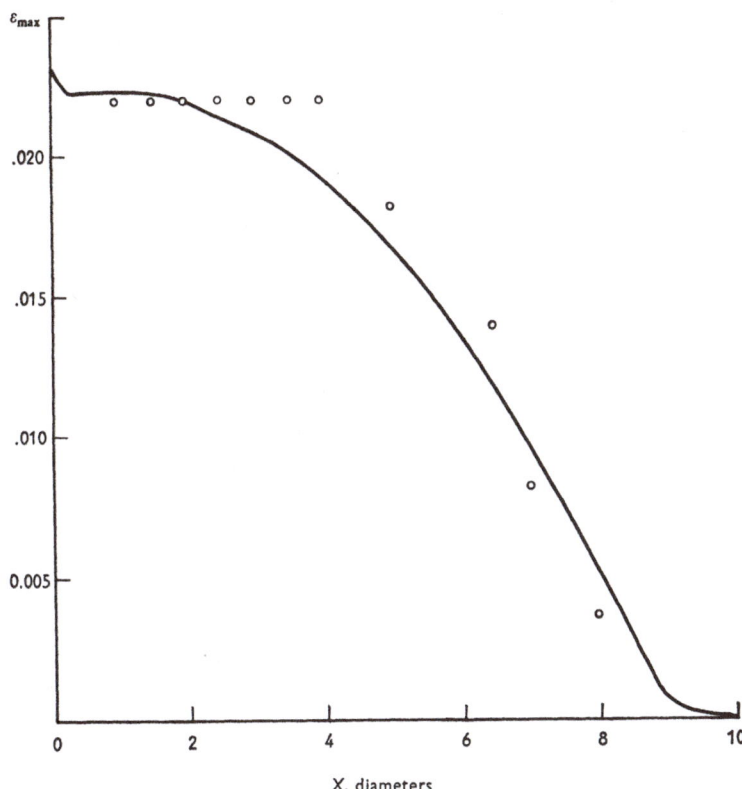

Fig. 11. Distribution of maximum strain along the bar (Example 2 and experiments).

paper (Cristescu [6]) that $\bar{\varepsilon}$ is the strain corresponding to the peak stress at the impacted end. In other words at $x = 0$ the lower part of the strain-time curve up to $\bar{\varepsilon}$ corresponds to the raising portion of the stress-strain curve, while the upper part of the strain-time curve (above $\bar{\varepsilon}$) corresponds to the relaxation of the stress (decreasing portion of the stress-strain curve). Thus all the parameters influencing the peak stress will influence $\bar{\varepsilon}$ as well. Numerical values for $\bar{\varepsilon}$ are given in Table II (see also Cristescu [6, 7]).

The perfectly elastic unloading mechanism seems to be in reasonably good agreement with the experimental data. This results from various aspects of the computed solutions as compared with experimental data. An overall picture of the shapes of the loading and unloading regions in the characteristic plane is given in Fig. 12 for example 2. Between the loading and unloading regions, denoted by L and U respectively, there is a strip

305

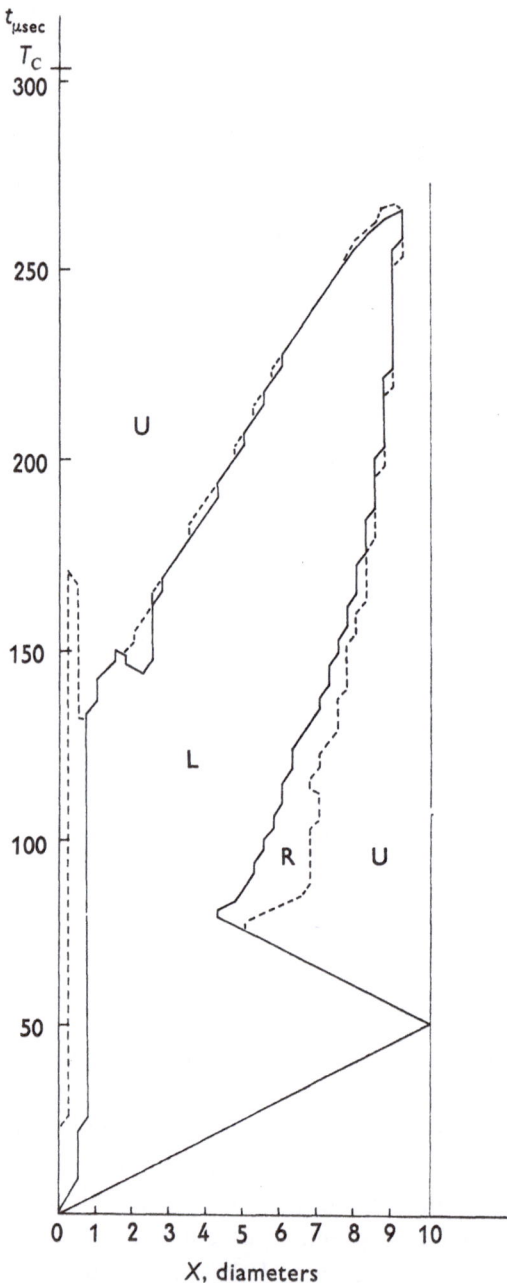

Fig. 12. Characteristic plane showing loading, unloading and relaxation regions (Example 2).

where relaxation takes place, i.e. stress is decreasing though strain increases. This region was denoted by R in Fig. 12. The unloading process starts from the free end of the bar. The first penetration of the unloading region into the loading region can again be observed experimentally. The coordinates of the tip of this first unloading region, denoted by x_U is given in Table II. x_U depends on various parameters but mostly on the yield stress, generally the higher σ_Y the smaller is x_U. The arrival times of the first unloading wave penetrating the loading region as computed assuming perfect elastic unloading is again in reasonable agreement with the experimental data. Figs. 7 and 8 are also showing a very reasonable agreement with the computed solution of the material response in the unloading domain close to the free end, where stresses and strains are relatively small. Generally the shape of the lower part of the unloading region near the free end of the bar as well as the overall behaviour of the solution near this end depend mainly on σ_Y and on k.

6. A generalized constitutive equation

The fact that constitutive equations of the form (3.1) can successfully describe the main phenomena which have been observed experimentally in one-dimensional dynamic plasticity tests is an encouragement for a tentative generalization for a general three dimensional case. There are various ways in which one may perform such a generalization, but it is only by considering experiments that the most reasonable generalization may be selected.

Here only one kind of such generalization will be presented but various other possible ways of generalization and their effect on the dynamic plastic response of the model will be discussed.

One generalized form of (3.1) was given in [3] Ch. X § 2, assuming small strains and rotations. Using the same notations as above this can be written in the form

$$\dot{e}_{ij} = \frac{\dot{s}_{ij}}{2G} + \frac{3k(\bar{\varepsilon})}{2E}\left[1 - \frac{f(\bar{\varepsilon})}{\sqrt{3}\sqrt{II_s}}\right]s_{ij} + \frac{3\Phi(\bar{\varepsilon}, \bar{\sigma})}{4II_s}s_{kl}\dot{s}_{kl}s_{ij}, \qquad (6.1)$$

where e_{ij} and s_{ij} are the strain and the stress deviators, while f and Φ are the same functions as defined by (3.3) and (3.7) respectively, where however, the strain is replaced by $\bar{\varepsilon} = (2/\sqrt{3})\sqrt{II_e}$, the stress by $\bar{\sigma} = \sqrt{3}\sqrt{II_s}$, and

$$II_e = \tfrac{1}{2}e_{ij}e_{ij}, \quad II_s = \tfrac{1}{2}s_{ij}s_{ij}. \qquad (6.2)$$

This is one of many possible ways of generalization of constitutive equation of the form (3.1) performed by following the classical way used in plasticity theory. Characteristic for this generalization is the fact that both relaxation boundary (RB) and the instantaneous response curve (IR) were generalized for the three dimensional case in a global (even isotropic) way. The problem of coupling of plastic waves for constitutive equations of the form (6.1) has been discussed by Cristescu [5]. It was shown in this paper that global IR will properly couple the plastic waves in a two-dimensional problem. For instance a realistic experiment of this kind is the combined stress experiment performed on thin walled tubes subjected to simultaneous tension and torsion. Thus if IR is global the waves involved in the problem will be at the same time shearing and longitudinal (compressive or tensile).

There are two ways in which from a physical point of view the two waves can interact when a constitutive equation of the form (6.1) is used. This is governed by the form of the functions entering in RB and in IR. It is known (Cristescu [5]) that proper coupling, i.e. coupling at the wave front, is possible only if IR is global. If RB is global, then interaction retarded in time between waves is the only possible interaction. If both RB and IR are global then both such interactions may occur simultaneously.

If we contemplate to define IR in a nonglobal form the proper coupling is impossible and the only possible interaction between the waves is a retarded in time one, through RB. Several intermediate cases when weaker but still proper coupling is present, are possible if some kind of piecewise definitions for IR is introduced. All these have been discussed by Cristescu [5].

The experimental data available today do not allow us to make a definite decision (even a qualitative one) concerning the way in which plastic waves are interacting (proper coupling, retarded in time, etc.). It is important to remember that the coupling problem is the dynamic counterpart of the problem of translation, deformation, etc. of the yield surface in static plasticity theory. Therefore, pending positive experimental data, one can contemplate various ways of generalizing the constitutive equation (3.1).

7. Conclusions

In the present paper it was shown that quasi-linear rate-type constitutive equations can successfully describe the main phenomena discovered by experimentalists in one-dimensional dynamic plasticity tests. Qualitatively all these phenomena can be described by such a constitutive equation even though for a more exact quantitative agreement some additional improvements of various constants and coefficient functions entering the constitutive equation may still be necessary. If the informations furnished by experimentalists are somehow contradictory, an analysis as the one shown may suggest what other experiments are still necessary.

The quasi-static stress-strain curve of finite form is giving poor results when used in dynamic plasticity problems (see Cristescu [6]). However, it is interesting to observe that if the relaxation boundary is used as a finite form stress-strain curve, the solution obtained gives a very reasonable overall agreement with the experimental data (see Cristescu and Bell [8]) though some of the phenomena such as peak stress, initial bump of the strain-time curve, the change of the curvature of the strain-time curves, etc., would be impossible to describe with a finite form stress-strain curve.

Acknowledgments

The author wishes to express his appreciation to Cornelia Cristescu for writing the program in FORTRAN and taking care of the computations. The computations were done at the Computing Center of the University of Florida for which the author is grateful to the Department of Engineering Science and Mechanics.

References

[1] Bell, J. F., Study of initial conditions in constant velocity impact. *J. Appl. Phys.*, vol. 31, *12* (1960) 2188–2195.

[2] Bell, J. F., *The Physics of Large Deformation of Crystalline Solids*. Springer Verlag, 1968.

[3] Cristescu, N., *Dynamic Plasticity*, North Holland, Amsterdam, 1967.

[4] Cristescu, N., The unloading in symmetric longitudinal impact of two elastic-plastic bars. *J. Mech. Sci.*, *12* (1970) 723–738.

[5] Cristescu, N., On the coupling of plastic waves as related to the yield condition. *Rev. Roumaine Sci. Techn., Ser. Mec. Appl.*, *16* (1971) 797–809.

[6] Cristescu, N., A procedure for determining the constitutive equations for materials exhibiting both time-dependent and time-independent plasticity. *Int. J. Solids Structures*, *8* (1972) 511–531.

309

[7] Cristescu, N., *Introduction to rate-dependent plasticity (a dynamical approach)*. Springer, Berlin (in press 1974).

[8] Cristescu, N. and J. F. Bell, On unloading in the symmetric impact of two aluminum bars; in *Inelastic behavior of solids*. McGraw Hill, 1970, pp. 397–419.

[9] Lee, E. H., The theory of wave propagation in anelastic materials; in N. Davids ed.: *Int. Symp. on Stress Wave Propagation in Materials*. Interscience Publ., 1960, pp. 199–228.

[10] Suliciu, I., private communication, 1972.

[11] Suliciu, I., L. E. Malvern and N. Cristescu, Remarks concerning the 'plateau' in dynamic plasticity. *Arch. Mech. Stosow, 24* (1972) 999–1011.

Superplasticity:
the behaviour and uses of ultra-fine grain alloys

J. L. Duncan

McMaster University, Hamilton, Ontario, Canada

Ultra-fine grain alloys have a grain size of 2 microns (m^{-6}) or less compared with 20 microns or greater in conventional fine grain materials. When hot worked at rates just below conventional forming speeds, many of these alloys display a low resistance to deformation which is independent of total strain and time but highly sensitive to strain rate. Generically, these are known as superplastic alloys and recently they have been produced on an industrial scale by thermo-mechanical processing. Components are formed from superplastic alloys in certain conventional metalworking processes and also by techniques previously applied only to thermoplastics and semi-molten glasses.

The deformation mechanisms which are thought to govern superplastic flow are described and a three dimensional mathematical theory of superplasticity is stated; this follows closely existing multiaxial creep theory and employs a constitutive relation similar to Norton's creep law. Tensile superplastic forming processes are essentially unstable in nature and it is suggested that the important areas of analysis relate to the investigation of geometric non-uniformities in the post-necking region. In compressive bulk forming processes, simple approximate analyses are required similar to those developed for strain hardening materials.

1. Introduction

An unusual behaviour in the cold-rolled Zn–Cu–Al tertiary eutectoid alloy was noticed by metallurgists in 1920 who observed that this allow 'behaved differently from ordinary crystalline materials such as aluminium but very similarly to ......pitch, glass, etc.' [1].[1]) It was subsequently shown that similar alloys could be stretched in tension to an enormous elongation, up to 2000%, almost as a glass rod can be drawn out in a hot flame, [2]. There was some lively debate about the basic structure of these materials. Were they amorphous, as suggested by their mechanical behaviour or were they poly-

[1]) Number in brackets designate References in Notes at end of paper.

311

crystalline as is usual in metals? Their grain structure was extremely fine and could only be imperfectly resolved by optical microscopy. This pheno-menon seems to have been regarded as a scientific curiosity and was largely forgotten in the 1930s. Further work appeared in the Russian literature [3] between 1944 and 1960 and the word 'superplasticity' was first used to describe the large tensile elongations observed. The Russian workers introduced the eutectoid Zn–Al alloy to the superplastic group and this has become the most widely used and studied of all the alloys. A review by Underwood [4] re-awakened the interest of Western workers and by 1965 the phenomenological character of superplasticity and the metallurgical conditions under which it occurs were clearly delineated [5]. The literature has grown rapidly and now, after fifty years, superplasticity is no longer considered a freak property but the normal behaviour of alloys in which a particular ultra-fine grain structure can be induced and retained in a stable form at elevated temperature.

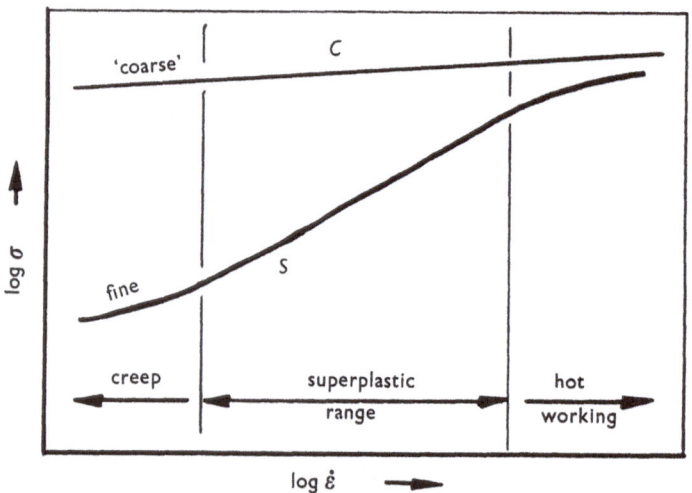

Fig. 1. Logarithmic schematic diagram of the typical stress, strain rate characteristic of a superplastic alloy, (curve S), compared with that of the same alloy, (curve C), in its conventional state.

Superplasticity is a hot-working phenomenon observed above a tempera-ture of about 40 % of the absolute melting point. The most important aspect of superplasticity can be observed by comparing the stress, strain rate characteristic of an alloy in its normal 'coarse' grain state with that of the same alloy in the ultra-fine grain, superplastic state, as in Fig. 1. At conventional forming speeds, both materials have a similar rate sensitivity and flow stress. At very low deformation rates – creep forming – the rate sensitivities are similar but the flow stress of the superplastic alloy is extremely low. There exists a transition range, often regarded as the 'superplastic range' where the strain rate sensitivity, for a metal, is unusually high. This range may extend over two or three decades of strain rate, and in a logarithmic diagram, the characteristic of many superplastic alloys is approximately linear and the properties can be described with the aid of the empirical relation,

$$\sigma = K\dot{\varepsilon}^m \tag{1}$$

where K and m are regarded as material parameters [6]. The index m, which is the slope of this line in the logarithmic diagram, is frequently called the strain rate sensitivity index following the more general definition of this property, namely

$$\frac{\dot{\varepsilon}}{\sigma} \cdot \frac{d\sigma}{d\dot{\varepsilon}} = \frac{d(\log \sigma)}{d(\log \dot{\varepsilon})} = m \tag{2}$$

In hot-working, conventional alloys have a definite yield point and their flow stress is dependent on strain as well as strain rate. Superplastic alloys, on the other hand, have no identifiable yield stress; they deform under any stress, however small, and their properties are surprisingly insensitive to strain. Their microstructure is virtually unchanged by large deformations and this, in itself, is an unusual characteristic for polycrystalline materials.

The classification of superplastic alloys presents some difficulties. They could be considered, perhaps pedantically, as fluids and, because their strain rate dependence is non-linear with an index in the range $0.25 < m < 0.8$, they would be classed as non-Newtonian fluids. Within this classification, the absence of a yield stress or 'threshold' stress required to initiate deformation, precludes their classification as Bingham plastics. Within the superplastic range, they fit the category of pseudo-plastic fluids, but this does not fit their behaviour at very low strain rates. There is probably little to be gained by attempting a rigid definition, [7]. Inasmuch as these alloys have a basically crystalline structure and can be handled in an apparently solid state, they are best regarded as solids. The term 'viscous solid' is a good description even if, in a strict sense, the words are contradictory.

A more useful classification can be made by comparing superplastic alloys with other materials which are hot-worked. Fig. 2 shows, schematically, the stress, strain rate characteristic for steel during forging and for a polymer in a thermoforming operation. The superplastic alloy is typically formed at a lower strain rate than either of the other two, but it may be seen that the level of stress and the strain rate sensitivity are both in between those of the typical metal and the thermoplastic.

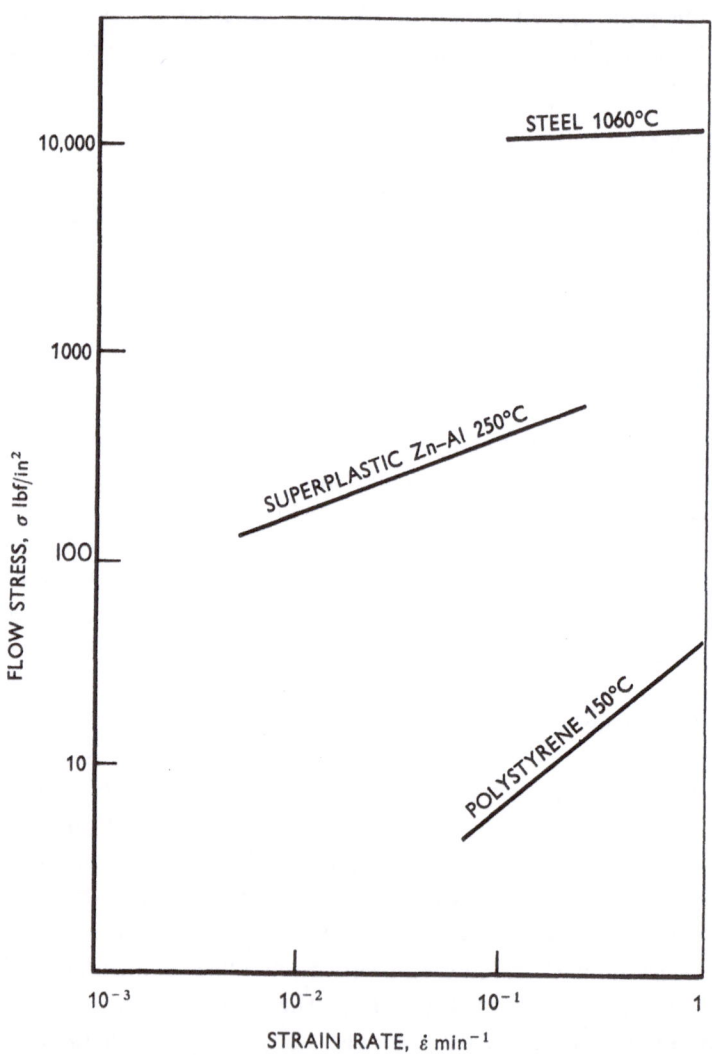

Fig. 2. Comparison of hot-working stress, strain rate relations for different materials.

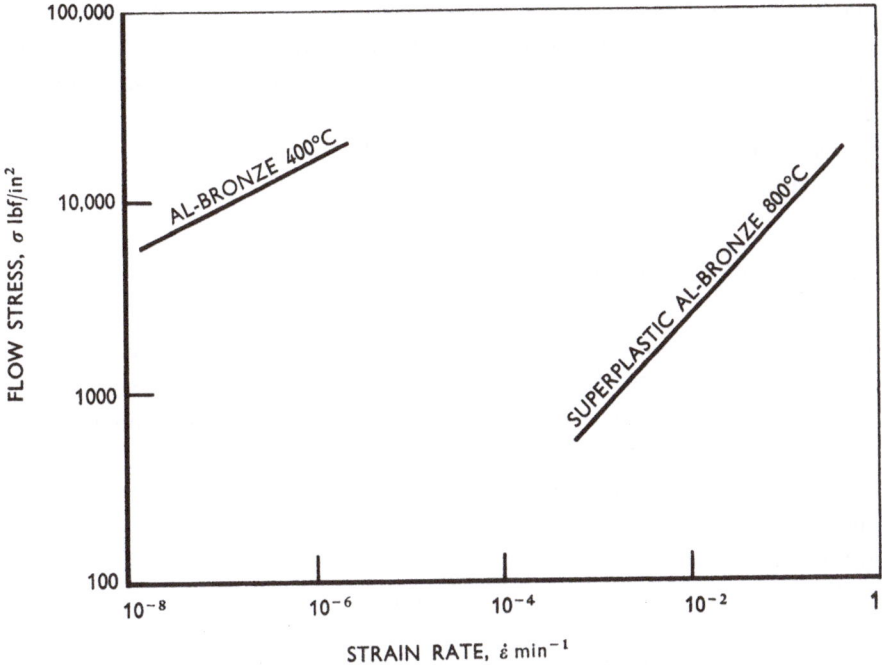

Fig. 3. Creep and superplastic deformation in aluminium-bronze.

Superplasticity is often regarded as an extension of normal creep behaviour in metals. There are, in fact, many affinities between creep and super-plasticity but, as shown in Fig. 3, the typical deformation rates in these processes differ by many orders of magnitude.

The overall physical manifestations of superplasticity are easily described and they are predictable. At low strain rates, superplastic alloys deform under the action of small stresses – a few hundred pounds per square inch, compared with a few thousand pounds in normal alloys. At higher strain rates, up to the lower limit of conventional hot-working speeds, superplastic alloys have a high strain rate sensitivity – typically 0.5 compared with 0.01–0.1 in most metals. This behaviour is sufficiently unusual and distinct to warrant being classed as a special deformation phenomenon, even though it appears to be the normal consequence of an ultrafine, micron size grain structure [8].

At the present time, the more striking aspects of superplasticity have largely been demonstrated in alloys in which, for special reasons, the ultra-fine microstructure can be obtained easily. As the interest in ultra-fine grain alloys grows and this structure is developed in more familiar alloys, it is

315

likely that superplasticity will be regarded as one region of the overall spectrum of material behaviour which can be exhibited generally by metals at elevated temperature. It is appropriate therefore to examine some of the fundamental changes in metal deformation mechanisms which can occur when the grain size is drastically reduced. This is a controversial, lively and confusing topic, but nonetheless interesting [9].

2. Deformation mechanisms

Metal physicists explain the familiar aspects of metal deformation very largely in terms of the behaviour of dislocations within the crystal lattice. Phenomena such as work-hardening, cyclic strain softening and thermal softening processes may all be explained by the existence, movement, inter-action and elimination of dislocations. The typical behaviour of metals, namely an identifiably rigid or elastic state, a definite yield stress and, at the most, only a weak dependence of stress on strain-rate are an outcome of these dislocation processes. Superplastic alloys, on the other hand, have

TABLE I

Conventional grain size classification

ASTM No.	Grain size micron	Range for commercial steels
1	254	
2	180	
3	127	
4	90	
5	64	
6	45	
7	32	
8	22.5	
9	16	
10	11	
11	7.9	
12	5.6	
13	4	
14	2.8	superplastic alloys

very different properties and yet retain the basically polycrystalline nature of metals. It would appear therefore that other deformation mechanisms must operate.

It has been mentioned that superplastic alloys all have an ultra-fine grain structure; the grain diameter is always less than 10 microns and generally less than 2 microns. It may be seen from Table I that superplastic grain sizes virtually start from the point at which the conventional classification of grain size ends. This table shows that superplastic grain sizes are at least an order of magnitude smaller than those in conventional 'fine-grain' materials. Another unusual aspect of the microstructure in superplastic alloys is the morphology. In normal alloys, individual grains are elongated in the direction of extension or, if the temperature is high enough, recrystallization occurs. In superplastic alloys, the grains do not elongate even though the material may be extended many hundred percent; during deformation the grains become, if anything, more equiaxed and develop a rounded appearance, as shown in Fig. 4.

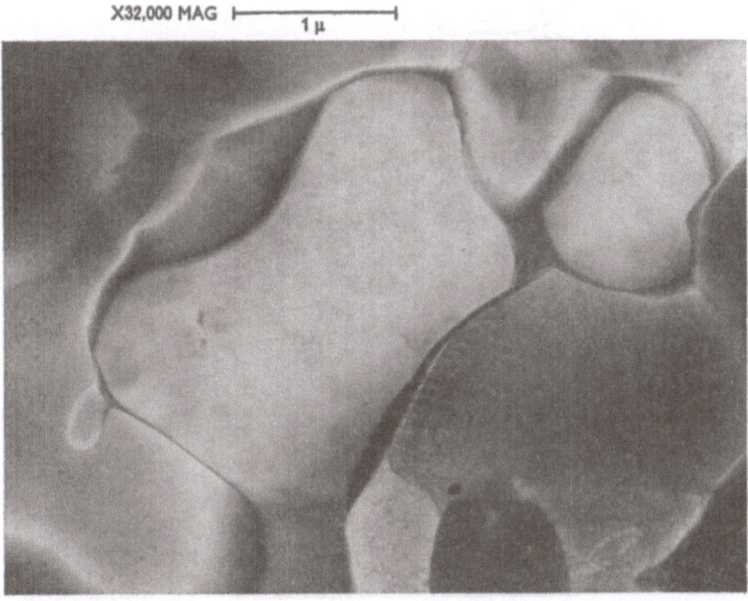

Fig. 4. Thin foil electron micrograph of heavily deformed superplastic Zn-Al. The average grain size is about 1 micron.

Metal physicists tend to agree on the mechanisms which could operate in superplastic flow, however they disagree on the relative importance of each mechanism. It is not proposed to enter this controversy here, but rather to list those mechanisms which could occur and to comment on the influence each mechanism may have on mechanical properties.

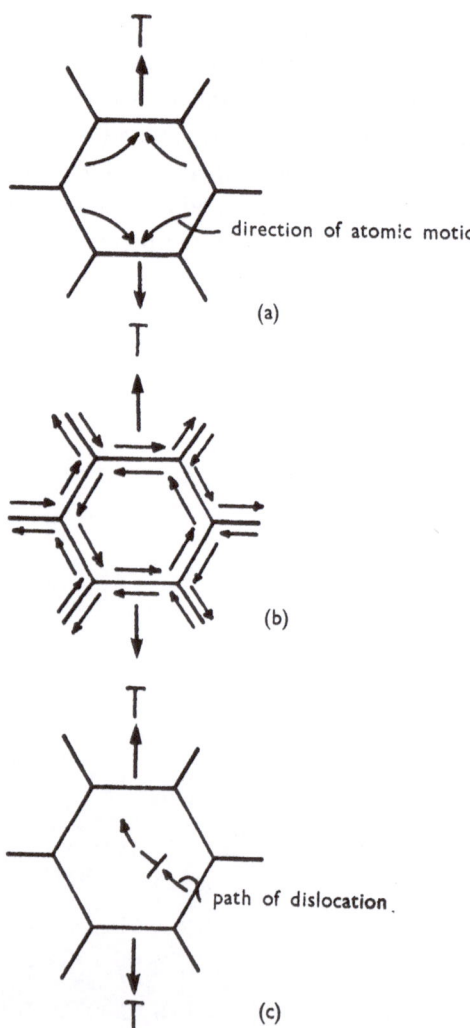

Fig. 5. Schematic representation of deformation mechanisms; (a) diffusion creep, (b) grain boundary shear and (c) recovery by dislocation climb.

Diffusion creep

It is postulated that in addition to dislocation imperfections in the lattice structure, vacancies or missing atoms also occur. Solute atoms may therefore diffuse or migrate as a result of these vacancies from regions of compressive stress in the grain to tensile regions, as shown in Fig. 5(a) [10]. Migration may be either through the body of the grain – Nabarro Herring mechanism – or along the grain boundary as proposed by Coble. The predicted strain rate for Nabarro Herring diffusion is,

$$\dot{\varepsilon} = C_1 \frac{\sigma}{L^2} \tag{3}$$

where L is the average grain size and C_1 a constant, and for Coble diffusion,

$$\dot{\varepsilon} = C_2 \frac{\sigma}{L^3} \tag{4}$$

In superplastic alloys it is found that

$$\dot{\varepsilon} \propto 1/L^a$$

where a is between 2 and 3 but the strain rates predicted by equations (3) and (4) are too low and both mechanisms would lead to elongation of individual grains. Further, the strain rate sensitivity index is not unity as predicted but tends to be about 0.3 or less at the low strain-rates at which diffusion mechanisms might be expected to dominate. It has been suggested that there is a 'threshold stress', σ_0, which must be exceeded for deformation to occur and that the stress, strain-rate relation should be,

$$\sigma = \sigma_0 + K\dot{\varepsilon}^m \tag{6}$$

This argument appears to be contrived, however, in order to permit a rate sensitivity index, m, approaching unity and, as mentioned, σ_0 has never been directly measured for a superplastic alloy.

Grain boundary sliding

Overall deformation in this mechanism results from relative motion of adjacent grains by shear along their boundaries, as in Fig. 5(b). Considerations of grain boundary viscosity lead to a theoretical strain-rate dependence of,

$$\dot{\varepsilon} = C_3 \frac{\sigma}{L} \tag{7}$$

but the rates predicted are much higher than those observed.

The hypothesis is generally favoured, however, as this mechanism would lead to the equi-axed, rounded grain structure which is typically observed after deformation. Obviously, though, grains cannot rotate relative to each other without causing voids in the structure unless there is some accommodating intergranular deformation. Accommodating mechanisms such as diffusion, slip or grain boundary migration can be invoked but the question then arises about which mechanism controls the deformation rate. Is this rate controlled by the grain boundary shearing process or by the accommodating processes which permit the grains to rotate? [11]

Crystallographic slip and recovery

In the higher strain rate region of the superplastic range, experimental evidence such as anisotropy of strain suggests that crystallographic slip associated with dislocations within the grain may be important [12]. One difficulty in supporting this thesis, however, is that superplastic alloys normally appear to be in a fully recovered state even after extensive deformation. It is well known that in dislocation creep, entanglement of the dislocations leads to formation of regular cells or sub-grains and the hypothesis is that in superplastic alloys, cell size would be greater than the actual grain size. As a consequence, dislocation entanglements would move toward the grain boundary and be removed by a recovery process such as grain boundary migration, dislocation climb, Fig. 5(c), or recrystallization. The observed rate sensitivity is attributed to these recovery processes.

Combined mechanisms

In ultra-fine grain alloys, the proportion of material existing in a disordered state at the grain boundaries is greater than in conventional alloys. It is not surprising, therefore, that a meaure of disorder exists in the metal physicists' explanation of superplasticity. It is clear, however, that a greater part of the deformation – 60% to 70% – results from grain boundary sliding, but it must be assumed that other mechanisms are involved and that these may be the ones which give rise to the characteristic behaviour shown in Fig. 1. At low strain rates, in the creep forming range, some diffusion process probably occurs while at higher rates approaching normal metalworking speeds, some crystallographic slip is likely. This, unfortunately, leaves a hiatus in the most important 'superplastic range' and it would be most ironic if, after all that has been written, this became regarded merely as a transitional mode of behaviour.

3. Development of a stable ultra-fine microstructure

Superplasticity is a comparatively unusual phenomenon because in many alloys an ultra-fine grain structure is not stable at a temperature of half the absolute melting point; grain growth causes a rapid reversion to the normal 'coarse' grain structure. It is worth commenting, therefore, on those factors which influence stability in the ultra-fine grain state.

The typical eutectic or eutectoid structure consists of very fine lamellae of two phases. This is not superplastic but the lamellae can be broken up and formed into ultra-fine equiaxed grains by drastic mechanical working. If the chemical composition of each phase is very different, grain growth can only occur by massive diffusion of solute atoms and even in micron size grains, this is a slow process. Most superplastic alloys are produced in this way and have a characteristic 'micro-duplex' structure consisting of two phases in approximately equal proportions. Certain variations exist such as those alloys in which the grain boundaries are 'pinned' by a precipitate. In a few alloys, such as the eutectoid Zn–Al, the ultra-fine structure can be obtained directly by a solid state decomposition, however even in these alloys, grain refinement by extensive working is preferred in industrial production. (It will be no surprise to the plastician that the wrought product is more reliable than the cast, heat-treated material.) Other mechanically produced alloys produced by powder compaction or electrodeposition have been studied in the laboratory.

The grain size of useful superplastic alloys is, at the moment, between about 0.5 and 5 microns. It is possible that techniques will be developed permitting further grain refinement and if, for example, a stable structure of 0.1 microns could be achieved, the material should have most attractive forming characteristics.

4. Superplastic alloys

Various lists of superplastic alloys appear in the literature and Table II is an example [13]. Most of these materials have been made in the laboratory only, but a few have been produced industrially. Some comment on these industrial alloys is worthwhile, although the remarks are ephemeral.

Eutectoid Zinc–Aluminium – This important alloy, 78 Zn–22 Al, has already been mentioned. The ultra-fine grain structure is easily obtained and is very stable. It can be readily worked in a variety of processes up to the transformation temperature 275°C. Variants of this alloy are produced as commercial alloys in both England and the USA [14].

Bronzes – Commercial aluminium bronze and cobalt-silicon-aluminium bronze can be produced in a superplastic form by quenching and warm rolling to produce an ultra-fine, two-phase, structure. The alloys are superplastic at about 800°C. They tend to show some cavitation after a few hundred percent elongation, but this is generally beyond the limits of extension required in technological processes. Their strength, corrosion and wear resistance and electrical conductivity at room temperature are very good [15].

High Temperature Alloys – Microduplex, 26 Cr–6 Ni, stainless steel and wrought Ti–6 Al–6 V titanium alloy show good superplastic properties at 900°C–1000°C [16] and have been used to produce commercial components.

5. Mathematical theory of superplasticity

The number of analyses of superplastic forming processes in the literature is quite small and even fewer papers present experimental data on the behaviour of superplastic alloys under multiaxial stress. In the work which does exist, however, it is assumed that the material is always isotropic and that it does not strain harden – these appear to be good assumptions for superplastic materials [17]. The mathematical theory adopted is very similar to that used for creep; using the von Mises hypothesis, one obtains, following Odqvist [18], the relation between the strain-rate tensor, $\dot{\varepsilon}_{ij}$, and the stress deviation tensor, s_{ij}:

$$\dot{\varepsilon}_{ij} = \frac{3}{2} \frac{dW}{d\sigma_e} \cdot \frac{s_{ij}}{\sigma_e} \tag{8}$$

TABLE II

Alloys in which superplasticity has been observed

Base metal	Alloy wt %	Temp °C	Grain size (microns = 10^{-6} metres)	Rate sensitivity index m
Aluminum	33 Cu	440–530	1–2	0.9
	12 Si 4 Cu	500	—	0.4
Cadmium	26 Zn	20	1–2	0.5
Chromium	27.5 Co	1200	—	—
Cobalt	10 Al	1200	0.4	0.3
Copper	10 Mg	700	—	—
	10–12 Al	500–700	3	0.6
	10 Al 1–4 Fe	800	10	0.8
	38–50 Zn	450–550	3	0.5
	40 Zn	600	3	0.7
	38 Zn Ti	600	5	0.7
	38 Zn 2 Fe	600	—	—
	2.8 Al 1.8 Si 0.4 Co	550	1	0.4
Iron	0.14 C 1.2 Mn 0.1 V	900	2	0.6
	0.34 C 0.47 Mn 2.0 Al	900	2	0.5
	0.42 C 1.9 Mn	730	1–2	0.6
	26 Cr 6.5 Ni) 30 Cr 6.0 Ni)	870–980	2	0.5
Lead	20 Sn	20	3	0.5
	5 Cd	0–100	1–10	0.6
Magnesium	0.5 Zr	500	20	0.3
	0 Zn 0.6 Zr	270–310	0.5	0.5
	23 Ni	450	—	—
	30 Cu	450	—	—
	33 Al	400	—	—
Nickel	nil	820	8	—
	39 Cr 8 Fe 2 Ti	980	2	0.5
Tin	5 Bi	20	1	0.5
	2–38 Pb	20	1–2	0.5
	33 Cd	20	1–2	0.5
Titanium	6 Al 4V	900–980	6	0.9
	5 Al 2.5 Sn	1000	18	0.7
Zinc	nil	0–20	1–2	—
	0.5 Al	20	1–2	—
	5 Al	200–360	1–2	0.7
	22 Al	200–260	1–2	0.5
	40 Al	250	1–10	0.5
Zirconium	—	900	12	0.5

where dW is the rate of work dissipation and σ_e is normally the stress invariant,

$$\sigma_e^2 = 3S_{ij} \cdot s_{ij}/2. \tag{9}$$

(In some cases a maximum shear stress criterion is adopted, following Tresca.) The term $dW/d\sigma_e$ is, of course, a scalar quantity. An effective strain rate, $\dot\varepsilon_e$, is introduced, where

$$\dot\varepsilon_e^2 = \tfrac{2}{3}\dot\varepsilon_{ij} \cdot \dot\varepsilon_{ij} \tag{10}$$

The constitutive equation in this theory is normally a generalization of equation (1) which, rigorously, should be written,

$$\left(\frac{\sigma_e}{\sigma_0}\right) = \left(\frac{\dot\varepsilon_e}{\dot\varepsilon_0}\right)^m \tag{11}$$

where σ_0 and m are material parameters and $\dot\varepsilon_0$ is an arbitrary constant which is often taken as 1 min^{-1} or 1 sec^{-1}. Eq. (11) is similar to the Norton creep law and fits experimental data well in the superplastic range. Care must be taken in drawing general conclusions from analyses in which this equation is employed; some solutions would indicate that different loading paths will produce identical changes in geometry. This arises because the 'shape' of the theoretical stress distribution is independent of the magnitude of the load. (One author has claimed this as an 'advantage' of the constitutive equation but one wonders whether the material will always be aware of this property of the mathematical model which it is supposed to obey.)

In the higher strain rate range, approaching that of conventional hot-working, the rate sensitivity of superplastic alloys decreases with increasing strain rate and a hyperbolic sine law may provide a better fit to the stress-strain rate curve; it would be preferable to use this law in, for example, an analyses of creep rupture of superplastic alloys.

6. Some superplastic forming processes

Uniaxial tension

The total elongation in superplastic alloys in simple tension, up to around 2000%, is most dramatic and unfortunately has overshadowed other aspects of superplasticity. These materials do not strain harden and the tensile process is therefore unstable from its inception; the load continually decreases and straining is non-uniform. The large elongations are a natural result of the strain rate sensitivity, as may be seen from Fig. 6. If a bar has an initial irregularity, dA, then locally the stress increases by dσ and the strain rate by d$\dot{\varepsilon}$, as shown in Fig. 6(a). If the material has a low strain rate sensitivity, Fig. 6(b), the local increase in strain rate is large and the irregularity develops quickly. If, however, the strain rate sensitivity is high, Fig. 6(c), the variation of strain rate along the bar is small and all sections will elongate considerably before an incipient neck has grown sufficiently to precipitate failure.

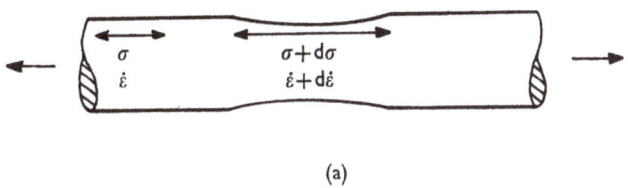

(a)

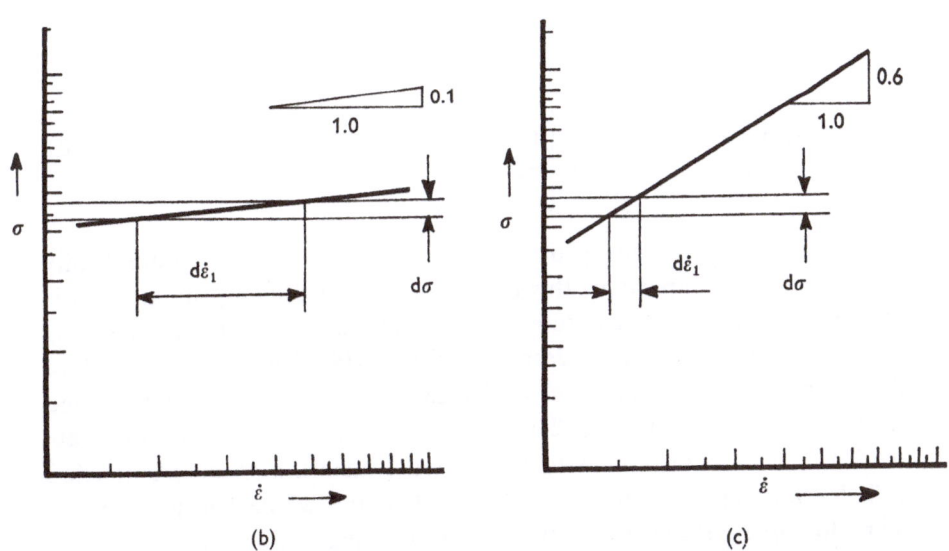

(b)   (c)

Fig. 6. Schematic representation of necking (a) in a weakly rate dependent material (b) and a superplastic alloy (c).

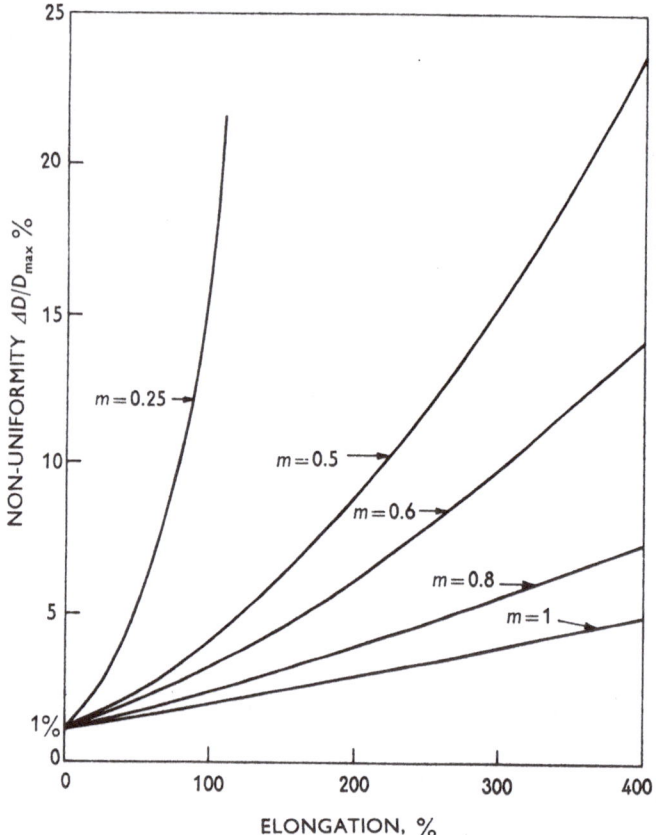

Fig. 7. Rate of growth of an initial 1% non-uniformity in tensile bars having different rate sensitivity index, m.

In superplastic forming processes, the way in which a non-uniformity develops is most important; this cannot be predicted by a general analysis because it depends on the nature of the initial inhomogeneity. Some results of a numerical analysis are presented in Fig. 7, [19]. The 'parallel' length of the test bar is assumed to taper, so that the difference in area from one end to the other is initially 1%. This diagram shows that with a strain rate sensitivity of 0.25 or less, as in conventional metals, a small imperfection will develop rapidly, whereas with $m = 0.5$, the whole bar may extend many hundred percent before the imperfection has become catastrophic.

It may be concluded from this discussion of the tensile test in superplastic alloys that the relevant feature of an analysis of tensile processes will not generally relate to the determination of the *onset* of instability but to the *growth* of non-uniformities which occur when a viscous material is stretched in an unstable process [20]. In technological processes, the limits of forming will be dependent not on the inherent ductility of the material but on the allowable variation of thickness in the formed part.

Biaxial stretching

The bulging of a superplastic sheet by lateral pressure is a typical example of one class of problems encountered in superplasticity. The change in geometry is very large indeed and both the stress and strain rate at a point in the sheet will vary with time. The axisymmetric problem has been solved by a finite difference solution but any extension of this to a more general problem would be extremely tedious. The results of the axisymmetric solution do, however, give some insight into the forming of superplastic sheet. If a constant pressure is suddenly applied to a flat circular diaphragm, the initial strain rate is high; it reaches a minimum as the shape develops and before thinning has become appreciable. Thereafter, the average strain rate accelerates and by the time the sheet has been formed into a hemisphere, the strain rate is several times greater than the minimum value. The analysis also indicates that the thickness is more uniform in the deformed part if the strain rate sensitivity of the material is high – as shown in Fig. 8, [21]. This is of obvious importance in the technological applications.

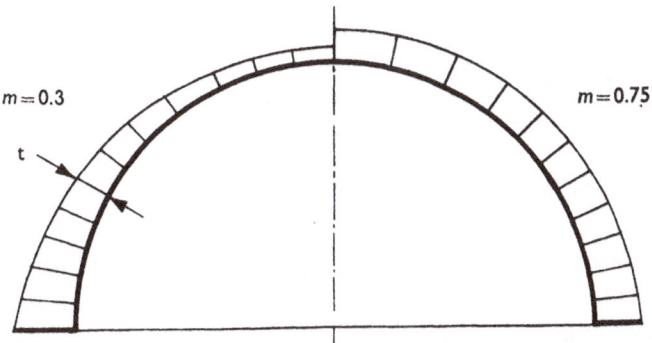

Fig. 8. Variation of thickness, t, in a hemisphere bulged from flat sheet of different rate sensitivity, m = 0.3 and m = 0.75.

Extrusion

The extrusion of superplastic alloys through a conical die has been studied by several authors. Transient effects are neglected and the stress and strain rate at a point which is fixed relative to the tooling do not vary with time. Formally, the problem is one of stationary creep; it is also similar to the extrusion of a strain-hardening, rate-independent material.

The existing superplastic solutions assume a particular stress distribution and are essentially 'slab' type analyses. In the author's opinion, these analyses are devoid of imagination and, technologically, are of limited value. What is required is a treatment similar to the semi-empirical technique now used to estimate extrusion pressures in strain-hardening materials [22]. In such an approach, one might estimate a mean flow stress from consideration of flow rates and volume of the deformation zone.

The powerful upper bound methods of conventional plasticity which employ velocity discontinuities cannot be used directly with highly rate-dependent materials, but possibly some analogue technique may be developed.

General considerations

Progress in the theoretical treatment of superplastic processes has been limited and from the viewpoint of fundamentals little has been contributed which did not already exist in the treatment of creep problems.

Many superplastic processes should be tractable because the microstructure typically does not change during large deformations and hence the material is insensitive to prior history. This assumption is often made in creep, where it is less frequently true, even as an approximation. The basic difficulty, however, is that changes in geometry can be enormous in superplastic processes and stationary problems such as extrusion are neither common, nor are they are the important one.

7. Industrial processes

Recent literature contains many examples of components formed experimentally from superplastic alloys. Because superplastic properties tend toward those of heated thermoplastics, the industrial objective has frequently been to produce superplastic metal parts using the inherently cheaper and simpler forming processes employed in the plastics industry. Parts have been produced by pressure forming sheet into a die or by blowing a cup-shaped extruded preform into a mould, as in a blow moulding process [23]. The advantages claimed are: lower cost of forming equipment, greater uniformity of thickness than in a conventional pressing and greater elongation or expansion than would be possible with a conventional metal. In certain structural components the superplastic part may be more economical than a plastic part because of the better creep resistance and higher strength of superplastic alloys compared with plastics at room temperature.

The majority of applications employ the zinc-aluminium superplastic alloy because of its low cost and reasonable room temperature properties. Other superplastic alloys such as the nickel-based and titanium alloys are attractive for higher strength components, particularly for aircraft engines. These are high cost items and reduction in tooling and manufacturing costs seem possible. In particular, a patented 'creep forging' process for these alloys has been most successful; this process bears some resemblance to injection moulding of plastics.

8. Conclusion

Superplasticity is the viscous behaviour of most alloys which have an ultra-fine grain structure when they are deformed slowly at an elevated temperature. Their mechanical behaviour can be described simply and with reasonable accuracy but the micro-mechanical mechanisms giving rise to this behaviour are not entirely clear. It appears that grain boundary sliding plays a large part and conventional slip a small role in overall deformation.

A large number of superplastic alloys have been studied and a few are available commercially. These are used in various metal forming processes, many of which resemble techniques used to form thermoplastics.

The present theory used in analyzing superplastic processes owes a considerable amount to the existing mathematical theory of creep. The alloys are exploited in very large strain processes and consequently the progress in the analysis of superplastic processes has not been great.

In the author's opinion, superplasticity is an interesting phenomenon and it is not without industrial importance. It presents a challenge to the plastician to develop analytical techniques suitable for studying the mechanics of superplastic processes. It may also entice plasticians away from some of the existing overworked fields and into the exciting and more refreshing areas of time-dependent plasticity.

Acknowledgements

The author would like to thank his colleagues, Dr. R. Sowerby and Messrs. A. R. Ragab and K. K. Jain, for their assistance and advice, and also the National Research Council of Canada, who has supported research into superplasticity at McMaster University.

Notes

There are two recent reviews dealing with metallurgical aspects of superplasticity: R. H. Johnson, *Met. Rev.*, *146* (1970) 115, and G. J. Davies et al, *J. Mat. Sci.*, *5* (1970) 1091. There does not appear to be an entirely satisfactory review of continuum aspects of superplasticity.

[1] From W. Rosenhain, J. K. Haughton and K. E. Bingham, *J. Inst. Metals*, *23* (1920) 261.

[2] In a remarkable paper, *J. Inst. Metals*, *54* (1934) 111, C. E. Pearson related these large elongations to the strain rate sensitivity, commented on the maintenance of the fine equiaxed grain structure and proposed a deformation mechanism – grain boundary sliding – which is still considered important in superplastic deformation.

[3] For example, A. A. Presnyakov and G. V. Starikova, *Fiz. Metal. Metaloved.*, *12* (1961) 873. Emphasis was given to the 'metastability' of the alloys, as in the quenched eutectoid Zn-Al. This is not now regarded as a condition for superplasticity, although in certain alloys an ultrafine structure can be obtained by decomposition of an unstable solid solution.

[4] E. E. Underwood, *J. of Metals*, *14* (1962) 914. Few writers of reviews see their effort rewarded as well as this.

[5] Principally by the excellent work of W. A. Backofen and his co-workers, D. H. Avery and D. L. Holt, published in *Trans A.S.M.*, 1964–66.

[6] Simple constitutive relations should not be used in a simple-minded manner; this warning is elegantly given by D. C. Drucker, *Int. to Mechanics of Deformable Bodies*, McGraw Hill, 1967. Difficulties which arise with the parameter m are mentioned in the paper by A. R. Ragab and J. L. Duncan in *Foundations of Plasticity*, Noordhoff, Leyden 1973, 271–285.

[7] Superplastic alloys would be non-strain hardening, rate-dependent solids or time-independent fluids!

[8] The term superplasticity has been applied to certain other phenomena observed in coarse grain materials. In certain environments which induce internal stresses – thermal cycling, irradiation or cycling about an allotropic phase transformation – very large tensile elongations can be obtained under the action of small stresses. This has been called environmental or transformation superplasticity and is described in, for example, D. Oelschlagel and V. Weiss, *Trans. A.S.M.*, *58* (1966) 143. Large tensile

elongations are not, by themselves, indications of superplasticity. If necking is suppressed artificially in a tensile test, elongations of several hundred percent can be obtained in many alloys, even in mild steel at room temperature as demonstrated by A. Taraldsen, *Materialprüfung*, 6 (1964) 189. In certain materials, deformation may lead to a refinement of the grain structure and a behaviour approaching superplasticity, as described by H. Naziri and R. Pearce, *J. Inst. Metals*, 97 (1969) 326. These may be genuine borderline cases between plasticity and superplasticity. For the purpose of this lecture, the phenomena referred to in this note are excluded.

[9] This introduction owes much to a penetrating survey by R. B. Nicholson, *Inst. of Metallurgists Review Course*, Series 2, No. 3, Eastbourne, U.K., 1969; he would not, however, wish to be held responsible for all of the comments.

[10] This diffusional creep model is supported in many of the papers by W. A. Backofen.

[11] See metallurgical reviews mentioned above for general discussion on grain boundary sliding. Cavitation at triple points is repeatedly observed in superplastic copper-based alloys (S. Sagat, P. A. Blenkinsop and D. M. R. Taplin, *J. Inst. Met.* (1972)) suggesting that grain boundary sliding occurs in some materials without adequate accommodating mechanisms.

[12] This is suggested by C. M. Packer, R. H. Johnson and O. D. Sherby, *Trans. TMS-AIME*, 242 (1968) 2485; A. Ball and M. M. Hutchinson, *Metal Sci. J.* 3 (1969) 1 and G. L. Dunlop and D. M. R. Taplin, *J. Aust. Ins. Met.*, 16 (1971), 195. Metallurgists seem to return to dislocations whenever the argument begins to falter.

[13] From D. M. R. Taplin, P. Rama Rao and V. V. P. K. Rao, *Proc. Silver Jubilee Conf. I.I.M.*, Delhi (1972).

[14] A series of alloys having the trade name 'Super·Z' are produced by the New Jersey Zinc Co., Bethlehem, Pa., USA and available at a price of approximately US$0.75 per pound. Their commercial literature on the use of these materials is very detailed. In England, Imperial Smelting Corp. Ltd. and its associates have been producing commercial alloys of this general type for some years.

[15] The superplastic behaviour of certain of its alloys is mentioned in specification sheets by the Olin Corp., USA.

[16] Forming experiments are described in G. C. Cornfield and R. H. Johnson, *Int. J. Mech. Sci.*, 12 (1970) 479.

[17] Torsion tests on superplastic Sn-Pb are described by A. Ghosh and J. L. Duncan, *Int. J. Mech. Sci.*, 12 (1970) 499 and further tests on Zn-Al by G. T. Roberts, M.Sc. Diss., Univ. of Manchester (UMIST) 1971.

[18] F. K. G. Odqvist, *Mathematical Theory of Creep*, Oxford, 1966. The appropriate theory is found in this and other books such as J. A. H. Hult, *Creep in Engineering Structures, Blaisdell*, 1966 and R. K. Penny and D. L. Marriott, *Design for Creep*, McGraw Hill, 1971.

[19] Following A. R. Ragab and J. L. Duncan, *loc. cit.*

[20] The tensile deformation of a time-dependent material is treated in a number of papers, e.g. N. J. Hoff, *Trans. ASME*, 20 (1953) 105, J. D. Campbell, *J. Mech. Phys. Solids*, 15 (1967) 359 and J. Klepaczko, *Int. J. Mech. Sci.*, 10 (1968) 296. The analysis of Hoff has been widely used in creep rupture studies. The behaviour of an actual test bar in tension (as opposed to the behaviour of a mathematical model purporting to describe it) is excellently presented by P. J. Wray, *J. Appl. Phy.*, 40, 8 (1970) 3352.

[21] Following G. C. Cornfield and R. H. Johnson, *loc. cit.*

[22] As given in W. Johnson and P. B. Mellor, *Plasticity for Mechanical Engineers*, Van Nostrand, 1962.

[23] Various papers relating to superplastic forming processes were published in *Int. J. Mech. Sci.* 1970–71; also J. F. Hubert and R. C. Kay, *Design Eng. Conf.*, *ASME*, April (1971).

Constitutive laws for granular media

Dragos Radenkovic

Ecole Polytechnique, Paris, France

Both principal idealisations of the behaviour of granular media: rigid-perfectly plastic scheme and models representing the subcritical deformation are discussed, each for its own sake and for its own merits. In the first case, different proposed flow-rules are analysed in connexion with limit-theorems for non-standard materials; in the second case, formulations based on the classical hardening rule are compared with an apparently better suited attempt to describe a loading process using hypoelastic constitutive law. It is concluded that principal problems concerning granular media still give matter for further research.

1. Introduction

The two salient features of the behaviour of granular media, shared by metals under usual working conditions – the existence of a *threshold of yielding* and a noticeably *rate-independent response* under stress – can be described more or less successfully by using the apparatus of the classical theory of plasticity.

Although the applications in the two domains (soils and metals) progressed simultaneously – the study of soils being for a long time even in a leading position (Coulomb 1773 – Tresca 1864, Kötter 1903 – Hencky 1923), the actual basic notions of the theory, especially concerning the 'hardening' i.e. phenomena which precede the yielding, were gradually built up from the observation of metal behaviour and often transposed to soils, sometimes without the necessary criticism.

Fig. 1 gives an idea of the similitude and the differences in the behaviour of metals and soils. Loading curves are shown in non-dimensional coordinates σ_{ax}/σ_0 (axial stress/yield stress) and $\Delta L/L_0$ % (stretch in %), the full curve corresponds to a typical steel, the broken lines represent the behaviour of sand in the triaxial apparatus under constant lateral stress: the initial density of sand was about 10% greater for the upper curve than for the lower one (data adapted from Bouthwell, cf. Stutz [47]).

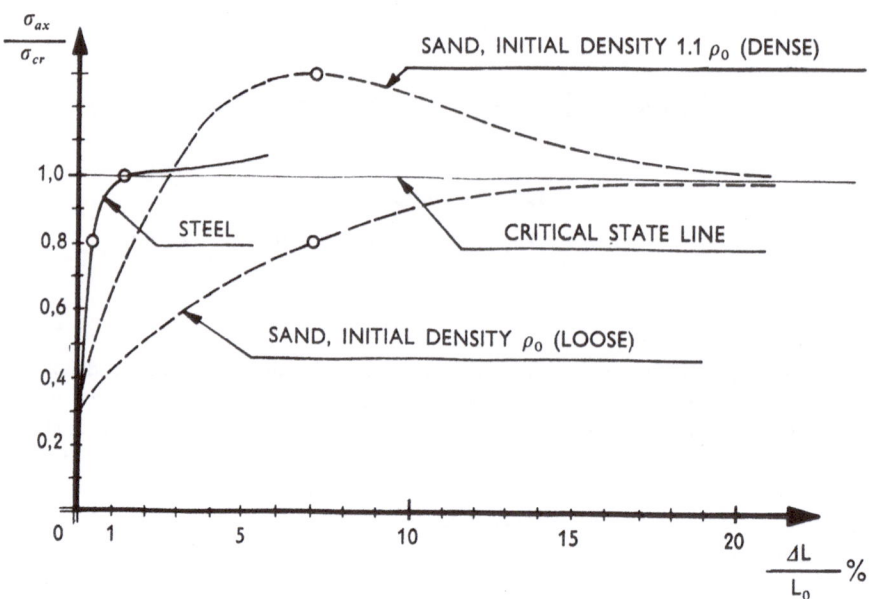

Fig. 1. Comparison of loading curves: metals-soils.

The behaviour observed is usually idealised, both for soils and for metals, in two different, almost opposite, directions. If attention is focussed on the, eventually asymptotic, horizontal part of the curves, forgetting the previous history, we obtain the *rigid-plastic* model. On the other hand, if the irreversible deformation is followed from the very beginning, the behaviour is represented by different models of *hardening*.

For metals, the situation is resumed schematically in Fig. 2, cf. Hill [18]. The hardening, which is essentially path-dependent, occurs beyond the initial *elastic frontier*. When the *yield frontier* (different from any actual elastic frontier!) is reached, the rate of deformation is of arbitrary magnitude under constant stress, but it is always perpendicular to the frontier in the corresponding point; we can say that such behaviour is not only rate-independent but also path-independent.

In the case of granular materials the picture seems to be, in a way, more intricate.

(i) The *rigid-plastic* model corresponds here rather to an asymptotic state than to 'initial' yielding (in the sense that this term is sometimes used). Fig. 3, based on the data from Fig. 1, shows that Mohr's circle for the critical state is indeed tangent to the envelope corresponding to the usual

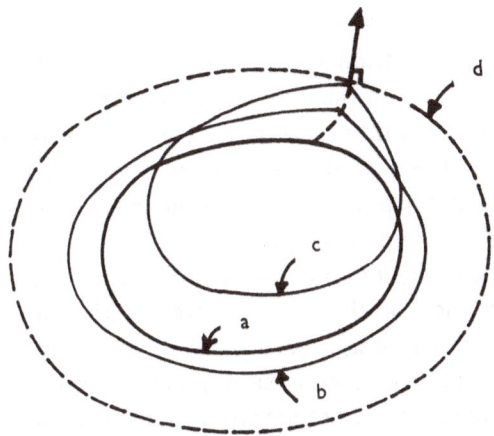

Fig. 2. Typical evolution of loading surfaces for metals.
a) Elastic frontier   b) c) Subsequent loading surfaces   d) Yield frontier

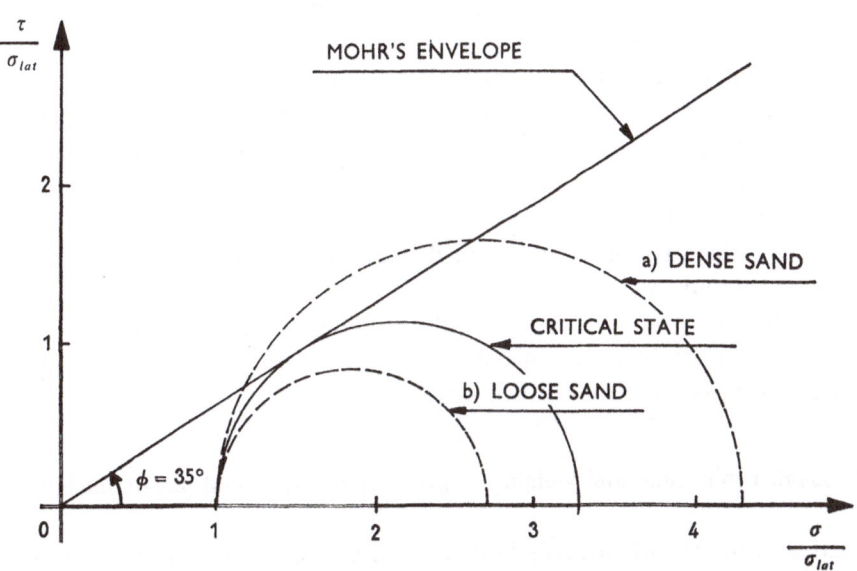

Fig. 3. Mohr's circles at differint stages of sand deformation a), b) states *at*
$\Delta L/L_0 = 7\%$.

angle of internal friction $\phi = 35°$; yet the circles corresponding to a relatively important deformation (7–8 %) are situated either above or below the envelope, depending on the initial density of the material.

The yielding occurs without change of volume, but only when a critical density is reached. The way this is achieved may depend on the previous history. So, we cannot expect a priori that the behaviour in yielding is path-independent as in the case of metals.

In spite of these difficulties, and of the fact that the changes in geometry due to the magnitude of the previous deformation may not always be negligible, the rigid-plastic approach is far from being hopeless. The estimates of the bearing capacity in many practical situations (slopes, foundations, excavations, etc...) are achieved quite successfully by calculating the corresponding *limit equilibrium* states.

(ii) Different models concerning the domain of *subcritical deformation* are doubly interesting. Their intrinsic aim is to follow the behaviour of a soil-superstructure system during the whole process of deformation preceding the failure. Hopefully this approach can also throw some light on the difficult and yet unsolved problem of finding a satisfactory approximation of the behaviour of granular media in the critical state.

The term subcritical deformation has been used here in order to show that usual models of metal plasticity are in this respect only of a relative value: the elastic domain is practically absent; the subsequent loading surfaces should be represented in a $\sigma - \rho$ stress-density space involving softening as well as hardening of the material; on the other hand the anisotropy, which is one of the main difficulties with metals, seems to play, if any, then only a secondary role.

In what follows, both principal idealisations (perfectly plastic and subcritical deformation of the media) will be discussed in greater detail, each for its own sake and for its own merits.

2. Limit equilibrium and yielding of granular media – rigid plastic models[1])

Statical fields of Sokolovski's kind, as well as kinematical mechanisms like Fellenius circles, are often used in practice. The engineer has intuitive confidence in such 'solutions', and most often he is not deceived.

---

[1]) Some additional remarks and details concerning the subject of this chapter are given in a later paper by Radenkovic [34].

Yet in order to be able to give a sound logical basis to the results obtained, and also to treat some essentially three dimensional problems, there is a need to know a *flow-rule* which would allow us to interconnect the statical and kinematical approaches to the problem of limit equilibrium. In fact, as is well known, 'solutions' which correspond to the rigid-plastic scheme obtain a definite physical meaning only through the so-called *limit theorems.*

In the case of standard materials (such as metals or clays), the associated flow rule, even if most often it is but a gross approximation to the real behaviour, readily opens the way to useful mathematical treatment. A flow rule for granular media is still a matter of controversy, but oddly enough limit theorems of a certain kind can even so be established, unfortunately far less powerful then those for standard materials.

2.1. *The flow-rule for granular media*

The first attempt to establish the equations which describe the velocity field corresponding to limit equilibrium is due to Mandel [23]. The idea springs from a natural interpretation of Coulomb's criterion, according to which the, isochoric, deformation consists of two slidings along the surface elements (tangent to stress characteristics) where the limit state $\tau = \sigma \operatorname{tg} \phi$ is just attained; the slidings are coupled with a rotation which has to be determined. Geniev [12] looked for a similar mechanism, but assuming that sliding occurs along only one family of stress characteristics. De Josselin de Jong [20], (cf. also [22]) developed independently the double sliding model. While the latter considers the accompanying rate of rotation to be free, Spencer [45] connects it to that of the principal axes of stress. In a later paper Mandel [24] proved that such a mechanism is incompatible if the rate of rotation is assigned to the axes fixed to the material element.

The corresponding geometrical relations are rather involved and, as a rule, they were not directly related to constituting relations in their usual form. Recently Mandl and Luque [25], noting that the permutation tensor *e* is hemitropic in the twodimensional case, have interpreted Geniev's model by:

$$d = \lambda As; \ \lambda \geqq 0 \quad \text{(bidimensional)}, \tag{2.1}$$

where the corresponding tensors represent: *d* – the rate of deformation, *s* the stress deviator, *A* a linear combination of one isotropic and one hemitropic fourth-order tensor.

337

Analysing Spencer's model from this point of view, we can notice that his basic equation (3.28) is identically satisfied, if we adopt as constitutive relations:

$$d = \frac{1}{2p} \overset{0}{s} \quad \text{(bidimensional)}, \tag{2.2}$$

where $\overset{0}{s}$ is the stress rate deviator and $2p = \sigma_{ii}$ – the mean stress. This would mean a kind of hypoelastic behaviour in yield, which sounds rather unusual but which is not impossible.

In the same way De Jong's free rotating model can be interpreted simply as:

$$d \text{ (isochoric) and } \sigma \text{ analytically } \textit{unrelated}, \tag{2.3}$$

but subjected to the thermodynamic restriction $\sigma \cdot d \geqq 0$.

The hypothesis of an isotropic relation between the tensors d and σ, which is more in the spirit of classical flow rules, is also often put forward. A fairly general, not necessarily quasi-linear, form of such a relation is given by:

$$d = \lambda \partial g; \quad \lambda \geqq 0 \text{ if } f(\sigma) = 0, \tag{2.4}$$

f being the yield criterion and g the so-called plastic potential.

Drucker and Prager [8] were the first to use relation of this kind for granular media with $f = g$ (standard material), which associated to Coulomb's criterion entails the increase of volume during the flow. Cf. also Shield [43]. Nowadays, if (2.4) is used, non standard materials $(f \neq g)$ are considered; most often, cf. Jenike and Shield[19], Haythornthwaite [15], the deformation is assumed to be isochoric, as was suggested long ago by Hill [16].

A discussion, in tensorial form, of different propositions (sliding models and isotropic relations between d and σ) is given by Zagainov [49].

Instead of (2.4) Sawczuk and Stutz [41] use the canonical form:

$$\sigma = \psi_0 \delta + \psi_1 d + \psi_2 d^2. \tag{2.5}$$

Applying a dimensional argument (rate-independence) it is possible to come back to (2.4) as a special case. This approach is interesting in view of possible generalizations – dependence of ψ_i on other scalar arguments than the invariants of d.

Now, reverting to Fig. 1, we may recall the known fact, cf. e.g. Roscoe [36], that in a very dense material pronounced non-homogeneous deformations appear as a consequence of local instabilities. The lines along which the sliding then occurs may well coincide with a family of stress characteristics (Geniev), but corresponding to the local state. On the contrary, a loose material seems to behave as a homogeneous isotropic medium.

Then, even if we discard extremal cases, we may ask to what extent the local asymptotic behaviour at yielding can be considered as path-independent; in other words: is it possible to claim a general validity for any of the relations proposed? In this respect, if the hypothesis (2.3) is not more disadvantageous in the discussion of limit theorems than any other assumption, it may in fact prove to be the most suitable.

As a further argument for such a conclusion we can add that experiments concerning non-homogeneous deformation, e.g. indentation by a flat punch, seem to contradict all theories; for instance cf. Stutz [46] and also the examples quoted by Spencer [45]. On the other hand Drescher [7] calculated the velocity fields corresponding to some statical solutions by using different flow rules of the kind (2.1) and (2.4); he found that in all cases considered the condition $\sigma d \geq 0$ (non-negative dissipation) was violated at least in a part of the field.

2.2. *Limit theorems*

Mathematical techniques used in order to establish limit theorems, and even their precise statement, are outside the scope of this paper, which is concerned with constitutive laws. Just some hints about what has been done, and what, eventually, could be done using the *flow-rules* available will be given in what follows.

The approach proposed by Drucker and Prager [8] based on the hypothesis of the associated flow-rule, tended to confirm the faith of engineers in Sokolowski's fields and Fellenius circles, but unfortunately it proved to be too optimistic. Drucker [9] found counterexamples considering systems of rigid bodies with Coulomb's interfaces for which there exist statically admissible fields beyond the limit of failure. This led him to propose [11] a tentative upper limit theorem.

Keeping the flow-rule (2.4) and distinguishing standard and non-standard materials, Radenkovic [31, 32] established upper and lower limit theorems, but with a gap between the two bounds rather too wide for practical purposes. Later, Palmer [29] obtained similar results. De Josselin de Jong [21] pointed out that a lower limit can also be given when d and σ are analytically unrelated. For an up-to-date presentation of limit theorems concerning standard and non-standard materials the paper by Salençon [38] may be consulted. The same author [39] discussing a practical problem of a foundation with Coulomb's friction, gives a counterexample, proving the non-validity of classical theorems, which is less artificial, thus more convincing, than Drucker's system of rigid bodies.

In a recent paper Nguyen Quoc Son and Radenkovic [28] have shown that the problem of *limit equilibrium* is related to *eigen-value* problems. In the latter, initial perturbations do not influence the value of critical loading (path-independence); on the other hand, the so called energy methods usually applied tend to determine the eigen-value, for which non-trivial solutions exist, and not the approximate form of these solutions. Similarly, when discussing limit equilibrium, it is the value of the limit loads that is interesting and not the local features of the deformation.

The counterexamples given by Drucker [9] and Salençon [39] show clearly that for non-standard materials, in the broadest sense of this term, we can not expect general results as powerful as those for media with the associated flow-rule. Still, we may hope that appropriate techniques would offer better approximations of the lower limit than those which are now available.

Very dense materials, where local instabilities can appear beyond Coulomb's asymptotic limit, may eventually present yet more difficulties, but at the other end of the scale, for very loose media – e.g. lunar soils, cf. Chung, Costes and Lee [4] – the changes in geometry are so important that the limit equilibrium approach is simply meaningless – cf. also Radenkovic and Salençon [33].

In such cases study of the subcritical deformation becomes necessary.

3. Subcritical deformation: elastoplastic and hypoelastic models

In any case, the rigid-plastic approach can only give information about the bearing capacity. If we need to follow the behaviour of a soil-superstructive system the subcritical deformation of the soil has to be taken into account. Leaving aside the long-time response of viscoplastic soils, we shall consider here only the rate-independent behaviour typical for granular media.

The pioneering and still most plentiful work in this direction was done by the Cambridge group; it started from a simple practical problem proposed by J. F. Baker to the late Professor Roscoe some twenty years ago; a broad review of the Cambridge research programm with numerous references to theoretical work and to corresponding experimental investigations was given by Roscoe [36].

A first attempt to consider the subcritical deformation as a kind of hardening process in the scope of the classical theory of plasticity is due to Drucker, Gibson and Henkel [10], who considered the short-time response of certain clays.

Working on the same lines, Poorooshasb and Roscoe [30] started from the idea that relative density (or void ratio) has to be taken as the parameter which determines hardening. In fact, the importance of this parameter in the subcritical deformation, as Roscoe [36] remarks, was stressed long ago by Hvorslev (1937) and Rendulic (1938).

The situation is schematically represented for the two-dimensional case in Fig. 4a. Let σ, τ be the coordinates in Mohr's plane and e the void ratio. The subsequent loading surfaces are defined by $f(\sigma, \tau, e_p) = 0$. The void ratio e_p under which plastic deformation occurs for a given σ, τ can be connected to an equivalent mean pressure p_0 by $g(p, p_0, e_p) = 0$. The maxima of e_p, say e_c, are situated on the critical state line $F(\sigma, \tau, e_c) = 0$ where the deformation is isochoric; the rule of normality of d to f having been adopted, for $e_p \neq e_c$ positive (or negative) dilatation occurs.

More generally we can represent loading surfaces by

$$f(\sigma, E) = 0, \tag{3.1}$$

where E is the parameter of hardening depending on initial density ρ_0 and on previous loading history

$$E = E[\rho_0, \overset{t}{\underset{0}{\sigma(z)}}]. \tag{3.2}$$

To a given loading surface we can associate the rate of plastic deformation

$$d^p = hdf|_E, \tag{3.3}$$

where $h = h(\sigma, E)$ is a symmetrical second order tensor, and $df|_E$ the gradient of f for E constant. The form of h can be derived, once (3.1) and (3.2) are given. A 'critical' surface may, if needed, be attached to the 'critical' values of E.

Elastic deformation can be added to (3.3), but, in fact, the deformation

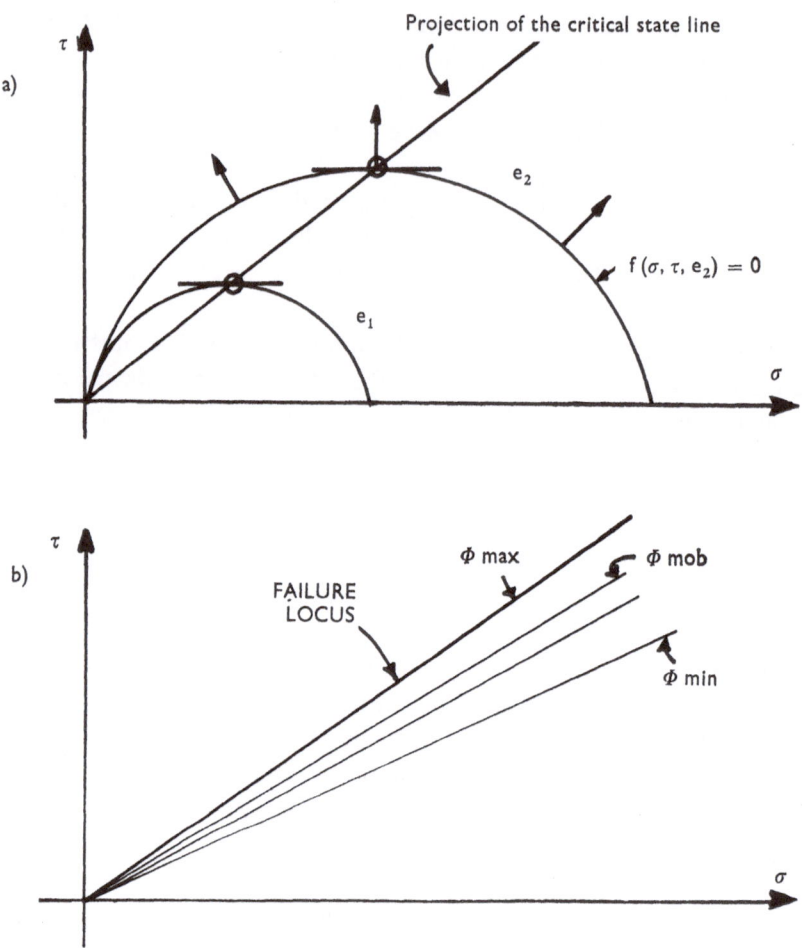

Fig. 4. Loading surfaces and failure locus for soils.
a) Loading surfaces with normality rule
b) Fan of Coulomb's lines for a given initial density

can be considered as reversible only for small cycles of stress; for instance, the elasticity has to be taken into account in some dynamic problems.

Possible versions of such a procedure are presented in more detail, for instance, in the papers by Smith and Kays [44] in view of direct numerical applications also by Gudehus [14] and Nikolaevskii [27]. (Cf. also [1], [2]).

The principal remark which should be made in respect to (3.3) is that **d** is *independent* of the direction of $\dot{\sigma}$ (whatever definition of the rate of stress $\dot{\sigma}$ is adopted); on the other hand the expression (3.3) is quasilinear in d and $\dot{\sigma}$. Both hypotheses spring from the classical theory of plasticity of metals and the question arises: to what extent *is their transposition to granular media legitimate*?

Another model of hardening is considered by Cole [5], who takes as loading surfaces a fan of Coulomb's straight lines – failure loci in Fig. 4b, where the 'mobilised' internal friction ϕ_{mob} depends on the density; the stress-dilatancy relation, Rowe [37], is not supposed to obey the normality rule.

An approach which parts from classical hardening rules is due to Sawczuk and Stutz [42]. They use hypoelastic type constitutive law, i.e. a rate-independent relation of the form $\dot{\sigma} = f(d, \sigma, \rho)$. In order to obtain manageable expressions they consider, as a first approximation, the quasi-linear expression:

$$\dot{\sigma} = a_0\delta + a_1 d + a_2\sigma, \tag{3.4}$$

where the coefficients a_i depend, in general, on ρ and on the invariants, and the joint invariants, of d and σ. The theory is presented in more detail by Stutz [48], so just a brief comment may be added here.

The relation (3.4) seems to suit the description of the subcritical deformation better than the hardening rule (3.3) since, in principle, d should be influenced by the direction of the rate of stress $\dot{\sigma}$. Yet such a law cannot be analytical in all directions of $\dot{\sigma}$ in order to avoid a known thermodynamic contradiction. A possible variant could to be imagine, for instance, a cone in the representative σ, ρ space limiting the domains of 'loading' and 'unloading', the coefficients a_i being different in either case. This would probably lead to a kind of hardening, but in a less classical form, cf. Hill [17]. On the other hand, as in the 'critical state' the deformation is supposed to occur under constant stress without change of volume, (3.4) reduces to $d = \lambda s$, analogous to (2.4); we have seen that such a flow-rule is controverted by many authors.

In this discussion we leave aside attempts which are rather too general, such as the paper by Goodman and Cowin [13]. The relations proposed by the authors are not rate-independent and so, in fact, do not concern our problem.

Finally, reviewing different proposed constitutive laws for granular media, we can notice that two questions, rather familiar in recent developments of the mechanics of continua, have received but little attention in our particular domain.

One of them is the problem of anisotropic behaviour raised sometimes by the experimentalists. In this respect granular medium would behave as a kind of rate-independent anisotropic fluids, which is certainly not easy to deal with.

The other question is that of Cosserat effects. The energy spent in the rotation of individual grains, or even of grain-packings, can not be taken into account if only global deformation is considered and we may ask ourselves to what extent this effect can be neglected. Its relative importance should probably be more pronounced in the subcritical domain than in yielding, but this is only a conjecture, cf. Satake [40], Nikolaevskii and Afanasiev [26]; in a recent paper Brown and Evans [3] put forward an argument which tends to show that this effect is in fact negligeably small.

4. Concluding remarks

As we are about to finish this discussion concerning the behaviour of granular media, it would perhaps be useful at least to mention some questions which have been omitted, although they are closely connected to our problem, such as:

(i) a survey of structural, called sometimes microrheological, investigations, both experimental and theoretical;

(ii) a critical examination of the many experimental results available;

(iii) examples of the analyses of limit equilibrium and, more recently, of elastic-plastic problems, performed in order to test different proposed theories;

(iv) dynamic problems.

As to the narrower problem of constitutive relations, we may sum up our review as follows.

A – Regarding the problem of *limit equilibrium* it is more or less universally admitted that the flow in the critical state occurs without change of volume, but the question of the analytical form of the flow-rule remains a point of controversy. Personally, I would tend to agree with the opinion that the axes of the rate of deformation tensor can be arbitrary. Still, we may ask:

a) can structural considerations, or a more careful study of subcritical deformation up to the critical state, clearly confirm this opinion or would they lead to another more definite flow-rule?

b) in any case, as the main problem in this respect is not the pattern of deformation, but a good approximation of the value of the *critical loading*, is it possible, even leaving a certain freedom to kinematic mechanisms, to obtain better lower limits than those now available?

B – Regarding the problem of *subcritical deformation*, there is certainly a difference between the hardening of metals and the non-elastic behaviour of granular media under small deviatoric stress. In this respect a hypoelastic law seems more satisfactory then the classical models. Again we ask:

a) how to assure that in a cycle of loading the work is not extracted from the matter, which is the known thermodynamic contradiction bound up with the hypoelastic scheme? To what extent are we not brought back to a kind of hardening if, say, a 'loading cone' is imagined in a sufficiently simple form?

b) the Cosserat effects probably play a more important role in the deformation of granular media than, for instance, in the case of metals; by introducing them shall we be able to clear up some old controversies, or on the contrary would this simply bring new, preferably avoidable, complications?

Now, after all these criticisms, questions and uncertainties, I feel the echo coming: 'L'art est difficile et la critique aisée...' and I quite agree, but I whould add that this also applies to the art of making a short general review of an involved problem.

References

[1] Berezhnoi, I. A., D. D. Ivlev, D. D. and V. B. Tchadov, On constructing models of cohesionless media by specifying the dissipation function. *Foundations of Plasticity* (Warsaw 1972). Noordhoff, Leyden 1973, 601–605.

[2] Brown, E. H., A theory for mechanical behaviour of sand. *Proc. 11th Int. Congr. Appl. Mech.* (Munich 1964), Springer, Berlin 1966, 183–191.

[3] Brown, C. B. and R. J. Evans, On the application of couple-stress theories to granular media. *Géotechnique*, 22 (1972) 356–361.

[4] Chung, T. J., N. C. Costes, and J. K. Lee, Boundary value problems with incremental plasticity in granular media. *Arch. Mech. Stos.* (to appear).

[5] Cole, E. R. L., The behaviour of soils in the simple shear apparatus. *Ph. D. thesis*, Cambridge 1967.

[6] Davis, E. H., Theories of plasticity and the failure of soil masses. In *Soil Mechanics: selected topics* (ed. I. K. Lee) (1968), 341–380.

[7] Drescher, A., A note on plane flow of granular media. *Problèmes de la Rhéologie*, Symp. Franco-Polonais, (Jablonna 1971), PWN, Warsaw 1973, 135–144.

[8] Drucker, D. C. and W. Prager, Soil mechanics and plastic analysis. *Quart. Appl. Math.*, *10* (1952) 157–165.

[9] Drucker, D. C., Coulomb friction, plasticity and limit loads. *J. Appl. Mech.*, *21* (1954) 71–74.

[10] Drucker, D. C., R. E. Gibson, and D. J. Henkel, Soil mechanics and work hardening theories of plasticity. *Trans. Amer. Soc. Civ. Eng.*, *122* (1957) 338–346.

[11] Drucker, D. C., On stress-strain relations for soils and load carrying capacity. *Proc. 1th Int. Conf. on Mech. of Soil Vehicle Systems*, 1961, 15–23.

[12] Geniev, G. A., Problems of the dynamics of granular media (in Russian), Gostechizdat, Moscow 1958.

[13] Goodman, M. A. and S. C. Cowin, A continuum theory for granular materials. *Arch. Rat. Mech. Anal. 44* (1972) 249–266.

[14] Gudehus, G., Elastic-plastic constitutive equations for dry sand. *Arch. Mech. Stos. 24*, 1972, 395–402.

[15] Haythornthwaite, R. M., Stresses and strains in soils. *Proc. 2d Symp. Nav. Struct. Mech.*, Pergamon Press, Oxford 1960, 185.

[16] Hill, R., The mathematical theory of plasticity, Cambridge Univ. Press, Cambridge 1950, p. 299.

[17] Hill, R., Some basic principles in the mechanics of solids without a natural time. *J. Mech. Phys. Solids 7* (1959) 209–225.

[18] Hill, R., The essential structure of constitutive laws for metal composites and poly-crystals. *J. Mech. Phys. Solids*, *15* (1967) 79–95.

[19] Jenike, A. W. and R. T. Shield, On the plastic flow of Coulomb solids beyond original failure. *J. Appl. Mech.*, *26* (1959) 599–602.

[20] De Josselin de Jong, G., Statics and kinematics in the failable zone of a granular material. Thesis, Delft 1959.

[21] De Josselin de Jong, G., Lower bound collapse theorem and lack of normality of strain rate to yield surface for soils. *IUTAM Symp. Rheol. and Soil Mech.* (Grenoble, 1964) 69–75. Springer, Berlin 1966.

[22] De Josselin de Jong, G., The double sliding, free rotating model for granular assemblies. *Géotechnique 21* (1971) 155–163.

[23] Mandel, J., Sur les lignes de glissement et le calcul des déplacements dans la déformation plastique. *C.R. Ac. Sc.*, Paris *225* (1947) 1272–1273.

[24] Mandel, J., Sur les équations d'écoulement des sols idéaux en déformation plane et le concept du double glissement. *J. Mech. Phys. Solids*, *14* (1966) 303–308.

[25] Mandl, G. and R. Fernandez Luque, Fully developed plastic shear flow of granular materials. *Géotechnique*, *20* (1970) 277–307.

[26] Nikolaevskii, V. N. and E. F. Afanasiev, On some examples of media with micro-structure of continuous particles. *Int. J. Solids and Struct. 5* (1969) 671–678.

[27] Nikolaevskii, V. N., Continuum theory of plastic deformation of granular media. *Foundations of Plasticity*, (Warsaw 1972). Noordhoff, Leyden 1973, 587—600.

[28] Nguyen, Quoc Son, and D. Radenkovic, La dualité des théorèmes limites pour une structure en matériau rigide-plastique standard. *Arch. Mech. Stos. 24*, 5–6, 991–998.

[29] Palmer, A. C., A limit theorem for materials with non-associated flow laws. *Journal de Mécanique 5* (1966) 217–222.

[30] Poorooshasb, H. B. and K. H. Roscoe, The correlation of the results of shear tests with varying degrees of dilatation. *Proc. 5th Int. Conf. Soil Mech., 1* (1961) 297–304.

[31] Radenkovic, D., Théorèmes limites pour un matériau de Coulomb à dilatation non-standardisée. *C.R. Ac. Sc.*, Paris *252* (1961) 4103–4104.

[32] Radenkovic, D., Théorie des charges limites. *Séminaire de Plasticité* (ed. J. Mandel), 1962, 129–142.

[33] Radenkovic, D. and J. Salençon, Equilibre limite et rupture en mécanique des sols. *Journées Françaises de Mécanique des Sols*. Bul. Liais. Labo. Rout. N° spécial 1971, 296–302.

[34] Radenkovic, D., Equilibre limite des milieux granulaires – modèles de comportement rigide – plastique. *Plasticité et Viscoplasticité*, Séminaire, (Paris 1972)

[35] Roscoe, K. H. and J. B. Burland, On the generalized stress-strain behaviour of wet clay. *Engineering plasticity* (ed. Heyman J., Leckie, F. A.), 535–609. Cambridge Univ. Press, Cambridge 1968.

[36] Roscoe, K. H., The influence of strains in soil mechanics. (10th Rankine Lecture). *Géotechnique, 20* (1970) 129–170.

[37] Rowe, P. W., The stress-dilatancy relation for static equilibrium of an assembly of particles in contact. *Proc. Roy. Soc., A 269* (1962) 500–527.

[38] Salençon, J., Ecoulement plastique libre et analyse limite pour les matériaux standards et non standards. *Annales de l'I.T.B.T.P.* 295–296 (1972) 91–100.

[39] Salençon, J., Un exemple de non-validité de la théorie classique des charges limites pour un système non standard. (private communication).

[40] Satake, M., Some considerations on the mechanics of granular materials. *IUTAM Symp. Mech. of Generalized Continua*, (Stuttgart 1967), 156–159. Springer, Berlin 1969.

[41] Sawczuk, A. and P. Stutz, On formulation of stress-strain relations for soils at failire. *ZAMP, 19* (1968) 770–778.

[42] Sawczuk, A. and P. Stutz, Contribution à l'étude de la loi rhéologique des milieux pulvérulents. *Problèmes de la Rhéologie*. Symp. Franco-Polonais, (Jablonna 1971), Warsaw 1973, 319–334.

[43] Shield, R. T., On Coulomb's law of failure in soils. *Journ. Mech. Phys. Solids, 4* (1955) 10–16.

[44] Smith, I. M. and S. Kay, Stress analysis of contractive or dilative soil. *J. Soil Mech. and Found.*, Div. *97* (1971) 981–988.

[45] Spencer, A. J. M., A theory of the kinematics of ideal soils under plane strain conditions. *J. Mech. Phys. Solids, 12* (1964) 337–351.

[46] Stutz, P., Contribution à l'étude de la loi de déformation plastique des sols. *Thèse Doct. Spec.*, Univ. Grenoble 1963.

[47] Stutz, P., Contribution à l'étude de la loi rhéologique des milieux pulvérulents. *Thèse*, Grenoble 1972.

[48] Stutz, P., Comportement élastoplastique des milieux granulaires. *Foundations of Plasticity*, (Warsaw 1972), 37–49. Noordhoff, Leyden 1973.

[49] Zagainov, L. S., Equations of plane steady-state notion of a granular medium. *Mech. of Solids* (Engl. transl.) *2* (1967) 130–134.

DISCUSSION

W. Szczepiński[1]). As an illustration to that part of the most interesting lecture presented by Professor Radenković in which the flow laws for granular media were considered I would like to show here a simple experimental test allowing us to decide quickly which flow law applies to a given granular material undergoing large deformations.

Two days ago Dr. Gudehus has shown a sophisticated experimental device, which allow him to perform experiments of great research value. However, it is well known that three-dimensional compression tests are complex and laborious. I think, therefore, that for preliminary tests a simple deformation mode may be useful.

Fig. 1 shows the solution regarding the motion of a granular medium pushed by the bucket of a loading machine (see Ref. [1]). The bucket moves with the velocity v_0 inclined at an angle β to the horizontal line. The network of stress characteristics covers the area of the curvilinear triangle ABC. The velocity solution has been obtained for the non-associated flow rule assuming constant volume of the deformed granular medium. Fig. 1b shows the mesh of velocity characteristics. The deforming region DEC is bounded by two lines of velocity discontinuity EC and DC. Material to the right of EC remains at rest, while the volume to the left of DC moves as a rigid block together with the bucket. The velocity hodograph is shown in Fig. 1c. Inner points of the curvilinear triangle $D'E'C'$ represent velocities of the respective particles in the deforming region DEC. This part of the hodograph is shown in enlarged scale in Fig. 1d.

Such a type of deformation may be applied to experimental testing of behaviour of granular media, because the deformation mode does not depend on the friction conditions along the contact surface between the tested medium and the pushing wall. If a cohesionless granular medium is considered and the stress-free upper edge is formed by a straight line, the

[1]) Institute of Fundamental Technological Research, Warsaw, Poland.

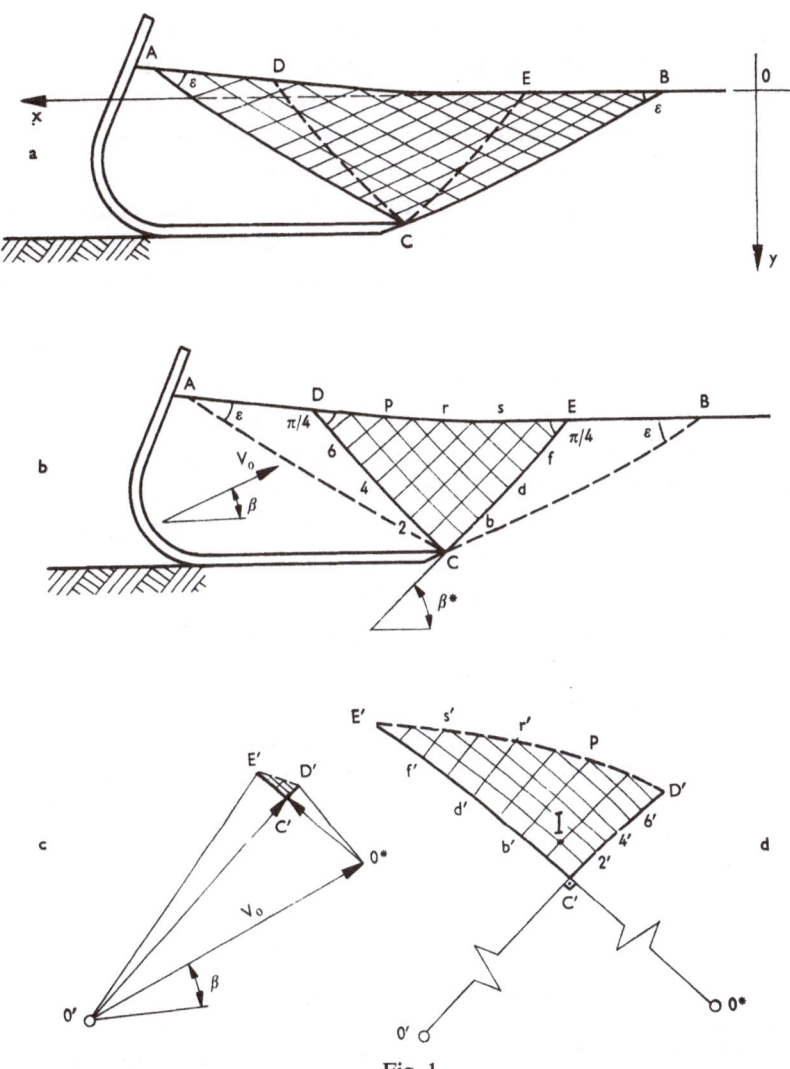

Fig. 1

patterns of stress and velocity characteristics are formed by two families of parallel straight lines. Theoretical solution for such a case can be obtained in an elementary way.

Fig. 2 shows the experimental deformation mode obtained with the use of glass rods of various diameters. One of the photographs was made when the container with the glass rods was fixed, while the model of the bucket

349

was shifted to the right. The second photograph was obtained when the bucket was fixed and the container was shifted to the left. The dead zones, the first one adjacent to the wall of the bucket and the second located to the right of the discontinuity line *EC*, are clearly visible. Both lines of velocity discontinuity intersect the stress-free surface at 45 degrees. The deforming region is shifted parallel to *EC*. Velocity jumps across the two lines of discontinuity are directed along the respective line. Thus, the actual velocity field coincides with the theoretical solution resulting from the non-associated flow rule mentioned above.

One of the results obtained for dry sand in a loose state is shown in Fig. 3. For the sake of comparison the theoretical lines corresponding to the lines of velocity jump *CD* and *CE* are shown in the photograph. Both actual discontinuity lines and direction along which the moving region is shifted coincide also in this case with the theoretical solution obtained for the non-associated flow rule. Analogous results have been obtained for other configurations of the stress-free boundary and for various values of the angle β [2]. Theoretical velocity fields resulting from the flow law associated with the Coulomb yield criterion depart qualitatively even from the actual fields.

For compacted dry sand the actual deforming regions at incipient flow were found to be slightly larger than those predicted by the flow rule assuming no change of the volume. However, even in this case the actual initial stages of motion were closer to the theoretical velocity fields resulting from the incompressibility and isotropy assumptions than to those obtained on the basis of the associated flow rule.

References

[1] Szczepiński, W., Some slip-line solutions for earthmoving processes, *Arch. Mech. Stos.*, Vol. 23 (1971) 885–896.
[2] Szczepiński, W. and H. Winek, On some practical problems of large flow of soils, *Proc. Symp. on Rheology* (Jabłonna 1971) PWN, Warsaw, 353–365.

A. Drescher[1]).   In the general lecture presented by Professor Radenković the attention was focused on theoretical problems arising in connection with constitutive equations for granular media. The decisive choice of a proper model, however, cannot be made without experimental evidence

[1]) Institute of Fundamental Technological Research, Warsaw, Poland.

Fig. 2

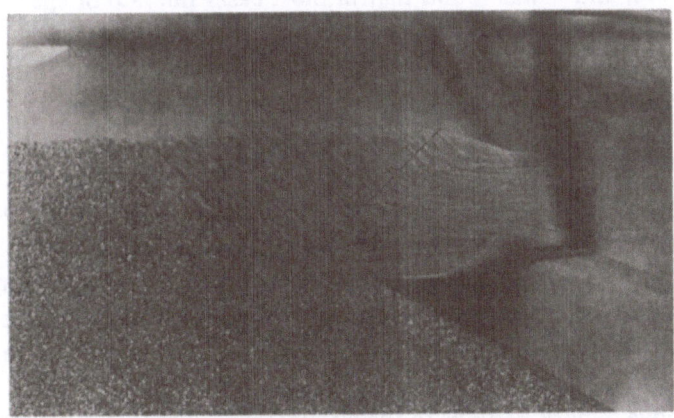

Fig. 3

as to the basic postulates involved in the theory. There are two generally accepted methods for experimental verification of these assumptions: (i) a direct one based on investigation of properties of a material sample, (ii) an indirect one, by comparing the theoretical solution of a boundary-value problem for a given model with experimental stress and/or strain (velocity) field. The second method has found a wide application for critical-state behaviour of granular media. The experiments were usually performed in plane-strain the velocity field being evaluated. The comparison of the theoretical and the experimental velocity fields, may give a valuable information only if the theoretical solution satisfies all the requirements imposed on the plastic field, i.e. if static and kinematic admissibility conditions and non-negativeness of the rate of specific plastic dissipation are fulfilled.

To illustrate this problem more in detail let us consider two of the commonly run plane-strain experiments: (i) horizontal penetration of a rigid wall into a granular medium (Fig. 4a), and (ii) indentation of a rigid punch into a semi-plane (Fig. 4b). In Fig. 4a the theoretical mesh of velocity characteristics corresponding to radial stress state for the non-associated and non-coaxial model is presented, while Fig. 4c shows the related hodograph. In Fig. 4b the net of velocity characteristics for the non-associated but coaxial flow rule and the stratically admissible solution of Prandtl type is drawn. In both cases the obtained velocity field satisfies all kinematical requirements except the condition of non-negativeness of the plastic dissipation. In [1] several boundary-value problems for non-associated flow rules were investigated, and it was found that in most cases the rate of specific plastic dissipation is negative within the field.

These results indicate on one hand, that the correct kinematical solutions based on the generally accepted statical solutions are for many problems very difficult to construct if non-associated flow rules are used. In other words, there are in some of the proposed flow rules strong internal constraints which restrict the applicability of these models to very particular problems. On the other hand, more criticism should be offered in drawing conclusions based on the test results obtained so far. It can be easy shown that many of the kinematical solutions, when compared with experiments, do not satisfy the 'thermodynamic' condition, and therefore cannot be accepted as physically realistic.

The above remark pertains to the classical plane-strain experiments where a natural granular material, usually sand, is used. It seems worthwhile

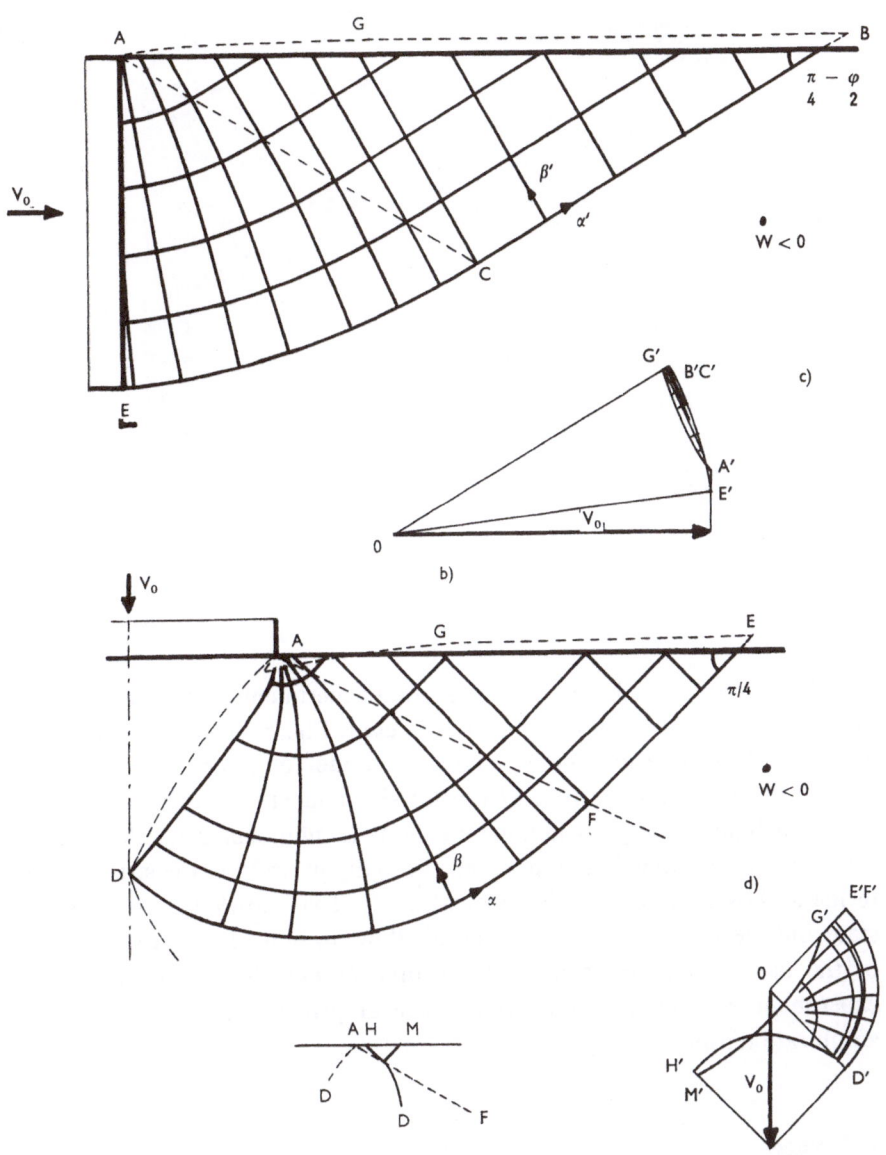

Fig. 4
a) Non-associated, non-coaxial flow law; b) Non-associated, coaxial flow law.

353

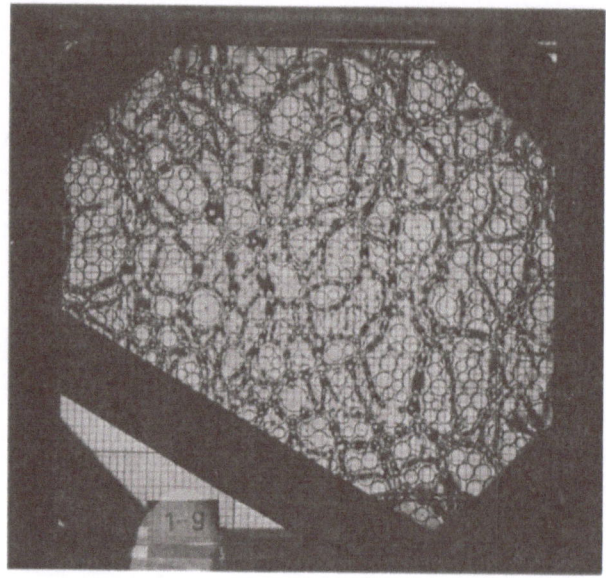

Fig. 5

to point out new possibilities of plane-strain experiments, if instead of sand, an optically sensible granular material is used. Fig. 5 shows the pattern of loading transition throughout a stressed model of a granular material, consisting of plane discs made from CR-39, and viewed in the circularly polarised light. Basing on familiar methods of photoelasticity the contact forces can be evaluated, and after the averaging procedure, the stress tensor inside a chosen region can be calculated [2]. The knowledge of the stress state and the strain state in the material allows to make a direct verification of the basic assumptions made in theoretical models. For instance, the experiment demonstrates non-coincidence of principal axes of stress and strain rate tensors.

References

[1] Drescher, A., *Arch. Mech. 24* (1972) 837–848.
[2] Drescher, A. and G. de Josselin de Jong, *J. Mech. Phys. Solids, 20* (1972) 337–351.

G. Gudehus[1]). A few points can be made in addition:

(i) In my opinion plastic anisotropy is of the same importance for granular materials as for metals. This view is supported both by microscopic (K. Wiendieck, Thèse, Grénoble, 1964, e.g.) and macroscopic experiments (author's discussion to Professor Phillips' lecture at this symposium).

(ii) One can draw another conclusion from the paper by Mandl and Fernandez Luque. As hemitropy and isotropy of second order tensor functions are not equivalent for a two-dimensional space, a 'two-dimensional' packing of parallel rods is not a relevant model for granular materials. Especially an observed difference of the principal axes of σ_{ij} and $\delta\varepsilon_{ij}^p$ in such a model does not prove that a similar difference holds for a granular material.

(iii) As the flow rule of granular materials is non-associated (at least its volumetric part) the existence of a plastic potential is not always warranted. Only in a two-dimensional stress space (as used for representing conventional triaxial tests) one can always find a potential. Otherwise a certain integrability condition must hold in order to enable the construction of a plastic potential; this would be an additional constitutive property.

[1]) University of Karlsruhe, Karlsruhe, GFR.

Discussion notes

to the international symposium on Foundations of plasticity

An axiomatic approach to the theory of thermomechanically constrained materials

F. Andreussi and P. Podio Guidugli

University of Pisa, Pisa, Italy

1. Introduction. In theories of purely mechanical behaviour of continuous media internal constraints are usually intended as scalar restrictions on the present value of the assumed deformation measure ([1], Sect. 30). As the class of possible motions becomes smaller, the material response must allow for arbitrary stresses which maintain the constraint. Incompressibility, rigidity and inextensibility are standard examples of purely mechanical constraints. In all of these cases, **one** is almost immediately led to appropriate constitutive assumptions for the arbitrary part of the stress, so that the resulting field equations can be written down and sometimes solved at various levels of generality.

In 1965 Truesdell and Noll [1] gave the theory of constrained materials an axiomatic formulation within the framework of their treatment of simple materials. In particular, they assumed a *reduced principle of determinism*, which consists in a suitable relaxation of the corresponding axiom for unconstrained materials and yields systematical predictions for the response to constraints. As the axiomatic approach proves especially valuable when unusual constraints are dealt with, even in a purely mechanical context, it is conceivable that it may be useful to study thermomechanical processes in thermomechanically constrained materials.

Actually, to have recourse to an *a priori* approach seems to be particularly cogent when, in addition to arbitrary stress fields, constraints may give raise to arbitrary thermodynamic fields.

In a previous paper [2] we discussed an extension, to thermomechanically constrained materials, of Truesdell and Noll's reduced principle of determinism and established some general results concerning the arbitrary parts of the constitutive functionals.

This note rests on the theoretical framework assembled in [2], but for a weaker characterization of the arbitrary fields. Consequently, results are less stringent than those obtained in [2]. Nevertheless, as it is shown by a typical example, present results are still definite enough to give the expected answer in many cases of interest.

2. Axiom of determinism. Let $x = x(X, t)$ be the motion and $\vartheta = \vartheta(X, t)$ the absolute temperature, which is assumed to be positive; X is the place in a reference configuration for the body and t is a monotonically increasing parameter. Some more notation is necessary to introduce the deformation gradient F, the stretching tensor D and the temperature gradient g. Further, let $\{T, \eta, \psi, h\}$ be an array of thermomechanical fields, namely, the (symmetric) Cauchy stress tensor, the entropy, the free energy and the heat flux vector, which are the values at X and t of some constitutive functionals depending on the histories of F, ϑ and g.

The ordered pair (x, ϑ) is called a *process* if the specific body force b and heat absorption r are so adjusted that the production of momentum and energy are null:

$$\text{div } T + \rho(b - \ddot{x}) = 0, \tag{2.1}$$

$$T \cdot D + \text{div } h + \rho(r - \dot{\varepsilon}) = 0.^1) \tag{2.2}$$

The choice of the constitutive functionals is restricted by the assumption that the production of entropy, denoted by γ, is non-negative in every process

$$\rho \vartheta \gamma = -\rho(\dot{\psi} + \vartheta \dot{\eta}) + T \cdot D + \frac{1}{\vartheta} h \cdot g \geqq 0. \tag{2.3}$$

As customary (*e.g.* [1], Sect. 79), the total entropy production can be interpreted as the sum of contributions due to internal dissipation δ ([3], Lect. 2) and to heat conduction:

$$\rho \vartheta \gamma = \rho \delta + \frac{1}{\vartheta} h \cdot g. \tag{2.4}$$

We now introduce the notion of a *simple thermomechanical internal constraint*, by which we mean an equation of the form

$$\mu(F, \vartheta, g) = 0 \tag{2.5}$$

---

^{1}) In Eq. (2.2) ρ is the density in the present configuration, $\varepsilon = \psi + \vartheta \eta$ is the (specific) internal energy and the dot stands for the usual inner product of second-order tensors.

in which μ is a scalar-valued function of the present values of F, ϑ and g. A process is said to be *admissible* in a constrained simple material if it obeys the constraint equation. As we have already noticed, if the class of possible processes is limited, the constitutive functionals must be such to allow the internal constraints to be maintained. Accordingly, we lay down the following *principle of determinism for simple materials subject to thermomechanical constraints*:

The fields T, η, ψ, h at time t are determined by the histories of the intermediate variables only to within 'added' fields $\bar{T}$, $\bar{\eta}$, $\bar{\psi}$, $\bar{h}$ such that[1])

$$T = \bar{T} + \hat{T}(F^t, \vartheta^t, g)$$

$$\eta = \bar{\eta} + \hat{\eta}(F^t, \vartheta^t, g) \, ;$$

$$\psi = \bar{\psi} + \hat{\psi}(F^t, \vartheta^t, g) \tag{2.6}$$

$$h = \bar{h} + \hat{h}(F^t, \vartheta^t, g)$$

moreover, the added fields affect neither the internal dissipation δ nor the conduction inequality:

$$\rho\bar{\delta} = -\rho(\dot{\bar{\psi}} + \vartheta\dot{\bar{\eta}}) + \bar{T} \cdot D = 0 \tag{2.7}$$

$$\bar{h} \cdot g = 0.$$

The former statement is modelled on and is plainly consistent with Truesdell and Noll's reduced axiom of determinism in the purely mechanical case. It asserts the following:

(i) The constitutive functionals decompose into the sum of two parts.

(ii) The '$\hat{}$' parts are fully determined by the process and are by no means affected by the '$\bar{}$' ones.

(iii) The constraint response '$\bar{}$' is ruled by conditions (2.7), which characterize the present theory among possible others. Conditions (2.7) impose the added fields not to influence Planck's and Fourier's inequalities, *i.e.*

$$\delta = \hat{\delta} + \bar{\delta} \geqq 0, \tag{2.8}$$

$$h \cdot g = (\hat{h} + \bar{h}) \cdot g \geqq 0. \tag{2.9}$$

[1]) We adopt the '$\bar{}$' and '$\hat{}$' notation of [4].

Clearly, if the added fields obey (2.7), they do not influence Clausius-Duhem inequality (2.3) as well. In other words, conditions (2.7) imply a 'frictionless' constraint response such that an admissible process is irreversible or not, irrespective of constraints. Moreover, (2.7) reduce to Truesdell and Noll's condition of null power in the mechanical limit.

Beside the formal presentation under form of a principle of determinism, the previous analysis makes patent two main ingredients of any axiomatic approach to the theory of constrained materials, namely, a notion of constraint and some characterizations relating the restrictions on the possible motions to the constraint response. In [2] constraints have the form (2.5), though characterizations (2.7) are replaced by the stronger requirement that the production of 'added' internal energy and entropy are null. On the other hand, in [4] a different notion of constraint is used, which in our notations looks like

$$A(C, \vartheta) \cdot \dot{C} + a(C, \vartheta)\dot{\vartheta} + a(C, \vartheta) \cdot \nabla \vartheta = 0 \tag{2.10}$$

and it is assumed that

$$\bar{\psi} = 0, \quad \bar{\gamma} = 0. \tag{2.11}$$

3. Some results. The material response to constraints is investigated by a standard procedure which immediately extends to any number of constraints equations.[1]) To meet the requirement of material objectivity we first express the constitutive functionals in terms of the new set of intermediate variables $C = F^T F$, ϑ and $\nabla \vartheta = F^T g$, so that Eq. (2.4) takes the form

$$\lambda(C, \vartheta, \nabla \vartheta) = 0. \tag{3.1}$$

We then consider the following vectors:[2])

$$\Gamma \equiv \left(\partial_C \bar{\psi} - \frac{1}{2\rho_R} \bar{S}, \partial_\vartheta \bar{\psi} + \bar{\eta}, \partial_{\nabla \vartheta} \bar{\psi} \right),$$

$$\Lambda \equiv (\partial_C \lambda, \partial_\vartheta \lambda, \partial_{\nabla \vartheta} \lambda), \tag{3.2}$$

$$\Xi \equiv (\dot{C}, \dot{\vartheta}, \nabla \dot{\vartheta}),$$

---

[1]) We refer to [2] for additional comments on the following developments.
[2]) Here ρ_R is the density in the reference configuration; $\bar{S} = (\rho_R/\rho)F^{-1}\bar{T}(F^{-1})^T$; ∂ stands for partial differentiation.

and in view of Eqs. $(2.7)_1$, (3.1) we impose the restriction

$$\Gamma \cdot \Xi = 0 \tag{3.3}$$

for all Ξ such that $\Lambda \cdot \Xi = 0$. In this manner we obtain the result

$$\Gamma = q(X, t)\Lambda, \tag{3.4}$$

or, equivalently, for all triples $(C, \vartheta, \nabla\vartheta)$ which satisfy the constraint equation

$$
\begin{aligned}
\frac{1}{2\rho_R}\bar{S} &= -q\partial_C\lambda + \partial_C\bar\psi, \\
\partial_\vartheta\bar\psi &= q\partial_\vartheta\lambda - \bar\eta, \\
\partial_{\nabla\vartheta}\bar\psi &= q\partial_{\nabla\vartheta}\lambda;
\end{aligned} \tag{3.5}
$$

moreover,

$$\bar{h} \cdot g = 0. \tag{2.7}_2$$

Eqs. (3.5), $(2.7)_2$ show to which extent the added fields are determined by the constraint equation in the present context. In particular, the stress is determined by the constraint equation only to within an arbitrary potential.

As a special example, we consider a material which is temperature-dependent extensible and perfectly conductive along the field of material directions e (cf. [5]). In this case the constraint equations take the form:

$$
\begin{aligned}
e \cdot Ce + f(\vartheta) &= 0, \\
e \cdot \nabla\vartheta &= 0
\end{aligned} \tag{3.6}
$$

and Eqs. (3.5), $(2.7)_2$ imply

$$
\begin{aligned}
\frac{1}{2\rho_R}\bar{S} &= -pe \otimes e + \partial_C\bar\psi, \\
\partial_\vartheta\bar\psi &= pf'(\vartheta) - \bar\eta, \\
\partial_{\nabla\vartheta}\bar\psi &= qe, \\
\bar{h} &= sFe.
\end{aligned} \tag{3.7}
$$

4. Concluding remarks. Within the axiomatic approach we developed, a central role is played by characterizations (2.7) of the constraint response.

When coupled with the time derivative of the constraint equation, conditions (2.7) yield the dependence of the added fields on the intermediate variables: arbitrary scalar coefficients appear and the '$^-$' quantities depend on C, ϑ, $\nabla\vartheta$ *via* the constraint equation. The indeterminacy concerning $\bar{\psi}$ suggests that one should impose further or different restrictions to achieve more definiteness.

Moreover, the analysis is limited to the class of constraints (2.5), which is quite special: extensions should include anholonomic, history dependent constraints and, of more interest, constraints in the form of inequalities.

References

[1] Truesdell, C. and W. Noll, The non-linear field theories of mechanics. *Handbuch der Physik III/3*, Ed. by S. Flügge, Springer-Verlag, Berlin (1965).
[2] Andreussi, F. and P. Podio Guidugli, *Bull. Acad. Pol. Sci., Sér. Sci. Tech.* (forthcoming).
[3] Truesdell, C., *Rational thermodynamics.* McGraw-Hill, New York (1969).
[4] Green, A. E., P. M. Naghdi and J. A. Trapp, *Int. J. Engng. Sci., 8* (1970).
[5] Trapp, J. A., *Int. J. Engng. Sci., 9* (1971).

Basic underwater explosive forming systems

T. Z. BLAZYNSKI

University of Leeds, Leeds, U.K.

The sequence of events associated with an underwater explosion has a decisive influence on the mode of the dynamic response of the target diaphragm. The detonation of a chemical high-explosive charge is accompanied by a high rate of propagation of the detonation wave through the charge, usually of the order of 7000 m/s. Consequently the physical state of the charge is altered in a very short period of time. A high-pressure, high-temperature gas bubble is created. The high pressure gives rise to a shock wave which, in turn, is transmitted, through the medium of water, to the diaphragm. A rapid increase in pressure, reaching a peak value almost instantaneously, is observed followed by an exponential type of decay. The expansion of the gas bubble continues, being accompanied by an outward accelerated flow of a mass of water. The decay of the pressure pulse to ambient and below ambient levels, together with the effect of the external pressure eventually causes the gas bubble to contract. Owing to energy losses, the contraction does not reach the original bubble size, but produces a somewhat larger diameter than that of the original bubble.

During the contraction of the bubble, the gas pressure begins to increase again, and, eventually, the increase results in the formation and propagation of a radial outward, distinctly non-shock, pressure wave. This wave is usually referred to as the primary bubble pulse, and, together with the original shock wave effect, constitutes a major source of energy flux that is delivered to the target. Depending on the geometry and properties of the system damped bubble pulsation may continue, but will not be of any consequence in the actual forming operation.

From the point of view of the target diaphragm, the incident primary shock wave imparts outward radial velocity to the sheet, which, if effectively secured to, for instance, a die, begins to deform. The reflection of this wave back into the medium, together with a rarefaction wave produced by the motion of the diaphragm, constitutes a secondary wave. The drop of the gas bubble pressure to a negative value causes the water to cavitate, and thus, temporarily, further loading of the sheet ceases. Phase one of the forming operation is thus completed.

As the contraction of the gas bubble begins, the inward flow of water

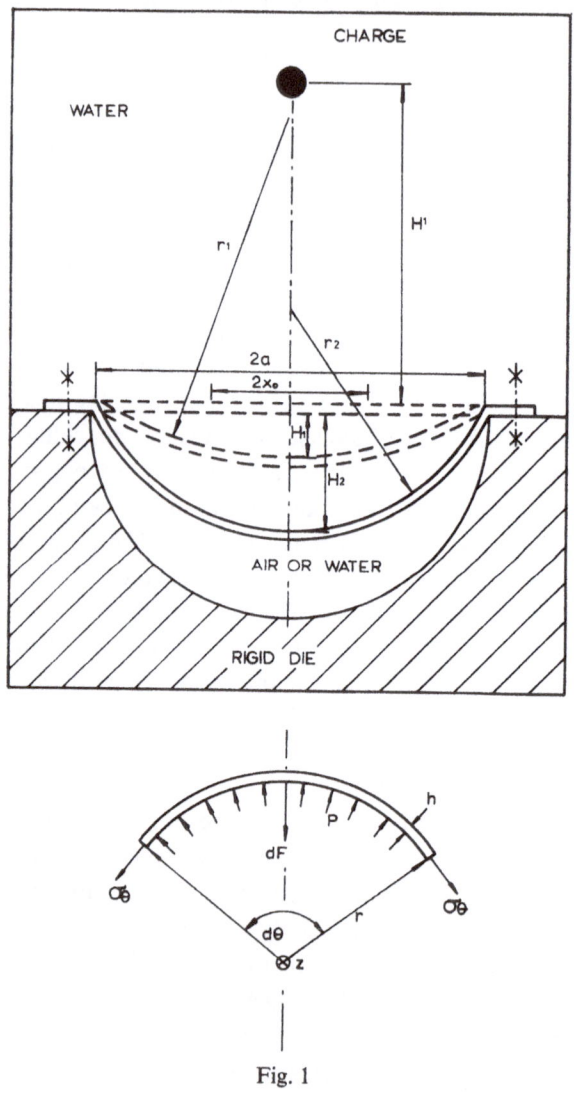

Fig. 1

associated with it, will cause the disappearance of cavitation, and, in due course, will alow the primary bubble energy pulse to reach the target. Further deformation of the diaphragm takes place.

In the more general case therefore, the deformation of the sheet may be said to have been produced by a double energy flux. The conditions required for this effect to occur are a large volume of transmitting medium that will alow the formation of the bubble, and the choice of a sufficiently large stand-off distance of the charge (H^1) Fig. 1. Considerable complications in the dynamic response of the sheet can be expected if the diameter of the primary bubble is less than (H^1), and, consequently, the bubble collapses against the target diaphragm.

Two basic forming systems can be used. The first, referred to as the 'air-cushion' system, is based on the use of water as the transmitting medium, the water being contained in a plastic bag (Fig. 1) positioned over the sheet-die assembly. The sheet is freely deformed against an air-cushion, the semi-cylindrical die being open at both ends. Full deformation, i.e. a complete conformity of the sheet with the die face would not normally be either aimed at, or achieved. The primary shock-wave generated by the charge does not only impart an energy flux to the diaphragm, but is also responsible for the destruction of the plastic container. In the absence therefore of

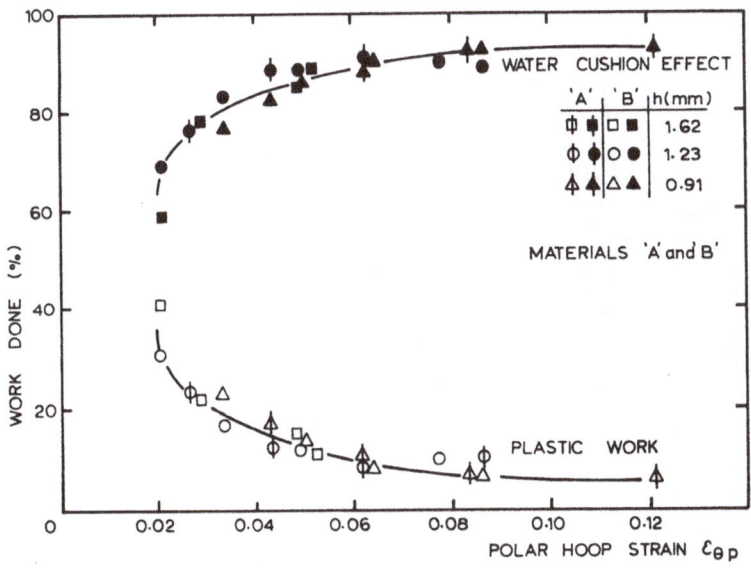

Fig. 2

a bubble pulse, the mechanism of sheet deformation is substantially modified. The shock-wave initially imparts high radial velocity to the sheet. On the cessation, however, of the pressure pulse, coincident with the end of the period of detonation of the charge, the energy associated with this velocity

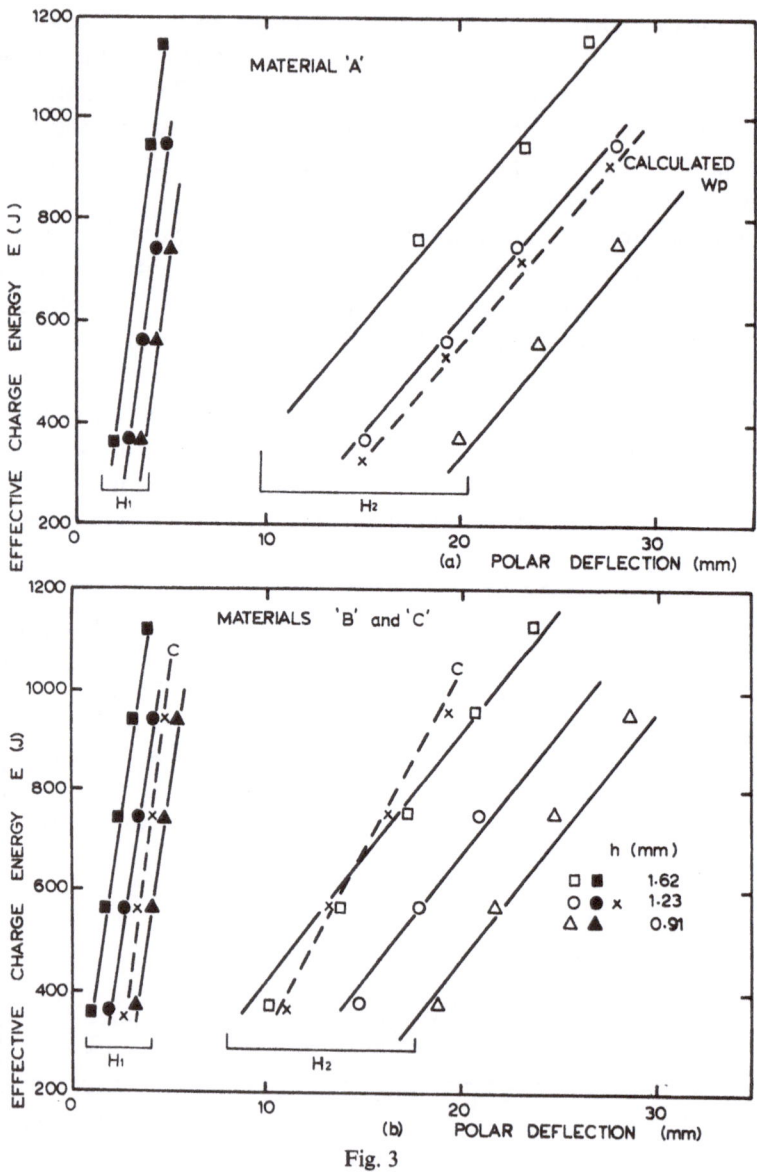

Fig. 3

is dissipated in overcoming the inertia effects of the moving target. The sheet is eventually brought to rest. The total polar deflection of the sheet (H_2) (Fig. 1) is thus obtained in two stages, i.e. during the period of the delivery of the shock pulse-(H_1), and the further increase in the time in which purely inertial effects are experienced.

In the second, 'water-cushion', system the sheet-die assembly is submerged in a metal tank filled with water, so that the deformation of the diaphragm takes place against a cushion of water contained between the sheet and the die (Fig. 1). The double energy effect becomes apparent. The effects of both the shock-wave and the primary bubble pulse are present, and are utilised on the one hand, in deforming the sheet, and on the other in producing the outflow of the mass of water originally present in the space between the die and the target.

Both situations can be analysed by using a quasi-static approach, that permits the calculation of the amount of plastic work and the estimation, where appropriate, of the water cushion effect.

The results for the water-cushion system are shown in Fig. 2. The close agreement between the calculated and measured (continuous line) values of H_2 are shown in Fig. 3 for the air-cushion system.

References

1] Johnson, W., A. Poynton, H. Singh and F. W. Travis, Experiments in the underwater explosive stretch forming of clamped circular blanks. *Internat. J. Mech. Sci., 8* (1966) 237.

[2] Sewailem, M. R. and T. Z. Blazynski, *Underwater explosive forming of rectangular metal sheet.* Proc. 8th Internat. MTDR Conference, Pergamon Press, Oxford, 1968, 1301.

[3] Blazynski, T. Z. and M. R. Sewailem, Air-Cushion effect in the explosive forming of metal sheet. *Engineer (London), 227* (1969) 58–62, 100–102.

[4] Bebb, A. H., Underwater explosions measurements. *Proc. Royal Soc. A, 20* (1951) 244.

Distributional aspects of the theory of plastic time dependent phenomena

Institute of Mining Research, Cracow, Poland

My contribution to the discussion is concerned with the specific rheological properties of certain class of real materials disregarded in usual theories. I mean here problems concerned with discontinuities of stress, strain and time relations observed in testing certain metals, especially, commercially pure annealed aluminium. According to my opinion their theoretical formulation may explain and clarify some aspects of classical theory of plasticity.

The properties in question were found and discussed by many experimentators (Portevin and Le Chatelier, Hanson and Wheeler, McReynolds, Bell and Stein, Kenig, Sharpe) and recently also described by Dillon [1], [2] and Bodner and Rosen [3].

Usually, the mentioned discontinuities are referred to the spontaneous effects being a result of a temporary lack of internal equilibrium during deformation process in room temperature. Thus, a considerable increase of strain in quasi instantaneous manner under constant or slightly increasing stress and jumping of stress under moderately increasing strain rate are examples of such phenomena.

On the stress-strain diagram the discontinuities appear in the hardening range of the curve and form a stepped relation the typical example of which is shown in Fig. 1 as found by Sharpe for aluminium. The size of steps depends on actual strain or strain rate and have a tendency to diminish with increasing strain.

As an introductory level to the stepwise hardening or discontinuous yielding one should consider the attainment of yield stress. This is a turning point in the change of material properties and the first step involving discontinuity.

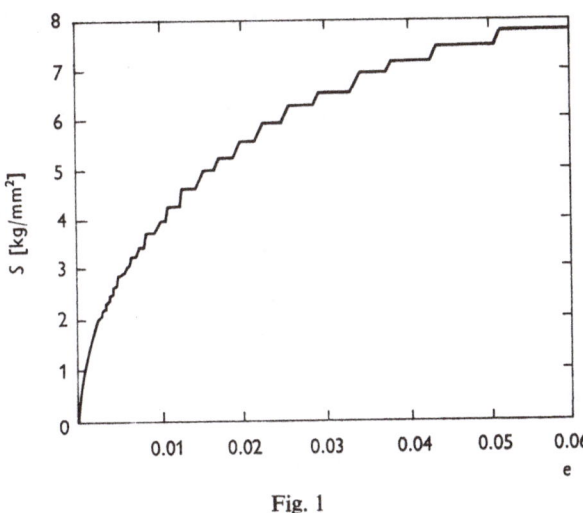

Fig. 1

It is clear that the stepwise relation between stress and strain provokes discontinuities of stress-time and strain-time relationships as indicating rapid creep and rapid relaxation in the material during comparatively short periods of time. Therefore, it seems reasonable to find a theoretical description of these phenomena on the rheological basis.

1. In considering the problem of instability of materials we start with an auxiliary concept of elasto-plastic behaviour which is of value in modeling a singular step of discontinuous yielding. According to my opinion this concept might be also of value in passing from elasto-plastic to perfectly plastic problems.

We regard ideal plastic material as a limiting case of an elasto-plastic body given by the relation between stress s and strain e as follows

$$s = 2s_0\pi^{-1} \operatorname{arc\,tg}[e \cdot g(e, \beta)], \quad \lim_{\beta \to 0} g(e, \beta) = \infty. \tag{1.1}$$

Thus, in the limit we obtain

$$\bar{s} = \bar{s}(\bar{e}) = \lim_{\beta \to 0} s = \begin{cases} s_0; & e > 0, \\ -s_0; & e < 0, \end{cases} \tag{1.2}$$

where s_0 denotes the yield stress and β is a physical parameter.

371

Consequently, the tangent moduli of the curves (1.1) and (1.2) are, respectively,

$$d_e s = 2s_0 (f - e d_e f)[\pi(f^2 + e^2)]^{-1}, \quad f = g^{-1}, \quad d_e = d/de, \tag{1.3}$$

and

$$d_{\bar{e}} \bar{s} = \lim_{\beta \to 0} d_e s = d_e [\theta(|\bar{e}|) \, \mathrm{sgn} \, \bar{e}] = \begin{cases} 0; & \bar{e} \neq 0, \\ \infty; & \bar{e} = 0, \end{cases} \tag{1.4}$$

where θ is the Heaviside distribution defined as

$$\theta(|\bar{e}|) = \begin{cases} 1; & \bar{e} \neq 0, \\ 0; & \bar{e} = 0. \end{cases} \tag{1.5}$$

With the above definition we simply have

$$d_{\bar{e}} \bar{s} = \bar{\delta}(\bar{e}), \quad \int_{-\infty}^{\infty} \bar{\delta}(\bar{e}) d\bar{e} = 1, \quad \bar{\delta}(\bar{e}) = (2s_0)^{-1} \cdot \delta(\bar{e}), \tag{1.6}$$

where δ is the Dirac distribution.

The concept given above may be interpreted in Fig. 2 in two ways. Firstly, it gives elasto-plastic complex behaviour (full line) and secondly, it represents a single step of instability (dotted line).

Furthermore, the given concept allows us to reckon and judge the proximity of elasto-plastic and rigid plastic solutions and a smooth passage from the former to the latter. As is known such possibility does not exist on the basis of the existing theories in general cases.

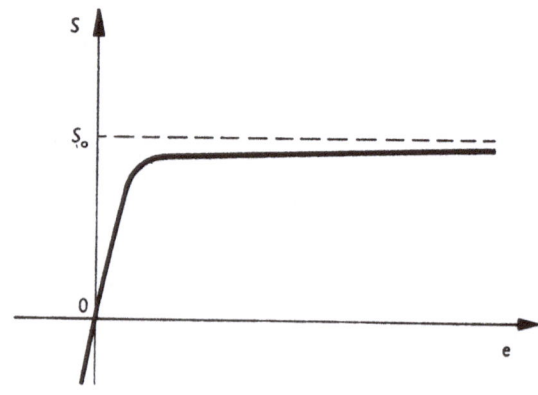

Fig. 2

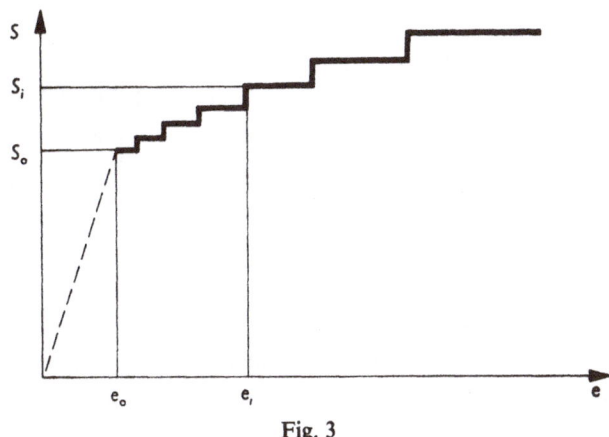

Fig. 3

2. In generalizing the theory of material instability we can start at first with the description of an ideal unstable material by a stepped function with horizontal platforms, Fig. 3.

If the yield stress s_0 is taken as the level of reference, then we can simply write stress-strain relation

$$s(e) = \lim_{n \to \infty} \sum_{i=1}^{n} \Delta s(e_i) = \int_0^s ds(e), \quad \Delta s(e_i) = (s_i - s_{i-1})\theta(e - e_i), \qquad (2.1)$$

in the form of the Stieltjes integral.

In general, the curve of instability shows the dependence of actual deformation state on the history of straining and straining rate as indicated by the magnitude of jumps and width of steps. In such case we postulate the form

$$s(\dot{e}, e) = \lim_{n \to \infty} \sum_{i=1}^{n} \mu(\dot{e}, e_i)\Delta p(e_i) = \int_0^e \mu(\dot{e}, e)dp(e), \quad \int_0^e |dp| < \infty, \qquad (2.2)$$

which indicates the mode of superposing stress. Here μ is distribution depending on actual strain and strain rate and p defines strain distribution of deformation process.

In order to illustrate the properties of unstable material described by the proposed theory we consider a discrete model consisting of sliding elements joined by strings put into the motion by applying a variable force or forcing velocity. The elements are characterized by different weights and friction coefficients, Fig. 4.

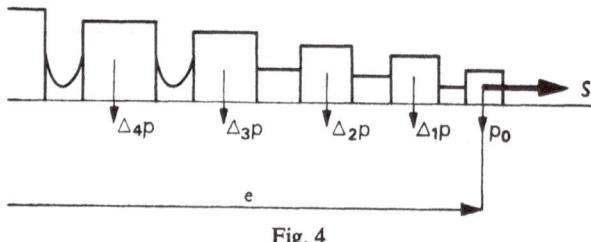

Fig. 4

By using the generalized superposition, it is possible to take into account the time dependence of instability [4]. Thus, using the reasoning leading to Eq. (2.2) we arrive at the law describing the stress increase in time

$$s(t) = \int_{t_0}^{t} \mu[t, \tau, \dot{e}(\tau), e(\tau)]\mathrm{d}_\tau p[e(\tau)], \tag{2.3}$$

where now μ plays the role of a hereditary function.

The equation (2.3) is of a relaxation type and gives information about stress change for a given strain rate. On the other hand, a counterpart of this equation written for superposed sudden or quasi instantaneous strain, takes the form

$$e(t) = -\int_{t_0}^{t} s(\tau)\mathrm{d}H[t, \tau, \dot{s}(\tau), s(\tau)], \tag{2.4}$$

where H is the generalized creep function.

In general both Eqs. (2.3) and (2.4) represent completely different phenomena which occur in physically different materials.

References

[1] Dillon, O. W., Jr., Experimental data on aluminium as a mechanically unstable solid. *J. Mech. Phys. Solids*, Vol. *11* (1963).

[2] Dillon, O. W., Jr., Waves in bars of mechanically unstable materials. *J. Appl. Mech.*, Vol. *33*, Nr *2* (1966).

[3] Bodner, S. R. and A. Rosen, Discontinuous yielding of commercially pure aluminium. *J. Mech. Phys. Solids*, Vol. *15* (1967).

[4] Bychawski, Z. and A. Fox, Some fundamental concepts on the theory of nonlinear viscoelasticity. *Arch. Mech. Stos.*, *16*, *6* (1966).

Boundary value problems with
incremental plasticity in granular media

T. J. Chung and J. K. Lee

The University of Alabama, Huntsville, Alabama, USA

N. C. Costes

Marshall Space Flight Center,
National Aeronautics and Space Administration, Huntsville, Alabama, USA

Constitutive theories associated with soil mechanics have always been criticized for their limitations in describing correctly the overwhelmingly complicated nature of the mechanical behavior of soil. In general, however, a theory based on concrete experimental evidence from a systematic soil testing program is considered to be more dependable than a theory based on purely theoretical speculation. In this respect, the constitutive theory proposed by Roscoe and his associates [1, 2] which is founded on the critical state concept [3] appears to be one of the most promising. Not only does the critical state theory depict reasonably well the soil behavior under combined stresses, but it also lends itself to mathematically consistent analytical methods based on the classical theory of plasticity [4]. The present study attempts to represent the critical state concept in terms of an incremental theory of plasticity and formulates governing equations which are convenient for a computational scheme using the finite element method.

Under conditions of triaxial compression, Roscoe and his co-workers showed that the deviatoric stress q, the mean normal stress p, and the void ratio e, form a surface called the state boundary surface (Fig. 1) which is unique for a given soil mass. They further theorized that a soil mass undergoing uniform shear distortion eventually reaches a critical state, whereupon it continues to distort without further change in its void ratio or in the effective stresses q and p. The projection of the curve representing such

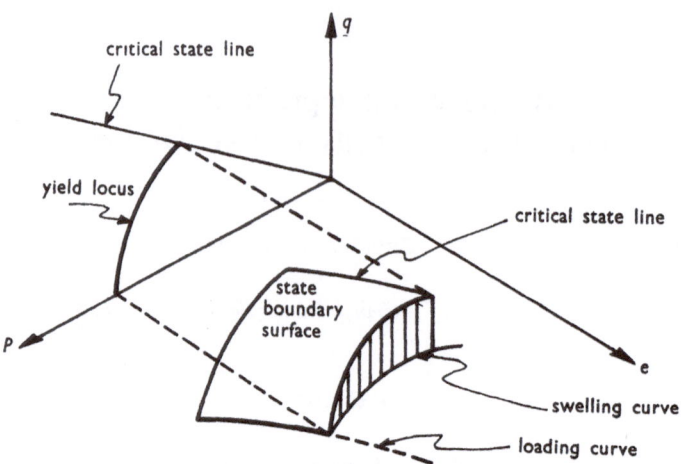

Fig. 1. State boundary surface (after Roscoe and Burland).

critical state conditions on the (p, q) plane is a straight line with a slope M

$$q = Mp. \tag{1}$$

When this curve is projected on to an $e - lnp$ plane, another straight line results of the form

$$e = e_a - \lambda lnp \tag{2}$$

where e_a is the void ratio at $p = 1$ and λ is a soil constant for loading at $q = 0$. Swelling curves, representing 'elastic' recoverable strains are also assumed to be straight and parallel lines of slope κ when plotted on an $e - lnp$ plot. This leads to the yield locus equation which is of the form

$$\frac{p}{p_0} - \frac{M^2}{M^2 + \eta^2} = 0 \tag{3}$$

where $\eta = q/p$, p_0 is the value of p at $\eta = 0$, and M is given by the Mohr-Coulomb criterion as

$$M = \frac{6 \sin \phi}{3 - \sin \phi}. \tag{4}$$

For three-dimensional boundary-value problems, expression (3) may be rewritten in terms of the second deviatoric stress invariant J

$$\frac{p}{p_0} - \frac{M^2}{M^2 + (3J/p^2)} = 0, \tag{5}$$

from which a plastic potential function can be defined by

$$F(p, J, p_0) = \bar{\sigma}^2 = 3J + p^2 M^2 - M^2 p_0 p \tag{6}$$

where $\bar{\sigma}$ is an equivalent yield stress. The parameter p_0 is a measure of the 'size' of the current yield locus and is treated as a variable which has a constant value at any given yield locus. As a result of a loading stress increment, the corresponding overall change in the void ratio of the soil de is associated with p_0 through the relations

$$de = de^{(e)} + de^{(p)} = -\lambda \frac{dp_0}{p_0} \tag{7}$$

$$de^{(e)} = -\kappa \frac{dp_0}{p_0}, \tag{8}$$

hence,

$$de^{(p)} = -(\lambda - \kappa) \frac{dp_0}{p_0} \tag{9}$$

in which d$e^{(e)}$ and d$e^{(p)}$ are, respectively, the recoverable (elastic) and non-recoverable (plastic) components of the total void ratio change de, and λ and κ are soil material constants, as defined before.

Following the procedure for the incremental theory of plasticity it can be shown after lengthy algebra that the stress increment d$\sigma^{\alpha\beta}$ may be written in the form,

$$d\sigma^{\alpha\beta} = D_{(e)}^{\alpha\beta mn} d\gamma_{mn}^{(e)} = D_{(e)}^{\alpha\beta mn}[d\gamma_{mn} - d\gamma_{mn}^{(p)}],$$

or

$$\tag{10}$$

$$d\sigma^{\alpha\beta} = [D_{(e)}^{\alpha\beta mn} + D_{(p)}^{\alpha\beta mn}]d\gamma_{mn}$$

in which $D_{(e)}^{\alpha\beta mn}$ and $D_{(p)}^{\alpha\beta mn}$ are fourth-order tensors representing elastic and plastic moduli, respectively; and dγ_{mn}, d$\gamma_{mn}^{(e)}$, and d$\gamma_{mn}^{(p)}$ are second order tensors, representing respectively the total, elastic, and plastic strain increments. $D_{(p)}^{\alpha\beta mn}$ takes the explicit form

$$D_{(p)}^{\alpha\beta mn} = -\frac{D_{(e)}^{\alpha\beta rs} B_{rs} R_{ij} D_{(e)}^{ijmn}}{1 + B_{pq} R_{tu} D_{(e)}^{pqtu}} \tag{11}$$

where

$$B_{mn} = (S_{mn} + a\delta_{mn})/b \tag{12}$$

$$S_{mn} = 3\frac{\partial J}{\partial \sigma^{mn}} \tag{13}$$

$$a = \frac{p}{3}[M^2 - (3J/p^2)] \tag{14}$$

$$b = \frac{(1+e)}{(\lambda - \kappa)}p^3[M^4 - (9J^2/p^4)] \tag{15}$$

$$R_{mn} = S_{mn} + a\delta_{mn}. \tag{16}$$

The elastic modulus $D_{(e)}^{\alpha\beta mn}$ can be derived from the set of equations

$$d\gamma_{mn}^{(e)} = d\psi_{mn}^{(e)} + \tfrac{1}{3}dv^{(e)}\delta_{mn} \tag{17}$$

where $d\psi_{mn}^{(e)} = \xi d\tau_{mn}$ and $dv^{(e)} = [\kappa/(1+e)p]dp = \omega dp$ are respectively the incremental elastic deviatoric and volumetric strains; $d\tau_{mn}$ is the deviatoric stress increment; and ξ is a scalar shear compliance to be determined experimentally.

The constitutive equation (10) is the familiar form readily applicable to the finite element equations of equilibrium

$$\sigma^{\alpha\beta}\frac{\partial\gamma_{\alpha\beta}}{\partial u^N}dv = F_N \tag{18}$$

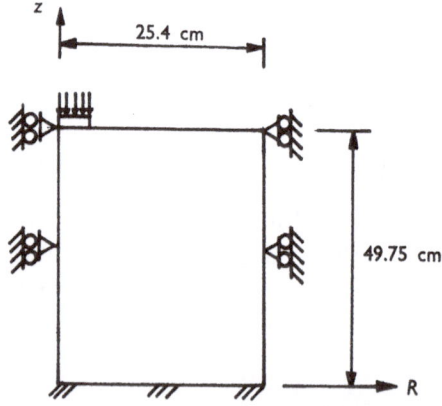

Fig. 2. Circular bearing plate on soil.

or, in incremental form,

$$\int_V d\sigma^{\alpha\beta} \frac{\partial \gamma_{\alpha\beta}}{\partial u^N} \, dv + \int_V \sigma^{\alpha\beta} d\left(\frac{\partial \gamma_{\alpha\beta}}{\partial u^N}\right) dv = dF_N. \tag{19}$$

Here u^N are nodal generalized displacements given by

$$u_i = \psi_{iN} u^N \tag{20}$$

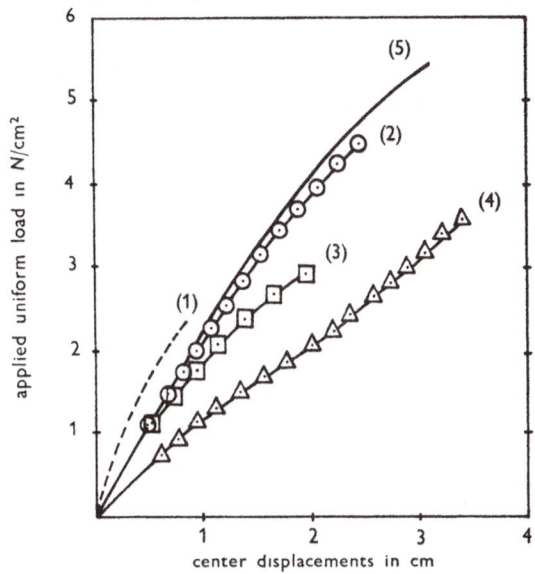

Fig. 3. Load-deformation curves for the bottom center of plate.

No.	Angle of friction	Void ratio	Density	Comp. index	Swelling index	Equivalent yield stress (initial)	Elast. shear compliance
	ϕ	e_0	γ	λ	κ	$\bar{\sigma}$	ξ
1	35°	0.875	0.0147 N/cc	0.05	0.01	0.4 N/cm²	10ω
2	35°	0.875	0.0147 N/cc	0.05	0.024	0.4 N/cm²	10ω
3	35°	0.875	0.0147 N/cc	0.1	0.024	0.4 N/cm²	10ω
4	35°	0.875	0.0147 N/cc	0.05	0.024	0.4 N/cm²	100ω
*5	35°	0.875	0.0147 N/cc	0.03–0.13	0.01	0.1–1.3	N/A

* Reference [7].

379

where u_i are the element displacement coordinates and ψ_{iN} is the normalized interpolation function [5, 6]. In matrix notation (19) may be rewritten in the form,

$$(K_{(L)} + K_{(G)} + K_{(P)})\mathrm{d}u = \mathrm{d}F. \tag{21}$$

Here $K_{(L)}$, $K_{(G)}$, and $K_{(P)}$ are the linear elastic, geometric, and plastic tangent stiffness matrices; and $\mathrm{d}F$ is the nodal load vector. Standard incremental loading techniques may be used to solve the nonlinear equations of (21).

A numerical example relating to a soil medium subjected to a surface loading through a rigid circular plate is shown in Fig. 2. Load-deformation

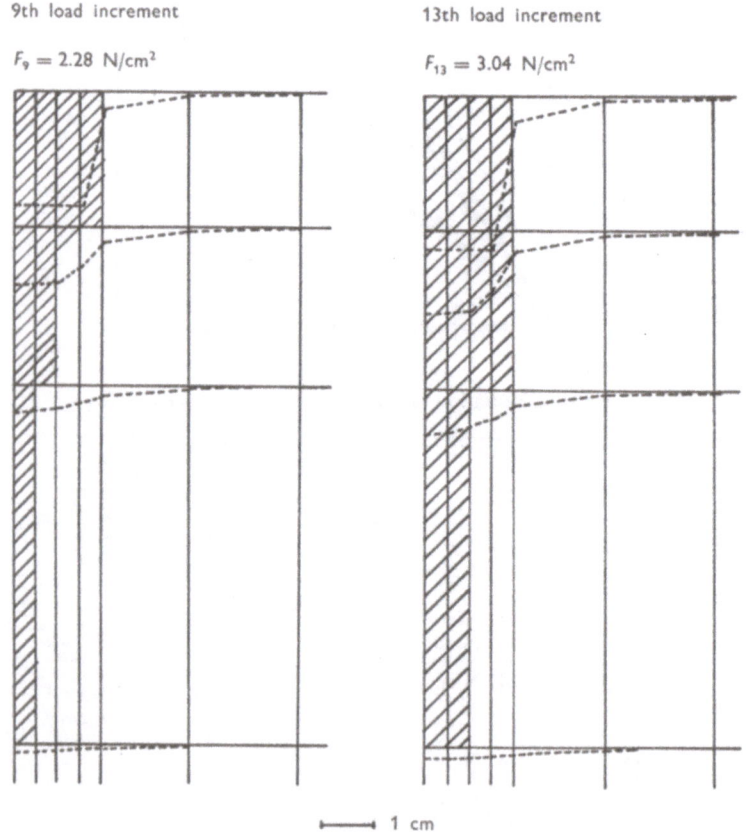

Fig. 4. Deformed geometry and yielded elements.

curves for the bottom center of the plate are shown in Fig. 3 and are compared with the experimental results of Namiq [7]. Deformed shapes and yielded regions are shown in Fig. 4.

In conclusion, the critical state concept with its representation by the classical incremental theory of plasticity appears to provide a powerful means for solving a wide variety of boundary value problems in soil media. The finite element model applied to this system is particularly appealing. The void ratio and density, compression and swelling indices, and angle of internal friction of the soil are accounted for in the analysis. Comparison of experimental results from load bearing plate tests with analytical results obtained for this theory under the same geometry, loading conditions, and material properties indicates good agreement between the present theory and experimental data.

References

[1] Roscoe, K. H., A. N. Schofield and C. P. Wroth, On the yielding of soils. *Geotechnique*, *Vol. 9* (1958) 71–83.
[2] Roscoe, K. H. and J. B. Burland, On the generalized stress-strain behavior of wet clay. *Engineering Plasticity*, Cambridge University (1968).
[3] Schofield, A. N. and C. P. Wroth, *Critical State Soil Mechanics*. McGraw-Hill, New York, 1968.
[4] Hill, R., *The Mathematical Theory of Plasticity*. Clearendon Press, Oxford, 1950.
[5] Zienkiewicz, O. C., *The Finite Element Method in Engineering Science*. McGraw-Hill, New York, 1971.
[6] Oden, J. T., *Finite Elements in Nonlinear Continua*. McGraw-Hill, New York, 1972.
[7] Namiq, L. I., *Stress Deformation Study of Simulated Lunar Soil*. Ph. D. Thesis, University of California, Berkeley, 1970.

On objective stress-rate in the constitutive relation for elastic-plastic solids

R. N. DUBEY

University of Waterloo, Waterloo, Canada

The constitutive relation for any material must be expressed in terms of an objective stress-rate which vanishes under rigid rotation. Many objective stress-rates have been proposed and used. They differ from each other by terms of the type stress strain-rate. The question naturally is whether there is any limitation on the choice of a particular stress-rate, or is it possible that the choice of certain stress-rate, in a boundary-value problem results in a conclusion contrary to experience. Here, I do not attempt to answer all these questions, but only to express some of the observations I made while analysing such problems.

Let us consider an incompressible elastic-plastic solid at some time t during a process of continued deformation. Suppose that the stress distribution σ_{ij} in the current state, together with other mechanical properties, is given. During a further infinitesimal deformation, the plastic component of the rate of deformation is given by

$$
\dot{\varepsilon}_{ij}^{p} =
\begin{cases}
\sigma_{ij}' \, \dfrac{\sigma_{kl}(\partial\sigma_{kl}/\partial t)}{H} & \text{for} \quad \sigma_{ij}' \dfrac{\partial\sigma_{ij}}{\partial t} > 0 \\[2em]
0 & \text{for} \quad \sigma_{ij}' \dfrac{\partial\sigma_{ij}}{\partial t} < 0
\end{cases}
\tag{1}
$$

where $\sigma_{ij}' = \sigma_{ij} - \frac{1}{3}\sigma_{kk}\delta_{ij}$ and H is a work-hardening parameter. The convected derivative of σ_{ij}, $\partial\sigma_{ij}/\partial t$, in (1) is objective and is related to the material derivative $\dot{\sigma}_{ij}$ by

$$
\frac{\partial\sigma_{ij}}{\partial t} = \dot{\sigma}_{ij} - \sigma_{ik}v_{j,k} - \sigma_{jk}v_{i,k}
\tag{2}
$$

where v_i is velocity and a comma followed by a suffix denotes spatial derivative. Let the yield condition be

$$J_2 - k^2 = 0; \quad J_2 = \tfrac{1}{2}\sigma'_{ij}\sigma'_{ij}. \tag{3}$$

Let us assume that the elastic components of the rate of deformation do not in any way influence the shape or the size of the yield surface. Under neutral loading, $\dot{\varepsilon}^p_{ij} = 0$ and we must have

$$\dot{J}_2 = \sigma'_{ij}\dot{\sigma}'_{ij} = \sigma'_{ij}\dot{\sigma}_{ij} = 0. \tag{4}$$

But, from (1) and (2),

$$\dot{J}_2 = \sigma'_{ij}\dot{\sigma}_{ij} = 2(\sigma_{ij}\sigma_{ik} - \tfrac{1}{3}\sigma_{mm}\sigma_{jk})\dot{\varepsilon}_{jk} \tag{5}$$

which may not necessarily vanish.

On the other hand, suppose we consider the co-rotational rate tensor, $D\sigma_{ij}/Dt$, as the objective stress-rate in (1). Using the relation

$$\frac{D\sigma_{ij}}{Dt} = \dot{\sigma}_{ij} - \sigma_{ik}\omega_{jk} - \sigma_{jk}\omega_{ik}$$

where ω_{ij} is the antisymmetric part of the velocity gradient, we obtain

$$\dot{J}_2 = \sigma'_{ij}\dot{\sigma}_{ij} = \sigma'_{ij}\frac{D\sigma_{ij}}{Dt}. \tag{6}$$

In view of the flow rule,

$$\varepsilon^p_{ij} = \begin{cases} \sigma_{ij}\dfrac{\sigma'_{kl}(D\sigma_{kl}/Dt)}{h} & \text{for} \quad \sigma'_{ij}\dfrac{D\sigma_{ij}}{Dt} > 0 \\[4mm] 0 & \text{for} \quad \sigma'_{ij}\dfrac{D\sigma_{ij}}{Dt} < 0 \end{cases} \tag{7}$$

we have $\sigma'_{ij}(D\sigma_{ij}/Dt) = 0$ for neutral loading. Hence $\dot{J}_2 = 0$. It is seen therefore that the neutral loading criterion $\sigma_{ij}(\partial\sigma_{ij}/\partial t) = 0$ is inconsistent with the requirement $\dot{f} = \dot{J}_2 = 0$. Similar remark pertains to other stress rates discussed, for instance, in the paper by Spencer and Ferrier [1], presented at this Symposium.

Next, let us consider an incompressible elastic-plastic plate, of dimension $2a \times 2b$ deforming, in a state of plane strain. Suppose that the stress distribution and stress deviator is given by

$$[\sigma_{ij}] = \begin{bmatrix} \sigma & 0 & 0 \\ 0 & 0 & 0 \\ 0 & 0 & \sigma/2 \end{bmatrix}, \quad [\sigma'_{ij}] = \begin{bmatrix} \sigma/2 & 0 & 0 \\ 0 & -\sigma/2 & 0 \\ 0 & 0 & 0 \end{bmatrix}. \tag{8}$$

We will write, in view of (8)$_2$, the unit normal to the yield surface, m_{ij}, as

$$[m_{ij}] = \frac{1}{\sqrt{2}} \begin{bmatrix} 1 & 0 & 0 \\ 0 & -1 & 0 \\ 0 & 0 & 0 \end{bmatrix} \tag{9}$$

and the constitutive relation in the form

$$\frac{\partial \sigma_{ij}}{\partial t} = 2\mu \left\{ \varepsilon_{ij} - m_{ij} \frac{m_{kl}\varepsilon_{kl}}{1 + H/2\mu} \right\} + \dot{p}\delta_{ij}. \tag{10}$$

The uniqueness criterion (Hill [2])

$$\int \dot{s}_{ij} v_{j,i} \, dV > 0 \tag{11}$$

with the help of

$$\dot{s}_{ij} = \frac{\partial \sigma_{ij}}{\partial t} + \sigma_{ik} v_{j,k} \tag{12}$$

can be expressed as

$$\int \left\{ 2\mu \left[\varepsilon_{ij}\varepsilon_{ij} - \frac{(m_{ij}\varepsilon_{ij})^2}{1+H/2\mu} \right] + \sigma_{ik} v_{j,k} v_{j,i} \right\} dV > 0. \tag{13}$$

We assume

$$v_1 = \sin pxf'(y), \quad v_2 = -p\cos pxf(y);$$

here $p = n\pi x/a$, is an integer; solve for $f(y)$ from the minimizing condition for (13), substitute in (13) to obtain, after some simplification,

$$\sigma + \frac{2H}{1+H/2\mu} > 0. \tag{14}$$

For strain-hardening material under tensile load, the inequality (14) always holds, hence there can be no symmetric (or necking) bifurcation of the boundary value problem. In deriving (14) we assumed $b/a < 1$. On the other hand, if we consider the constitutive relation in terms of $D\sigma_{ij}/Dt$, then

$$\frac{D\sigma_{ij}}{Dt} = \dot{p}\delta_{ij} + 2\mu\left[\varepsilon_{ij} - m_{ij}\frac{m_{kl}\varepsilon_{kl}}{1+h/2\mu}\right]$$

then after similar routine analysis, we obtain the following uniqueness criterion

$$\frac{2h}{1+h/2\mu} - \sigma > 0$$

from (11). Hence the boundary-value problem may have non-unique solution under tensile load, if $\sigma = 2h/(1+h/2\mu)$.

References

[1] Spencer, A. J. M. and J. E. Ferrier, *Foundations of Plasticity*. Noordhoff Int. Publ. Leyden (1973), 9–24.
[2] Hill, R., *J. Mech. Phys. Solids*, 7 (1959) 209.

Extremum principles in the dynamics of
rigid-plastic bodies and mathematical programming

M. I. ERKHOV

Central Building Research Institute, Moscow, USSR

Hitherto no adequate theory describing dynamic response of plastic structures has been developed nor suitable computational schemes exist to treat
numerically the initial boundary value problems. In the present paper extremal principles of dynamics of rigid-plastic bodies are formulated together with the dynamic theory of limit equilibrium. The general mathematical problem of dynamics of rigid-plastic bodies is reduced to the
problem of linear 'LP' or non-linear 'NP' programming. The LP method
is subsequently used to solve dynamic problems for square and circular
plastic plates.

When solving dynamic problems the following conditions should be
satisfied:

a) Equations of motion and boundary conditions on the surface S of the
   body;
b) The yield condition $f(\sigma_{ij}) \leqq 0$;
c) Relation between the components of the acceleration $\ddot{u}_i$ and velocity
   vectors $\dot{u}_i$;
d) Incompressibility requirement and kinematic boundary conditions;
e) Initial conditions for displacements, velocities and accelerations.

In what follows σ_{ij} and $\dot{\varepsilon}_{ij}$ denote respectively components of the stress and
strain rate tensors, γ is mass density, X_i and p_i are components of body
force and surface traction vectors, V denotes volume of the body, t is time
and $i, j = 1, 2, 3$. According to the associated flow rule the yield function
$f(\sigma_{ij})$ is taken as the plastic potential.

A stress field σ_{ij}° satisfying the conditions a) and b) is called a statically
admissible stress field. A velocity field ($\dot{u}_i^*$, $\dot{\varepsilon}_{ij}^*$ or $\dot{u}_i^{**}$, $\dot{\varepsilon}_{ij}^{**}$) etc. satisfying the

386

condition d) is called a kinematically admissible velocity field. To each velocity $\dot{u}_i^*$, $\dot{u}_i^{**}$ etc. corresponds an acceleration $\ddot{u}_i^*$, $\ddot{u}_i^{**}$ etc. The quantities without upper indices denote the actual solution.

The principle of virtual velocities states that

$$\int_V \sigma_{ij}^\circ \dot{\varepsilon}_{ij}^* dV + \int_V \gamma \ddot{u}_i^* \dot{u}_i^* dV - \int_S p_i \dot{u}_i^* dS - \int_V X_i \dot{u}_i^* dV = 0. \tag{1}$$

where the functions p_i, σ_{ij}°, $\ddot{u}_i^*$, X_i satisfy the condition a).

Using the principle of maximum plastic work Eq. (1) can be transformed to give

$$\int_V \sigma_{ij}^* \dot{\varepsilon}_{ij}^* dV + \int_V \gamma \ddot{u}_i^* \dot{u}_i^* dV - \int_S p_i \dot{u}_i^* dS - \int_V X_i \dot{u}_i^* dV \geqq 0. \tag{2}$$

where σ_{ij}^* is related to $\dot{\varepsilon}_{ij}^*$ through an associated flow rule.

The minimum principle: for an actual solution the functional (2) attains minimum and is equal to zero.

Taking a different value for the pressure term p_i in (1), $p_i^{\circ*} \neq p_i$ one can always choose such $\dot{u}_i^*$ that an equality sign will hold in (2). The expression (1) is replaced now by the inequality

$$\int_V \sigma_{ij}^\circ \dot{\varepsilon}_{ij}^* dV + \int_V \gamma \ddot{u}_i^* \dot{u}_i^* dV - \int_S p_i^{\circ*} \dot{u}_i^* dS - \int_V X_i \dot{u}_i^* dV \leqq 0. \tag{3}$$

The following *maximum principle* can be stated: *for an actual solution the functional (3) attains maximum and is equal to zero.* Here σ_{ij}°, X_i, $\ddot{u}_i^*$ and $p_i^{\circ*}$ satisfy the condition a). In the expressions for minimum and maximum principles (2) and (3) a different acceleration $\ddot{u}_i^{**}$ can be used instead of $\ddot{u}_i^*$ keeping the value of the velocity unchanged $\dot{u}_i^*$. Some alternative forms of the extremum principles (2) and (3) can also be formulated. With the acceleration terms equal to zero inequalities (2) and (3) yield a familiar form of the extremum principles for statically loaded rigid-plastic bodies.

The exact solution or an approximation to it is obtained by taking an appropriate system of functions and minimizing the functional (2) or maximizing the functional (3).

The resulting solution to dynamic problems is unique. Integrating (2) with respect to time from 0 to t a *minimum principle* is obtained which does not involve accelerations

$$\int_0^t \int_V \sigma_{ij}^* \dot{\varepsilon}_{ij}^* dV dt + \tfrac{1}{2} \int_V \gamma \dot{u}_i^* \dot{u}_i^* dV -$$

$$- \int_0^t \int_S p_i \dot{u}_i^* dS dt - \int_0^t \int_V X_i \dot{u}_i^* dV dt \geqq 0, \qquad (4)$$

where $u_i^* = \dot{u}_i^* = 0$ for $t = t_0$.

Consider the time interval $t \geqq t_k$ where t_k is the so called 'time to rest'. Then, by virtue of the condition $\dot{u}_i^*(t_k) = 0$, another form of the minimum principle is obtained where the second term in (4) vanishes. The resulting expression is linear and thus convenient in further applications.

The maximum principle (3) can be integrated in a similar way to get appropriate linear and quadratic forms.

The extremum principles discussed above constitute the basic relationships of the dynamic theory of limit equilibrium. Evaluation of the absolute value of the functionals in question can serve as a certain criterium of the accuracy of the solution. For these purposes however the following forms of the functionals appear to be more suitable

$$\int_V \gamma \ddot{u}_i^{**} \dot{u}_i^* dV \geqq \int_V \gamma \ddot{u}_i^* \dot{u}_i^* dV, \quad \int_S p_i^* \dot{u}_i^* dS \geqq \int_S p_i^{\circ *} \dot{u}_i^* dS, \qquad (5)$$

in which $\ddot{u}_i^{**}$ and $\ddot{u}_i^*$ are respectively an upper and lower bound for $\ddot{u}_i$ in an average sense. Similarly p_i^* and $p_i^{\circ *}$ are upper and lower bound for p_i (according to the second expression (5) holds either $p_i^* = p_i$ or $p_i^{\circ *} = p_i$).

The system of functions σ_{ij}°, $\ddot{u}_i^{**}$, p_i, entering the first formula (5), is in equilibrium whereas the functions $\ddot{u}_i^*$, σ_{ij}^*, p_i satisfy the equality (1) if σ_{ij}^* is substituted for σ_{ij}° (in this case the sign equal to zero holds but (1) is no longer a virtual velocity principle). Similarly, the system of functions σ_{ij}°, $\ddot{u}_i^*$, $p_i^{\circ *}$ appearing in the second formula (5) is in equilibrium whereas the functions $(\sigma_{ij}^*, \ddot{u}_i^*, p_i^*)$ are determined by (1) provided that σ_{ij}^* and p_i^* are substituted respectively for σ_{ij}° and p_i (in this case (1) can no longer be interpreted as a principle of virtual velocities).

The expressions (5) provide extremum principles and also upper and lower bounds for $\ddot{u}_i$ and p_i which can be easily applied in practical computations (for example in the dynamically loaded shells close bounds on the actual solution can be obtained by considering inscribed and circumscribed yield conditions).

The extremum principles formulated in the present note enable to use more effective methods of mathematical programming to treat dynamic problems. For $t > t_k$, the functional to be minimized is

$$Z = \int_0^{t_k} \int_V d \, dVdt - \int_0^{t_k} \int_S p_i \dot{u}_i^* dSdt, \quad d = \sigma_{ij}^* \dot{\varepsilon}_{ij}^*. \tag{6}$$

In plate and shell problems the dissipation is expressed in terms of generalized stress E_j, and strain rates e_j rather than σ_{ij} and $\dot{\varepsilon}_{ij}$. In the case of piece-wise linear yield surface the equality $D = E_j^* e_j^*$ is replaced by a set of inequalities

$$D \geqq E_j^{(m)} \dot{e}_j^* \tag{7}$$

where m denotes the number of the corner points on the yield surface. Discretizing the problem in space and time by means of finite differences the problem of linear programming can be stated as follows: to find a minimum for the functional Z (6) subject to the restriction of the equality type a) for E_j^o and $\ddot{u}_i^*$, restrictions of the inequality type b) for E_j^o and (7) for D.

The problem of quadratic programming consists in minimizing the functional (4) at each time step under the same restrictions as in the preceding case.

The method of linear programming has been applied to the solution of a simply supported plate loaded by a uniformly distributed rectangular pressure pulse of the duration t_1. The radius of the plate R was divided on 5 equal intervals. The dimensionless permanent central deflections $wR\gamma/\sigma_s h^2 t_1^2$ was compared with the exact solution of the same problem in Table I.

TABLE I

r/R	0	0.2	0.4	0.6	0.8	1
Computed values	31.16	28.2	21.09	14.08	7.03	0
Exact values	31.50	28.10	22.20	14.55	7.40	0

Here r denotes the radial coordinate, w is the vertical deflection, σ_s is a yield stress in tension, $2h$ denotes the thickness of the plate and the assumed value of the load intensity was $p = 18(\sigma_s h^2/R^2)$. The computed value of the response time $t_k = 3t_1$ was equal to the exact solution. A high accuracy of the method of linear programming is evident.

The same method was also used to solve a simply supported rectangular plate loaded by a uniformly distributed pressure acted in the interval $0 \leq t \leq t_1$. The length of the plate edge $2L$ was divided on 10 intervals. The linearized Tresca yield condition was used, the resulting static load carrying capacity was found to be $p' = 5.716(\sigma_s h^2/L^2)$. In the dynamic problem the assumed load intensity was $p = 3p'$. The computed permanent deflections of the plate in function of the rectangular coordinates x and y with the origin of the plate centre are presented in Table II.

Further examples, discussion of the proposed principles and references can be found in [1–3].

TABLE II

y/L	x/L	$wL^2\gamma/\sigma_s h^2 t_1^2$	y/L	x/L	$wL^2\gamma/\sigma_s h^2 t_1^2$
	0	30.710		0.4	19.529
	0.2	27.529		0.6	13.508
	0.4	21.018	0.4	0.8	6.767
0	0.6	14.054		1.0	0
	0.8	7.009		0.6	10.641
	1.0	0	0.6	0.8	5.578
	0.2	26.258		1.0	0
	0.4	21.018		0.8	2.790
0.2	0.6	14.049	0.8	1.0	0
	0.8	6.991	1.0	10.	0
	1.0	0			

References

[1] Erkhov, M. I., On extremal principles in the dynamics of rigid-plastic solids (in Russian), *Stoit. Mekh. Rasch. Soruzh. No 5* (1970).

[2] Erkhov, M. I., The dynamic limit analysis (in Russian), *Izv. AN SSSR, Mekh. Tverd. Tela No. 2* (1971).

[3] Erkhov, M. I., Extremum principles in the dynamics of rigid-plastic bodies and mathematical programming, *Arch. Mech. 25* (1973) 69.

Plasticity analysis in soil mechanics problems

Royal Military College, Kingston, Ontario, Canada

Plastic failure zones are developed in many classes of soil mechanics problems, particularly in sands. These problems have commonly been examined in laboratory tests at model scale, and the results expressed in terms of dimensionless force coefficients related in effect to the initial condition of the sand deposit. Engineers generally consider however that the coefficients obtained from these studies tend to suggest failure loads which are much higher than could reasonably be expected from previous experience with full-scale structures, (Kerisel, [1]).

The critical condition for sand stability is one of limiting shear stress described by a maximum value of the ratio of shear to normal stress τ/σ_n. The Mohr-Coulomb failure criterion assumes that τ/σ_n is constant, and defines an angle of shearing resistance ϕ for the sand which is apparently independent of stress levels, but highly dependent on the initial density of the soil.

Plasticity solutions for ultimate loadings in sands commonly use the forward integration procedures of the method of stress characteristics described by Sokolovskii [2], assuming that ϕ remains constant. Force coefficients from these solutions are again higher than would normally be acceptable in engineering applications. The purpose of this discussion is to review the causes for these discrepancies from normal engineering experience, and to indicate the directions in which improvements can be made.

Boundary conditions in many soils problems are often very complex and changes in initial boundary assumptions have a strong influence on computed failure loads (Graham, [3]; Graham and Stuart, [4]). It has been shown for example (James and Bransby, [5]) that the angle of soil-structure contact friction δ, mobilized down the face of a wall which is failing into a mass of sand, can decrease from ϕ at the top of the wall, to almost zero at the bottom. When $\phi = 35°$, this latter boundary specification reduces the calculated K_p coefficient by 34% from the commonly calculated figure of 8.57 for $\delta = \phi = $ constant (Table I). Furthermore, the displacements required to mobilize full shearing resistance in sand are much larger than in most other branches of materials science. As a result, structural displacements preceding 'failure' are often large enough to significantly alter the stress distribution in the failure zone and hence also to change the ultimate loads on the structure. Table I shows for example, that K_p increases by about 17% as the passive wall moves through 5° of rotation about its base.

TABLE I

Values of passive wall force coefficient K_p

$\phi = 35°$, $\delta = \phi$ at top of wall

Wall rotation preceding failure	Contact friction at bottom of wall	
	$\delta = \phi = 35°$	$\delta = 0$
0°	8.57	5.63
5°	10.03	6.56

In a similar fashion, the ultimate bearing pressure of strip footings increases by about 35% as the foundation settles through 10% of its width, (Graham and Stuart, [4]).

It is now known that the assumption that ϕ is a constant strength parameter for a given initial sand density is only true over the limited range of stresses used in most strength testing programmes. When the range of stresses is extended over several orders of magnitude, the limiting shear stress condition τ/σ_n is not constant, and a curved failure envelope replaces the straight Mohr-Coulomb line. This is represented in Fig. 1 by a typical relationship between ϕ and log (mean principal stress pressure), (De Beer, [6]; Bishop, [7]). In any real failure zone therefore, ϕ cannot remain constant, and must vary systematically throughout the failable region. It is relatively easy to incorporate this variability into the method of stress characteristics in terms of the average stress levels in each of the characteristic node calculations (Berezantsev and Kovalev, [8]; Graham and Pollock, [9]).

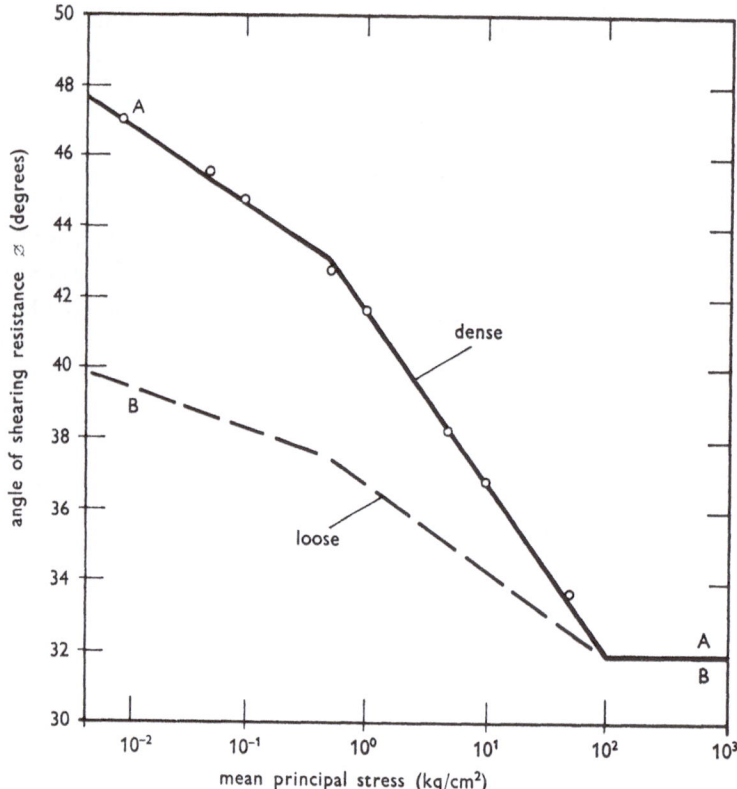

Fig. 1. Typical ϕ vs. pressure relationships.

The stresses attained in failure zones in sand depend not only on the initial density of the soil, but also on the size of the structure. The distribution of mobilized shearing resistance and the resulting 'dimensionless' force coefficients are therefore also size-dependent. It is not possible to include the ϕ vs. pressure dependency in conventional limit-analysis solutions, and the whole problem of scale effect in sands has hitherto escaped analytical solution. The author has however been able to apply the pressure-dependent solution to a number of problems in earthworks engineering, and has produced trends of behaviour similar to those found in model and field tests. As an example of this work, Fig. 2 shows stress-characteristics calculated for surface footings 0.28 m and 3.72 m wide respectively, using the ϕ vs. pressure relationship AA in Fig. 1 for uniform dense angular sand. The variation of ϕ in one of the computed failure zones is about 7°, and the

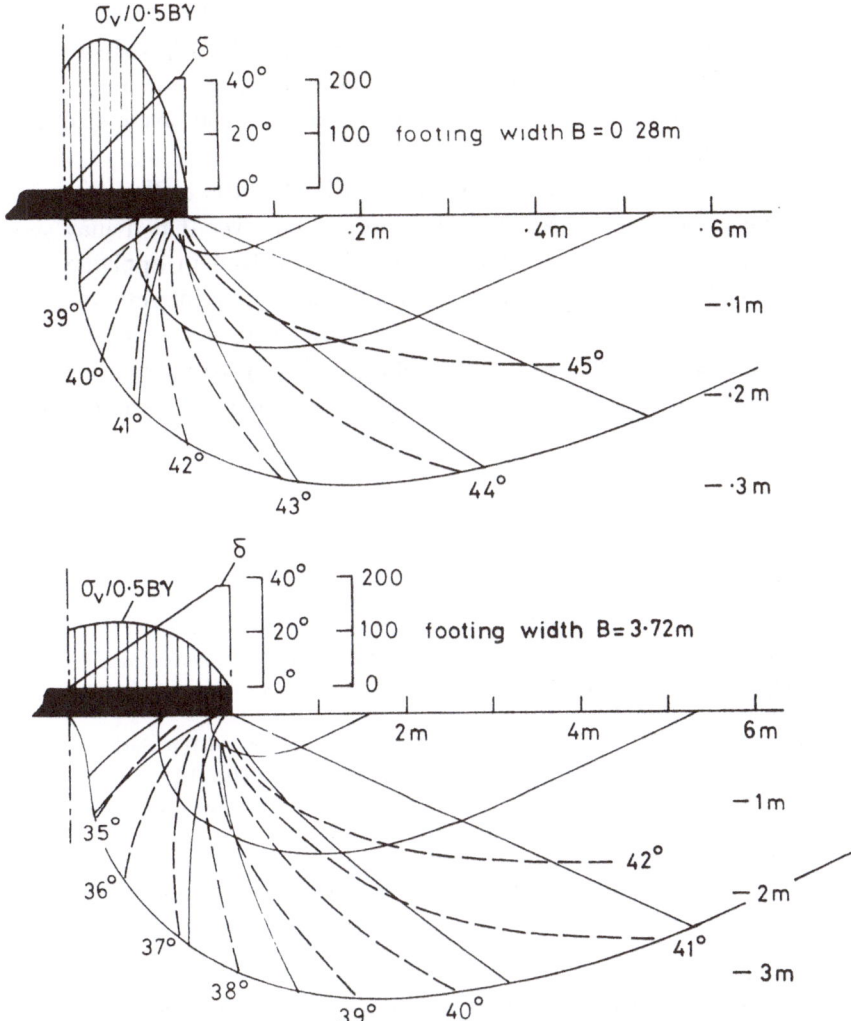

Fig. 2. Scale dependent stress characteristic fields for the surface footing problem.

difference in ϕ-mobilization at geometrically similar points in the two fields is about 3.5° to 4°. This corresponds to a 52% decrease in the force coefficient N_γ for a ten-fold increase in footing width. In engineering terms, changes of this magnitude are obviously of major importance.

Results of carefully controlled tests on model footings of different sizes are not readily available. Some preliminary tests carried out by the author at Queen's University, Belfast, suggest that an eight-fold increase in width produces a 41% reduction in N_y for $\phi = 35.5°$, and 34% reduction for $\phi = 38.2°$. The agreement between these experimental results and computations based on partly tentative strength-pressure relationships, is considered encouraging.

It is appreciated that this Discussion contributes little to the theoretical development of plasticity analysis. It is hoped however that it may serve some purpose in pointing out the difficulties involved in applying techniques already existing to some problems of engineering interest where the specification of boundary and field conditions becomes of major importance. Acceptable results can only be obtained by very careful mathematical modelling of the problem following detailed examination of the trends of laboratory behavior.

References

[1] Kerisel, J., Le language des modèles en mécanique des sols, Proceedings 5th European Conf. on Soil Mechanics and Foundations Engineering, Madrid 1972, 9–30.

[2] Sokolovskii, V. V., Statics of granular media. Pergamon Press, New York, 1965.

[3] Graham, J., Calculation of passive pressure in sand. *Canadian Geotechnical Journal, Vol. 8* (1971) 566–578.

[4] Graham, J. and J. G. Stuart, Scale and boundary effects in foundation analysis. *Journal of SM and FE Division, ASCE, Vol. 97, No. SM11* (1971) 1533–1548.

[5] James, R. G. and P. L. Bransby, Experimental and theoretical investigations of passive earth pressure problems, *Geotechnique, Vol. 20* (1970) 17–37.

[6] De Beer, E. E., The scale effect in the transposition of the results of deep-sounding tests on the ultimate bearing capacity of piles and caisson foundations. *Geotechnique, Vol. 13* (1963) 39–75.

[7] Bishop, A. W., The strength of soils as engineering materials. *Geotechnique, Vol. 16* (1966) 91–128.

[8] Berezantsev, V. G. and I. V. Kovalev, Consideration of the curvilinearity of the shear graph when conducting tests on model foundations. *Soil Mechanics and Foundation Engineering, No. 1* (Jan.–Feb. 1968) 3–8.

[9] Graham, J. and D. J. Pollock, Scale-dependent plasticity analysis for sand. *Civil Engineering and Public Works Review, London, Vol. 67* (1972) 245–251.

A note on the behaviour of hardening – softening granular media

Tomasz Hueckel

Institute of Fundamental Technological Research, Warsaw, Poland

Introduction. The purpose of this note is to discuss some models of mechanical behaviour of granular media. We shall restrict our attention to the models relying on the theory of dilatational plasticity. By dilatational plasticity we mean a plastic flow accompanied by irreversible volume changes. The theory assumes existence of yield surface and plastic potential flow rule enabling description of dilatation as well as compaction of the material.

Theory. In general, both loading and unloading should be considered in the theory, since granular materials are known which exhibit equally pronounced reversible and irreversible response to straining. The strain increment is assumed to be composed of recoverable and irrecoverable parts $d\varepsilon'_{ij}$ and $d\varepsilon''_{ij}$, respectively. One of the simplest sets of physical relations for a non-linear, inelastic, time-independent behaviour may be written in the form:

$$d\varepsilon_{ij} = d\varepsilon'_{ij} + d\varepsilon''_{ij} \tag{1}$$

$$d\varepsilon'_{ij} = A_{ijkl}d\sigma_{kl}; \quad d\varepsilon''_{ij} = d\lambda \frac{\partial g(\sigma_{ij}, \rho'')}{\partial \sigma_{ij}} \tag{2}$$

$$f = f(\sigma_{ij}, \rho'') = 0, \tag{3}$$

subject to conditions

$$d\lambda > 0, \quad \text{if} \quad f = 0, \ df = 0.$$

$$d\lambda = 0, \quad \text{if} \quad f < 0 \quad \text{or} \quad f = 0 \quad \text{and} \quad df < 0, \tag{4}$$

397

where: $g = 0$ denotes a plastic potential, $f = 0$ is a yield condition, A_{ijkl} stands for the material moduli tensor (depending on the state of deformation), $d\lambda$ is a non-negative multiplier, and ρ'' denotes the irreversible part of the density change with respect to the assumed reference state.

In principal stress space the function $f = 0$ can be visualized as a single parameter family of surfaces. It is assumed that the surfaces $f = 0$ and $g = 0$ are closed in the stress space.

The assumption that $f = 0$ is closed admits the plastic deformation due to isotropic compression, whereas a closed $g = 0$ allows for both compaction ($d\varepsilon_{ii}'' < 0$) resulting in hardening, and dilatation ($d\varepsilon_{ii}'' > 0$), resulting in softening of the material. The incompressible flow takes place for such points of $g = 0$ where $d\varepsilon_{ii}'' = 0$. The set of these points constitutes the critical state line. For $f = g = 0$, equation (2b) is reduced to the associated flow law (Fig. 1). This was discussed by Mróz [3, 4] and applied to some boundary value problems [5, 6]. The critical state theory was developed by Roscoe et al. [7].

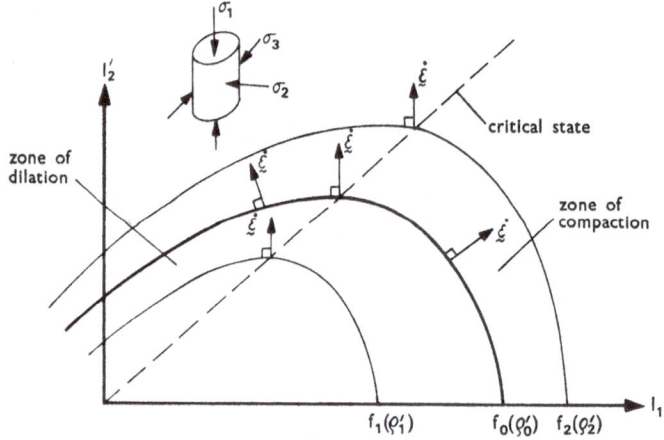

Fig. 1. Instantaneous yield surface for triaxial test; $I_2' = -\frac{1}{3}(\sigma_1 - \sigma_2)$; $I_1 = -\frac{1}{3}(\sigma_1 + 2\sigma_2)$.

Problems. There are three main problems of special interest in the theory: i) flow law, ii) material instability, which may appear whenever softening occurs, iii) coupling of reversible and irreversible strains.

Flow law. The main point of the problem is to determine the potential function $g = 0$, and the yield condition $f = 0$. The former function gives

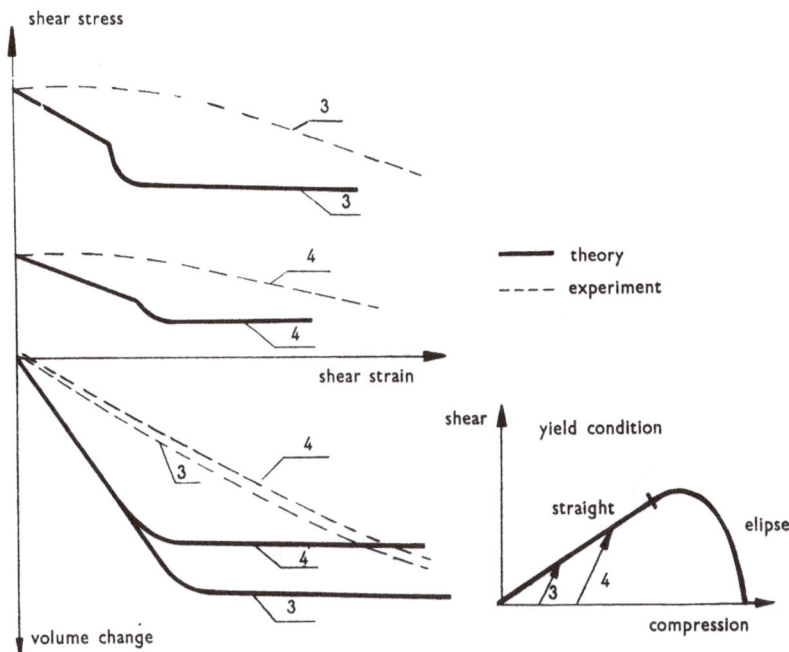

Fig. 2. Triaxial test results for dilatation; 3, 4-loading paths (after [1]).

the information on the relation between shear and volumetric strain increments, while the latter describes the extend of the elastic range. The appropriate tests were carried out [1] on triaxial compression apparatus and
elaborated for a prescribed yield condition $f = g$. The theory agrees qualitatively with experiments in the dilatation range, whereas in compaction
significant discrepancies occur (Fig. 2). This leads to the conclusion that, in
general, surfaces f and g do not coincide, althrough they converge in the
dilatation range.

Instability. Instability and the resulting non-uniqueness of solution of a
boundary value problem may occur when a part of the body undergoes
dilatation, associated with softening. For example, both in expansion and
compression of the thick-walled tube of granular media the solutions [5, 6]
can either be stable or unstable (Fig. 3). Apart from the symmetric solution,
some other solutions might be found which correspond to localized asymmetric flow. The question thus arises, what is the deformation mode actually developing in a real material. The problem seems to be of more general

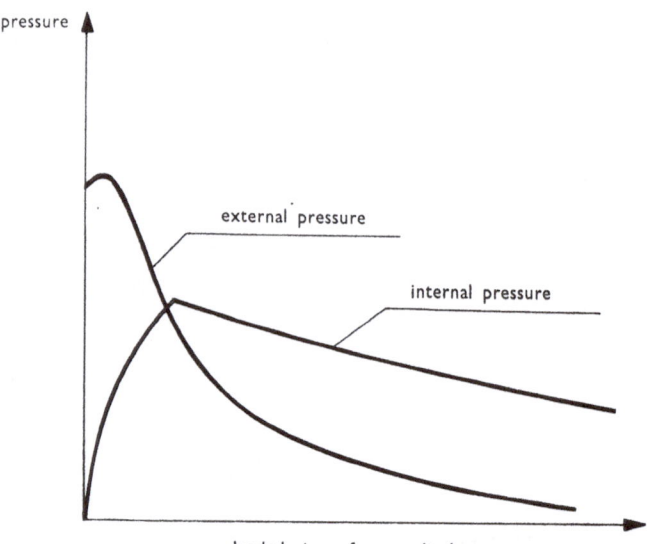

Fig. 3. Thick-walled tube pressure versus displacement.

nature, and has not been solved so far. The phenomena of the local defor-mations are observed in practice (silos, foundations) and force the engineers to propose some *ad hoc* theories.

The non-uniqueness for the corners of the yield condition on the critical line do not appear in the model, since the points on this line belong to the subsequent yield surfaces.

Strain coupling. This problem is due to high non-linearity during unloading and reveals when the specific form of the tensor A_{ijkl} and its dependence on strains are sought.

It should be assumed in general that the reversible strain increment may depend on permanent deformations in granular material. It is observed that the stress-strain curves during subsequent unloading differ one from another due to a different irreversible deformation.

It is therefore suggested that the tensor A_{ijkl} might depend not only on ε'_{ij} but also on ε''_{ij}. This results in coupling of the strain parts. Two simple examples may be given when the bulk modulus K and shear modulus G are functions of the density change with respect to the assumed reference state and in particular of i) the total density change $\tilde{\rho} = \rho' + \rho''$, $K = K(\tilde{\rho})$;

$G = G(\tilde{\rho})$, and ii) the reversible part of the density change ρ' only, i.e. $K = K(\rho')$, $G = G(\rho')$ [2]. In the latter case, no coupling is assumed. In order to detect whether or not the coupling exists for a specific material, the appropriate experiments were carried out under uniaxial strain [2]. The tests were performed on wheat grains and rice grains which exhibit relatively large reversible part of strain. The moduli K and G and the plastic hardening function for the cases i) and ii) may be determined from experiments recording the material response to loading and several unloading. Solutions of uniaxial strain problem, with the real material constants, were compared with the experimental curves for wheat in the case of existance or absence of coupling (Fig. 4). The comparison indicates that the coupling for the wheat grains is not so significant as it was suggested in the case i), and a rule with weaker coupling should be chosen. This approach,

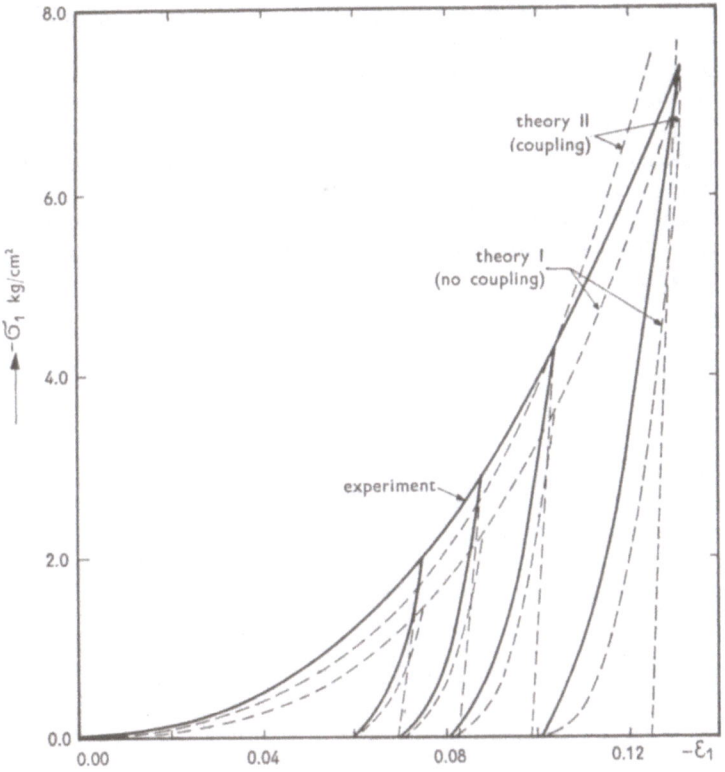

Fig. 4. Wheat grains. Uniaxial strain test.

was merely intended to show some fundamental topics of the coupling problem and to present some simplest examples. It should be stressed that the range of validity of the variable density model is limited. Because of the lack of elastic potential, the closed cycles can generate or dissipate the energy within the yield surface. The model is thus limited to the description of the behaviour of the granular media in the virgin loading and subsequent unloading.

References

[1] Drescher, A., Experimental verification of the density-hardening media (in Polish) *Rozpr. Inż. 20* (1972).

[2] Hueckel, T. and A. Drescher, Non-linear description for inelastic behaviour of granular media, pending publ.

[3] Mróz, Z., On a theory of density-hardening media, *Acta Mech.*, in press.

[4] Mróz, Z. and A. Drescher, Limit plasticity approach to some cases of flow of bulk solids, *J. Engng for Industry, 91,* 2 (1969).

[5] Mróz, Z. and T. Hueckel, Some boundary-value problems for variable density materials, Proc. Polish-French Symp. 'Problèmes de la Rhéologie' (1971) PWN, Warsaw (1973) 173–191.

[6] Mróz, Z. and K. Kwaszczyńska, Some boundary-value problems for granular density-hardening media (in Polish) *Rozpr. Inż. 19* (1971) 15–42.

[7] Roscoe, K. H. and H. B. Poorooshasb, A theoretical and experimental study of strains in triaxial compression tests on normally consolidated clays, *Geotechnique, 13* (1963) 12–38.

Some remarks on the strain-rate sensitive behavior of shells

NORMAN JONES

Massachusetts Institute of Technology, Cambridge, Mass. USA

An experimental investigation into the dynamic plastic behavior of various shells has been reported recently in [1] and [2]. The inner surfaces of the test models were subjected to uniformly distributed impulsive velocities which produced extensive plastic deformation. An examination will now be made of the viscoplastic characteristics of some of these models.

In 1957, Cowper and Symonds [3] proposed an appealingly simple constitutive relation to describe the uniaxial behavior of a strain-rate sensitive material. The Cowper-Symonds relation can be generalized in the form

$$\frac{\sigma_e}{\sigma_0} = 1 + \left(\frac{\dot{\varepsilon}_e}{D}\right)^{1/p} \tag{1}$$

where $\sigma_e = (\frac{3}{2}s_{ij}s_{ij})^{\frac{1}{2}}$, $\dot{\varepsilon}_e = (\frac{2}{3}\dot{\varepsilon}_{ij}\dot{\varepsilon}_{ij})^{\frac{1}{2}}$, σ_0 is the static uniaxial yield stress, and D and p are constants which must be obtained experimentally for a given material. The Cowper-Symonds relation with $D = 40.4 \sec^{-1}$ and $p = 5$ provides a fair representation of the mean of the experimental data which has been obtained from a large number of dynamic studies into the viscoplastic behavior of hot rolled mild steel specimens loaded uniaxially. Unfortunately, the results from dynamic uniaxial tests on aluminum 6061 T6 specimens are contradictory. Maiden and Green [4] and Lindholm and Yeakley [5] reported that aluminum 6061 T6 exhibited no strain-rate sensitivity in uniaxial compression, at least for strain-rates up to $10^3 \sec^{-1}$, approximately. On the other hand, Turnbow and Ripperger [6], Mok (see [3]) Hoge [7] and Hoggatt and Recht [8] observed various amounts of strain-rate sensitivity for aluminum 6061 T6. However, the strain-rate sensitivity of aluminum 6061 T6 in all these tests was less than that predicted by equation (1) with the constants $D = 6500 \sec^{-1}$ and $p = 4$ which were proposed by Cowper and Symonds on the basis of the earlier experimental

403

results of Evans (1942), Whiffen (1948) and Klinger (1950) [3]. It is observed that a curve with $p = 4$ and $D = 1,288,000 \text{ sec}^{-1}$, approximately, passes through the average of the widely scattered data of Mok [3] and of references [6] to [8]. It must be emphasized that these values of D and p for aluminum 6061 T6 are tentative due to the poor agreement between the experimental results of different authors and the absence of any data for strain-rates above 7,000 sec^{-1}.

Perrone [9, 10] has shown for some simple rigid-plastic structures loaded impulsively, that the influence of material strain-rate sensitivity on the structural response can be adequately catered for with a time independent yield stress which is evaluated from equation (1) for the initial strain-rates. This method must be modified if the influence of geometry changes is important (e.g. large deflection viscoplastic response of plates [11]). However, it is evident from the experimental results which are presented in [1] and [2], that geometry changes are not important for the test models at least for permanent transverse displacements up to about twice the corresponding shell thickness.

The fully clamped hemispherical shells which were examined in [1] had a mean radius R, wall thickness H and were subjected to a uniformly distributed impulsive velocity V_0. Thus, the initial in-plane strain-rates were $\dot{\varepsilon}_\phi = \dot{\varepsilon}_\theta = V_0/R$, while for volume conservation the initial transverse strain-rate was $\dot{\varepsilon}_z = -2V_0/R$. It is straightforward to show that $\dot{\varepsilon}_e = 2V_0/R$ and $\sigma_e = \sigma_0^1$, where σ_0^1 is the initial dynamic yield stress. Equation (1), therefore, becomes $\sigma_0^1/\sigma_0 = 1 + \{2V_0/(RD)\}^{1/p}$, where $D = 40.4 \text{ sec}^{-1}$ and $p = 5$ for hot rolled mild steel, and $D = 1,288,000 \text{ sec}^{-1}$ and $p = 4$ for aluminum 6061 T6. The experimental values of the maximum permanent transverse displacements (W_0) which were reported in [1] are replotted in Fig. 1 with the non-dimensionalized impulse parameter $\lambda^1 = 4\rho V_0^2 R^2/(\sigma_0^1 H^2)$ as an abscissa (ρ is density of specimens).

The cylindrical shell panels of axial length $2L$, mean radius R and wall thickness H, which were examined in [2], were fully clamped along the two longitudinal edges to give an included angle of 2α (90 degrees, approximately) and were free along the two circumferential edges. The panels were loaded throughout a central rectangular zone of area A_L with a uniformly distributed impulsive velocity V_0. Thus, the initial in-plane strain-rates within the central zone were $\dot{\varepsilon}_\theta = V_0/R$ and $\dot{\varepsilon}_x = 0$, while for volume conservation $\dot{\varepsilon}_z = -V_0/R$. Equation (1) can now be written $\sigma_0^1/\sigma_0 = 1 + \{2V_0/(\sqrt{3} RD)\}^{1/p}$. Therefore, the initial yield stresses throughout the loaded and

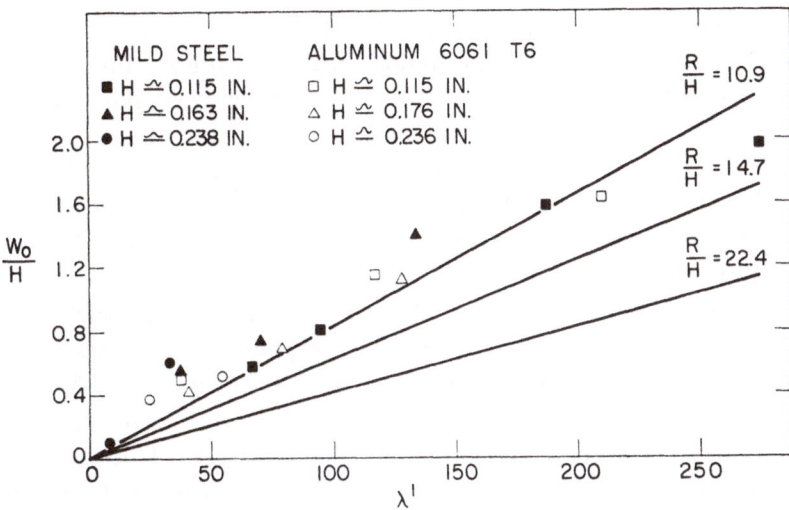

Fig. 1. Maximum permanent transverse deflections (W_0) for fully clamped hemispherical shells loaded impulsively ($R \simeq 2.58$ in.) (See reference 1 for further details.)

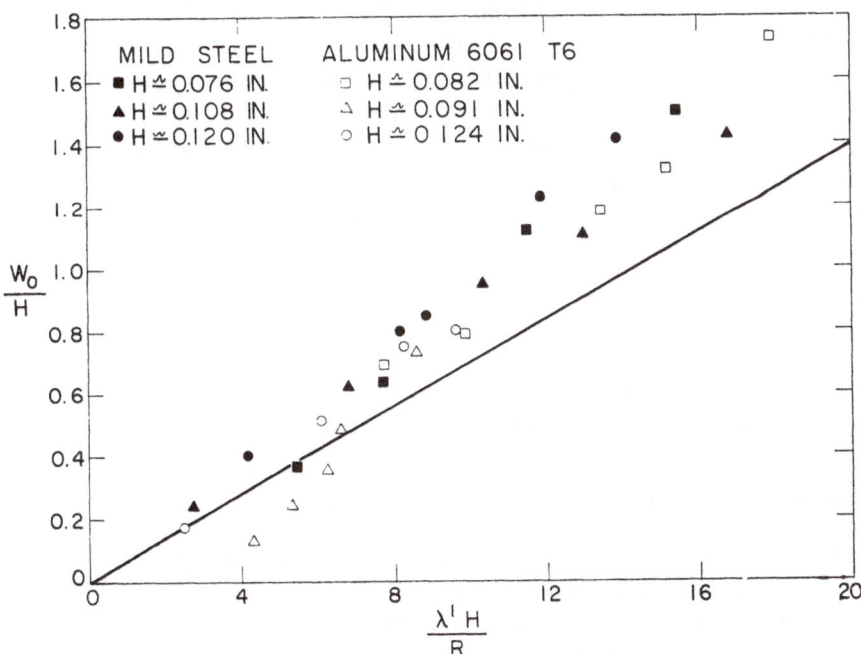

Fig. 2. Maximum permanent transverse deflections (W_0) of cylindircal shell panels loaded impulsively ($R \simeq 2.05$ in.) (See reference 2 for further details.)

405

unloaded zones were σ_0^1 and σ_0, respectively. In order to simplify the calculations, it was assumed that an average yield stress for an entire panel was $\{1+(\sigma_0^1/\sigma_0-1)A_L/A_T\}\sigma_0$, where $A_T = 4\alpha RL$. Strictly speaking, the coefficient of σ_0 should be evaluated on the basis of the energy dissipation rates in A_L and $A_T - A_L$ rather than simply on the areas A_L and $A_T - A_L$. The experimental values of the maximum permanent transverse displacements (W_0) which were reported in reference [2] are replotted in Fig. 2 with respect to $\lambda^1 H/R$.

The predictions of the approximate theoretical method which was developed in [12] and [13] for the dynamic response of shells made from a rigid perfectly plastic material are also presented in Figs. 1 and 2.

The most striking feature of Figs. 1 and 2 is that the mean curves of the hot rolled mild steel results and of the aluminum 6061 T6 results almost coincide. Thus, it appears that a theoretical analysis for a strain-rate insensitive material could be used to describe the behavior of a strain-rate sensitive one when using Perrone's suggestion and a generalized form of the Cowper-Symonds constitutive relation. As remarked previously this procedure is restricted to impulsive loadings and can only be used without modification when geometry changes do not occur during deformation. Finally, it should be remarked that insufficient experimental data is available currently for an accurate description of the dynamic uniaxial material properties of aluminum 6061 T6, while the accuracy of equation (1) in its general form cannot be ascertained since scarcely any tests have been conducted on metals which are subjected to either dynamic biaxial stress states or variable stress histories.

Acknowledgement

The author wishes to thank the Structural Mechanics Branch of O.N.R. for their support of this work under contract N00014–67–0204–0032.

References

[1] Jones, N., J. G. Giannotti and K. E. Grassit, An experimental study into the dynamic inelastic behavior of spherical shells and shell intersections, *Archiwum Budowy Maszyn* (Arch. of Mech. Eng.), *Vol. 20, 1* (1973) 33–46.

[2] Jones, N., J. W. Dumas, J. G. Giannotti and K. E. Grassit, The dynamic plastic behavior of shells, *Dynamic Response of Structures* (1972) 1–29, Ed. G. Herrmann and N. Perrone, Pergamon Press.

[3] Symonds, P. S., Viscoplastic behavior in response of structures to dynamic loading, Behavior of Materials Under Dynamic Loading (1965) 106–124. Ed. N. J. Huffington, A.S.M.E.

[4] Maiden, C. J. and S. J. Green, Compressive strain-rate tests on six selected materials at strain-rates from 10^{-3} to 10^4 sec^{-1}, *Journal of Applied Mechanics, Vol. 33* (1966) 496–504.

[5] Lindholm, U. S. and L. M. Yeakley, High strain-rate testing: tension and compression, *Experimental Mechanics, Vol. 8* (1968) 1–9.

[6] Turnbow, J. W. and E. A. Ripperger, Strain-rate effects on stress-strain characteristics of aluminum and copper, 415–440. Proc. 4th Midwestern Conference on Solid Mechanics, 1959.

[7] Hoge, K. G., Influence of strain-rate on mechanical properties of 6061 T6 aluminum under uniaxial and Biaxial states of stress, *Experimental Mechanics, Vol. 6* (1966) 204–211.

[8] Hoggatt, C. R. and R. F. Recht, Stress-strain data obtained at high rates using an expanding ring, *Experimental Mechanics, Vol. 9* (1969) 441–448.

[9] Perrone, N., On a simplified method for solving impulsively loaded structures of rate-sensitive materials, *Journal of Applied Mechanics, Vol. 32* (1965) 489–492.

[10] Perrone, N., Impulsively loaded strain-rate-sensitive plates, *Journal of Applied Mechanics, Vol. 34* (1967) 380–384.

[11] Jones, N., Finite-deflections of a rigid-viscoplastic strain-hardening annular plate loaded impulsively, *Journal of Applied Mechanics, Vol. 35* (1968) 349–356.

[12] Walters, R. M. and N. Jones, An approximate theoretical study of the dynamic plastic behavior of shells, *International Journal of Non-Linear Mechanics, Vol. 7, No. 3* (1972) 255–273.

[13] Jones, N. and R. M. Walters, A comparison of theory and experiments on the dynamic plastic behavior of shells, *Archives of Mechanics, Vol. 24* (1972) 701–714.

On shakedown criteria

JAN A. KÖNIG

Institute of fundamental Technological Research, Warsaw, Poland

Introduction. The aim of this note is to confront the usual intuitive notion of shakedown with physical criteria of material failure and with structural stability under fluctuating loads.

The theory of shakedown is a natural extension of limit analysis. However if an elastic-plastic structure is subjected to repeated loads there exists a danger of failure before the limit load interaction is reached.

The problem was noticed by Grüning [1] as early as in 1926 and static shakedown theorems were formulated a few years later by Bleich [2] and Melan [3]. None of those writers and the majority of their followers as well gave however a precise definition of shakedown in terms of criteria of material failure or unserviceability conditions.

A mathematical definition of shakedown was given in [4]. It states that the shakedown is ensured if the energy dissipated during an arbitrary process remains finite:

$$w_p = \int_0^t \sigma_{ij}\dot{\varepsilon}_{ij}^P dt < \infty. \tag{1.1}$$

Some inaccuracies (repeated by various authors) occuring in formulations and demonstrations of shakedown theorems have been discussed by Rychlewski [5] who proposed a less stringent criterion, namely

$$\bar{w}_p = \frac{1}{V}\int_V \int_0^t \sigma_{ij}\dot{\varepsilon}_{ij}^P dt dV < w_0 \tag{1.2}$$

where w_0 is a constant.

The properly formulated Bleich-Melan theorem for the case of framed structures, based on the criterion (1.2) is due to Konieczny [6].

408

In the paper presented to the Symposium by Maier [15] the following criterion:

$$\max_{i,j} \lim_{t \to \infty} |\varepsilon_{ij}^P(x, t)| = \varepsilon_{\infty}^P(x) < \infty \tag{1.3}$$

was applied succesfully in a finite element approach. This requirement is equivalent to that given by the formula (1.1).

All the above criteria approximate the actual criterion of material failure under fluctuating loads. They contain also, to some extent, a safety criterion against incremental collapse. However in the latter case no local criterion is adequate. An integral analysis of the whole structure is necessary in a more refined approach.

Incremental collapse. To formulate adequate theorems regarding shakedown it is necessary to have reliable shakedown criteria. Therefore the intuitive criteria such as (1.1), (1.2) and (1.3) should be checked against either criteria of material failure or those of structural unserviceability.

The former will be done in the next section. The latter, at the present state of knowledge, has to be limited to an estimation of maximum deflections occuring in a shakedown processes. Some results have been obtained quite recently in this field both in finite element and continuum description of structures [13–18]. The most general approach was presented by Maier in his paper presented at this Symposium. It has been demonstrated that the necessary conditions of shakedown are also the sufficient ones for certain cases of geometric nonlinearity.

In connection with this result the following remark will be made: In a linearization of the deflection upper bound problem the inequality (3.7) of the next section may be useful:

$$\left(\frac{1}{\kappa} w_p\right)^{\frac{1}{2}} = \left(\frac{1}{\kappa} \int_0^t \sigma_{ij} \dot{\varepsilon}_{ij}^P dt\right)^{\frac{1}{2}} \geq \left(\int_0^t \dot{\varepsilon}_{ij}^P \dot{\varepsilon}_{ij}^P dt\right)^{\frac{1}{2}} \geq$$

$$\geq (\varepsilon_{ij}^P \varepsilon_{ij}^P)^{\frac{1}{2}} \geq \max_{i,j} |\varepsilon_{ij}^P| \tag{2.1}$$

and in this way the dissipated energy inequality becomes linearized without any linearization of the yield condition.

Low-cycle fatigue. For symmetric cyclic plastic strains with the constant amplitude $\delta\varepsilon^P$ the Coffin-Miner formula [7, 11] holds:

$$(\delta\varepsilon^P)N^{1/m} = C. \tag{3.1}$$

Here N is a number of cycles to collapse, m stands for a material constant, taken usually as 2. The formula remains valid for the case of simple tension if one puts $N = \frac{1}{4}$.

Formula (3.1) can be extended to multi-dimensional stress states if the amplitude $\delta\varepsilon^P$ is replaced by the following expression [9, 10]:

$$\Delta\varepsilon^P = (\tfrac{2}{3})^{\frac{1}{2}}[(\delta\varepsilon_1^P - \delta\varepsilon_2^P)^2 + (\delta\varepsilon_2^P - \delta\varepsilon_3^P)^2 + (\delta\varepsilon_3^P - \delta\varepsilon_1^P)^2]^{\frac{1}{2}} =$$

$$= (\tfrac{2}{3})^{\frac{1}{2}}(\delta\varepsilon_{ij}^P \delta\varepsilon_{ij}^P)^{\frac{1}{2}}. \tag{3.2}$$

If subsequent cycles are of different amplitudes, the criterion may be approximated by

$$\sum_{i=1}^{N} \frac{1}{N_i} = 1 \tag{3.3}$$

where N_i denotes failure number related to the amplitude at the ith cycle [11].

In a processes of plastic deformations the criterion assumes the form as follows (Cf. [12]):

$$(\varepsilon_{ij}^P \varepsilon_{ij}^P)^{\frac{1}{2}} \int_0^t (\dot{\varepsilon}_{ij}^P \dot{\varepsilon}_{ij}^P)^{\frac{1}{2}} dt = B \tag{3.4}$$

where B is a material constant. This formula follows from the previous one by replacing the number of cycles by the length of trajectory in the plastic strain space:

$$L = \int_0^t (\dot{\varepsilon}_{ij}^P \dot{\varepsilon}_{ij}^P)^{\frac{1}{2}} dt.$$

If plastic deformations are initially zero i.e. $\varepsilon_{ij}^P(0) = 0$ then, according to the Cauchy inequality applied to integrals, we obtain:

$$[\varepsilon_{ij}^P(t)\varepsilon_{ij}^P(t)]^{\frac{1}{2}} \leq \int_0^t [\dot{\varepsilon}_{ij}^P(\tau)\dot{\varepsilon}_{ij}^P(\tau)]^{\frac{1}{2}} d\tau. \tag{3.5}$$

Thus, a safe estimation for (3.4) is given by the following inequality

$$\int_0^t [\dot\varepsilon_{ij}^P(\tau)\dot\varepsilon_{ij}^P(\tau)]^{\frac{1}{2}}d\tau \leqq B^{\frac{1}{2}}. \tag{3.6}$$

If the yield condition $F(\sigma_{ij}) = K$ contains the origin of coordinate system $\sigma_{ij} = 0$ then there exist two constants κ and μ such that $0 < \kappa \leqq \mu$ and

$$\kappa(\dot\varepsilon_{ij}^P\dot\varepsilon_{ij}^P)^{\frac{1}{2}} \leqq \sigma_{ij}\dot\varepsilon_{ij}^P \leqq \mu(\dot\varepsilon_{ij}^P\dot\varepsilon_{ij}^P)^{\frac{1}{2}}. \tag{3.7}$$

For the Huber-Mises yield condition $\mu = \kappa = (\frac{2}{3})^{\frac{1}{2}}\sigma_0$, where σ_0 denotes the yield stress in simple tension. For Tresca yield condition in plane stress $\mu = 2^{\frac{1}{2}}\sigma_0$, $\kappa = \sigma_0/2^{\frac{1}{2}}$.

The following safe estimate of the failure criterion can be derived from (3.4) and (3.6):

$$w_p = \int_0^t \sigma_{ij}\dot\varepsilon_{ij}^P dt \leqq B^{\frac{1}{2}}. \tag{3.8}$$

By comparing this result with the shakedown criteria (1.1), (1.2) and (1.3) we see that the criterion (1.2) is an average over the whole body volume and the criteria (1.1) and (1.3) follow from (3.8) if $B = \infty$.

Conclusions. Many experimental data agree with the predictions of the classical shakedown theorems. Nevertheless some discrepancies have been observed. They are explained by strainhardening and geometric nonlinearity effects. It is feasible however, that also the differences between theoretical and physical criteria of shakedown can be responsible for those discrepancies.

References

[1] Grüning, M., Die Tragfähigkeits statisch unbestimmten Tragwerke aus Stahl bei belibig häufig wiederholter Belastung. Springer 1926.

[2] Bleich, H., Über die Bemessung statisch unbestimmter Stahlwerke unter Berücksichtigung des elastisch-plastischen Verhaltens des Baustoffes. *Bauingenieur 13* (1932) 261.

[3] Melan, E., Die Spannungszustand eines Mises-Henckyscher Kontinuums bei verändlicher Belastung. *Sitz. Ak. Wiss. Wien IIa, 147* (1938) 73.

[4] König, J. A., Theory of shakedown of elastic-plastic structures. *Arch. Mech. Stos. 18* (1966) 227–238.

[5] Rychlewski, J., On basic theorems of shakedown of elastic-plastic bodies. XI Polish Conference on Solid Mechanics, Bielsko-Biała, September 1967.

[6] Konieczny, L., Shakedown theory of beams (in Polish). *Mech. Teor. Stos. 8* (1970) 257–276.

[7] Coffin, L. F. and J. P. Tavernelli, Cyclic straining and fatigue of metals. *Trans. Met. Soc. AIME 215* (1959) 794–807.

[8] Serensen, S. V., Problems of limit analysis for low number of loading cycles. (in Russian), Nauka, Moscow (1969) 6–25.

[9] Taira, S., T. Inoue and M. Takahashi, Low cycle fatigue under multiaxial stresses. Proc. 10-th Japan Congr. Testing Materials, Kyoto (1967) 18–23.

[10] Mattavi, J. L., Low cycle fatigue behaviour under biaxial strain distribution. *J. Basic. Engng. 91* (1969) 23–31.

[11] Miner, R., Cumulative damage in fatigue. *J. Appl. Mech. 12* (1945) 159–164.

[12] Novozhilov, U. V. and O. G. Rybakina, On prospects of constructing a strength criterion for general stress. Strength for low number of loading cycles. Nauka, Moscow (1969) 71–81.

[13] Vitiello, E., Upper bounds of plastic strains in shake-down of structures subjected to cyclic loads. Meccanica 1972 (to appear). .

[14] Ponter, A. R. S., An upper bound to the small displacements of plastic, perfectly plastic structures. ASME paper No 72-APM-V.

[15] Maier, G., A shakedown matrix theory allowing for work-hardening and second-order geometric effects. *Foundations of Plasticity*, Noordhoff Int. Publ., Leyden 1973, 417–433.

[16] Brzezinski, R. and J. A. König, Deflections of elastic-plastic framed structures at shakedown. *Journal of Structural Mechanics 2* (1973) 211–228.

[17] Brzezinski, R. and J. A. König, Evaluation of shakedown deflections of framed structures by linear programming. Symposium on Plastic Analysis of Structures, Jassy, September 1972, 101–116.

[18] König, J. A., Shakedown deflections. A finite element approach. *Theoretical and Applied Mechanics* (Bulgarian) *3* (1972) 65–69.

Comment on elastic and plastic rotations

J. KRATOCHVÍL

Institute of Solid State Physics, Prague, Czechoslovakia

Kinematics of recently proposed finite-strain theories of inelastic behavior is usually based on the resolution of the deformation gradient F into a plastic part P followed by an elastic part E

$$F = EP. \tag{1}$$

To justify (1), three configurations are introduced: the reference configuration κ, the actual configuration χ, and the intermediate local configuration κ', such as F: $\kappa \to \chi$, P: $\kappa \to \kappa'$, E: $\kappa' \to \chi$. The resolution (1) has a meaning only if κ' is clearly defined.

Professor Onat in his lecture at this symposium choose to interpret κ' as a configuration reached from χ by unloading. Such interpretation, as any other interpretation based on the concept of stress, defines κ' only to within a rotation (in isotropic materials the rotation can be arbitrary). To fix the choice of κ' Onat assumed that E is equal to the right stretch tensor of unloading deformation, i.e. the rotational part of E is the unit tensor. Similar assumptions were introduced by Lee [1]. Precise interpretation of κ' defined by unloading and valid also for anisotropic materials was given by Mandel [2].

The purpose of this comment is to describe briefly an alternative approach [3–5] applicable to crystalline materials which is based on interpretation of elastic and plastic deformation proposed by Kröner [6]. The approach could be useful e.g. in the description of preferred orientation phenomenon, so called texture, which accompanies inelastic deformation in polycrystalline materials and can considerably influence the deformation process.

From the microscopic point of view elastic deformation is related to changes of distances between atomic planes and to rotation of crystallo-

413

graphic directions. On the other hand, plastic deformation, being a mode of deformation which just moves atoms into crystallographically equivalent positions, does not cause such changes. Crystal defects which appear during plastic deformation disturb locally the periodicity of the crystal lattice. These characteristic features of deformation modes should be reflected in a phenomenological theory.

To distinguish the elastic and plastic parts of deformation, X-ray diffraction technique can be helpful especially in single crystals. Standard X-ray methods are able to indicate the changes of distances between atomic planes and the rotation of crystallographic directions (in polycrystals X-ray measurements refer to certain average changes of interplane distances and rotation of directions of preferred orientations of crystalline grains). The intermediate configuration κ' may be interpreted as the reference configuration of the X-ray measurement, i.e. a configuration where the geometrical arrangement and the orientation of crystallographic directions with respect to the fixed axes are kept constant for all time; (in this sense the suggested approach is a special case of Mandel's theory [2]). Hence we can specify both stretch and rotational parts of E, and the resolution (1) has unique meaning. Unlike the case of unloading interpretation of κ', the suggested kinematics is not based on the concept of stress.

In the theory of inelastic behavior of crystalline materials [3–5], where the proposed interpretation of the relation (1) is used, the elastic and plastic rotations play an important role.

Material time derivative of (1) yields

$$L = L_E + L_P, \tag{2}$$

where $L = \dot{F}F^{-1}$, $L_E = \dot{E}E^{-1}$, and $L_P = E\dot{P}P^{-1}E^{-1}$. In the theory it is assumed that not only the plastic stretching tensor $D_P = \frac{1}{2}(L_P + L_P^T)$ but also the plastic spin tensor $W_P = \frac{1}{2}(L_P - L_P^T)$ is governed by the constitutive equation. For rate-dependent materials we have following reduced forms of these constitutive equations

$$D_P = ED_P(E^T E, \theta, E^T \operatorname{grad} \theta, \alpha)E^T, \tag{3}$$

$$W_P = EW_P(E^T E, \theta, E^T \operatorname{grad} \theta, \alpha)E^T, \tag{4}$$

where D_P is a symmetric and W_P a skew-symmetric tensor function, θ means the temperature, and α is a set of parameters $\alpha^{(i)}$, $i = 1, ..., n$, which describe structural changes caused by lattice defects. If $\alpha^{(i)}$ are tensors, then

α in (3), (4) must be replaced by a set of convected quantities, e.g. $E^T\alpha^{(i)}E$ in the case of second order tensors $\alpha^{(i)}$. According to the proposed interpretation the constitutive relation (4) controls the rotation of distinguished directions of anisotropy in the material.

For isotropic materials the equations (3) and (4) become

$$D_P = D_P(EE^T, \theta, \text{grad } \theta, \alpha), \tag{5}$$

$$W_P = W_P(EE^T, \theta, \text{grad } \theta, \alpha), \tag{6}$$

where now D_P and W_P are isotropic functions. It is interesting to note that if W_P is independent of grad θ and $\alpha^{(i)}$ are scalars, Wang's representation theorem [7] yields $W_P = 0$. Then, the skew part of (2) is reduced to the trivial statement that $W = W_E$ and instead of (2) it is sufficient to consider

$$D = D_E + D_P. \tag{7}$$

D, W and D_E, W_E are the symmetric and skew-symmetric parts of L and L_E respectively.

However, if W_P is sensitive to grad θ or $\alpha^{(i)}$ are not scalars we cannot deduce from Wang's theorem that in (6) $W_P = 0$ (physically it means that certain distinguished directions are present in the material). Then not only (7) but also the skew part of (2) can be important in solving boundary value problems even for isotropic materials.

The same statements concerning W_P are valid also for rate-independent materials as shown by Trávníček [8].

References

[1] Lee, E. H., Elastic-plastic deformation at finite strains. *J. Appl. Mech. 36* (1969) 1-6.
[2] Mandel, J., Relations de comportement des milieux elastiques-plastiques et elastiques-viscoplastiques. Notion de repere directeur. *Foundations of Plasticity* (edited by A. Sawczuk) Noordhoff, Leyden, 1973, 387–400.
[3] Kratochvíl, J., Finite-strain theory of crystalline elastic-inelastic materials. *J. Appl. Physics, 42* (1971) 1104–1108.
[4] Kratochvíl, J., On a finite strain theory of elastic-inelastic materials. *Acta Mechanica 16* (1973) 127–142.
[5] Kratochvíl, J., Finite-strain theory of inelastic behavior of crystalline solids. *Foundations of Plasticity* (edited by A. Sawczuk) Noordhoff, Leyden, 1973, 400–415.
[6] Kröner, E., Kontinuums-theorie der Versetzungen und Eigenspannungen. Springer, Berlin, 1958.

[7] Wang, C.-C., A new representation theorem for isotropic functions: An answer to Professor G. F. Smith's criticism of my papers on representations for isotropic functions, Part 2. *Arch. Rational Mech. Anal. 36* (1970) 198.

[8] Trávníček, L., Finite deformation of elastic-plastic solids. Thesis, Charles University, College of Math. and Physics, Prague, 1972. (XIII. Int. Congress of Theoretical and Applied Mechanics, Moscow, 1972).

A minimum principle in dynamics of elastic-plastic continua at finite deformation

L. H. N. Lee and C. M. Ni

University of Notre Dame, Notre Dame, Ind., USA

An absolute minimum principle in dynamics of elastic-plastic continua subject to finite deformation is developed. The principle is based on the concept of employing finite differences in accelerations in formulating a variational principle by Gibbs in classical mechanics [1]. The concept is particularly well suited to the study of non-holonomic systems. A parallel minimum principle in dynamic plasticity has been developed by Tamuzh [2] for rigid-plastic bodies involving infinitesimal deformations.

The absolute minimum principle is expressed in terms of Lagrangian strains, E_{KL}, and Piola-Kirchhoff stresses, S_{KL}. Green and Naghdi assume that the Lagrangian strain can be divided into two parts, elastic and plastic parts, E'_{KL} and E''_{KL}. The justification of the assumption is given in References [3, 4]. It is assumed here that the constitutive relationship is not influenced by the strain acceleration but may be influenced by the plastic strain velocity, $\dot{E}''_{MN}$, such as

$$S_{KL} = S_{KL}(E'_{MN}, E''_{MN}, \dot{E}''_{MN}, \theta) \tag{1}$$

where θ is the temperature.

The true displacement, U_M^+, and acceleration, $\ddot{U}_M^+$, fields of a body, which has an initial volume V, bounded by surface A, are distinguished from all possible ones by satisfying the equations of motion in the Lagrangian coordinates:

$$[S_{KL}^+(\delta_{ML} + U_{M,L}^+)]_{,K} + \rho_0(F_M - \ddot{U}_M^+) = 0 \tag{2}$$

where ρ_0 is the initial mass density and F_M is the body force per unit mass. The true Piola-Kirchhoff stresses, S_{KL}^+, satisfy the boundary conditions

$$S_{KL}^+(\delta_{ML} + U_{M,L}^+)N_K = T_M \tag{3}$$

417

on that part of the initial surface area, A_T, where the surface force per unit area, T_M, is prescribed, N_K is the outward unit normal to A.

It has been shown, by Eqs. (1), (2), and (3) and Reference [5], that the true acceleration field of the body, which has predetermined displacement and velocity fields at time t, is distinguished from all kinematically admissible ones by having the absolute minimum value of the following functional

$$J = \int_V \frac{\rho_0}{2} \ddot{U}_M^2 dV + \int_V S_{KL} \ddot{E}_{KL} dV -$$

$$- \int_V \rho_0 F_M \ddot{U}_M dV - \int_{A_T} T_M \ddot{U}_M dA +$$

$$+ \int_\sigma [S_{KL} N_K (\delta_{ML} + U_{M,L}) \ddot{U}_M] d\sigma. \tag{4}$$

The last integral of Eq. (4) is extended over all internal surfaces σ having directions N_K and where the accelerations are discontinuous. The quantities enclosed by the boldface bracket indicate the jump. If Piola-Kirchhoff stresses depend on strain accelerations, the minimum principle remains valid for a limited class of acceleration fields subject to the kinematic boundary constraints and the requirements that

$$\ddot{E}_{KL}(S_{KL}^+ - S_{KL}) \leqq 0 \text{ in } V$$

and

$$S_{KL} = S_{KL}^+ \text{ on } A_T. \tag{5}$$

Ordinarily, it is sufficient to use the first variation of J with respect to the acceleration, $\delta_{acc.} J = 0$, to establish governing equations or to solve a problem by a direct method of variational calculus. The present minimum principle circumvents certain difficulties in treating constitutive relationships by other variational principles involving displacement or velocity variations.

For relatively simple problems, such as non-linear vibrations of elastic beams and impulsive loading of a rigid-plastic beam with axial constraints, approximate closed form solutions, which agree with available analytical results, have been obtained directly by the minimum principle [5]. For relatively complex problems, numerical methods have been developed. Incremental numerical procedures based on Kantorovich's method [6] and the minimum principle have been developed and applied to the determination of the responses of inelastic beams [7, 8], rectangular plates [9] and cylindrical shells [10] to impulsive loadings. An incremental finite difference numerical procedure has also been developed and employed to investigate the dynamic behavior of a cylindrical shell panel [10].

Acknowledgement. This research was supported by the National Science Foundation of the United States of America under Grant GK-11034 to the University of Notre Dame.

References

[1] Pars, L. A., A treatise on analytical dynamics, Heinemann, London, (1965) 200–201.

[2] Tamuzh, V. P., On a minimum principle in dynamics of rigid-plastic bodies, *PMM* *26, 4* (1962) 715–722.

[3] Green, A. E. and P. M. Naghdi, A thermodynamic development of elastic-plastic continua, Proc. of Symposia of IUTAM on the Irreversible Aspects of Continuum Mechanics, Springer-Verlag, New York (1968) 117–131.

[4] Green, A. E. and P. M. Naghdi, Some remarks on elastic-plastic deformation at finite strain, *Int. J. Engrg. Science, 9,* 12 (1971) 1219–1229.

[5] Lee, L. H. N. and C. M. Ni, A minimum principle in dynamics of elastic-plastic continua at finite deformation, *Archives of Mechanics, 25, 3* (1973) 457–468.

[6] Kantorovich, L. V. and V. I. Krylov, Approximate methods of higher analysis, Interscience Publisher, New York (1958) 240.

[7] Ni, C. M. and L. H. N. Lee, Dynamic behavior of inelastic beams at finite deformation, *Proc. CANCAM71* (1971) 455–456.

[8] Ni, C. M. and L. H. N. Lee, Finite earthquake-response of an inelastic structure, *J. Engrg. Mech. Division, ASCE, 98, EM6* (1972) 1529–1546.

[9] Sureshwara, B., L. H. N. Lee and T. Ariman, Impulsive loading of rectangular plates with finite plastic deformations, *Proc. SECTAM VI* (1972) 553–579.

[10] Ni, C. M., Dynamic behavior of inelastic cylindrical shells at finite deformation, a Ph. D. thesis, under the direction of L. H. N. Lee, submitted to the Graduate School of the University of Notre Dame, 1971.

A note on displacement bounding techniques for dynamically loaded inelastic structures

W. J. MORALES

Westinghouse, Tampa, Florida, USA

The determination of the response of inelastic structures subjected to a variety of dynamic loadings consisting of a combination of impulsive and time-dependent blast loading has received a great deal of attention in recent years. It was recognized from the start that due to non-linearities introduced by plastic deformation, rate sensitive yielding, and large geometry changes the simple extension of the elegant analytical approaches developed in linear elastic problems would not be possible. Progress has been achieved along two distinct but equally important approaches.

One such approach consists primarily on numerical computations. Great advances have been made in a relatively short time as more sophisticated finite difference codes such as MIT PETROS III [1] and the Karman N. Code [2]. The PETROS III code has been written to include all the necessary types of constitutive relations and can handle any configuration of arbitrary shape under complex dynamic loadings with geometry changes occurring during the deformation process. While such codes should form part of the arsenal of computational tools of the design engineer, one must be aware of the enormous amount of machine time required to describe the transient response of dynamically loaded structures. An example described later in the test illustrates this point. The development of finite element techniques to solve this class of non-linear problems falls short to the versatility of finite difference techniques [3, 4] while little gain if any is to be expected in the computational costs.

The second line of approach has been more traditional, employing analytical methods to characterize the structural response. This alternative approach includes converging approximate solution or mode approximation technique and displacement bounding methods. The mode approximation technique originally developed by Martin and Symonds [5] for rigid-perfectly plastic continua under impulsive loading has been extended to time-dependent inelastic structures [6] and transient blast loadings [7]. Similarly, the development of the displacement bounding approach has followed the same order of refinement as the mode approximation method. The bounding method consists of the bracketing of the inelastic transient structural response through the application of two theorems requiring simple and relatively crude assumptions of some portion of the solution, either the structural response or the state of stress due to an assumed point load resulting in lower and upper bounds of the permanent deformation respectively. The upper bound theorem first introduced by Martin [8] for impulsively loaded rigid-perfect plastic structures has since been extended to time-dependent continua [9], blast loading [10] and large deformations [11]. The development of the lower bound theorem has proceeded in a similar fashion.

From the initial formulation of the theorem to bound the deformation of rigid-perfectly plastic structures [12], the theorem has been extended to

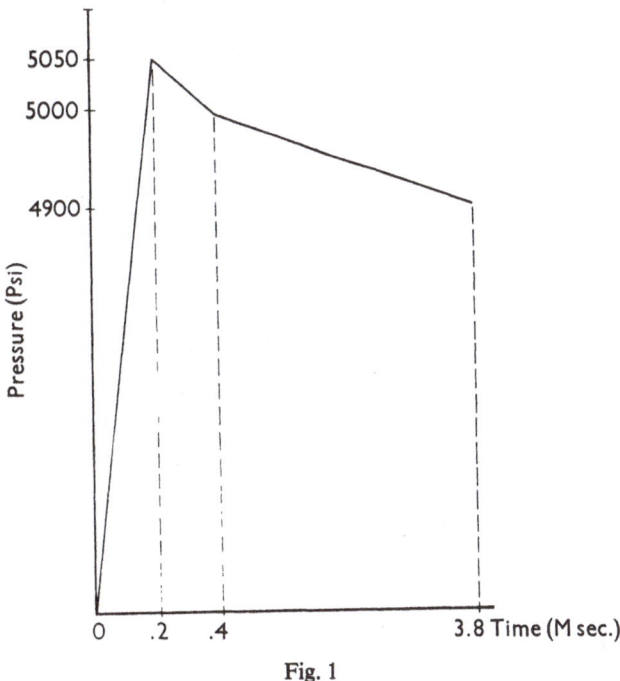

Fig. 1

visco-plastic continua [13] and time-dependent dynamic loading [14]. Recently, the theorem has been formulated to include the effect of large displacements making the bounding technique an extremely effective tool for preliminary design as will be illustrated in the example shown below.

Without denying the usefulness of the mode approximation technique in preliminary design, the bounding approach constitutes a more powerful method for, while the former approach might approximate the exact solution closely (very questionable for complex geometries) it gives no indication as to its relative magnitude compared to the exact solution. In the other hand, the upper bound theorem yields a consistent conservative estimate while the lower bound allows for the close bracketing of the exact structural response specially when optimization techniques are employed. The potential of this technique has been recognized by the AEC which has employed a coded version of these theorems to provide preliminary design calculations for a sodium-water reaction occurring in a sodium cooled steam generator of a fast breeder reactor.

The bounding theorems were applied to a sample problem consisting of a simply supported cylindrical vessel under an axisymmetric transient blast loading as shown in Fig. 1. The structure was assumed to respond visco-plastically with the following material and geometric constants:

$$n = 5, \quad D = 40.4 \text{ sec}^{-1}, \quad \sigma_0 = 50,000 \text{ psi}$$

where

$$\sigma = \sigma_0(1 + (\dot{\varepsilon}/D)^{1/n})$$

density $= 0.00074$ lb-sec^2-in^{-4}, length $= 80$ in;
inside radius $= 15.87$ in; thickness $= 0.50$ in.

The results of the analyses utilizing the bounding method and the 'exact' solution obtained by the application of the PETROS III finite difference code are summarized in Fig. 2. Average machine running time for each bounding solution was about 9 cpu seconds while for each PETROS III run about 180 cpu seconds were needed. It is to be noted that although no attempt was made to optimize the results, consistent bounds and reasonable engineering solutions were obtained at the expense of little machine time.

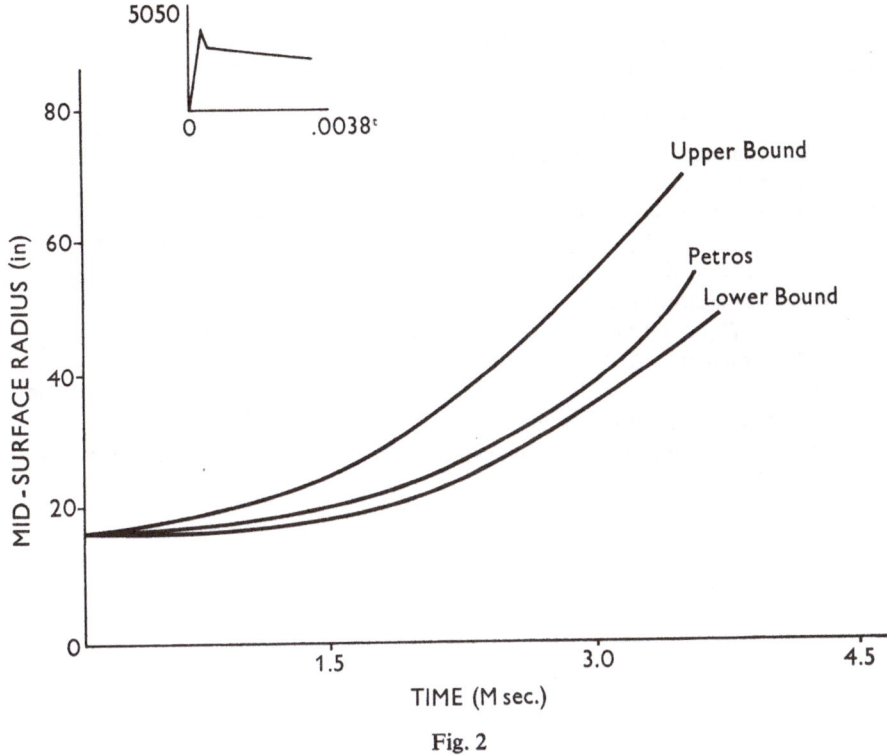

Fig. 2

References

[1] PETROS 3: A finite-difference method and program for the calculation of large elastic-plastic dynamically-induced deformations of multilayer variable-thickness shells, Aeroelastic and Structures Research Laboratory, MIT, 1972.

[2] Bothell, L. and W. Hubka, Development of a method which may be used for predicting the inelastic response of cylinders subjected to strong blast loads, Karman Nuclear, Colorado, 1970.

[3] MARC-CDC: Non-linear finite element analysis, MARC Analysis Research Corporation, Providence, 1972.

[4] Farhoomand, I. and E. Wilson, A non-linear finite element code for analyzing the blast response of underground structures, Structural Engineering Laboratory, Berkeley, California, 1971.

[5] Martin, J. B. and P. Symonds, Mode approximations for impulsively-loaded rigid-plastic structures, *ASCE, EM5, V. 92* (1966) 43–66.

[6] Martin, J. B. and J. D. O'Keeffe, Convergence approximations for dynamically loaded rigid-plastic structures, NSRDC, Contract N00189-68-C-0157, Brown U., 1969.

[7] Kaliszky, S., Approximate solutions for impulsively loaded inelastic structures and continua, *Int. J. of Non-Linear Mechanics*, 5 (1970) 143–158.

[8] Martin, J. B., Impulsively loading theorems for rigid-plastic continua, *ASCE, EM5*, 90 (1964) 27–42.

[9] Martin, J. B., Time and displacement bound theorems for viscous and rigid-visco-plastic continua, *Developments in Theoretical and Applied Mechanics*, 3 (1968) 1–22, ed. by W. A. Shaw.

[10] Robinson, D. N., A displacement bound principle for elastic-plastic structures subjected to blast loading, *J. of Mechanics and Physics of Solids*, 18 (1970) 65–80.

[11] Martin, J. B. and A. Ponter, Bounds on large deformations of impulsively loaded elastic-plastic structures, U. of Leicester, Rpt. 70, 1970.

[12] Morales, W. J. and G. E. Nevill, Lower bounds on deformations of dynamically loaded structures, *AIAA J.*, 8 (1970) 2043–2047.

[13] Morales, W. J., A lower bound theorem for dynamically loaded rigid-viscoplastic structures, to be presented in the Archives of Mechanics, 1972.

[14] Morales, W. J., A displacement bound theorem for blast loaded inelastic structures, *ASCE, EM3* (1972).

Optimal design criteria for reinforced plates and shells

ZENON MRÓZ

Institute of Fundamental Technological Research, Warsaw, Poland

Consider a perfectly plastic structure such as plate or shell for which the generalized stresses Q and strain rates q and rates of displacements w define the static and kinematic fields. Assume that the middle surface is prescribed and both support conditions and type of loading are specified. Our aim is to find an optimal design which minimizes the cost function

$$I = \int_A F(\Phi_i) \mathrm{d}A, \qquad i = 1, 2, ..., m \tag{1}$$

for prescribed safety factor of plastic collapse. This condition is expressed as follows

$$C = \int_A D(q, \Phi_i) \mathrm{d}A - \mu \int_A p \cdot w \mathrm{d}A = 0 \tag{2}$$

where $D(q, \Phi_i)$ denotes the specific dissipation power per unit area of the middle surface, μ is the prescribed safety factor and p is the specified loading. The design variables are assumed to be unknown functions of position on the middle surface. The number of design variables is arbitrary. In particular, Φ_i may represent fiber densities in reinforcing layers, their angles of orientation, variable plate thickness, etc. In this note we derive the optimality criterion without going into details. A more complete presentation will be given elsewhere.

Since (2) constitutes a constraint on (1), let us consider the functional $J = I - \lambda C$ where λ denotes the Lagrange multiplier. Considering the first variation of J, we have

$$\delta J = \int_A \frac{\partial F}{\partial \Phi_i} \delta \Phi_i \mathrm{d}A - \lambda \int_A \left(\frac{\partial D}{\partial q} \cdot \delta q + \frac{\partial D}{\partial \Phi_i} \delta \Phi_i \right) \cdot \mathrm{d}A -$$

$$- \mu \int_A p \cdot \delta w \mathrm{d}A = 0, \tag{3}$$

and since

$$\frac{\partial D}{\partial q} = Q, \quad \int_A Q \cdot \delta q \mathrm{d}A = \mu \int_A p \cdot \delta w \mathrm{d}A, \tag{4}$$

the stationarity conditions take the form

$$\frac{\partial F}{\partial \Phi_i} = \lambda \frac{\partial D}{\partial \Phi_i}, \qquad i = 1, 2, \dots, m, \tag{5}$$

and are represented by a set of functional relations from which the design variables Φ_i can be determined in terms of the generalized strain rates q.

In order to derive the conditions for global minimum of I, consider any other design corresponding to the same limit load and characterized by Φ_i', w', q', Q'. Let us take the difference

$$I' - I = \int F(\Phi_i') \mathrm{d}A - \int F(\Phi_i) \mathrm{d}A =$$

$$= \int \frac{\partial F}{\partial \Phi_i} (\Phi_i' - \Phi_i) \mathrm{d}A + \int \left[F(\Phi_i') - F(\Phi_i) - \frac{\partial F}{\partial \Phi_i} (\Phi_i' - \Phi_i) \right] \mathrm{d}A \tag{6}$$

and since

$$\int D(q, \Phi_i) \mathrm{d}A - \mu \int p \cdot w \mathrm{d}A = \int D(q', \Phi_i') \mathrm{d}A - \mu \int p \cdot w' \mathrm{d}A,$$

$$\int \frac{\partial D}{\partial q'} \cdot (q - q') \mathrm{d}A = \mu \int p \cdot (w - w') \mathrm{d}A, \tag{7}$$

there is

$$\int \frac{\partial D}{\partial \Phi_i} (\Phi_i' - \Phi_i) \mathrm{d}A = \int \left[D(q, \Phi_i') - D(q', \Phi_i') - \frac{\partial D}{\partial q'} \cdot (q - q') \right] \mathrm{d}A -$$

$$- \int \left[D(q, \Phi_i') - D(q, \Phi_i) - \frac{\partial D}{\partial \Phi_i} (\Phi_i' - \Phi_i) \right] \mathrm{d}A, \tag{8}$$

and in view of (5), the equality (6) can be presented in the form

$$I' - I = \int \left[F(\Phi'_i) - F(\Phi_i) - \frac{\partial F}{\partial \Phi_i} (\Phi'_i - \Phi_i) \right] dA +$$

$$+ \lambda \int \left[D(q, \Phi'_i) - D(q', \Phi'_i) - \frac{\partial D}{\partial q'} \cdot (q - q') \right] dA -$$

$$- \lambda \int \left[D(q, \Phi'_i) - D(q, \Phi_i) - \frac{\partial D}{\partial \Phi_i} (\Phi'_i - \Phi_i) \right] dA. \tag{9}$$

It is obvious from (9) that $I' > I$ if the first two integrands of (9) are positive-definite whereas the last one is negative-definite. Thus the sufficient stationarity conditions (5) correspond to global optimum provided:

(i) the specific cost is a convex function of design variables,
(ii) the specific dissipation power is a saddle function of q and Φ_i i.e. a convex function of q and a concave function of Φ_i.

The conditions (i) and (ii) are *sufficient conditions* for the global minimum of I. To illustrate these conditions, consider the case of pure flexure and only one design variable which can be identified with plate thickness, $\Phi = h$. Then $F = a + bh$, (a, and b are constants), $D = D^0(q)h^n$, where $n = 1$ for sandwich sections with variable sheet thickness h and $n = 2$ for solid sections. The stationarity conditions (5) give

$$D^0(q)h^{n-1} = \text{const.} \tag{10}$$

Fig. 1 illustrates the sufficient optimality conditions. While the convexity of cost functions with respect to Φ_i and the dissipation function with respect to q occur, there is no general concavity property of D with respect to Φ_i. (see Fig. 1c). In fact, for $n \leq 1$ the dissipation function is concave with respect to h and the global minimum of I occurs. On the other hand, for solid sections we have $n \geq 1$ and the dissipation function is convex with respect to both q and h. Then the stationarity condition may not correspond to an optimal design. This point has been clearly illustrated in [1].

Let us now introduce geometric constraints. First, assume that the design variables remain within the intervals $\Phi_i^- \leq \Phi_i \leq \Phi_i^+$, Fig. 1a. The stationarity conditions (5) can easily be modified by introducing penalty costs for designs lying outside this interval. Increasing slopes of penalty cost functions, we finally force the design variables to lie within prescribed intervals. Instead of (5), we now have

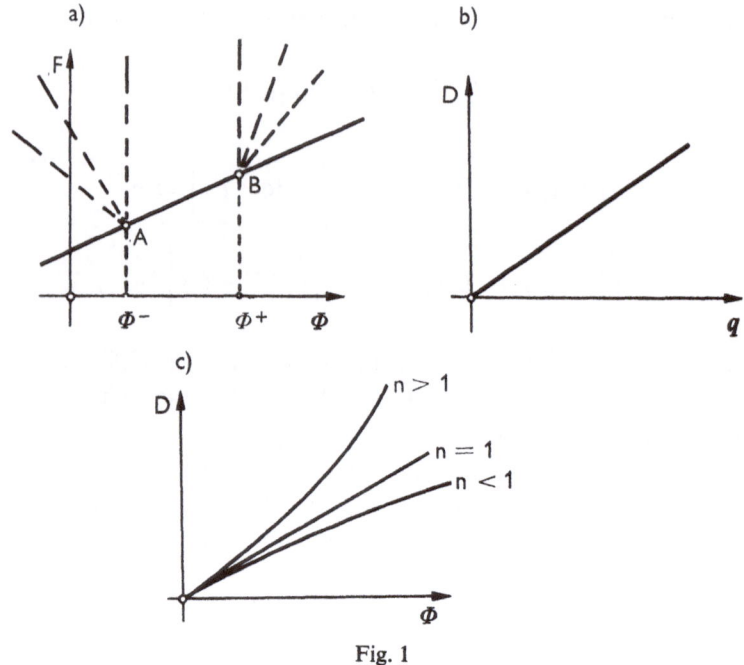

Fig. 1

$$\left.\frac{\partial F}{\partial \Phi_i}\right\}\begin{array}{c}\geqq\\=\\\leqq\end{array}=\lambda\frac{\partial D}{\partial \Phi_i}\left\{\begin{array}{l}\Phi_i=\Phi_i^-\\\Phi_i \text{ unconstrained}\\\Phi_i=\Phi_i^+\end{array}\right.\tag{11}$$

Consider now the parametric constraints. Let the design variables be speci-fied in terms of assumed shape functions and unknown design parameters. Thus $\Phi_i = \Phi_i(a_k)$, $k = 1, 2, ..., s$. The stationarity conditions now take the form

$$\int \frac{\partial F}{\partial \Phi_i}\frac{\partial \Phi_i}{\partial a_k}\,\mathrm{d}A = \lambda\int \frac{\partial D}{\partial \Phi_i}\frac{\partial \Phi_i}{\partial a_k}\,\mathrm{d}A, \qquad (k = 1, 2, ..., s).\tag{12}$$

Consider, for instance, the case when the median surface is divided into s sub-regions over each of which the variable thickness is given by $h_j = \beta_j(x)a_j$ where a_j denotes the design parameters and $\beta_j(x)$ are assumed shape func-tions. The total cost and the dissipation power are given by

$$I = \sum_{j=1}^{s}\int \beta_j(x)a_j\mathrm{d}A_j, \quad D = \sum_{j=1}^{s}\int D^0(q)\beta_j^n a_j^n\mathrm{d}A_j\tag{13}$$

and the stationarity conditions (12) take the form

$$\frac{a_1^{n-1} \int D_1^0 \beta_1^n \mathrm{d}A_1}{\int \beta_1 \mathrm{d}A_1} = \frac{a_2^{n-1} \int D_2^0 \beta_2^n \mathrm{d}A_2}{\int \beta_2 \mathrm{d}A_2} = \dots \tag{14}$$

For $n = 1$ (sandwich sections), (14) becomes

$$\frac{\int D_1^0 \beta_1 \mathrm{d}A_1}{\int \beta_1 \mathrm{d}A_1} = \frac{\int D_2^0 \beta_2 \mathrm{d}A_2}{\int \beta_2 \mathrm{d}A_2} = \dots = \frac{\int D_i^0 \beta_j \mathrm{d}A_j}{\int \beta_j \mathrm{d}A_j}. \tag{15}$$

In other words, the stationarity condition requires the mean dissipation powers in each sub-region to be equal. This condition generalizes previous criteria derived for piecewise constant cross-section of a structural member.

The present approach starting from the dissipation function possesses certain advantages when compared with static approach used by Rozvany [2] and previously by Marçal and Prager [3]. It is assumed that the cost function is explicitly expressed in terms of generalized stresses, $F = F(Q)$. However, in many cases it may occur that the number of design variables may exceed that of generalized stresses and in order to eliminate design variables from the cost function, some local optimization should be carried out. In the present approach, the optimality condition follows directly from the set of relations which express Φ_i in terms of q. Further, the optimality conditions, such as (14) and (15) are characterized by simple mechanical interpretation. It seems therefore that these two approaches, although equivalent, should be more carefully investigated and their relative advantages should be studied.

References

[1] Kozłowski, W. and Z. Mróz, Optimal design of solid plates, *Int. Journ. Solid. Struct.* *5* (1969) 781–794.
[2] Rozvany, G. I. N. and S. R. Adidam, Recent advances in plastic optimal design, *Foundations of Plasticity*, Noordhoff, Leyden, 1973, 201–217.
[3] Marçal, P. V. and W. Prager, A method of optimal plastic design, *J. de Méc. 3* (1964) 509–530.

Coupled thermo-plastic problems

B. E. POBEDRIA

Moscow State University, Moscow, USSR

According to Ilyushin's deformation theory of plasticity [1] stress-strain relations in an active loading process are

$$s_{ij} = 2G[1 - \omega(\varepsilon_u, T)]e_{ij}$$
$$\sigma = K(\theta - 3\alpha\vartheta) \tag{1}$$

where

$$s_{ij} = \sigma_{ij} - \sigma\delta_{ij}; \quad e_{ij} = \varepsilon_{ij} - \tfrac{1}{3}\theta\delta_{ij};$$
$$\sigma = \tfrac{1}{3}\sigma_{kk}; \quad \theta = \varepsilon_{kk}; \quad \sigma_u = \sqrt{s_{ij}s_{ij}}; \quad \varepsilon_u = \sqrt{e_{ij}e_{ij}}$$

and G, K stand for the elastic moduli, ϑ – denotes temperature increase, ω is Ilyushin's plasticity function. The stress-strain relations in unloading are simply given by the Hooke law. Substituting relations (1) in the equations of motion and taking into consideration the Cauchy relations for small deformations three equations are obtained for components of the displacements vector u and temperature T

$$\sigma_{ij,j}\{u, T\} + \rho F_i = \rho \ddot{u}_i(0). \tag{2}$$

According to the first law of thermodynamics.

$$\rho C_p \dot{T} - (\lambda T_{,i})_{,i} = -3[\alpha T_0 \sigma]^{\cdot} + \rho q + W^* \tag{3}$$

where q – stands for the heat source per unit mass, W^* denotes the dissipation function.

It is proved, that in the deformation theory of plasticity W^* is given in [2]

$$W^* = \sigma_u[\omega\varepsilon_u]^{\cdot}.$$

430

The following coupled problem of thermo-plasticity is formulated: to solve the system (2), (3) with appropriate boundary and initial conditions. The existence and uniqueness of the generalized solution in the Sobolev space $\mathring{W}_2^{(1)}$ can then be proved.

Three examples illustrating the effect of coupling were considered by the writer.

1. A simple stretching of the thermally isolated beam by the loading $P(t)$ increasing with time. The linear hardening materials were considered for which

$$\omega = (1-A)\left[1 - \frac{\varepsilon_s}{\varepsilon_u}\right]h(\varepsilon_u - \varepsilon_s)$$

where $0 < A \leqq 1$, h – the Heaviside function. The temperature of the specimen due to stretching was determined. It is found that the steel 38XA specimen under loading equal 80 percent of its ultimate strength σ_b warms up about 10°C, while in purely elastic stretching the specimen cools by 0,5°C.

The following approximate formula is obtained for the temperature increase of the specimen (in °C) if the specimen is loaded to rupture

$$\vartheta(t) = \kappa \frac{1-A}{A}.$$

For steel $\kappa \approx 0,8$.

2. A torsion of a circular bar. Having found a torque for the known parameters of the problem the heat conduction equation was solved with known heat source, which is the dissipation function W^*.

The resulting equation was solved by the Green function method.

3. A spherical cavity in an elastic-plastic medium under internal pressure. The plastic properties of the material were assumed to depend on temperature. The problem is solved by the successive approximation method and its convergence of the solution was proved.

References

[1] Ilyushin, A., Plasticity. Pergamon Press, New York, 1950.
[2] Pobedria, B. E., On coupled problems of continuum mechanics (in Russian). In *Uprugost i nieuprugost*, No. 2, MGU, Moscow 1971.

Un exemple de non validité de la théorie classique des charges limites pour un système non standard

JEAN SALENÇON

Ecole Polytechnique, Paris, France

Introduction. Le calcul à la rupture couramment utilisé en résistance des matériaux et en mécanique des sols repose sur les théorèmes suivants:

Considérant une structure S soumise à un système de charges proportionnel à un paramètre $\lambda > 0$, soit $\lambda\Phi$[1]) alors:

Théorème 1 – si un système de charges $(\lambda_s\Phi)$ est tel que l'on puisse trouver un champ de contraintes qui l'équilibre sans excéder nulle part la limite de rupture, la structure résistera à ce chargement (i.e. $\lambda_s \leqq \lambda_1$ valeur limite).

Théorème 2 – si on connaît un mode de rupture admissible pour la structure, dans lequel un système de charges $\lambda_c\Phi$ est suffisant pour produire la ruine (i.e. puissance des forces motrices excédant la puissance des forces résistantes), la structure ne résistera pas à ce chargement (i.e. $\lambda_1 \leqq \lambda_c$).

Ces théorèmes qui peuvent paraître intuitifs et conformes au bon sens expriment[2]) les résultats de la théorie classique des charges limites (cf. [4, 10]) et on sait que ceux-ci ne peuvent être établis que dans l'hypothèse du *principe du travail maximal*[3]), qui doit être vérifié par les lois du comportement plastique des matériaux constitutifs, et par les conditions de frottement aux interfaces entre les solides de la structure étudiée [2, 9, 10].

Rappelons que lorsqu'il n'en est pas ainsi, des théorèmes ont été donnés [2, 7, 8, 6, 10] permettant d'encadrer la zone possible pour les charges limites, l'unicité de λ_1 ne pouvant plus être établie.

Les cas se produit couramment en mécanique des sols où l'on a affaire à des matériaux non standards. Les difficultés d'application des théorèmes

---

[1]) λ apparaît comme le coefficient de sécurité de S par rapport à Φ, [1].

[2]) Dans le cas d'un système de charges proportionnel à un paramètre.

[3]) Rappelons que cette condition est équivalente à la convexité du critère de plasticité et à la 'normalité' de la loi de comportement (matériau dit standard [7, 8]).

limites classiques apparaissent alors dans l'utilisation de la méthode ciné-matique (méthode des cercles de glissement etc.): l'évaluation de la puis-sance des forces résistantes fait l'objet de nombreux errements, la puissance dissipée n'étant pas une fonction univoque de la vitesse de déformation.

On étudie ici un problème où le principe du travail maximal est satisfait par la loi de comportement du matériau constituant le système, tandis que la condition de frottement à l'interface est non standard.

Le problème posé. On étudie en déformation plane, la force limite de poin-çonnement d'un demi-espace en matériau de Tresca standard non-pesant de cission limite k, par un poinçon de largeur $2a$ placé dans une entaille rec-tangulaire de largeur $2a$ et de profondeur h (problème de la fondation peu profonde). A l'interface de contact $(BAA'B')$ entre le poinçon et le demi-plan, la liaison est unilatérale avec frottement de Coulomb; avec les nota-tions de la Fig. 1, on a donc la règle suivante:

$$\left.\begin{array}{l} \sup\{\sigma,\ \tau^2-\sigma^2\operatorname{tg}^2\varphi,\ \tau^2-k^2\}<0 \quad : \quad [u_n]=[u_t]=0 \\[2mm] \left.\begin{array}{l}\sup\{\sigma,\ \tau^2-\sigma^2\operatorname{tg}^2\varphi,\ \tau^2-k^2\}=0\\[1mm]\sigma<0\end{array}\right\} : \left\{\begin{array}{l}[u_n]=0,\\[1mm]\tau\cdot[u_t]\geqq 0\end{array}\right. \\[4mm] \sigma=0,\quad \tau=0 \qquad\qquad : \quad [u_n]\geqq 0,\ [u_t]\ \text{arbitraire} \end{array}\right\} \tag{1}$$

On supposera dans la suite,

$$\operatorname{tg}\varphi > 0{,}39. \tag{2}$$

La Fig. 1a représente cette condition de frottement: critère de glissement et loi de glissement associée. On voit que la règle de normalité n'est pas satisfaite [10]: cette condition n'est pas standard.

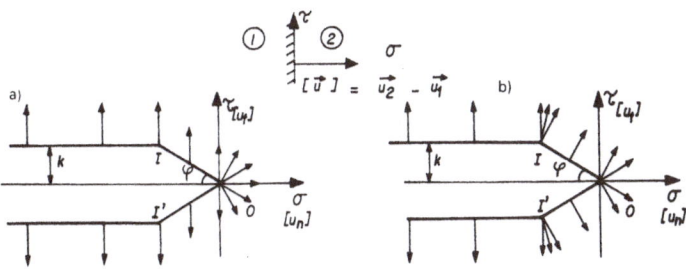

Fig. 1

(La Fig. 1b représente pour le même critère de glissement, ce que serait la loi de glissement pour un interface standard: discontinuité de vitesse purement tangentielle uniquement pour $|\tau| = k$; discontinuité de vitesse avec décollement inclinée à φ sur l'interface pour $|\tau| = -\sigma\,\mathrm{tg}\,\varphi$, points coniques, en 0 d'ouverture $(\pi - 2\varphi)$ seulement et en I et I').

Solution statique. Le champ de contraintes symétrique dont la partie droite de la Fig. 2 représente la moitié, fournit une solution statique du problème, (analogue au champ proposé dans [12]).

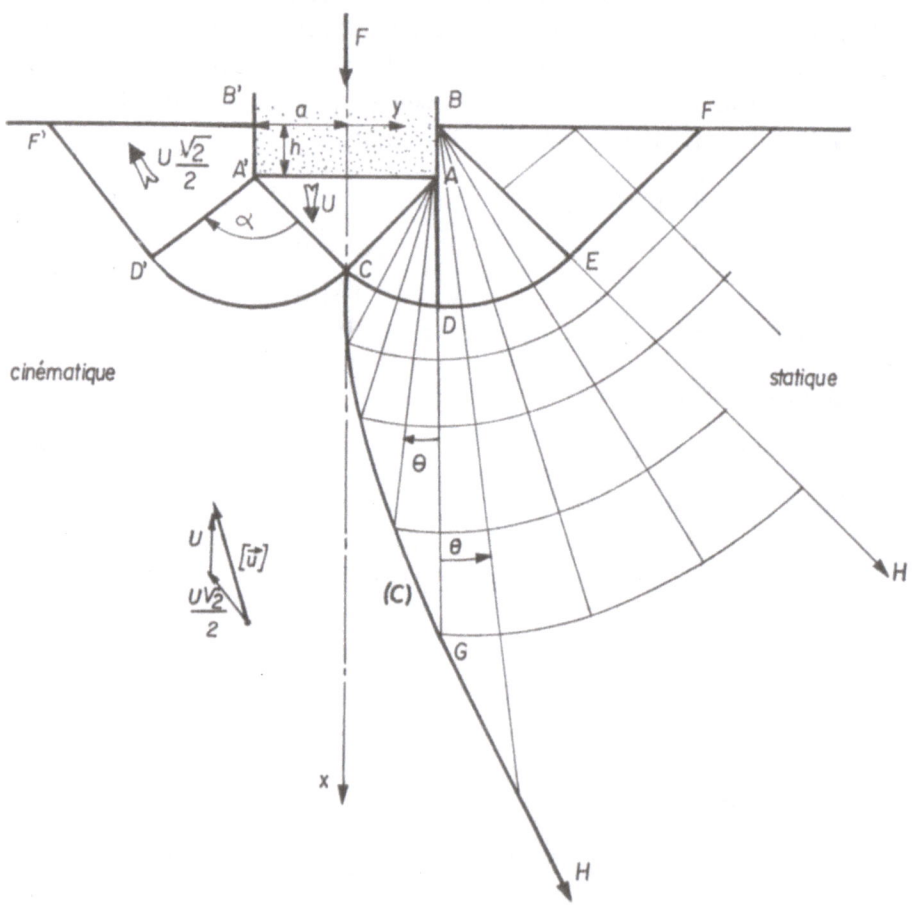

Fig. 2

Dans *BEF* le champ est homogène; *BDE*, éventail de Prandtl de sommet *B* et d'ouverture $\pi/4$; *AOC*, éventail de Prandtl de sommet *A* et d'ouverture $\pi/4$; *A'AC*, champ homogène.

La contrainte sur *A'A* est: $\tau = 0$, $\sigma = -(\pi+2)k$ qui est admissible; sur *AB* on a: $\tau = k$, $\sigma = -(\pi/2+1)k$ qui est admissible pour tg $\varphi > 0,39$.

Au-dessous de *CDEF* le champ de contraintes statiquement et plastiquement admissible est obtenu par la méthode de Shield [11]: on détermine une ligne de discontinuité du champ de contraintes (*C*), issue de *C*, telle que l'on obtienne un prolongement licite en utilisant au-dessus de (*C*) les mêmes champs homogènes, semi-homogènes, que dans *ACD*, *BCD*, *BDE*, *BEF* et au-dessous de (*C*) le champ statiquement admissible, vérifiant $\tau_{xy} = 0$, défini par la continuité de la contrainte sur (*C*) et la condition $(\sigma_x - \sigma_y) = 2k$ sur (*C*) dans ce champ.

Ici la courbe (*C*) se compose de deux arcs:

– l'arc *CG* du prolongement de Shield classique pour un champ de Prandtl centré en *A* et de rayon $a\sqrt{2}$, dont l'équation en coordonnées polaires de centre *A* et d'axe *AD* est ([9]):

$$\rho = a\sqrt{2}\bigg/\left(\sin\left(\frac{\pi}{4} - \theta\right)\cdot\text{tg}\left(\frac{\pi}{8} - \frac{\theta}{2}\right)\right) \tag{3}$$

pour $-(\pi/4) \leqq \theta < 0$;

– l'arc *GH* homothétique (centre *G*, rapport $\overline{GB}/\overline{GA}$) de l'arc d'équation (3) pour $0 \leqq \theta \leqq \pi/4$, et qui a ainsi pour équation en coordonnées polaires (*B*, *BD*):

$$\rho = \left(a\sqrt{2}+h\left(1 - \frac{\sqrt{2}}{2}\right)\right)\bigg/\left(\sin\left(\frac{\pi}{4} - \theta\right)\cdot\text{tg}\left(\frac{\pi}{8} - \frac{\theta}{2}\right)\right) \tag{4}$$

$0 \leqq \theta \leqq \pi/4$;

ces deux arcs se raccordent tangentiellement en *G*.

Les valeurs de σ_x, σ_y sur (*C*) sont, pour chaque valeur de θ, les mêmes que dans le prolongement de Shield classique: on est donc assuré que dans la zone située au-dessous de (*C*) le champ de contraintes est plastiquement admissible.

On dispose ainsi d'un champ de contraintes statiquement et plastiquement admissible. L'intégration des contraintes sur *B'A'AB* fournit la valeur de la force de poinçonnement correspondant à cette solution statique:

$$F_s(h/a) = 2ak(\pi + 2 + h/a).\qquad(5)$$

Remarquons au passage que la solution statique considérée ici est celle qui donne la plus forte valeur de $F_s(h/a)$, parmi toute la classe schématisée sur la Fig. 3 (où le champ de contraintes au-dessous de *CDEF* est encore obtenu par la méthode de Shield [9]).

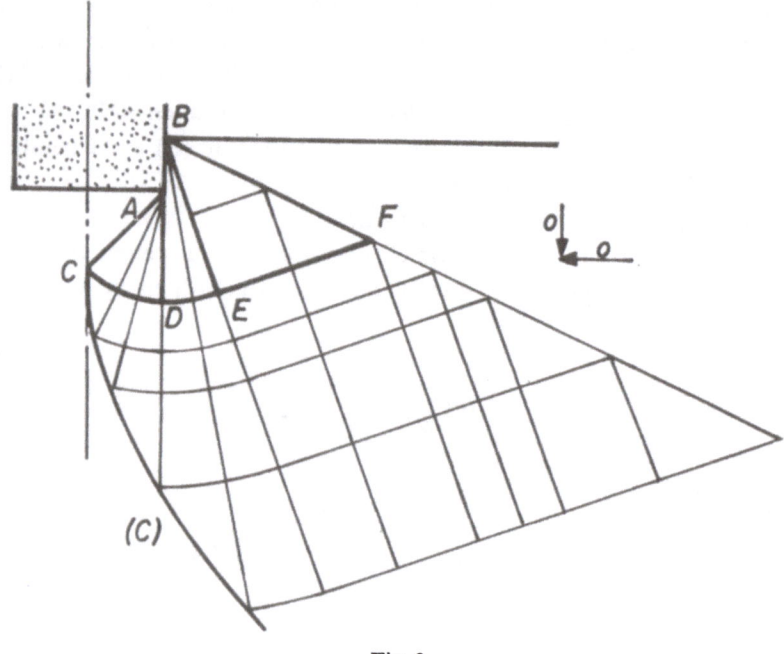

Fig. 3

Solution cinématique. Le schéma d'écoulement symétrique dont la partie gauche de la Fig. 2 représente la moitié fournit une solution cinématique du problème. Il s'agit d'un mode de déformation classiquement utilisé ([5, 3] par exemple): il y a enfoncement du bloc $A'AC$ à la vitesse verticale U du poinçon; déformation plastique dans l'éventail $A'CD'$ d'ouverture α; discontinuité de la vitesse $(= U/\sqrt{2})$ le long de $ACD'F'$, et $A'D'F'B'$ glisse en bloc le long de $D'F'$; la discontinuité de vitesse entre sol et poinçon le long de $A'B'$ se compose donc de $U(\sqrt{2}/2)$ selon $D'F'$ et de U selon la verticale (cf. Fig. 2): il y a décollement.

Pour l'application de la méthode cinématique classique, le calcul de la puissance des forces résistantes dans ce mode de déformation ne pose aucun problème: le matériau étant standard la puissance dissipée a une expression univoque dans la zone déformée et le long des lignes de discontinuité $A'C$ et $CD'F'$; enfin le long de $A'B'$ où la discontinuité de vitesse correspond à un décollement, on a $\sigma = \tau = 0$ d'après (1), (cf. Fig. 1a), et la puissance dissipée y est *nulle* sans aucune ambiguïté.

Egalant la puissance des forces motrices à la puissance des forces résistantes, on obtient pour ce mécanisme:

$$F_c(h/a, \alpha) = 2ak\left((1+2\alpha+\text{tg}(3\pi/4-\alpha) + \frac{h}{a}\,\frac{\sqrt{2}}{2}\,\frac{1}{\cos(3\pi/4-\alpha)}\right). \quad (6)$$

On voit que:

$$F_c(h/a, \pi/2) = 2ak(2+\pi+h/a) = F_s(h/a). \quad (7)$$

D'autre part, l'étude de $F_c(h/a, \alpha)$ sur $0 \leqq \alpha \leqq 3\pi/4$, à h/a fixé montre que $F_c(h/a, \alpha)$ passe par un minimum pour $\alpha = \alpha_0$;

$$\alpha_0 = 3\pi/4 - \arcsin\left(-\frac{h}{a}\,\frac{\sqrt{2}}{8} + \frac{\sqrt{2}}{8}\sqrt{h^2/a^2+16}\right) \quad (8)$$

croît de $\pi/2$ à $3\pi/4$ quand h/a croît à partir de 0.

La solution correspondant à $\alpha = \alpha_0$ est donc la meilleure de la classe considérée, puisque fournissant la valeur de F_c la plus faible.

La Fig. 4 représente $F_c(h/a, \alpha_0)$ en fonction de h/a.

Il résulte évidemment de (7) que l'on a:

$$F_c(h/a, \alpha_0) < F_s(h/a) \quad \text{pour} \quad h/a > 0. \quad (9)$$

Ainsi les résultats de la théorie classique des charges limites sont en défaut puisque le chargement obtenu par la méthode statique est supérieur à celui obtenu par la méthode cinématique.

Cela provient évidemment du caractère non-standard de la condition de frottement sur $BAA'B'$. On vérifie facilement par exemple que dans l'hypothèse, sans signification pratique, d'un interface standard de même critère (Fig. 1b), la puissance dissipée le long de $B'A'$ n'est plus toujours nulle, bien qu'il y ait décollement: suivant les valeurs relatives de α et φ, la discontinuité de vitesse peur appartenir au cône des normales extérieures en I

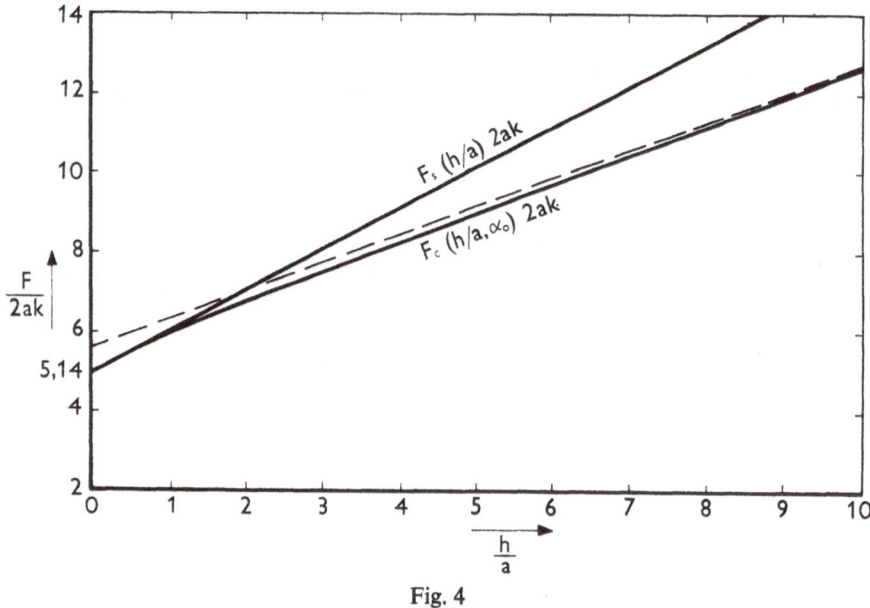

Fig. 4

et il y a dissipation qui conduit à modifier l'expression de $F_c(h/a, \alpha)$ par rapport à (6) en:

$$F_c(h/a, \alpha) = 2ak\left[1 + 2\alpha + \text{tg}\left(\frac{3\pi}{4} - \alpha\right) + \frac{h}{a}\,\frac{\sqrt{2}}{2}\,\frac{1}{\cos(3\pi/4 - \alpha)}\right] +$$

$$+ 2ak \times \text{Max}\left\{0, 1 + \frac{\sqrt{2}}{2}\cos\left(\frac{3\pi}{4} - \alpha\right) - \frac{\sqrt{2}}{2}\,\frac{\sin(3\pi/4 - \alpha)}{\text{tg}\,\varphi}\right). \quad (10)$$

Il s'ensuit, compte-tenu de la condition (2) sur φ, que la plus faible valeur de $F_c(h/a, \alpha)$ ainsi obtenue est supérieure à $F_s(h/a)$.

Conclusion. La présente note attire l'attention sur les dangers de l'application des théorèmes limites classiques, dans le cas de matériaux satisfaisant le principe du travail maximal et d'interfaces non standards: en particulier, sans paraître à aucun moment poser de problèmes, une telle utilisation peut conduire à des paradoxes.

L'exemple présenté n'a rien de pathologique, à la différence de ceux donnés dans [2]; il s'agit d'un problème banal dont on ne connaît pas de solution complète, et les schémas que nous avons utilisés dans l'étude sont tout à fait habituels.

Jean Salençon

Le praticien de la mécanique des sols, pour qui il s'agit là de la force portante d'une fondation peu profonde, notera sur la Fig. 4 que dans la gamme des valeurs usuelles ($h/a \leq 2$ par exemple), les résultats donnés par les deux solutions diffèrent peu. Mais la question reste posée de façon générale de savoir quelle approche on devra choisir et quel résultat on devra adopter; en particulier l'usage de la méthode statique n'est plus une garantie de sécurité.

References

[1] Courbon, J., Plasticité appliquée au calcul des structures. Cours E.N.P.C., Paris 1971.
[2] Drucker, D. C., Coulomb friction, plasticity and limit loads. *Jl. Appl. Mech. Trans. A.S.M.E., vol. 21, 1* (1954) 71–74.
[3] Hansen, Bent, Bearing capacity of shallow strip footings in clay, *C.R. 7ème congr. int. Mec. Sols, Mexico, Vol. 2* (1969) 107–113.
[4] Mandel, J., Cours de Mécanique des milieux continus, Tome II Annexe 20. Gauthier-Villars, Paris, 1966.
[5] Meyerhof, G., The ultimate capacity of foundation, *Géotechnique, Vol. 2, no 4* (1950/1951) 301–332.
[6] Palmer, A. C., A limit theorem for materials with non-associated flow-laws, *J. de Mécanique, Vol. 5, 2* (1966) 217.
[7] Radenkovic, D., Théorèmes limites pour un matériau de Coulomb à dilatation non-standardisée. *C.R. Ac. Sc. Paris, 252* (1961) 4103–4104.
[8] Radenkovic, D., Théorie des charges limites in *Séminaire de Plasticité, ed.* J. Mandel, P.S.T., no 116, 1962.
[9] Salençon, J., La théorie des charges limites dans la résolution des problèmes de plasticité en déformation plane. Thèse Doc. ès-Sc., Paris, 1969.
[10] Salençon, J., Théorie de la Plasticité pour les applications à la Mécanique des Sols. Cours E.N.P.C., Paris, 1972.
[11] Shield, R. T., *J. Appl. Mech. 21* (1954) 193 et *cf.* A. Philipps, Introduction to plasticity, Ronald Press, N.Y. (1956) 158.
[12] Sokolovski, V V., Statics of Soil Media. Butterworths – London (1960) 118.

Elastic-plastic transition

B. R. SETH

Birla Institute of Technology, Ranchi, India

In a number of papers like those of J. Kratochvíl and T. Tokuoka presented at the symposium the current tendency to treat plasticity as a separate subject could be noticed. The usual assumptions of incompressibility and yield conditions are made in almost all papers. The fully plastic state is a limiting state like that of ideal fluid flow. It should be obtained from the transition state, in which both linear and plastic effects dovetail into each other in a non-linear manner, by assuming the response coefficients to approach a particular value.

In an elastic medium, undergoing deformation, the strain and rotation are at first small and the strain is pretty well linear. As the parameter of deformation increases the strain changes from linear to non-linear in character. In cases like torsion secondary normal stresses, absent in the linear theory, set in. With a further increase the material starts to yield, which may be called the transition state. The properties of the material change and new constitutive relations come into existence.

Analytically, it should be possible to identify the yielding with the critical points of the differential system defining the medium. If these points are not obvious in one system it should be possible, by continuous mappings to recognise them in some state plane.

Geometrically, the reciprocal deformation ellipsoid can be expected to indicate the transition (or critical points) when one or more of its principal axes tend to become zero or infinite. In linear theories the non-linear terms in the strain components are neglected. These terms, being rotational in character, should be retained, as rotation or spin effects are very significant during transition.

During transition the physical quantities like the stresses and strain take on asymptotic values. Such values have to be determined at the transition points. The new constitutive equations can then be deduced from them. It now becomes unnecessary to assume a yield condition to determine the elastic-plastic boundary.

As regards yield conditions, we should use the strain tensor field which is defined for any type of medium-homogeneous, heterogeneous, isotropic and aelotropic. A change from elastic to plastic deformation can be interpreted as a mapping of one state into the other and the yield an asymptotic subspace. In other words, at yield the macroelement breaks down, and the transformation matrix becomes singular. In terms of the invariants of the strain-tensor field, J_1, J_2, J_3 it is possible to show that the yield condition is of the form

$$8J_3 - 4J_2 + 2J_1 = 1.$$

In a large number of cases including that of aelotropic materials this reduces to a generalized form of tresca yield condition given by –

$$\tau_{11} - k_0\tau_{33} = k_1,$$

where $\tau_{11} > \tau_{22} > \tau_{33}$ are the principal stresses and k_0, k_1 functions of the response coefficients.

A number of problems including those of spherical and cylindrical shells, rotation of cylinders, sheet bending and collapse of thick cylinders have been successfully treated by the transition method.

To sum up –

(i) Elastic-plastic transition should not be replaced by a yield condition, which may or may not exist.
(ii) When it exists it should be derivable from the field equations as a limiting concept.
(iii) The yield condition in current use neglect the Bauchinger effect, so that the magnitude of the yield stress in tension and compression is the same, which is incorrect.
(iv) The plastic state, being an ideal one, should be derivable as a limiting state from a transition one; it should not be treated separably from the elastic one.
(v) Consistent with the general theory of transition fields it should be possible to show that elastic-plastic transition gives rise to sub-harmonic fields.

Discussion notes

Bibliography

[1] Seth, B. R., Transition theory of elastic-plastic deformation, creep and relaxation, *Nature, 195* (1962) 896–897.

[1a] — Simple case of transition phenomenon, Proc. Eight Army Math. Conf. 1962, Madison, (1963) 409–447.

[2] — Elastic-plastic transition in torsion, MRC Tech. Rep. 302, 1962; *Zeit. Angew. Math. Mech. 44* (1964) 229–233.

[3] — Transition theory of creep and relaxation, MRC Tech. Rep. 391, 1962; 11-1-23.

[4] — Elastic-plastic transition in shells and tubes under pressure MRC Tech. Rep. 295, 1962; *Zeit. Angew. Math. Mech. 43* (1963) 345–351.

[5] — Transition theory of sheet-bending, MRC Tech. Rep. 326, 1962; *Prikl. Mat. Mek. 27* (1963) 380–382.

[6] — On the problem of transition phenomenon – *Bull. Inst. Politechnic, Iasi, Roum. 10* (1964) 255–262.

[7] — Transition phenomenon in physical problems – *Bull. Cal. Math. Soc. 56* (1965) 83–89.

[8] — Generalized strain and transition concepts for elastic-plastic deformation, creep and relaxation Proc. XI Int. Cong. of Applied Mechanics, Munich, 1964 383–389.

[9] — Continuum concepts of measure. Presidential Address, Xth Cong. of Theo. and Appl. Mechanics., Madras, 1965, 1–16.

[10] — Measure concept in mechanics. *Int. Jour. of Non-linear Mechanics, 1* (1966) 35–40.

[11] — Irreversible transition in continuum mechanics, IUTAM Symposia, Vienna, 1966, 359–366.

[12] — Plane transitions. *Ind. J. Math. 9* (1967) 499–504.

[13] — Transitions-basic concepts. *Ind. J. Mech. and Math. Part I.* (1968) 11–16.

[14] — Space transisions – Problems of hydrodynamics and continuum mechanics, Sedov anniversary volume, Moscow, 1969, 453–458; SIAM 1966, 632–638.

[15] — New concepts in continuum mechanics, Trans. XIV Conf. Army Math., 1969, 283–303. *Bull. Cal. Math. Soc. 62* (1970) 49–58.

[16] — Aelotropic plasticity, Novozhilov anniversary volume, Moscow, 1970, 419–425.

[17] — Transition conditions – The yield condition. *Int. J. Non-linear Mech. 5* (1970) 279–285.

[18] — Transition analysis of collapse of thick cylinders, *Zeit. Angew. Math. Mech. 50* (1970) 617–621.

[19] — Creep rupture. IUTAM symposium on creep in structures, Gothenburg, 1970, Springer-Verlag, 167–169.

[20] — Creep transition, Muskhelishvilli, anniversary volume Moscow, 1971, 463–468.

[21] Hsu, T. C., S. R. Davies and R. Royles, A study of the stress-strain relationship in the work-hardening range, ASME, Paper No. 66-WA/Met. 1.
Hsu, T. C., Definition of the yield point in plasticity and its effects on the shape of the yield locus. *Jour. Strain Analysis, 1* (1966) 331–338.

[22] Purushothama, C. M., Elastic-plastic transition, *ZAMM, 45* (1965) 401–408.

[23] Hulsurker, S., Transition theory of creep of shells under uniform pressure, *ZAMM*, *46* (1966) 431–437.

[24] Narasimhan, M. N. L. and K. S. Sra, Generalized measures of deformation-rates in viscoelasticity and their applications to rectilinear and simple shearing flows. *Ind. J. Non-linear Mech.*, *4* (1969) 361–372.

[25] Narasimhan, M. N. L. and R. N. Knoshaug, Generalized measures of deformation-rates in secondary flows of viscoelastic fluids between rotating spheres. *Ind. J. Non-linear Mech.*, *7* (1972) 161–174.

[26] Borah, B. N., On the yield condition and transition fields in elastic-plastic deformation of solids. *Ind. J. Pure and App. Math.*, *2* (1971) 335–343.

[27] — Thermo-elastic-plastic transition tubes under uniform pressure and steady state temperature, *Jour. Math. Phy. Sci. 4* (1970) 288–301.

[28] — Thermo-elastic-plastic transition of shells under uniform pressure and steady state temperature. *Ind. J. Pure and App. Maths. 3* (1972) 82–91.

[29] Seth, B. R., Transition theory of elastic plastic deformation, Extension lectures, Osmania University, Hyderabad, 1964, pp. 1–64.

[30] — Transition problem of aelotropic yield and creep rupture, CISM, course and lectures, No. 47, Udine, Springer, 1970, 1–43.

The constitutive aspects of aging metals

Donald C. Stouffer and Alvin M. Strauss

University of Cincinnati, Cincinnati, Ohio, U.S.A.

Introduction. This discussion is directed toward introducing a constitutive theory for metals describing the effect of aging on deformations in the plastic range [1]. The general approach presented here is phenomenological and axiomatic, and is directed towards establishing a firm theoretical foundation for all non-linear aging simple materials. The generalized results can then be specialized for application to plastically deformed metals.

Most of the classical theories of plasticity are time-translation invariant and consequently aging effects are not included. It is well known, however, that aging effects may play an important role in the deformation process of metals. For example, Lazan [2] reported that in metals significant intrinsic changes in the microstructure can occur in time and that this 'aging' can affect the mechanical properties of the material. Clearly these effects can be important and invalidate mathematical models that are assumed to remain unchanged in time.

Aging simple materials. In an aging constitutive theory it is necessary to clearly define the time variables used in its development. Let t_c represent the beginning of the aging phenomenon in the material, let t represent the present time, and let τ represent any time between or including t_c and t; that is, $\tau \in [t_c, t]$. The time t_c is defined so that no event prior to t_c is considered relevant to the constitutive theory. Using these definitions, it can be shown from the time-translation portion of the principle of material objectivity, [3], that the elapsed time $t - t_c$ is an appropriate measure of the 'age' of the material. Thus on selecting the origin of the time scale at t_c (i.e.; $t_c = 0$) the time measure t assumes the special dual definition of the age of the material and the present time.

Consider now a simple material that admits aging effects. Mathematically this implies that at some time t the Kirchoff stress tensor $S(t)$ can be determined by the history of the Lagrangian stain tensor $E(\tau)$ for all $\tau \in [0, t]$, and that the stress depends explicitly upon the age of the material t. Thus the objective constitutive equation for a simple aging material can be written as

$$S(t) = \underset{\tau=0}{\overset{t}{Q}}[E(\tau), t] \tag{1}$$

where Q is a functional that completely characterizes the mechanical properties of the material.

To analyze the influence of the aging in the constitutive equation (1) it is convenient to assume that the non-zero part of the plastic deforming history is very short compared to the aging time scale so that the aging effect can be neglected during the deforming process. Let $\hat{E}(\tau)$ be a strain history initiated at $\tau = 0$. Thus $\hat{E}(\tau) = 0$ for $\tau \in (-\infty, 0)$ and $\hat{E}(0) \neq 0$ for τ belonging to the small time interval $[0, \hat{t}]$. The stress $\hat{S}(\hat{t})$, calculated from equation (1) using $\hat{E}$, is the stress in an 'unaged' material or a material with a standard amount of aging. Designate $\hat{E}$ and $\hat{S}$ as reference or base histories and assume

$$\hat{S}(\hat{t}) = \underset{\tau=0}{\overset{\hat{t}}{Q}}[\hat{E}(\tau); \; -] \equiv \underset{\tau=0}{\overset{t}{Q^{(1)}}}[\hat{E}(t-\tau)] \tag{2}$$

where $Q^{(1)}$ is the functional for the non-aging response of the material.

Consider another strain history E that is identical to $\hat{E}$ but displaced by an interval $t^* = t - \hat{t} \geqslant 0$ along the time scale. The stress history $S(t)$ can be determined from equation (1) as

$$S(t) = \underset{\tau=0}{\overset{t}{Q}}[E(\tau), t] = \underset{\tau=t^*}{\overset{t}{Q}}[E(\tau), t]. \tag{3}$$

The stress history $S(\tau)$ will differ from the reference stress history $\hat{S}(\tau)$ by the perturbation history

$$\Sigma(\tau)|_{t^*}^t = [S(\tau) - \hat{S}(\tau - t^*)]|_{\tau = t^*}^{\tau = t} \tag{4}$$

for aging materials. If the material is non-aging $\Sigma(\tau) = 0$ for every choice of t^* and the material is called time-translation invariant.

These results will allow the constitutive functional Q in equation (1) to be resolved into two functionals; $Q^{(1)}$ as previously defined and $Q^{(2)}$ a functional for the perturbation stress $\Sigma(\tau)$ that results from a change in the material properties due to aging. If t^* is selected as the origin of the time scale, then equation (1) can be replaced by

$$S(t) = Q^{(1)}[E(t-\tau)] + Q^{(2)}[E(\tau), t-t_c]. \qquad \qquad (5)$$
$$\tau=0 \qquad \qquad \tau=0$$

The functional $Q^{(2)}$ will vanish if $t-t_c$ is very small on the aging time scale of if the material is non-aging.

An integral representation for rate independent materials. The constitutive functionals $Q^{(1)}$ and $Q^{(2)}$ can be represented by a series of multiple integrals, [4], provided they satisfy certain continuity requirements. To obtain a convenient working relationship for engineering problems let us assume $Q^{(1)}$ and $Q^{(2)}$ can be adequately represented by the first term of these multiple integral expansions.

The constitutive functional $Q^{(1)}$ can be further specialized for rate-independent materials by introducing an arc-length parameterization. Let

$$p = \gamma(\tau) = \int_0^\tau [E'(\xi) \cdot E'(\xi)]^{\frac{1}{2}} d\xi, \qquad \qquad (6)$$

where ( )′ represents differentiation with respect to time.

Assume γ has a Lipschitz inverse (or is absolutely continuous) and define the inverse as, [5],

$$\tau = \Gamma(p). \qquad \qquad (7)$$

Using the Pipkin-Rivlin [6] definition of rate-independence, which Owen and Williams [7] demonstrated was identical to that of Truesdell and Noll [8], the constitutive functional $Q^{(1)}$ can be replaced by

$$Q^{(3)} = Q^{(3)}[\bar{E}(P-p)] \qquad \qquad (8)$$
$$p=0$$

where $P = \gamma(t)$ and $\bar{E}(p) = E[\Gamma(p)]$. Combining the above results allows the constitutive equation (5) to be rewritten as

$$S = \int_0^P G[P-p]\bar{E}(p)dp + \int_0^t H(\tau, t-t_c)E(\tau)d\tau. \qquad \qquad (9)$$

The stress in (9) is a function of P, t and $t-t_c$. It is possible to eliminate P

by using (6) and (7) in (9) to obtain

$$S(t, t-t_c) = \int_0^t \{G[\gamma(t)-\gamma(\tau)][E'(\tau) \cdot E'(\tau)]^{\frac{1}{2}} +$$

$$+ H(\tau, t-t_c)\} E(\tau) d\tau. \tag{10}$$

The constitutive equation is now, once again, completely in the time domain.

A somewhat more convenient form can be developed if the strain in equation (1) is replaced by the strain rate history $E'(\tau)$. Repeating the above development using E' gives

$$S(t, t-t_c) = \int_0^t \{K[\gamma(t)-\gamma(\tau)] + L(\tau, t-t_c)\} E'(\tau) d\tau. \tag{11}$$

If the material is isotropic in the undeformed state, each of the fourth order tensor valued material functions G, H, K, and L can be replaced by scalar valued functions. If the metal responds elastically to hydrostatic loads, as in usually true even for extremely large loads, the dilatation terms can be further simplified. For complete details [1].

References

[1] Stouffer, D. C. and A. M. Strauss, A phenomenological theory of aging effects in metals, *J. Engineering Materials and Technology, Vol. 1* (1973) 107–111.

[2] Lazan, B. J., *Damping of Materials and Members in Structural Mechanics*, Pergamon Press, Oxford, 1968.

[3] Stouffer, D. C. and A. S. Wineman, A constitutive representation for linear, aging environmental-dependent viscoelastic materials, *Acta Mechanica, Vol. 13* (1972) 31–53.

[4] Green, A. E. and R. S. Rivlin, The mechanics of non-linear materials with memory, *Archive for Rational Mechanics and Analysis, Vol. 1* (1957) 1–21.

[5] Pepe, W. D. and A. M. Strauss, Theory of viscoplastic response, *Archives of Mechanics, Vol. 23.3* (1971) 405–412.

[6] Pipkin, A. C. and R. S. Rivlin, Mechanics of rate-independent materials *Journal of Applied Mathematics and physics (ZAMP) Vol. 16* (1965) 313–327.

[7] Owen, D. R. and W. O. Williams, On the concept of rate-independence *Quarterly Applied Mathematics, Vol. 26.3* (1968) 321–329.

[8] Truesdeel, C. and W. Noll, Handbuch der Physik, Vol. III/2, edited by S. Flügge, Springer-Verlag, Berlin, 1965.

Formulation of some homogeneous
thermodynamic processes as variational inequalities

PIERO VILLAGGIO

University of Pisa, Pisa, Italy

1. Introduction. Many problems of continuum mechanics have a particular non-classical formulation in that they require the simultaneous solution of equations and inequalities. These inequalities express the effect of certain restrictions imposed, for instance, by unilateral constraints or by certain conditions of plasticity.

The most general approach to these problems is given by the theory of variational inequalities. The central problem of this theory is to find a vector field u, defined in a subset K of a reflexive Banach space X, such that

$$(A(u), v-u) \geqq 0 \qquad \forall v \in K. \tag{1.1}$$

where A is a non-linear map of X into its dual X' and $(,)$ denotes the pairing between X and X'. If A is monotone and continuous on finite dimentional subspaces of X, then there exists a solution of the problem (1.1), and this solution is unique if A is strictly monotone (Stampacchia [1]).

In this context the evolution of a thermo-elastic body $\mathscr{B}$ under prescribed initial and boundary conditions can be considered. The unknowns of the problem are the displacement of field u relative to an assigned reference configuration $\kappa(\mathscr{B})$ and the temperature field ϑ. The equations of the problem are balance of momentum (Cauchy's equations) and balance of energy. An *a priori* restriction on possible solutions is imposed by the Clausius-Duhem inequality. It is important to note that this point of view differs from that of Coleman and Noll [2]. Indeed, while this classical approach studies those restrictions that must be placed on constitutive equations in order that the Clausius-Duhem inequality hold for every process, the formulation of the problem as a variational inequality renders plausible the existence of processes in circumstances for which the constitutive equation does not satisfy the restrictions imposed by Coleman and Noll's theory.

Whether or not this problem is well-posed rests substantially on the regularity of the operator[1]) and the convexity of the subset K. Since the first property is usually satisfied, it is interesting to study the implications of the second property on the structure of the constitutive equations.

2. Simple homogeneous processes. At present the theory of variational inequalities is not sufficiently wide to include, in a uniform way, all kinds of problems, because the techniques depend crucially on the nature of the specific problem[2]); therefore we consider only a particular class of homogeneous simple processes, that is those described by only one geometrical variable Y and the temperature ϑ, both functions of the time t.

The basic quantities for our study are the *thermodynamic force* $\omega = \hat{\omega}(Y)$, the *internal energy* $\varepsilon = \hat{\varepsilon}(Y, \vartheta)$ and the *entropy* $\eta = \hat{\eta}(Y, \vartheta)$. The assumption that ω depend only on Y is made for simplicity in the application of the theory.

Strong formulation. The equations of the problem are:
(1) The equilibrium equation

$$\ddot{Y} - \hat{\omega}(Y) = B(t), \tag{2.1}$$

where $B(t)$ is the *external force*;
(2) the first law

$$\dot{\varepsilon} + \hat{\omega}(Y)\dot{Y} = Q(t), \tag{2.2}$$

where $Q(t)$ in the *heating*;
(3) the Clausius-Duhem inequality (Truesdell [4])

$$\dot{\psi} + \hat{\omega}(Y)\dot{Y} + \eta\dot{\vartheta} \leq 0, \tag{2.3}$$

where $\psi = \varepsilon - \eta\vartheta$ is the *free energy*.
The boundary conditions are supposed assigned as follows:
(1) The motion is considered in a bounded interval $[0, T]$ and

$$Y(0) = Y(T) = 0; \tag{2.4}$$

---

[1]) That is on its monotonicity and continuity on finite dimensional subspaces.
[2]) According to whether the operator A is hyperbolic, parabolic or elliptic (Lions [3]).

(2) at the initial instant the temperature is prescribed, that is

$$\vartheta(0) = \vartheta_0 > 0. \tag{2.5}$$

Observation. It is evident that the assumption (2.4) is rather restrictive, because it limits the results to the class of motions defined on finite intervals of time which homogeneous displacements at the endpoints.

The hypothesis that ω be independent of ϑ allows us to solve (2.2) with respect to ϑ and to express (2.3) in terms of Υ only, the equation (2.2) can be eliminated and the problem is reduced to finding a function $\Upsilon(t)$ such that the equation (2.1) holds with the restriction

$$\dot{\bar{\psi}} + \omega \dot{\Upsilon} + \eta \dot{\bar{\vartheta}} \leqq 0, \tag{2.6}$$

where $\bar{\psi}$ and $\bar{\vartheta}$ are the new expressions of ψ and ϑ obtained by the solution of (2.3).[1]).

Weak formulation. Let us consider the reflexive Banach space X of the functions $\Upsilon(t)$, which, together with their first derivatives, are square integrable on $[0, T]$. This space is endowed with the norm

$$\| \Upsilon \|^2 = \int_0^T (\Upsilon^2(t) + \dot{\Upsilon}^2(t)) dt. \tag{2.7}$$

Further V is the linear subspace of X defined by functions $\Upsilon(t)$ satisfying (2.4), while K is a subset of V described by functions $\Upsilon(t) \in V$ satisfying the inequality (2.6).

Then, putting

$$A(\Upsilon) = -\ddot{\Upsilon} + \hat{\omega}(\Upsilon) + B(t) \tag{2.8}$$

we see that $\Upsilon(t)$ is a solution of the general problem (1.1).

Conditions of solvability. In addition to the condition of continuity of $A(\Upsilon)$ on finite dimensional subspaces, the solvability of the problem (1.1) is essentially implied by the following two conditions (Stampacchia [1]):

I) There exists a constant $\kappa > 0$ such that

$$(A(\Upsilon_2) - A(\Upsilon_1), \Upsilon_2 - \Upsilon_1) \geqq \kappa \| \Upsilon_2 - \Upsilon_1 \|^2 \qquad \forall \Upsilon_1, \Upsilon_2 \in K. \tag{2.9}$$

[1]) Of course all the conditions ensuring the existence of this process are supposed to be satisfied.

Postulate I. Assuming that, for every $\Upsilon \in V$, the partial derivative $\hat{\omega},_\Upsilon(\Upsilon)$ satisfies the inequality

$$\hat{\omega},_\Upsilon(\Upsilon) \geqq \alpha > 0, \tag{2.10}$$

where α is a constant, the inequality (2.9) automatically holds.

II) The subset $K \subset V$ is convex and closed.

Postulate II. The surface of the equation

$$\dot{\psi} + \omega \dot{\Upsilon} + \eta \dot{\vartheta} = 0 \tag{2.11}$$

is convex in the space of the variables Υ and $\dot{\Upsilon}$. Then, observing that the condition

$$\vartheta = \frac{1}{\varepsilon,_\vartheta} (Q(t) - \varepsilon,_\Upsilon \dot{\Upsilon} - \omega \dot{\Upsilon})$$

follows from (2.3), (2.11) assumes the form

$$\left[\psi,_\Upsilon + \omega - \frac{1}{\varepsilon,_\vartheta} (\psi,_\vartheta + \eta)(\varepsilon,_\Upsilon + \omega) \right] \dot{\Upsilon} + \frac{1}{\varepsilon,_\vartheta} (\psi,_\vartheta + \eta)Q(t) = 0, \tag{2.12}$$

which is convex if and only if (Villaggio [5])

$$\left. \begin{aligned} \psi,_\Upsilon + \omega - \frac{1}{\varepsilon,_\vartheta} (\psi,_\vartheta + \eta)(\varepsilon,_\Upsilon + \omega) &= \kappa_1, \\[2mm] \frac{1}{\varepsilon,_\vartheta} (\psi,_\vartheta + \eta) &= \kappa_2, \end{aligned} \right\} \tag{2.13}$$

where κ_1 and κ_2 are two constants.

3. Conclusion. The equations (2.13) show that, even for certain homogeneous simple processes, the requirement that the thermomechanical problem be well-posed yields certain conditions on the constitutive equations which are less restrictive than those imposed by classical thermodynamics.

References

[1] Stampacchia, G., Variational inequalities. Theory and appl. of monotone operators. Proc. NATO Adv. Study Inst. Ed. Oderisi, 1968.

[2] Coleman, B. D. and W. Noll, The thermodynamics of elastic materials with heat conduction and viscosity. *Arch. Rat. Mech. Anal. 13* (1963) 167–178.

[3] Lions, J. L., Quelques méthodes de résolution des problèmes aux limites non linéares. Dunod et Gauthier-Villars, Paris, 1969.

[4] Truesdell, C., The elements of continuum mechanics. Springer, Berlin, 1965.

[5] Villaggio, P., Formulation of some homogeneous thermodynamic processes as variational inequalities, *Archiwum Mechaniki Stosowanej* (in press).

Stable growth of a crack in a rate-sensitive Tresca solid

MICHAEL P. WNUK

South Dakota State University, Brookings S.D., USA

An extension of McClintock and Rice theory of stable crack growth is proposed for the tensile mode of fracture and small scale yielding condition. Time-dependent phenomena are incorporated so that the combined effect of plastic and viscous deformation may be taken into account. In the first original paper published on this subject in 1958 by McClintock (J. Applied Mechanics, Trans. ASME, Dec. 1958, pp. 528–588) it has been noted that some creeping in biaxially stressed aluminum foils tested for subcritical crack growth, was indeed observed. McClintock's comment was that because of time effects present, variations in testing rate might have contributed to the scatter.

Consider a linear viscoelastic matrix which contains a moving crack. With the exception of high stress regions, where yielding occurs according to the Tresca plasticity condition, the matrix is described by the constitutive equations

$$s_{ij} = \int_{-\infty}^{t} G_1(t-\tau) \frac{\partial e_{ij}(\tau)}{\partial \tau} \, d\tau, \quad s = \int_{-\infty}^{t} G_2(t-\tau) \frac{\partial e(\tau)}{\partial \tau} \, d\tau \qquad (1)$$

where $G_1(t)$ and $G_2(t)$ are the relaxation moduli in shear and isotropic compression respectively, s_{ij} and e_{ij} denote the deviatoric parts of stress and strain tensors, while $\delta_{ij}s$ and $\delta_{ij}e$ are the hydrostatic parts of these tensors. The extended correspondence principle (cf. Graham, Rep. of North Carolina State U., Nov. 1966; and Wnuk and Knauss, Int. J. Solids Structures, 1970, Vol. 6, pp. 995 to 1009) allows one to reduce the viscoelastic displacement in the crack plane to the form

$$u_y(x, t) = u_y^0(x, t) + \int_{t_0}^{t} \frac{\dot{\Omega}(t-\tau)}{\Omega(0)} u_y^0(x, \tau) d\tau \qquad (2)$$

Discussion notes

where $u^0(x, t)$ is the elastic solution to the same boundary value problem (we drop the index y), x is the space coordinate and t denotes current time. Note that t_0 is the instant at which point at x first 'senses' a non-zero displacement due to the approaching crack front. The kernel $\Omega(t)$ is related to the relaxation moduli as follows

$$\Omega(t) = \mathscr{L}^{-1}\left[\frac{2(2G_1^*(s)+G_2^*(s))}{s^2(G_1^*(s)+2G_2^*(s)G_1^*(s))} ; \quad s \to t\right]. \tag{3}$$

Stars symbolize the Laplace transforms and $\mathscr{L}^{-1}$ is the inverse transform. If the Poisson ratio can be regarded constant, then $\Omega(t)$ is shown to reduce to the product of $2(1-v)$ and the creep compliance $D(t)$ available from any standard creep test. We shall denote the normalized compliance $D(t)/D(0)$ by $\psi(t)$.

The following function is introduced as a measure of the damage accumulated at a given material point P, located at $x = x_P$

$$\Phi(\dot{R}, l, t) = \int_{t-\delta t}^{t} S[x = x_P, \tau]\dot{u}[x = x_P, \tau]d\tau \tag{4}$$

Here $S(x_P, \tau)$ is the stress restraining separation of crack faces, δt is the time used by the crack front to traverse its own process zone, $R(t)$ is the length of plastic zone and $l(t)$ is crack half-length. Note that the lower limit of the integral in eq. (4) corresponds to the instant just prior to failure at point P, while the upper limit denotes the time at which P collapses. The length Δ is on the order of a characteristic micro-structural size, compare McClintock and Irwin in 'Plasticity Aspects of Fracture Mechanics', ASTM STP 381, 1964, pp. 84–113. Since Δ is small versus crack size (it is exactly zero only for a perfectly brittle continuum), one may apply the quasi-steady approximation of Glennie and Willis (J. Mech. Phys. Solids, 1971, Vol. 19, pp. 11–30) and divide the unsteady motion of an accelerating crack into a number of constant speed segments, each of which occupies the time interval $\delta t = \Delta/l$.

If the restraining stress is assumed constant ($= Y$) and the RHS of eq. (4) is chosen as $2Yu_0$, in where u_0 denotes the initiation opening displacement, then eq. (4) degenerates into the 'final stretch' criterion. This criterion turns out to be in a much better agreement with the experimental data for the subcritical range of the stress intensity level than the COD criterion, (cf. Vincent in Polymer, 1971, Vol. 12, pp. 534–546). Noteworthy is also the fact that under certain assumptions ($R/\Delta \gg 1$) the final stretch criterion can be derived from McClintock's ε_f-criterion. This is possible if one assigns the strains within the Dugdale zone as follows

$$\varepsilon(x_1) = \varepsilon_0 + \frac{\Delta\varepsilon_f^p}{u_0}\{-|\operatorname{grad} u(x_1)|\}, \quad 0 \le x_1 \le R(x_1), \quad \varepsilon_0 = Y/E \quad (5)$$

where $u = u(x_1, R(x_1))$ is the displacement within the plastic zone and x_1 denotes the distance from the crack tip. Since the crack moves while the control point is stationary, one may regard x_1 as a time-like variable.

Combining all the basic equations listed above we arrive at the following equation of motion

$$R - \sqrt{R(R-\Delta)} + \Delta\,\frac{dR}{dl} + \frac{\Delta}{2}\log\frac{\sqrt{R}+\sqrt{R-\Delta}}{\sqrt{R}-\sqrt{R-\Delta}} +$$

$$+ \frac{\delta\psi}{\Delta}\,R^2\int_0^\varepsilon\left\{\sqrt{\rho(s)(\rho(s)-s)} - \frac{s}{2}\log\frac{\sqrt{\rho(s)}+\sqrt{\rho(s)-s}}{\sqrt{\rho(s)}-\sqrt{\rho(s)-s}}\right\}ds = R_0 \quad (6)$$

in where $\varepsilon = \Delta/R(t-\delta t)$, $\rho(s) = 1 + (\varepsilon-s)dR/dl$, $\delta\psi = \psi(\delta t) - \psi(0)$ and all R's are taken at the instant $t - \delta t$. Two cases are now considered:

1°) elastic-plastic matrix with no rate sensitivity, then $\delta\psi = 0$, and equation (6) subject to the final stretch criterion reduces to a differential equation governing the slow growth of McClintock's type

$$\Delta\,\frac{dR}{dl} = R_0 - R + \sqrt{R(R-\Delta)} + \frac{\Delta}{2}\log\frac{\sqrt{R}-\sqrt{R-\Delta}}{\sqrt{R}+\sqrt{R-\Delta}} \quad (7)$$

or, for $R/\Delta \gg 1$

$$\Delta\,\frac{dR}{dl} = R_0 - \frac{\Delta}{2}\log\frac{4R}{\Delta}. \quad (7a)$$

The initial slope $(dR/dl)_0$ expressed in terms of the initial plastic zone size $r_0 = R_0/\Delta$ or in terms of ductility parameter $\alpha = \varepsilon_f^p/\varepsilon_0$ is

$$\left(\frac{dR}{dl}\right)_0 = \begin{cases} \sqrt{r_0(r_0-1)} + \frac{1}{2}\log\dfrac{\sqrt{r_0}-\sqrt{r_0-1}}{\sqrt{r_0}+\sqrt{r_0-1}} \\[2mm] \sqrt{\alpha(1+\alpha)} + \frac{1}{2}\log\dfrac{\sqrt{1+\alpha}-\sqrt{\alpha}}{\sqrt{1+\alpha}+\sqrt{\alpha}}. \end{cases} \quad (8)$$

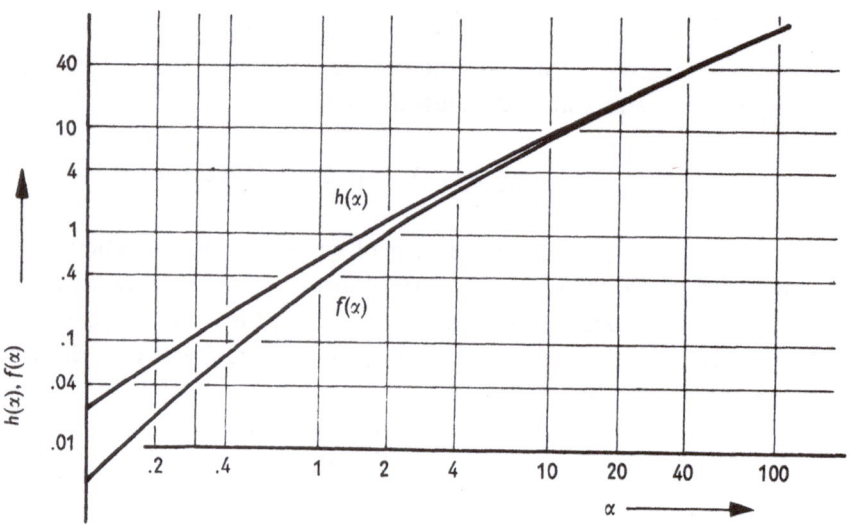

Fig. 1. Initial slope of the R-curve vs. ductility parameter α; $h(\alpha)$ denotes the present solution, $f(\alpha)$ is Rice's solution.

It compares closely to Rice's solution for mode III derived from the incremental theory of plasticity, $(dR/dl)_0^{\text{Rice}} = \alpha - \log(1+\alpha)$, see Fig. 1. Fig. 2 shows a few graphs which resulted from integration of eqs. (7) and (7a). The latter turns out to be integrable in a closed form. Both describe the 'universal' R-curve, which depends only on the threshold value of R and on the initial crack size, but it does not depend on geometry.

The steady-state limit of R results from the following transcendental equation

$$r_0 - r_\infty + \sqrt{r_\infty(r_\infty - 1)} + \tfrac{1}{2}\log\frac{\sqrt{r_\infty} - \sqrt{r_\infty - 1}}{\sqrt{r_\infty} + \sqrt{r_\infty - 1}} = 0 \qquad (9)$$

where $r_\infty = R_\infty/\Delta$. Crack begins to move at $r = r_0$, while the transition to rapid propagation occurs at r_f, determined readily from the criterion $dR/dl = \partial R/\partial l$. Of course $r_0 \leqq r_f \leqq r_\infty$.

2°) The second case is that of a rate-sensitive matrix with Dugdale plastic zones present. Here the complete equation (6) has to be tackled. It simplifies considerably if one assumes $R/\Delta \gg 1$. Then one has

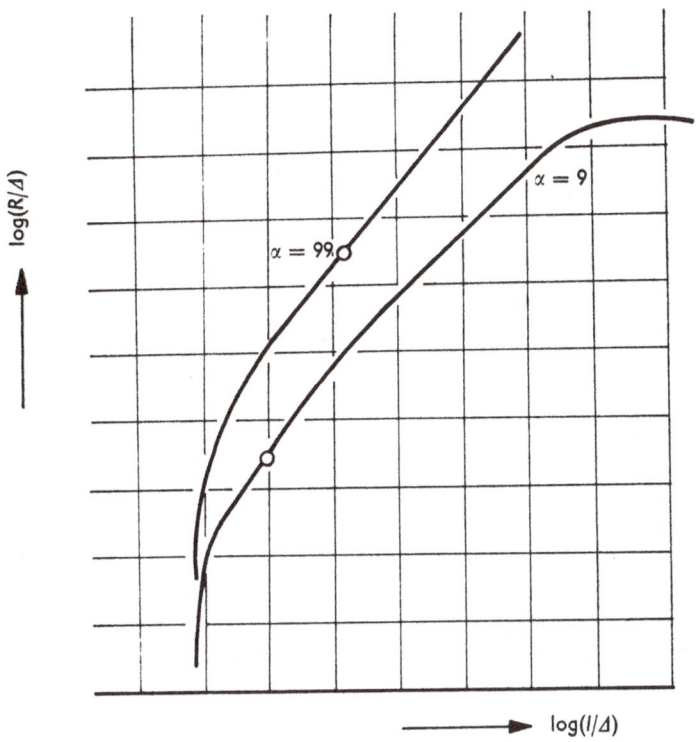

$\log(R/\Delta)$

$\alpha = 9$

$\alpha = 99$

$\log(l/\Delta)$

Fig. 2. Slow growth of McClintock's type at two different ductilities $\alpha = 9$ and $\alpha = 99$. Circles denote the final instability points. Initial crack size equals 100Δ.

$$\frac{\Delta}{2}\log\left(\frac{4R}{\Delta}\right) + \Delta[1 + CR(\partial R/\partial Q)^{-1}]\frac{dR}{dl} -$$

$$- \Delta C(\partial R/\partial Q)^{-1}R\frac{\partial R}{\partial l} = R_0 \qquad (10)$$

as the governing equation of motion. Here $[(dR/dl) - (\partial R/\partial l)]C\Delta(\partial R/\partial Q)^{-1}$ has been substituted for $\delta\psi(= B\Delta/\dot{l})$; B denotes the slope of creep compliance at time zero, Q is the loading parameter and $C = B/\dot{Q}$. Eq. (10) has been integrated numerically for the case of a finite crack in an infinite plate, and the results are shown in Fig. 3. The figure illustrates the effect of the rate sensitivity of the material and the rate of loading on the shape of the R-curve. It is seen that not only the slope of the curve is affected, but also

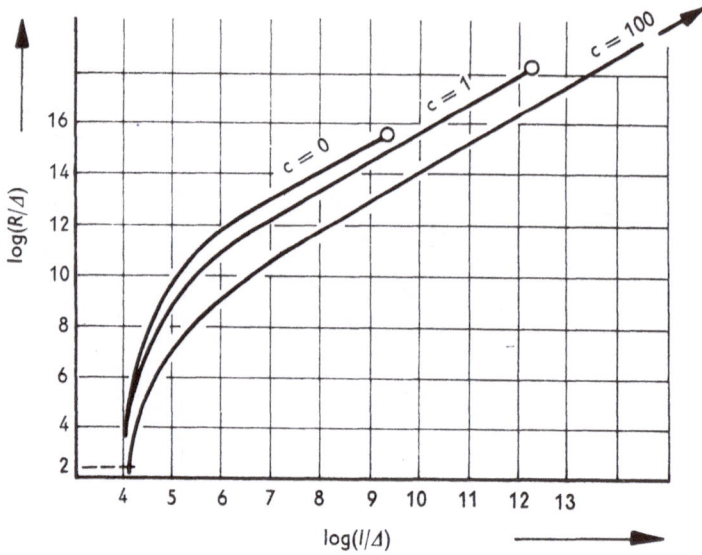

Fig. 3. Slow growth of a crack in a ductile rate dependent solid. Ultimate instability is reached when the slope of the R-curve plotted on log-log scale equals unity (marked by circles). Initial crack size is 100Δ.

pronounced changes in location of the ultimate instability point are observed. Examples of integration of the equation of motion when the viscoelastic dissipation is dominant, are considered in more detail by Wnuk in Int. J. Fracture Mechanics, 1971, Vol. 7, No. 4, pp. 383–407.

Discussions to papers contributed

to the international symposium on Foundations of plasticity

J. H. Argyris and A. S. L. Chan

Static and dynamic elasto-plastic analysis by the method of finite elements in space and time

Foundations of Plasticity
pp. 147–175

DISCUSSION

E. H. Lee.[1])   I wish to comment on the solution of the forging problem, of a cylindrical billet struck by a hammer. On first impact a plane wave ABC of one-dimensional strain (no lateral displacement) will emanate from the impact surface, this condition being modified by the waves CD and EA emanating from the free cylindrical surface as indicated in Fig. A. The condition inside $EABCD$ will be pure axial compression, below ABC the material will be undisturbed, and more complicated motion will occur behind the waves AE and DC. In view of this motion, it does not seem to

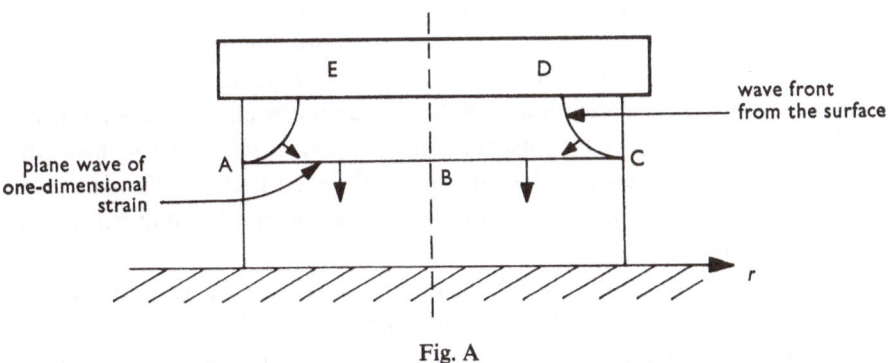

plane wave of
one-dimensional
strain

wave front
from the surface

Fig. A

¹) Stanford University, Stanford, USA.

me that the uniformly strained disc elements used in the calculation will provide a satisfactory representation. Of course, later in the motion a more uniform distribution of strain may arise which can be more accurately represented by the disc finite elements.

S. Valliappan.[1]) The authors are to be commended for extending the elasto-plastic analysis by finite element method from static to dynamic problems. The writer agrees with the authors that any nonlinear analysis involves a large computation time and an efficient iterative scheme is necessary to solve real problems. The writer [V.1] and Zienkiewicz et al [V.3] have developed such an efficient iterative approach which has been applied successfully to many real problems. [V.2, V.4]

The authors have not clearly described in the text, their iterative scheme except providing a flow chart. The flow chart shows that at each step, the stiffness matrix for each element is calculated, the structure stiffness matrix is assembled again and the system of simultaneous equations is solved. This may be the reason for such a large computer time (about 60 min. on CDC 6600) involved in solving the problems given as demonstrations of the method proposed by the authors. The 'initial stress method' described by the writer in [V.2] gives the solution of problems such as 'Approximation to Machining Process' and 'Notched Specimen' (Fig. 5 and 10 of [V.2]) in about 40 min. on IBM 360 (which is equivalent to 5 min. on CDC 6600).

The writer agrees with the authors that in cases like the 'Rectangular Strip in Tensile Test' it is more realistic to prescribe increments of displacements rather than loads since the pressure is not uniformly distributed. The same effect can be clearly observed in the results given in Fig. 6 of the writer's paper [V.2] for the problem on 'Approximation to Machining Process'. The writer has solved several problems in geomechanics of which one example 'Strip Foundation on Soil' shows that by applying the uniform displacements, the pressure distribution obtained is not uniform beneath the foundation. Moreover it has also been observed that the ultimate load obtained by prescribing uniform displacements rather than uniform loads is quite different.

Finally the writer wishes to ask the authors whether they have tried more refined elements than triangular elements for the elastiplastic analysis. If so, what are their comments on the efficiency of such elements? The writer is

[1]) University of New South Wales, Kensington, Australia.

inclined to think that even though more complex elements will produce more accurate results with fewer elements, the time involved in the computation will also be greater.

[V. 1] Valliappan, S., *Nonlinear stress analysis of two-dimensional problems with special reference to rock and soil mechanics*, Ph.D. Thesis, University of Wales (1968).

[V. 2] Valliappan, S., Elasto-plastic analysis of anisotropic work-hardening materials, *Arch. Mech. Stos. 24* (1972) 465–481.

[V. 3] Zienkiewicz, O. C., S. Valliappan and I. P. King, Elasto-plastic solutions of Engineering problems, initial stress, finite element approach, *Int. J. Num. Meth. Eng., Vol. I* (1969) 75–100.

[V. 4] Zienkiewicz, O. C. and S. Valliappan, analysis of real structures for creep, plasticity and other complex constitutive laws, Int. Conf. Struc., Solid Mech., Eng. Design, Southampton (1969).

A. S. L. Chan.[1]) In connection with Professor Lee's comment I wish to add that the present finite element model may not be exactly accurate in the very first instance of impact, but with a radial and hoop strain varying linearly in the axial direction (although not radially) for every element, it seems to give an adequate approximation in the later stages. The analysis may be improved by the use of a more refined element if so wished; however, the method will still be applicable.

Dr. Valliappan was interested in the structure stiffness matrix. The large computer time involved in the calculation is entirely due to the necessity of assembling the plastic stiffness matrix by a numerical integration for every element after each iteration. The elastic calculation of exactly the same problem by the same procedure takes less than 3 minutes (CDC 6600) from first impact to separation (bounce-off). It seems therefore that a stiffness matrix based on a prescribed stress distribution on the finite element may have an advantage.

The authors agree with Dr. Valliappan that the use of the refined elements does require more computing time than using more of the simpler elements if the stiffness matrix has to be calculated by numerical integration. However, this may not be true if a mixed model (specifying both stress and displacement variations) is used so that the plastic stiffness may be calculated quickly.

[1]) Imperial College of Science and Technology, London, England.

G. Augusti and A. Baratta

Theory of probability and limit analysis of structures under multi-parameter loading

Foundations of Plasticity
pp. 347–364

DISCUSSION

N. C. Lind and S. R. Parimi.[1]) The authors should be congratulated to directing the attention towards a timely and exiting subject. A new light is cast over the familiar concepts of plastic limit analysis under uncertainty. The study is also important from the viewpoint of practice because it uncovers some profound implications of applying a particular deterministic theory in what is actually a probabilistic context. It is regrettable, therefore, that the authors have chosen to limit the study to random strengths and given loads, for in practice (excepting certain special structures, e.g., pressure vessels) the load dispersion is two to five times greater than the strength dispersion, both expressed in terms of the coefficient of variation. One may well be justified in sometimes neglecting strength dispersion, but load dispersion is almost always important.

The probabilistic theorems of limit analysis proposed by the authors can be stated in alternative forms that seem at once clearer and easier to justify. Consider a structure with random strength and random loading. Plastic collapse may conceivably take place by motion of one or more mechanisms, constituting a subset S_1 of the space, S, of possible mechanisms. The probability of collapse conditional upon it being by this subset is necessarily less than er equal to the probability of failure. Hence $P_\psi \leq P_c$. Conversely, the structure may withstand collapse by the activation of one or more possible stress fields, constituting a subset of all conceivable equilibrium stress fields. Thus, the reliability is greater than or equal to the probability

[1]) University of Waterloo, Ontario, Canada.

464

that the structure does not collapse by reasons of a subset of equilibrium stress fields. Hence, $P_c \leqq P_\psi$.

These theorems are a direct consequence of the calculus of events – the notion of statical and kinematic admissibility play no role in the proofs but tend to create confusion in the definitions of events $\{C\}$, $\{D\}$, etc. It is perhaps to be expected, then, that these theorems do not provide as much information as the deterministic theorems of limit analysis. For example, the kinematic theorem in one version states: The collapse load factor is the smallest kinematically admissible load factor. The latter is a set function; for (at least) one element of the space, S, of all kinematically admissible fields the load factor equals the collapse load factor. Also P_γ is a set function on the space, S, by contrast, there is generally no single element for which P_γ equals P_c. Similar remarks apply to P_ψ. As a practical consequence of this difference. it is not possible to determine P_c exactly merely by an exhaustive study of all admissible fields, (i.e., determining the probabilities associated with the admissibility of each of the elements in S), while the collapse load factor may always be found if such a study is made in the deterministic case. Also, the equality signs in Eqs. (8) and (11) do not hold true unless the probability of a joint event (either the intersection of kinematic fields or the union of equilibrium stress fields) is implied in the above equation. For the example given, it is possible to construct two families of Γ_p and Ψ_p curves for various values of P, analogous to those shown in Fig. 3 of the paper. Γ_p and Ψ_p curves passing through the same point in $W_1 - W_2$ space (A in Fig. A), obviously would have different probabilities associated. Hence it is only possible to bound P_c, such that the bounds do not fall into a certain range containing P_c. For structure A in Fig. A, $P_A^- < P_{CA} < 0.1$ and for structure B, $0.01 < P_{CB} < P_B^+$.

For these reasons the theorems are not the probabilistic counterparts of the classical limit theorems. However, if the probability of a compound event (i.e., two or more fields being admissible) is implied in P_γ (or P_ψ) it would be possible to bound P_c as closely as one wishes.

It would be useful if the authors could state with precision what meaning should be attached to the basic concepts of limit analysis when loads and strengths are probabilistic. It seems that a field can only be said to be (for example: statically) admissible with such and such probability. What, then, is the probability of statical admissibility to be associated with a set of two such fields, several or infinitely many fields, or a continuum of fields?

Quite apart from the conceptual difficulties in calculation of probabilities

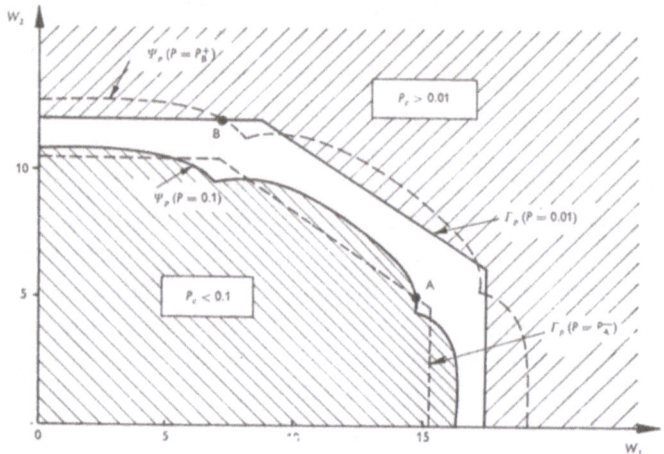

Fig. A. Γ_p, Ψ_p curves in load space for various probabilities.

of such compound events as $\{C\}$, $\{D\}$, etc., in this paper, there may be practical difficulties if the sets of admissible fields are large or do not have discrete elements, finite in number. For example, how could P_γ be calculated for the plane strain subset of the kinematically admissible motions of a three-dimensional body?

J. Murzewski.[1]) 1. The authors have calculated the conditional lower bound probabilities $P_{\gamma j}[W]$, $j = 1, 2, 3, \ldots, N$, by separate considerations of N specific collapse mechanisms for the load system W_i, $i = 1, 2, \ldots, n$. The overall lower bound probability $P_\gamma[W]$, i.e. the probability that any mechanism from the number N is active, might be a much better approximation in the authors' opinion, but the calculations would be rather cumbersome and they have not been performed in the numerical example. This opinion seems to be a little exaggerated. Significant improvement in the approximation may be obtained for such load proportions $\alpha = W_2/W_1$ for which the limiting straight lines Γ_{pj} and Γ_{pj+1} intersect and also in the vicinity of such points. For the small P, the joint probability of two neighbouring mechanisms may be taken into account only at the intersection point,

$$P_\gamma \approx P_{\gamma j, j+1}.$$

[1]) Technical University, Cracow, Poland.

466

The calculation for the normally distributed limit moments $\tilde{M}_{ol}$ is rather simple. This will be shown below and a derivation of the overall lower bound limit curve Γ_p for $P = 0.01$ may supplement the numerical example of the paper.

The correlation coefficients for the collapse mechanisms are calculated as follows,

$$\rho_{j,j+1} = \frac{\overline{(\mathscr{L}_j \mathscr{L}_{j+1})} - \overline{\mathscr{L}_j}\overline{\mathscr{L}_{j+1}}}{\sqrt{(\mathrm{Var}(\mathscr{L}_j)\,\mathrm{Var}(\mathscr{L}_{j+1}))}} \quad \text{for } j = 1, 2.$$

The authors' formulae (35) and the distribution parameters $\overline{M}_{ol} = 10\ tm$, $\mathrm{Var}(M_{ol}) = 1\ (tm)^2$ give

$$\rho_{12} = \frac{6}{\sqrt{60}} = 0.774, \quad \rho_{23} = \frac{4}{\sqrt{40}} = 0.632.$$

The load proportions $\alpha_{j,j+1} = W_2/W_1$, corresponding to the intersection points of the limiting lines Γ_{pj} and Γ_{pj+1}, are calculated from the inequalities (38b) taking equality sign. One obtains

$$\alpha_{12} = \frac{52.64 - 34.30}{34.30}\frac{2}{3} = 0.356, \quad \alpha_{23} = \frac{35.34}{52.64 - 35.34}\frac{2}{3} = 1.36.$$

Values of two-dimensional normal distribution $F(x, y, \rho_{xy})$ are tabulated [M.4] and there is only one unknown W_1 in the equation

$$F(\lambda_j, \lambda_{j+1}, \rho_{j,j+1}) = 0.01,$$

where

$$\lambda_j = (\mathscr{L}_j - \overline{\mathscr{L}}_j)/\sqrt{(\mathrm{Var}(\mathscr{L}_j))},$$

$$\mathscr{L}_j = 2W_1, \quad \mathscr{L}_{j+1} = (2 + 3\alpha_{12})W_1 \quad \text{for } j = 1,$$

$$\mathscr{L}_j = (2 + 3\alpha_{23})W_1, \quad \mathscr{L}_{j+1} = 3\alpha_{23}W_1 \quad \text{for } j = 2.$$

Other proportions $\alpha = W_2/W_1$, close to the values α_{12}, α_{23}, are also taken into account in order to get more points of the curve Γ_p. (See Fig. B.)

2. The authors determine approximate values of the collapse probability $P_c[W]$ at each point in the n-dimensional load space W_i, $i = 1, 2, \ldots, n$, for the random strengths $\tilde{M}_{ol}$ and they suggest to apply the multiple integral (47) for the evaluation of the structural reliability R. However, this is only

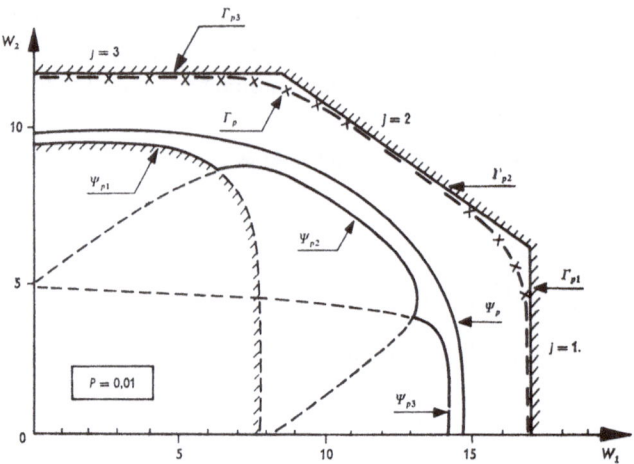

Fig. B. A corrected upper bound Γ_p for the safe domain of load factors. Probability
$P = 0.01$.

one possibility of solution. The other one is to determine the collapse
probability $P_c[M_0]$ in the m-dimensional strength space M_{0l}, $l = 1, 2, 3, \ldots$
$\ldots, m$, for the random loads $\tilde{W}_i$ and to calculate a multiple integral

$$R = 1 - \int_{-\infty}^{+\infty} \cdots \int_{-\infty}^{+\infty} P_c[M_0] p_M[M_0] \, \mathrm{d}M_{0l} \ldots \mathrm{d}M_{0m}.$$

The first method should be simpler for computations only for $n < m$. Both
methods become cumbersome for multi-parameter loading and multi-parameter strength, even if the computers are applied. This is probably the
reason why the authors have not calculated the integral (47) in their numerical example. The calculations of the reliability R is not necessary if we
accept the probability of disqualification instead of the probability of
failure as the measure of structural safety. Such a theory represents better
the actual social and economical conditions in building [M.2]. According
to the theory, the design loads for a given structure can be determined from
the solution presented by the authors.

3. The authors assume that the strengths at m points of the structure are
independent random variables and no failures at other points are possible.
For example, $m = 11$ selected cross-sections of the frame on Fig. 1, are
numerically analyzed in the paper. However, the selection criterion is not

468

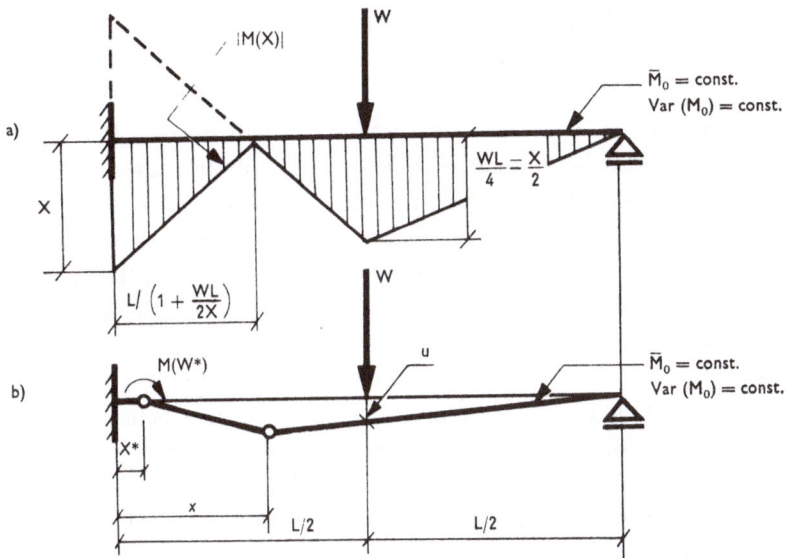

Fig. C. The statically admissible moment diagrams (a) and the kinematically admissible collapse mechanisms (b) for a stochastically non-homogeneous redundant beam. Parameters: a) $X \geqq 0$, b) $0 \leqq x^* \leqq L$, $0 \leqq W^* \leqq W$.

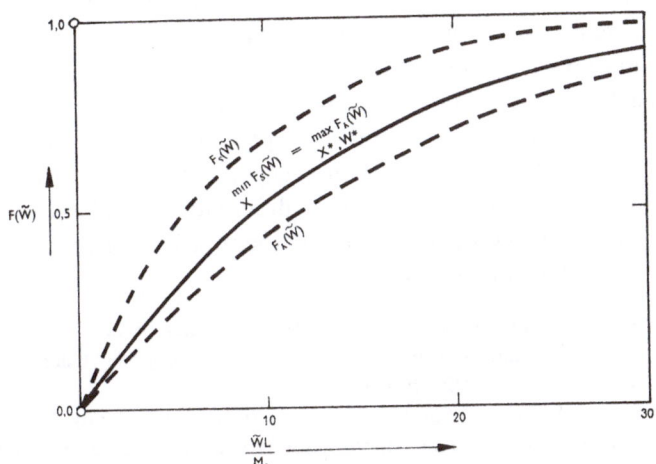

Fig. D. Cumulative distrivution function of the limit load $\tilde{W}$ for the redundant beam from Fig. C. $\tilde{M}_0$ – the Weibull random variable, $\text{Var}(M_0) = (\overline{M}_0)^2$. $F_S(\tilde{W})$: stratic solution, $F_K(\tilde{W})$: kinematic solution.

quite clear. Usually as random variables the strengths of the 'critical cross-sections' are taken (in the sense of the deterministic limit analysis), so it would be $m = 5$ for the frame under consideration (Nos. 1, 4, 6, 8 and 11 on Fig. 1). The limit cases are $m = 1$ and $m \to \infty$. If $m = 1$, the actual structure can be thought of as a sample structure taken out from a population of nominally identical structures. The case $m \to \infty$ means that every cross-section of the frame exhibits a random strength stochastically independent of the other strengths. The probabilistic structural analysis in this case is similar to the analysis of a size-effect. A stochastic theory of the size-effect for brittle bodies, initiated by W. Weibull, has already been extended for the ductile statically determined structures [M. 3]. Some concepts regarding exact elasto-plastic solutions for stochastically non-homogeneous redundant structures have also been suggested [M. 2], but the calculations were found to be too difficult. One may expect that approximations in the form of lower and upper bounds would be useful for designers. Recently, it has been proved [M. 1] that the diagram of admissiblie moments for the probabilistic static theorem as well as the collapse mechanisms for the probabilistic kinematical theorem are those of the deterministic analysis of perfectly homogeneous rigid-plastic structures, provided that the coefficient of variation of the stochastically variable strength is small enough. It is important, both for a finite and infinite number m, to take into consideration all statically admissible moment diagrams and all kinematically admissible collapse mechanisms. Under this conditions the maximum lower bound and the minimum upper bound of the probability of the carrying capacity converge to each other and the probabilistic solution may be recognized as a complete one. In our considerations [M. 1] the convergent lower and upper bounds for a stochastically non-homogeneous ($m \to \infty$) redundant beam (Fig. C) were determined by means of differential calculus from a set of one-parameter moment diagrams and two-parameters collapse mechanisms. The results are shown in Fig. D.

[M. 1] Machowski, A. and J. Murzewski, Carrying capacity of steel beams and frames from a stochastically non-homogeneous material, Regional Conference on Tall Buildings (Warsaw, 1972) (in print).

[M. 2] Murzewski, J., *Safety of building structures* (in Polish), Arkady, Warsaw 1970.

[M. 3] Murzewski, J. and J. Sojka, The probability characteristics of the limit load of a structure made of a quasi-homogeneous ductile material (in Polish), *Rozpr. Inż.* 15 (1967) 259–627.

[M. 4] Smirnov, N. V. and L. N. Bolshev, *Tables for computation of the two-dimensional normal distribution function* (in Russian), AN SSSR, Moscow 1962.

G. Augusti[1]**) and A. Baratta.**[2]**)** The authors want to thank the discussers, whose contributions enhance the value of their work, and also afford the opportunity of clearing up some points that were apparently obscure in the original text.

First of all, it is worth remarking – once more – that this paper and its parent works [1] [2] concern theoretical investigations on the randomness of the ultimate strength of ductile structures; the distinct, and probably more important, problem of load uncertainty and dispersion is *not* investigated, but is well present in the authors' minds. The definitions of (probability of) statical admissibility and kinematical sufficiency, and the basic theorems, were formulated for *given* loads, i.e. in the form of *conditional probability* problems. This is a sort of standard way of defining and presenting random strength problems in the structural engineering literature: of course, the complete solution of a *structural reliability* problem requires combining the results on the strength variability with some stochastic definition of loads (perhaps in the form suggested by Ferry-Borges and Castanheta [R. 1]), as very briefly indicated in the concluding section of the paper.

Murzewski suggests the introduction of a collapse probability $P_c[M_0]$ defined in (generalized) stress space rather than in load space: this appears an interesting alternative, and it would easily be possible to re-formulate the bounding theorems and obtain procedures to limit $P_c[M_0]$ on both sides. However, it seems that in usual problems the number m of significant stresses is seldom smaller than the number n of independent load parameters: therefore, Murzewski's m-fold reliability integral would seldom imply a computational advantage. Further complications would arise when more than one stress component should be taken into account in each point (section) of the structure. In either formulation, the calculation of the manyfold reliability integral might take advantage of the known stochastic techniques for numerical integration [R. 2].

It is hoped that the final printed form of the paper already gives an answer to some of the queries by Lind and Parimi, who had apparently examined the first proofs. Their presentation of the probabilistic theorems is definitely more concise (but is it clearer?) than the original one. However, it is not true that 'the notion of statical and kinematical admissibility plays no role in

[1]) Università di Firenze, Florence, Italy.
[2]) Università di Napoli, Naples, Italy.

the proofs'. For instance, to say that the structure does not collapse if it can 'withstand collapse by activation of one or more possible stress fields', is perfectly equivalent to state the static theorem of limit analysis; and like the formal theorem, such a statement does not hold true if the existence of a statically admissible stress field is not enough to warrant the survival of the structure (e.g. because of instability).

The authors do think that an analogy exists between the 'classical' and the 'probabilistic' limit analysis theorems: the formulation of the latter very clearly included 'the probability of a joint event'. In the numerical example (which was introduced in the paper only to demonstrate the practical applicability of the proposed procedures, and kept concise also because of time and length limitations), such joint probabilities were explicitly considered for the stress fields only (i.e., in the determination of the Ψ_p domains); Murzewski's discussion now includes some analogous calculation for the mechanisms, and the consequent improvement of an outer (Γ_p) domain. Consideration of more and more fields and their joint probabilities, involving only numerical and not conceptual difficulties, would lead to the determination of closer and closer bound to the true solution: the question of the rigorous *convergence*, given the continuity of the admissible stress field set, is currently being investigated [R. 3]. However, the advantage of the suggested procedures, which allow to calculate alternatively outer and inner bounds, lies just in the possibility of halting the calculations as soon as the difference between the bounds is deemed narrow enough in comparison with the advantage expected from, and the effort required by, further refinements. The authors never thought that this point was even approached in their numerical example, and this is now made self-evident by Lind and Parimi's Fig. A which shows that the region $P_c > 0.01$ (defined by a Γ_p domain) and the region $P_c < 0.1$ defined by a Ψ_p domain) do not intersect, and therefore do not provide in any point a well-defined interval for the probability of collapse P_c.

Finally, with reference to Murzewski's paragraph No. 3, the Authors underline that in this paper they did not intend to discuss at all the major problems of a convenient choice of the sections of possible yielding, let alone to present a 'criterion of selection'. Apparently, a compromise will have to be struck in some way between the two – equally unreasonable – extremes of limiting them to the 'critical sections' of deterministic limit analysis, and of increasing their number so much that the other, computationally essential, assumption of stochastic independence between the strengths of each such section, becomes completely unrealistic. With regard to these problems, Murzewski's works sound of great interest, and the authors hope that they will soon become available in a language that will allow them a wider circulation.

[R. 1] Ferry-Borges, J. and M. Castanheta, *Structural safety*, L.N.E.C. Course No. 101, 2nd. Ed., Lisbon 1971, 144–148.

[R. 2] Shreider, Yu. A. (Editor), *Methods of statistical testing*, Elsevier Publ. Co., N.Y. 1964.

[R. 3] Baratta, A., *The static approach in probabilistic limit analysis* (in preparation).

H. D. Bui, A. Zaoui et J. Zarka

Sur le comportement élasto-plastique et viscoplastique des monocristaux et polycristaux métalliques de structure cubique à faces centrées

Foundations of Plasticity
pp. 51–75

DISCUSSION

H. Zorski.[1])   How did you calculate the interaction forces between dislocation lines?

E. Kröner.[2])   Since I have followed the work of Zarka in recent years I would like to give the following comment which also is a kind of evaluation of the present work of Bui, Zaoui and Zarka.

In the symposium paper Zarka specifies the dislocation state in terms of straight dislocation segments which at the beginning of the deformation are distributed at random as well in space as in length. He then applies Peach and Koehler's formula, a basic equation of dislocation theory, in order to calculate the elastic interaction between the dislocation segments. Within the framework of this model Zarka is able to calculate the motion of the segments and the formation of new segments, thus the progress of deformation under the applied load. This calculation which includes contributions of all possible glide systems leads to a single crystal plastic stress-strain law which is also valid under triaxial stress. The latter fact is very important if one wishes to apply the results to polycrystalline behaviour. In fact, the so-called self-consistent theory, originally proposed by myself and further developed and applied by Budiansky, Wu and Hutchinson implies an averaging over the *in situ* stress-strain curves of the single crystal-

---

[1]) Institute of Fundamental Technological Research, Warsaw, Poland.
[2]) University of Stuttgart, Stuttgart, G.F.R.

474

lites. These curves are practically not accessible to experiments; hence any other way to obtain them should be of great value.

Based on these results, Bui, Zaoui and Zarka were able to calculate the plastic stress-strain behaviour of polycrystalline materials in certain special situations. In doing this they even succeeded in removing some of the restrictions in the self-consistent method which were assumed by the previous workers in order to simplify the calculations.

Notwithstanding the fact that the model of dislocation segments misses certain features of the real dislocation state I consider this work an important step towards the integration of physical research on the microscale into more macroscopic descriptions of the deformation behaviour of crystalline solids.

J. Zarka.[1]) It is difficult to answer in a few minutes the very important question posed by Professor Zorski. I shall thus just sketch the procedure I used when evaluating the interactions. All the relevant computations can be found in [Z. 1].

According to the mathematical theory of dislocations (cf. [Z. 2]) the internal stresses at any point of the crystal can be evaluated once a distribution of dislocations is given. Calculations of this kind, however, involve volume integrals and are cumbersome. I preferred to introduce certain approximations in order to reach simple explicit formulas.

1) Since, generally the dislocation lines are composed of small straight segments which stretch out in some particular directions, I grouped in one system the set of the segments which are parallel and which have the same Burgers vector. I assumed that these segments were randomly distributed and I calculated the shortest mean distance, d, between two segments of one system and the shortest mean distance, D, between two planes which contain some segments of one system. In the quantities d and D there appear, the mean number of segments per unit volume and their mean length.

2) I carried out the computations of the interactions in admitting the classical hypothesis of straight infinitely long dislocations in an infinite isotropic elastic medium. I considered a dislocation of one system and I wanted to be sure that this dislocation could be moved through the other systems; for that I took the maximum absolute value of all the actions.

[1]) Ecole Polytechnique et Centre National de la Recherche Scientifique, Paris, France

However I weighted these actions in taking into account the finite length of the dislocations when I introduced the quantities d and D.

Such a procedure of course, gives only an approximation of the interactions. Experimentally, it was found that the results obtained were sufficiently accurate, [Z. 3].

As regards Professor Kröner's remarks I wish only to thank him for his very kind comment.

[Z. 1] Zarka, J., Sur la viscoplasticité des métaux, *Memorial de l'Artillerie Française*, 2ème fasc. (1970) 213–292.

[Z. 2] Kröner, E., *Kontinuums Theorie der Versetzungen und Eigenspannungen*, Springer, Berlin 1958.

[Z. 3] Zarka, J., Etude du comportement des monocristaux métalliques. Application à la traction simple du monocristal C.F.C. *Journal de Mécanique, Vol. 12, N° 2* (1973) 275-318.

P. K. Fung, D. J. Burns and N. C. Lind

Yield under high hydrostatic pressure

Foundations of Plasticity
pp. 287–299

DISCUSSION

J. Litoński.[1]) In connection with the reported tests by Fung, Burns and Lind I would like to present some results regarding influence of hydrostatic pressure on strain hardening of copper. In my tests conical specimen were used, [L. 1]. It was shown in [L. 2] that conical specimen were suitable for the purpose because the strain hardening exponent $n = \mathrm{d} \ln \tau / \mathrm{d} \ln \mathrm{tg}\, \gamma$ as a function of the deformation $\mathrm{tg}\, \gamma$ can be found once the distribution of the permanent shear deformation on the lateral surface of the plastically twisted conical specimen is known (τ is the shear stress and γ is the angle of the plastic shear). For conical specimens, twisted at the pressures 1, 7700, 12000 and 15000 kgcm^{-2}, the relations between n and $\mathrm{tg}\, \gamma$ were derived. The curves of Fig. A correspond to the average of two or three test cones twisted at the same hydrostatic pressure. It can be concluded from the curves that in the pressure range from 7700 to 15000 kgcm^{-2} the strain hardening exponent n is slightly smaller than at 1 kgcm^{-2}. The decrease of the mean value of n in the range $0.04 \leqq \mathrm{tg}\, \gamma \leqq 0.43$, is of about 6 to 13 percent.

Experiments in which uniform elongation of cylindrical copper specimens was measured at the hydrostatic pressure of 1 and 10000 kgcm^{-2} were also made. The results indicate a small decrease of the uniform elongation at 10000 kgcm^{-2} which corroborates the conclusion drawn from the torsion tests.

The change of the strain hardening exponent is connected with the change of the stress-strain curve. From Fig. 5 of the authors' paper, where the initial parts of stress-strain curves are shown at the normal and 4200 kgcm^{-2}

[1]) Institute of Fundamental Technological Research, Warsaw, Poland.

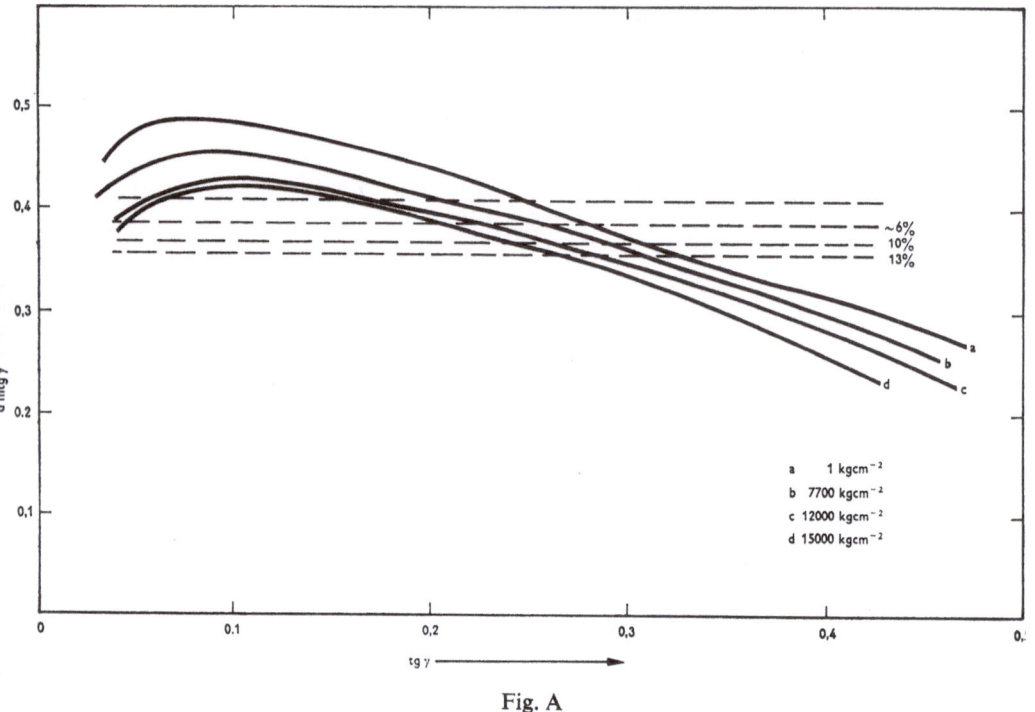

Fig. A

hydrostatic pressure, respectively it can be concluded that the curve (b), after the initial yield differs up to about 10 percent from the curve (a). Could the authors comment to what extent this difference in strain hardening may be due to the high pressure effect?

[L. 1] Litoński, J., IBTP Report No. 45, Warsaw 1971.
[L. 2] Litoński, J., *Rozpr. Inżyn. 19* (1971) 485–500.

P. Germain and E. H. Lee

Plane waves in elastic-plastic solids

Foundations of Plasticity
pp. 257–270

DISCUSSION

J. Najar.[1])   The paper shows that for a dissipative solid governed by non-linear differential equations any attempt to consider a shock wave as a jump surface usually fails in view of a lack of a complete set of jump conditions.

There are at least *three possibilities* to overcome this difficulty, two of them being mentioned in this paper. The first and apparently the simplest idea is to complete the algebraic jump conditions with an *algebraic constitutive relation* for a shock wave in the medium considered. Eq. (40) gives an example of such a relation. There are a number of questions which have to be clarified in derivation of such shock laws. The most important one is to show how constitutive invariance requirements should be formulated. The way of deriving the missing condition as suggested in the paper, where a finite difference relation (40) is proposed by analogy to the differential equation (10), is not always obvious. The desired shock laws may be derived by means of a generalized functions method.

The second possibility is of a more general character than that used in the commented paper. One can namely consider a narrow layer in the vicinity of the shock wave front and suppose that the layer is governed by a *specific constitutive law*. Eqs. (42–44) may be regarded as an example of such a law. This approach is somewhat similar to the well-known Prandtl's boundary layer method for viscous fluids; an explicit distinction in the physical properties of the fluid in the outer flow and in the boundary layer is fundamental for this method.

A third possibility, following from Prandtl's idea is analysed in [N. 1].

[1]) Institute of Fundamental Technological Research, Warsaw, Poland.

The shock wave in this approach forms a narrow region in the neighbourhood of the shock front. This region is regarded now as *a gradient layer*, where the gradients of some field functions (e.g. pressure or velocity components) are high enough to enable linearizations of the governing set of equations. As a small parameter of this linearization the dimensionless thickness of the gradient layer can be chosen. This parameter should be small compared with the reciprocal of the dominating gradient magnitude.

A comparison of the last two approaches shows that, while the starting point of both is the same, the results are in general different. Instead of mathematical simplifications obtained by the last method, the second approach brings further complication when taking into account the viscosity effects. While it is a matter of discussion what range of problems is more suitably described by each of these attempts, there is no doubt that for a wide variety of problems the gradient layer method gives a simpler description, which can be readily used for solving some important practical problems.

An example of such a solution is given in [N. 2], where the problem of collision of two metal plates is studied from the viewpoint of perfect plasticity. It is supposed that a high pressure, occurring at the instant of the collision on the contact surface, decreases rapidly inside the plates, giving rise to a gradient layer in the vicinity of the surface. Assuming a relative slip of both plates due to the tangential component of the velocity of collision it is proved that an instantaneous instability of the contact surface occurs. This instability might be considered as a factor causing the wavy contact surface often observed in explosive welding of metal plates.

[N. 1] Najar, J., Gradient methods in the dynamics of strong discontinuities and their applications to the theory of plasticity, Proc. Polish-French Symp. 'Problèmes de la Rhéologie' (1971) P.W.N., Warsaw 1973, 265–276.
[N. 2] Najar, J., Surface waves in a perfectly plastic solid (in Russian), *Mekh. Tverd. Tela, No. 5* (1969) 187.

P. Germain[1] and E. H. Lee.[2])   We thank Dr. Najar for his suggestion of another approach to shock structure analysis. It will be interesting to see whether this yields a more convenient basis for analysis than the method suggested in the paper.

[1]) University of Paris, Paris, France.
[2]) Stanford University, Stanford, USA.

K. S. Havner

An analytical model of large deformation effects in crystalline aggregates

Foundations of Plasticity
pp. 93–106

DISCUSSION

T. H. Lin.[1]) This analysis of multicrystal elastic-plastic deformation is interesting. This method seems to be able to consider anisotropy of crystals without much additional complexity of calculations. In the calculation of deformation textures of polycrystals, it is necessary to calculate the change of crystal orientations at large deformation. This proposed method seems to be adaptable to calculate this deformation texture.

[1]) University of California, Los Angeles, USA.

Y. Horie

On the thermodynamic states of plastically deformed solids

Foundations of Plasticity
pp. 219–233

DISCUSSION

J. T. Fong.[1])   The purpose of this brief note is to show the major similarities and differences between the results reported by Horie [F. 1] and Fong and Simmons [F. 2] at the same symposium.

1. *Decomposition of the energy differential.* In Fig. A I show the decomposition of the energy differential as a sequence of arrows with broken ones

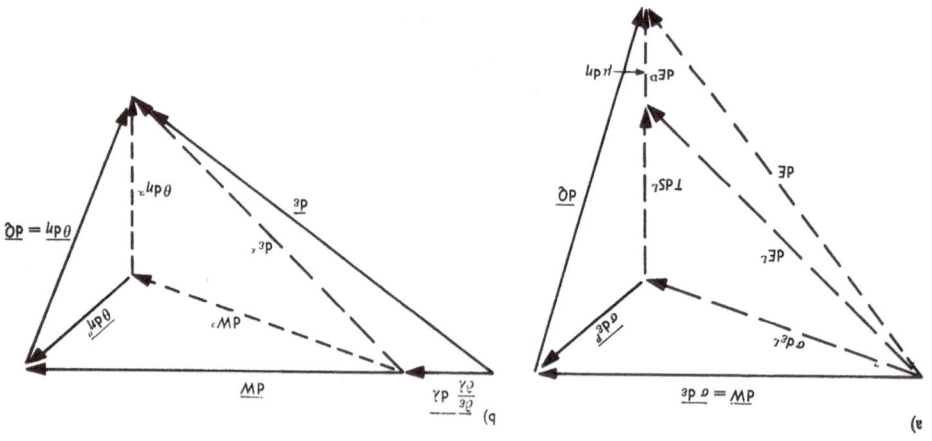

Fig. A.   A pictorial representation of the decomposition of the energy differential in thermodynamics
a) Horie's theory [F. 1].   b) Rate-independent case of Fong-Simmons' theory [F. 2].

[1]) U.S. National Bureau of Standards, Washington, USA.

denoting path-independent differentials and full arrows path-dependent differentials. The symbols attached to each collection of arrows correspond to those defined in [F. 1, F. 2]. The similarities between Horie's formulation and the rate-independent case of Fong and Simmons theory is not surprising, since both theories were motivated by the work of Bridgman [F. 3]. The major differences between the two theories are:

(a) Horie's dislocation energy differential dE^D is path-independent and of a thermal origin, whereas the counterpart $(\partial \varepsilon / \partial \lambda) d\lambda$ in Fong-Simmons' theory is path-independent and mechanical in origin.

(b) Horie left the plastic work differential $\sigma d\varepsilon^p$ unspecified, whereas Fong and Simmons used a single-integral to represent the equivalent differential $\theta d\zeta''$.

2. *Additional remarks.* Horie's result on proving Drucker's 'path independence in the small' is very interesting since Fong and Simmons derived a similar result using adiabatic consideration in a recent paper [F. 4]. The use of a single-integral leads to many interesting possibilities which is mentioned in Fong-Simmons' paper at this symposium.

[F. 1] Horie, Y., On the thermodynamic states of plasticially deformed solids, *Foundations of Plasticity*, Noordhoff Int. Publ., Leyden 1973, 219–233.

[F. 2] Fong, J. T. and J. A. Simmons, A non-equilibrium thermodynamic theory of simple materials based on a single-integral entropic functional, *Archives of Mechanics*, 23 (1972) 789.

[F. 3] Bridgman, P. W., The thermodynamics of plastic deformation and generalized entropy, *Rev. Modern Physics*, 22 (1950) 56.

[F. 4] Fong, J. T. and J. A. Simmons, A decomposition theorem on the scalar potential of a compressible BKZ fluid (submitted to a Technical Journal).

Y. Horie.[1]) Comments such as Fong's seem to strengthen the view that despite the fact that there is no definitive thermodynamic theory of plasticity, such an approach will expose fundamental problems in plasticity and provide adequate solutions to many important practical problems.

Path-independent dE^D is, in my formulation, a special case of the incremental energy change. It will be path-dependent if μ is a function of $\boldsymbol{n}$.

[1]) North Carolina State University, Raleigh, N.C., USA.

J. Klepaczko

Some experimental investigations of the elastic-plastic wave propagation in bars

Foundations of Plasticity
pp. 451–462

DISCUSSION

H. Fukuoka.[1])   I would like to ask Dr. Klepaczko to explain what does he mean by the strain-rate history effect.

J. Klepaczko.[2])   The yield stress is not uniquely specified by the strain and strain-rate. The strain-rate history effects have been studied in [K. 1, K. 2].

[K. 1] Klepaczko, J., Effects of strain-rate history on the strain hardening curve of aluminium, *Arch. Mech. Stos., 19* (1967) 211.

[K. 2] Klepaczko, J., Strain-rate history effects for polycrystalline aluminium and theory of intersections, *J. Mech. Phys. Solids, 16* (1968) 255.

[1]) Osaka University, Toyonaka, Japan.
[2]) Institute of Fundamental Technological Research, Warsaw, Poland.

J. Kratochvíl

Finite-strain theory of inelastic behavior of crystalline solids

Foundations of Plasticity
pp. 401–415

DISCUSSION

E. H. Lee.[1])   What is the significance of the order of matrix multiplication $F = EMP$. This choice influences the expressions for L_E, L_M and L_P. Do these have there different physical significances?

J. Kratochvíl.[2])   The theory suggested in [K. 1] is closely related to the order of visco-plastic, structural and elastic deformation gradients (denoted by P, M and E respectively) in the decomposition of the deformation gradient F

$$F = EMP \tag{1}$$

Any change of the order of E, M and P in (1) has to be accompanied by modifications of the transformation properties of E, M and P cf. Eqs. (2.10)–(2.12), (3.3), (3.4) of [K. 1]), as well as by changes of the definitions and the relations for the kinematical quantities $E_{(t)}$, M_t, P_t, L_E, L_M and L_P.

The most important alteration induced by a reordering in (1) is, however, a change in physical meaning of E, M, P and the related quantities. This fact was remarked also by Clifton [K. 2] who analyzed consequences of the decompositions $F = EP$ and $F = \bar{P}\bar{E}$. The change in physical meaning of E, M and P can deeply influence the procedure for establishing effective constitutive equations for actual materials.

In [K. 1] I attempted to justify the meaning of E, M and P by the solid state physics. From the microscopic point of view an elastic deformation,

[1]) Stanford University, Stanford, California, USA.
[2]) Institute of Solid State Physics, Prague, Czechoslovakia.

485

denoted in a phenomenological theory by E, is related to changes of distances between atomic planes and to rotations of crystallographic directions. The structural deformation M is connected with local disturbances of the crystal lattice and with changes of the crystallographic structure during phase-transformations. Visco-plastic[1]) deformation P, being a mode of deformation which just moves atoms into crystallographically equivalent positions, causes no changes in the crystal lattice. Solid state physics methods are able to distinguish these different modes of deformation and in some cases even to register them during a thermo-mechanical process (e.g. diffraction techniques are helpful especially for single crystals).

Two new kinematical concepts, the elastic motion and the quasi-elastic motion, introduced in [K. 1] are closely related to the described interpretation. The elastic motion $\hat{\chi}_t(\tau)$, $\tau \in \langle t, t_0 \rangle$, is supposed to be a motion of the body (generally only in the local sense) from the present configuration $\chi(t)$ during which elastic deformation is changing to the value this deformation had at the reference time t_0; the moving local elastic reference configuration $\hat{\varkappa}_t$ is then reached, i.e. $E^{-1}: \chi(t) \to \hat{\varkappa}_t$. Similarly the quasi-elastic motion requires to return simultaneously both elastic and structural deformations to their values at t_0; we have then the moving local quasi-elastic configuration $\varkappa'_t$; i.e. $(E')^{-1}: \chi(t) \to \varkappa'_t$. Alternatively $\varkappa'_t$ can be reached from $\hat{\varkappa}_t$ changing structural deformation to the level it had at t_0, i.e. $M^{-1}: \hat{\varkappa}_t \to \varkappa'_t$.

The elastic motion and the intermediate configurations $\hat{\varkappa}_t$ and $\varkappa'_t$ can be experimentally realized (at least in some special cases or in idealized or approximate sense)[2]). To produce the elastic motion and the configuration $\hat{\varkappa}_t$ we can utilize the property of solids that the response to a sudden change of deformation and temperature is elastic. An attempt to analyze the interrelation of sudden changes, unloading processes as well as the concept of elastic motion has been described in [K. 3]. To attain the intermediate configuration $\varkappa'_t$ both densities of crystal defects and phase-transformations which cause structural deformation are changed in a controlled way by suitable thermo-mechanical treatment leaving visco-plastic deformation constant.

---

[1]) The term 'visco-plastic' used in [K. 1] and here is not the most convenient expression. Many types of viscous effects are caused by rearrangement of crystal defects and hence fall into the category of structural changes.

[2]) The concept of quasi-elastic motion is convenient in the formulation of the theory, but its practical realization is probably very complicated.

If another decomposition of F is considered, e.g.

$$F = \bar{P}\bar{M}E, \tag{2}$$

first the intermediate configuration $\bar{\varkappa}_t$ have to be specified, $\bar{P}^{-1} \colon \chi(t) \to \bar{\varkappa}_t$; hence we face the difficult problem how to produce plastic deformation without elastic or structural changes. We encounter similar difficulties when attempting to realize intermediate configurations related to other types of the decomposition of F.

Summarizing, it is not difficult to formulate theories analogical to that proposed in [K. 1] which are based on different decompositions of F, but from the physical point of view the decomposition (1) has a certain advantage. It seems to me that the decomposition (1) has, at least at present, the greatest chance to be experimentally realized.

[K. 1] Kratochvíl, J., Finite-strain theory of inelastic behaviour of crystalline solids, *Foundations of Plasticity*, Noordhoff Int. Publ., Leyden 1973, 401–415.

[K. 2] Clifton, R. J., On the equivalence of $F^e F^p$ and $F^p F^e$, *Trans. ASME, J. Appl. Mech. 39* (1972) 287.

[K. 3] Kratochvíl, J., On a finite strain theory of elastic-inelastic materials, *Acta Mechanica 16* (1973) 127.

Th. Lehmann

On large elastic-plastic deformations

Foundations of Plasticity
pp. 571–585

DISCUSSION

P. Germain.[1])   When you write $g_{ik} - \mathring{g}_{ik}$ the result may be regarded to be either the components of the Lagrange-Green strain tensor or the components of the Euler-Almansi strain tensor. When you want to compute material derivatives such a distinction is important. How do you interprete $g_{ik} - \mathring{g}_{ik}$ of the formula (8)?

Th. Lehmann.[2])   The quantity $g_{ik} - \mathring{g}_{ik}$ as well as the quantity $\mathring{g}_{ik} - g_{ik}$ may be interpreted in different manner as Sedov [L. 1] shows. In my paper all these quantities are related to the base of the deformed body-fixed coordinate-system. The material derivative has to be calculated with respect to this reference system (see foot-note to Eq. (5)).

[L. 1] Sedov, L. I., Foundations of the nonlinear mechanics of continua, Pergamon, Oxford, 1966.

[1]) Université Paris VI, Paris, France.
[2]) Ruhr-Universität, Bochum, FRD.

T. H. Lin

Microstress fields of slip bands and inhomogeneity of plastic deformation of metals

Foundations of Plasticity

pp. 77–91

DISCUSSION

K. S. Havner.[1])   Professor Lin is to be congratulated on a most interesting paper, which is a significant step toward an understanding of the formation and growth of discrete slip bands in metals. He has demonstrated once again (as in earlier papers) that for representing physical behavior at a scale of 10^{-4} mm $(0.1\ \mu)$, continuum inhomogeneous slip models are equally as viable as models based upon continuous distributions of dislocations. The grossly inhomogeneous nature of deformation at this level, well-known from experiment and demonstrated in the paper through analytical calculations, also has been pointed-out by the discusser [H. 1] and previously emphasized by Hill [H. 2].

The paper is quite well-done, but I have one or two questions, and there appear to be a few inconsistencies which should be checked. The latter are undoubtedly only copying errors, however, and certainly do not detract from the overall high quality of the paper. In brief:

(a) The right-hand sides of both (12) and (13) are twice the right-hand side of (11) evaluated along the respective coordinate axes. It seems either a 2 is missing from the numerator of (11) or else all subsequent expressions for τ_R are twice the correct values.

(b) From (21) and (22), the sign preceding $3Ae^p$ in both (23) and (24) should be positive, not negative. In fact, were it negative the inequality following (23), $\tau_{I_0} - \tau_{I_y} > 0$, would not be strictly correct.

[1]) North Carolina State University, Raleigh, N.C., USA.

In the last set of calculations there is defined 'a fine-grain, aluminum polycrystal of cubic shape and 40μ in linear dimension'. I believe Professor Lin intends *each* crystal grain to be cubic and of dimension 40μ, with the dimensions of the polycrystal considered very large in comparison. These calculations are among the most interesting in the paper, giving an approximate determination of the highly heterogeneous plastic strain within the grain as a sequence of slip bands forms and demonstrating how certain previously active bands become inactive as loading progresses.

It would be of great value if the slip-band model and analysis were extended to include consideration of the actual elastic anisotropy of metal crystals. Perhaps Professor Lin could comment upon the availability (or feasibility) of solution for a point body force in an infinite, elastic, *orthotropic* medium.

[H. 1] Havner, K. S., An analytical model of large deformation effects in crystalline aggregates, *Foundations of Plasticity*, Noordhoff Int. Publ., Leyden 1973, 93–106.
[H. 2] Hill, R., *Surveys in Mechanics* (G. I. Taylor 70th Anniversary Volume), Cambridge University Press, (1956) 7–31.

F. McClintock.[1]) Can the method be extended to the practically important case of precipitate particles of the order of 200–300 Angströms in size, where slip bands are much shorter and finer?

M. P. Wnuk.[2]) How does your model provide for explanation of plastic strain accumulation under the cyclic load?

T. H. Lin.[3]) I want to thank Professor Havner for his kind comments. In answering his specific question (a) and (b), the '2' in Eq. 11 should be deleted, and the sign preceding $3Ae^p$ in both Eqs. 23 and 24 should be positive instead of negative, as pointed out by Professor Havner. The final version includes these corrections.

In the last set of calculations, 'a fine-grain aluminum polycrystal of cubic shape and 40μ in linear dimension', I mean the individual crystals are assumed to be cubic and of a linear dimension of 40μ, with the dimensions of the polycrystal very large in comparison to that of the individual crystal.

[1]) Massachusetts Institute of Technology, Cambridge, Mass., USA.
[2]) South Dakota State University, Brookings, S.D., USA.
[3]) University of California, Los Angeles, Calif., USA.

490

I do not know of any close-form solution of the stress field caused by a point body force in an infinite, elastic orthotropic medium. However solution of the problem is available in series form, see 'On the construction of the Green Tensor for the Basic Equation of the theory of Elasticity of an Anisotropic Infinite Medium' by Lifshitz, I. M. and Rozentzweig, L. N., *Journal of Experimental and Theoretical Physics*, U.S.S.R. 17:783–791, 1947.

Referring to Professor McClintock's question I may only add that the method should be able to be extended to slip bands much shorter and finer. The analogy between body force and plastic strain gradient is applicable to slip bands of different shapes.

Professor Wnuk's question requires some comments. Lattice imperfections exist in all metals and give rise to an initial heterogeneous microstress field. Consider two *closely located* thin slices P, Q in a most favorably oriented crystal located at a free surface of a polycrystal. A small initial resolved shear stress is assumed to be positive in slice P and negative in slice Q. To simplify the calculation, this initial resolved shear stress is further assumed to vanish elsewhere. Under the initial tensile loading, the algebraic sums of the initial and the applied resolved shear stress in P is larger than that in Q and hence P will slide first giving plastic strain. The plastic strain remains after unloading and the stress field caused by the plastic strain is called the residual stress field. The resolved residual shear stress field is *continuous*. Hence slip in P relieves the positive resolved shear stress not only in P but also in Q. This helps to keep Q from reaching the positive critical shear stress in the tensile loading and also increases the negative resolved shear stress that helps slip in Q during the reversed loading. When the negative resolved shear stress in Q reaches the critical shear stress during the reversed loading, plastic strain occurs in Q and a new residual stress field is produced. The new residual resolved shear stress field increases the positive resolved shear stress not only in Q but also in P. This increase of positive resolved shear stress in P helps to keep the resolved shear stress in P from reaching the critical shear stress during the reversed loading and also helps P to slide in the subsequent forward loading. This process is repeated for every loading cycle and hence gives plastic strain accumulation under cyclic load.

G. Maier

A shakedown matrix theory allowing for workhardening and second-order geometric effects

Foundations of Plasticity
pp. 417–433

DISCUSSION

M. Życzkowski.[1])   A general theory, proposed by Maier, may be illustrated by simple examples showing the importance of geometrical effects in the plastic shakedown problems. Consider a system consisting of a slender rigid block and of two deformable, stable in compression, perfectly elastic-plastic bars, subject to the vertical force P_1 and the horizontal force P_2 varying in prescribed limits, Fig. A. The behaviour of such a system under quasistatic loadings has been discussed in [Z. 3]. Similar systems, but subject to one compressive force only, have been used in the theory of elastic-plastic stability under conservative, [Z. 1], and non-conservative loadings [Z. 2].

The system is externally and internally statically determinate and the classical Melan's theorem cannot be applied since the residual stresses simply cannot develop. Disregarding geometrical effects we find the reactive forces

$$R_1 = \frac{P_1}{2} + \frac{L}{a} P_2, \quad R_2 = \frac{P_1}{2} - \frac{L}{a} P_2. \tag{1}$$

The elastic carrying capacity (equal here to the limit carrying capacity) is described by a square in the dimensionless coordinates $p_1 = P_1/\sigma_0 F$, $p_2 = LP_2/a\sigma_0 F$ (σ_0 being the yield point and F cross-sectional area of the deformable bars):

$$p_2 = \pm (p_1 \pm 1). \tag{2}$$

[1]) Technical University, Cracow, Poland.

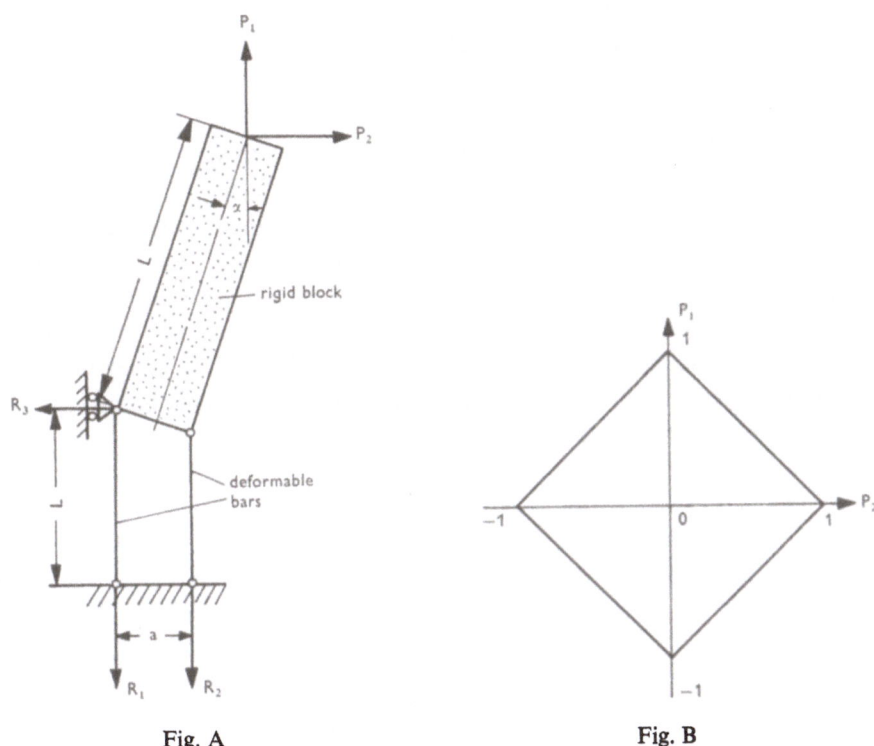

Fig. A   Fig. B

Only the inside of the square corresponds here to the shakedown region, Fig. B.

This conclusion changes in an essential manner if the geometrical effects were taken into account. At first, instead of a square we obtain the limit curve in the form of a curvilinear tetragon, partially convex and partially concave. The equations of equilibrium in the deformed state (under the assumption of small displacements)

$$R_1 + R_2 = P, \quad R_2 a = -P_2 L + P_1\left(\frac{a}{2} + \alpha L\right),$$ (3)

and Hooke's law applied to both bars

$$(R_1 - R_2)\frac{l}{EF} = a\alpha$$ (4)

(where α stands for the deflection angle of the rigid block) determine the reactive forces R_1 and R_2. The equations $R_1 = \pm\sigma_0 F$ and $R_2 = \pm\sigma_0 F$ describe now the limit curve as follows

$$p_2 = \pm(p_1 \pm 1)(\mu p_1 + 1), \tag{5}$$

where $\mu = 4Ll\sigma_0/a^2 E$ is a parameter connected with the slenderness and elasticity of the system.

Further, considering shakedown of the system under variable repeated forces, we may determine a one-parameter family of shakedown regions. Residual stresses cannot exist, but residual strains are possible. They result in a residual mean elongation Δl_0 of the deformable bars and in a residual deflection angle α_0 of the rigid block. The quantity Δl_0 is of no influence on the further behavior of the system, but α_0 appears in the modified Eq. (4) of subsequent elastic changes, namely

$$(R_1 - R_2)\frac{l}{EF} = a(\alpha - \alpha_0). \tag{6}$$

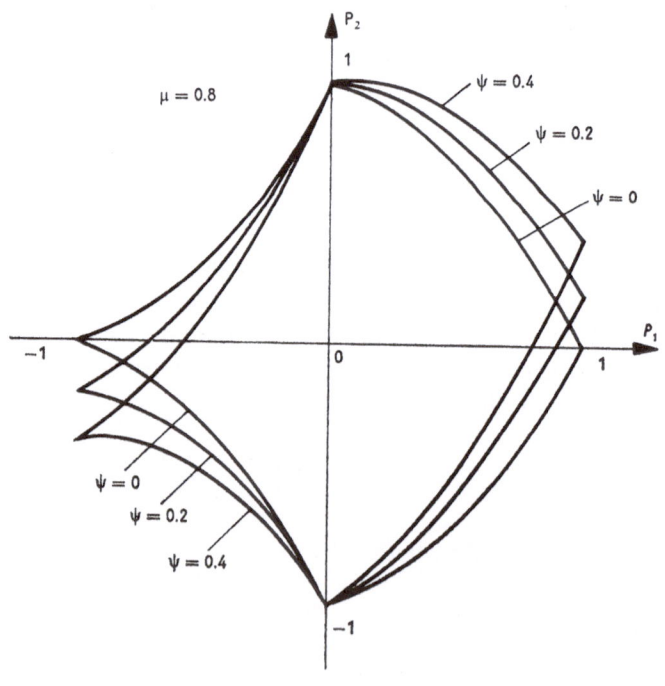

Fig. C

The modified limit curves, determining the shakedown regions, are as follows

$$p_2 = \pm (p_1 \pm 1)(\mu p_1 + 1) + \psi p_1, \tag{7}$$

where the parameter ψ equals $\psi = (2L/a)\alpha_0$; they are shown in Fig. C. Thus the geometrical effects make it possible to discuss the family of shakedown regions in the problem which is quite trivial if the effects are not taken into account.

[Z. 1] Panovko, Inzh. Sbornik *20* (1954), 160–163.

[Z. 2] Podstrigach, Ya. S. and Yu. A. Tchernukha, Prikladnaya Mekhanika *8* (1972), 9, 3–9.

[Z. 3] Życzkowski, M., Complex loadings in plasticity (in Polish) PWN, Warszawa 1973.

G. Maier.[1])   The discussion by Zyczkowski and the note by König [M. 1] are welcome supplements to the paper. The importance of paying attention to the shakedown criteria is rightly emphasized by König. The safety criterion against local failure by low cycle fatigue as expressed by inequality (3.8) of his note, clearly implies safety also against incremental collapse as understood traditionally (configuration changes unbounded in time or 'divergent'); however it cannot rigorously guarantee that excessive deflections occur eventually and cause unserviceability. The same holds for the less stringent criterion (1.3) of bounded plastic strains; this condition, therefore, is less realistic and accurate than the above criterion only as far as low cycle fatigue is concerned. Some structures, like e.g. frictional systems, may be reasonably thought of as capable of unlimited local dissipation. For such systems incremental collapse may be regarded as the only failure phenomenon and the relevant safety criterion should ensure that some meaningful displacements never exceed a given 'critical' value, or, at least, that a suitable norm of the displacement field is bounded in time. Also this kind of shakedown criteria, besides those discussed by König, might be the basis for a physically sound and sometimes technically useful theory of the adaption under variable repeated loads.

[M. 1] König, J. A., On shakedown criterion, this volume, 408–412.

[1]) Polytechnic of Milan, Italy.

Z. Mróz

A description of workhardening of metals with application to variable loading

Foundations of Plasticity
pp. 551–570

DISCUSSION

O. Bruhns.[1]) In connection with Professor Mróz' contribution to this symposium I would like to present a new concept of describing cyclic deformation processes.

In the past ten years an extensive research has been carried out concerning elasto-plastic behavior and low-cycle fatigue of metals under cyclically varying loads. Experiments regarding complex loading histories have shown that the agreement between experimental and theoretical results is not satisfactory within the usual assumptions: a single yield criterion, the associated flow rule based and either isotropic or kinematic hardening law. In a cyclic uniaxial tension-compression test with fixed total strain range, the stress range increases or decreases according to the employed material tending to a fixed value corresponding to a steady cycle. In theoretical considerations, however, the isotropic workhardening model predicts an elastic behavior after a certain number of cycles, whereas the kinematic workhardening rule predicts a steady plastic cycling after the first cycle of loading. Predictions of both theories of hardening are not in agreement with experimental observation.

In order to describe the behavior of metals under cyclic loading models consisting of assemblages of simple elasto-plastic sub-elements have been proposed by Besseling [B. 1], Iwan [B. 2]. Mróz [B. 3] developed the so called model of 'Workhardening moduli'. Backhaus [B. 4] allows for de-

---

[1]) Ruhr-Universität, Bochum, G.F.R.

formation of the initial yield surface in addition to expansion and translation of this surface in the stress space.

In this note I propose a more general nonlinear hardening rule allowing to describe qualitatively some typical phenomena observed in uniaxial and multiaxial stress or strain cycling. All considerations are made within a finite elasto-plastic deformation theory. The deformations are assumed to be isothermal and depending only on the stress or strain history. Furthermore, we assume the plastic part of the deformations to be incompressible.

Elastic $\bar{d}_k^i$ and plastic $\tilde{d}_k^i$ strain rates satisfy the relation

$$d_k^i = \bar{d}_k^i + \tilde{d}_k^i \tag{1}$$

Let ε_k^i denote the strain tensor, d_k^i the strain rate tensor, and $(\ )|_0$ covariant derivation with respect to time, according to Zaremba and Jaumann so that

$$\varepsilon_k^i|_0 = \frac{\partial}{\partial t}\,\varepsilon_k^i + d_r^i\varepsilon_k^r - d_k^r\varepsilon_r^i, \tag{2}$$

σ_k^i and τ_k^i denote the stress tensor and the stress deviator respectively. All quantities are referred to a material-fixed coordinate system.

The constitutive relations consist of stress-strain rate relations, a yield condition and a hardening law

a) A deformation law in the form

$$d_k^i = \lambda\,\frac{\partial F}{\partial \sigma_k^i} + \varkappa\tau_k^i|_0 + \bar{d}_k^i \tag{3}$$

combines the two commonly used theories. The first part of (3) corresponds to a deformation law of the 'incremental theory' and together with the elastic part is known as the Prandtl-Reuss law. The second term together with the elastic strain rate represents an incremental form of the 'deformation theory'. The elastic part of the deformation is

$$\bar{d}_k^i = \frac{\lambda}{2G}\,\tau_k^i|_0 + \frac{1}{9K}\,\sigma_r^r|_0\delta_k^i \tag{4}$$

where G denotes the shear modulus and K stands for the bulk modulus. The parameter $\varkappa$ enables to vary the contribution of the 'deformation theory' in the deformation law. Unloading is assumed to be elastic.

b) The yield condition is taken as a special form of the general relation

$$F(\sigma_k^i;\ldots) = \underset{0}{A} + \underset{1}{A_k^i}\tau_i^k + \underset{2}{A_{ks}^{ir}}\tau_i^k\tau_r^s + \ldots = 0 \tag{5}$$

namely

$$F(\sigma_k^i; \ldots) = (\tau_k^i - \alpha_k^i)(\tau_i^k - \alpha_i^k) - k^2(w) = 0, \tag{6}$$

This class of yield conditions developed by Melan, Prager, Ziegler, Shield, Kadashevitch and Novozhilov includes isotropic as well as kinematic hardening, but does not account for deformations of the shape of the yield locus. Let

$$\alpha_k^i|_0 = c\bar{d}_k^i$$

denotes the translation of the yield surface while $k^2(w)$ in (6) gives a measure for its expansion. The intensity of kinematic hardening is described by the parameter c, and w stands for the plastic work.

c) Usually a hardening rule is assumed in such a manner, that the constitutive equations lead to linear hardening in the simple tensile test. It is here where the new concept will be introduced that reflects a nonlinear material behavior in the tensile test:

$$k^2(w) = k_0^2 \left\{ \left(1 + \frac{Bw}{k_0^2} \right)^{1/m} \left[1 - \zeta \left(1 + \frac{Bw}{k_0^2} \right)^{(m-2)/m} \right] + \zeta \right\}^2 \tag{7}$$

Eq. (7) includes the case of linear hardening for $m = 2$. Here and after $0 < \zeta \leq 1$. The deformation law (3) eventually becomes

$$d_k^i = \frac{\left[(1 - c\varkappa)(\tau_s^r - \alpha_s^r) - \frac{dk^2}{2dw} \tau_s^r \varkappa \right] \tau_r^s|_0}{ck^2 + \frac{dk^2}{2dw}(\tau_s^r - \alpha_s^r)\tau_r^s} (\tau_k^i - \alpha_k^i) + \varkappa \tau_k^i|_0 + \bar{d}_k^i$$

In a tensile test this constitutive law leads to a nonlinear hardening independent of the values of ζ and $\varkappa$, except for the case $m = 2$. The influence of these parameters shows only in other loading processes, as for example in simple shear.

I shall demonstrate the application of the proposed constitutive laws in the simple cyclic tension-compression test between fixed limits of deformation. The results concern a brass alloy 70/30 with

$$E = 780 \text{ Mp/cm}^2, \quad \sigma_0 = .9 \text{ Mp/cm}^2 \quad \text{and} \quad B = 52.9 \text{ Mp/cm}^2,$$

the resulting exponent is $m = 2.83$.

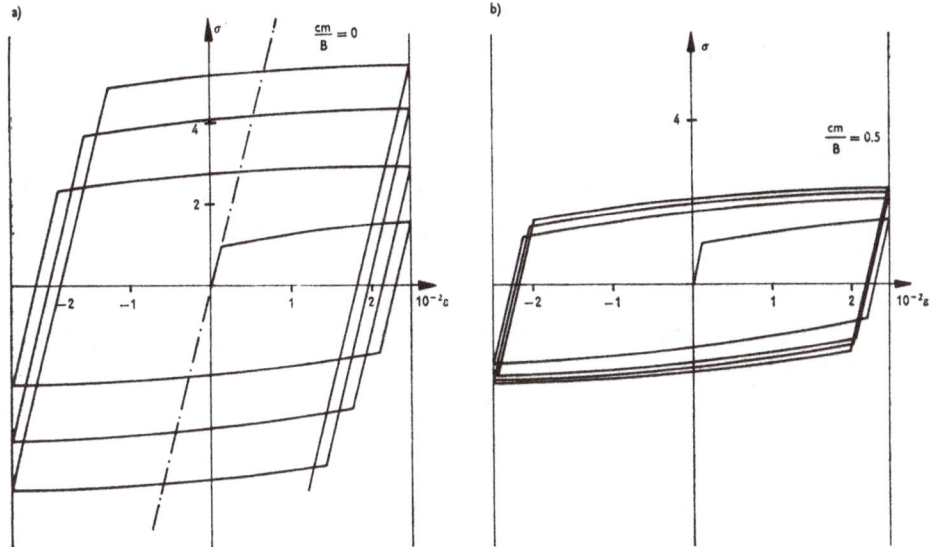

Fig. A. Uniaxial cycling, $m = 2.83$.

In Fig. A the stress σ is plotted versus the strain within fixed limits. Fig. Aa concerns a purely isotropic hardening ($cm/B = 0$). This model leads to a pure elastic behavior (dash-dotted line) after a transistory period of permanent strain increase. The same is observed for a linear hardening model. But every change of the parameter cm/B, (Fig. Ab, $cm/B = .5$) makes the loading cycle tending to a limit state after an initial phase of increase. It can concluded therefore that the proposed nonlinear model better fits experimental observations than the simple theories of pure isotropic or kinematic hardening with linear workhardening rule.

[B. 1] Besseling, J. F., A theory of plastic flow for anisotropic hardening in plastic deformation of an initially isotropic material, *Natn, Aero. Res. Inst. Rep. S-410*, 1953.

[B. 2] Iwan, W. D., On a class of models for the yielding behaviour of continuous and composite systems, *Journ. Appl. Mech. 34* (1967) 612–617.

[B. 3] Mróz, Z., On the description of anisotropic workhardening, *Journ. Mech. Phys. Sol. 15* (1967) 163–175.

[B. 4] Backhaus, G., Zur Fliessgrenze bei allgemeiner Verfestigung, *Z. Angew. Math. Mech. 48* (1968) 99–108.

Z. Mróz.[1]) The model discussed by Dr. Bruhns is a combination of kinematic and isotropic hardening with non-linear stress-strain charac-

[1]) Institute of Fundamental Technological Research, Warsaw, Poland.

teristics. Such a model has already been discussed in the literature for small plastic strains. However, author accounts for geometry changes and uses convective stress rate. Moreover, he introduces the non-associated flow rule (3) which combines both flow and deformation theory of plasticity. Applicability of such a rule should thus be limited to loading paths lying in a certain conical region near the normal to the yield surface. For loading paths tangential to the yield surface, the plastic strain rate does not vanish and continuity condition between elastic and plastic regions is not satisfied. It is interesting that this law can be used to predict cyclic strain accumulation and to describe plastic hysteresis loops, but its general applicability should be limited.

G. I. N. Rozvany and S. R. Adidam

Recent advances in optimal plastic design

Foundations of Plasticity
pp. 201–217

DISCUSSION

S. Kaliszky.[1]) Authors give a compact summary of recent advances in plastic optimal design and, in addition, extend the existing theorems to non-convex cost functions, alternate load systems and multi-component systems.

The illustrative examples represent valuable contributions to the subject. Some of them, however, lead to extreme solutions and therefore cannot form a direct basis for practical design. The reason is that in the specific cost function ψ a minimum volume (minimum thickness or reinforcement etc.) requested from structural point of view is not taken into consideration, Fig. A. Apparently, one can easily overcome this difficulty by introducing

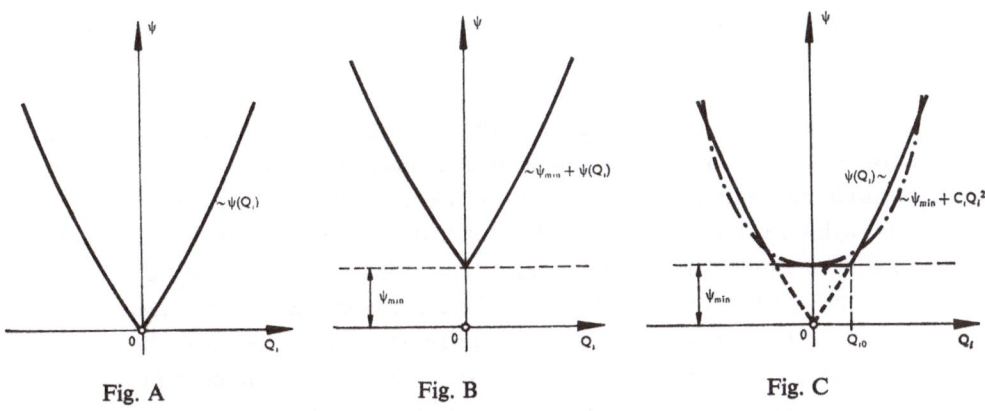

Fig. A   Fig. B   Fig. C

[1]) Technical University, Budapest, Hungary.

501

an additive constant ψ_{min} in the specific cost function, which represents the necessary minimum volume (thickness, reinforcement etc.) (Fig. B):

$$\psi = \psi_{min} + \psi(Q_i).$$

Since from the point of view of variational calculus an additive constant is irrelevant, this change does not influence the optimal solution. It means, that the minimum volume can be applied arbitrarily and independently from the determination of internal forces. This way of solution, however, is not economical because in the calculation the contribution of the minimum volume to the balancing of internal forces is not taken into consideration. Consequently, we loose the cost of the minimum volume.

One can utilize the necessary minimum volume by introducing two specific cost functions as below (Fig. C):

$$\psi = \psi_{min} \ \text{if} \ |Q_i| \leqq Q_{i0},$$

$$\psi = \psi(Q_i) \ \text{if} \ |Q_i| > Q_{i0}.$$

In such a manner we can obtain more economical and realistic solutions but, on the other hand, the consideration of two cost functions leads to more complicated calculations. The difficulty can be overcome by approximating the actual functions by a single smooth cost function. In some cases (e.g. reinforced concrete beams, slabs and shells; sandwich beams) good approximation can be obtained by using a quadratic specific cost function:

$$\psi = \psi_{min} + \sum C_i Q_i^2.$$

This approximation reduces the optimal plastic design to a special elastic solution.

W. Szczepiński.[1]) Congratulating the authors on this interesting work, I would like to make one complementary remark. In the presented paper are considered relatively simple structural elements like beams, frames, plates and cylindrical shells. It might be worthwhile to mention here that plastic design methods can be applied also to another types of structures. As an example I shall show how one of these methods, namely the method of piecewise-homogeneous statically admissible stress fields, may be useful in mechanical engineering design. From the limit-design theorems of the theory of plasticity the following conclusion results:

[1]) Institute of Fundamental Technological Research, Warsaw, Poland.

A safe estimate of the shape and dimensions of the structural element may be obtained on the basis of an appropriate statically admissible stress field. The boundaries of such a stress field specify the external surface of the element. Selecting the field with the minimum area one obtains the smallest volume of the element in the considered class of statically admissible stress fields.

Numerous examples of practical application of this technique have been described in [S. 1, S. 2]. Three examples are shown below.

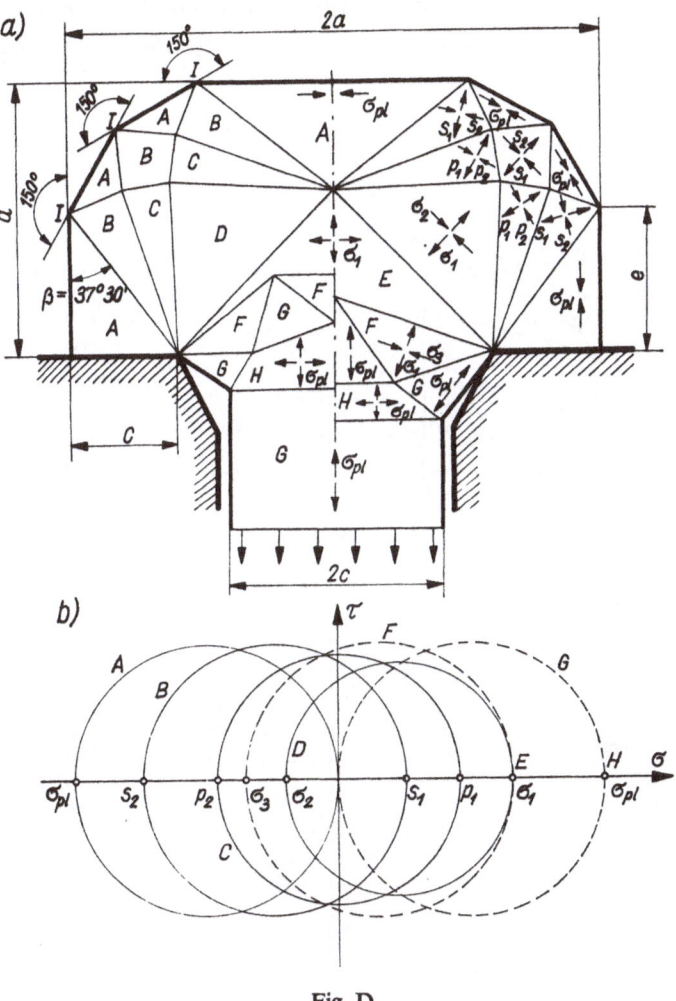

Fig. D

Consider the design of the head of a tension member shown in Fig. D. Assume that the thickness of the whole element is constant and that the ratio a/c is given. It may be shown that the solution presented in the figure, constructed for the Tresca yield criterion, is valid both for plane strain and plane stress conditions. Thus it may be applied for elements with arbitrary thickness.

The yield load for the lower part is $P = 2c\sigma_{pl}$ per unit thickness. The safe estimate of the shape of the head may be obtained by means of any statically admissible stress field corresponding to this value of the load. The upper part of the stress field forming the contour is composed of two symmetrical fields each with the axis of symmetry, making an angle of 45°

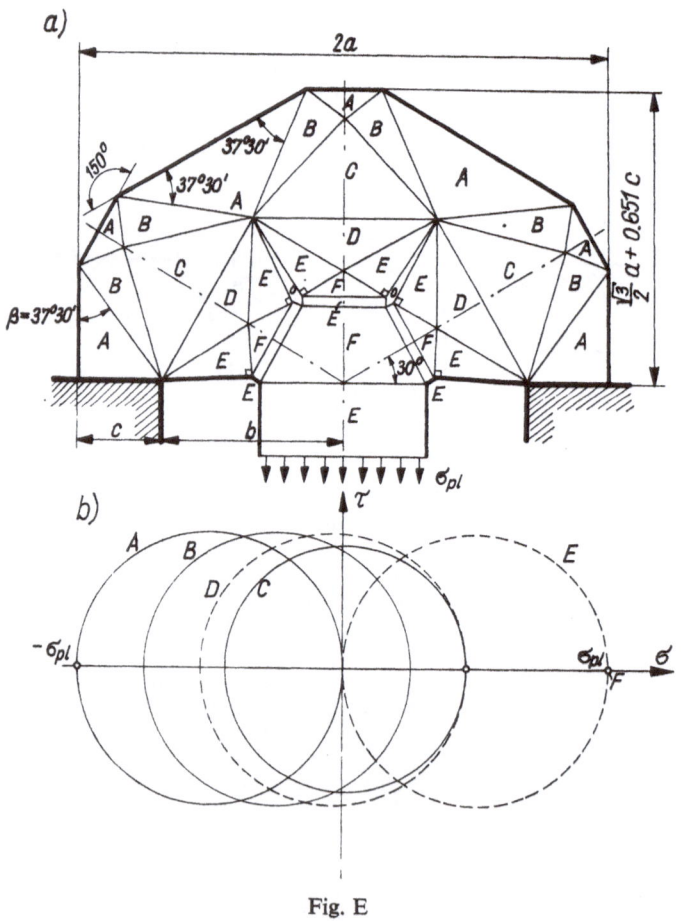

Fig. E

with the vertical axis. The angles at all corners I are assumed to have the same value of 150°. If the triangles A and B are to be in the plastic state, the lines of stress discontinuity separating them must form an angle $\beta = 37°30'$ with the corresponding sectors of the outer contour. The regions C, D and E are stressed below the yieldpoint.

For the lower part of the element, two possible solutions are shown on

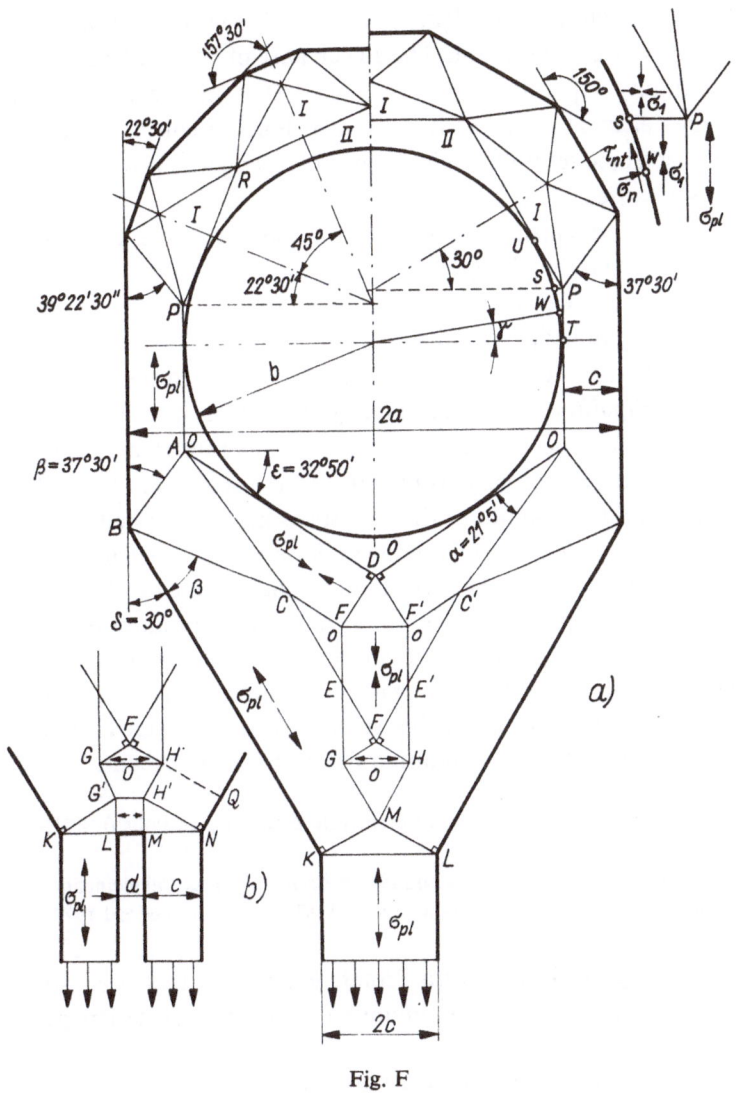

Fig. F

505

both sides of the vertical axis, Fig. Da. The left-hand side solution gives more economical contour of the connecting region between the head and the body. The regions F and G are plastic. The region H is subjected to isotropic tension.

For larger values of the a/c ratio, a more economical contour of the head may be obtained using the stress field shown in Fig. E. The upper part of the field is composed of three elementary fields whose axes of symmetry form angles of 60°. Regions A and B are plastic, while the stress in the triangles C remains below yield the point. The lower part of the field below the lines of discontinuity separating regions C and D is also composed of three elementary fields. The regions E are under uniaxial tension. The regions F are under isotropic tension, while the narrow wedge-shaped regions marked by 0 are stress-free.

In Fig. F is presented a stress field for the head of a tension member connected with the remaining elements of the structure by means of a bolt located in the hole. Consider first the upper part of the field. On both sides of the axis of symmetry two possible solutions are presented, the left-hand solution giving a more economical contour. For the lower part of the element two configurations are presented (out of many possible). In the left-hand solution the rod is composed of two separate strips. The distance between both strips may vary from zero to the value equal to the length of the sector GH. Such a solution with two strips may be applied if the rod has an I-section, and if the thin wall joining both strips is assumed to be stress-free.

Strictly speaking the method used above should not be classified as one of the methods of optimum design in terms of the optimum design theory. However, the statically admissible stress field technique has been proved very effective for structural elements of complex shape for which other methods cannot be applied. In this sense this method can be treated as a preliminary approach to the optimum design of such elements.

[S. 1] Szczepiński, W., Plastic Design of Structural Elements, (in Polish), PWN, Warszawa 1968.

[S. 2] Szczepiński, W., Limit analysis and plastic design of structural elements of complex shape, Progress in Aerospace Sciences, Vol. 12, Pergamon, Oxford 1972.

G. I. N. Rozvany.[1]) I agree with Professor Kaliszky that it is often necessary to introduce a specified minimum value for certain design parameters

[1]) Monash University, Clayton, Victoria, Australia.

and then the corresponding modified cost function can be approximated by a quadratic function. The first derivatives of the latter being linear, the optimal solution on the basis of the criteria outlined by the authors would also give the stress field for a linearly elastic structure, confirming earlier results by Professor Kaliszky [R. 1]. Alternatively, the specific cost function can be replaced by a tri-linear function which can be handled by the methods outlined in the paper under discussion. This approach was discussed in a previous paper, [R. 2].

Professor Mróz's note [R. 3] is found to be particularly valuable since it presents a formulation that is expressed in terms of design variables which are not necessarily generalized stresses. In [R. 4] it is assumed that a specific cost component $\bar{\psi}_l$ is a given function of the stress vector Q. If the local geometrical properties (section properties) are not given uniquely by the stress vector Q then a local optimization is, indeed, necessary. This is not necessarily a disadvantage because, in effect, the optimization procedure is partitioned into two separate operations.

Splitting the specific cost function into separate components, $\psi = \sum \psi_l$ (when possible) does not make the formulation more general but simplifies the solution of the problem considerably.

As Professor Save has pointed out at this symposium, the formulation given in [R. 4] is more general than the one proposed by Mróz, since it is not restricted to minimum weight design and can be applied to problems in which the specific cost is *not* a linear function of the specific volume. Professor Mróz's first equation under (13) clearly states that in his formulation the specific cost *is* the specific volume. This explains that he has obtained a criterion in terms of uniform mean energy dissipation.

The first author would like to add the following two comments on Professor Mróz's formulation:

1) It appears that optimality criteria obtained by Prager and Shield and in [R. 4] are more explicit because they take the form of a strain-stress relation in which the strains are given by the gradient of the cost function. Using this approach, a problem of optimization reduces to a problem of analysis. If the optimality condition is expressed in the terms of uniform energy dissipation, as in various papers by Drucker, Shield, Mróz and Save, the same optimal solution can be obtained, but the uniform energy dissipation statement must first be converted into a strain-stress relation.

2) To prove the sufficiency of a necessary condition, I do not find it necessary to use the procedure outlined in Eqn. (6) through (9) by Professor

507

Mróz. In my derivations, Euler-Lagrange equations are used for obtaining
necessary optimality criteria for convex cost functionals. Since the minimum
is always unique in terms of the cost value for such functionals, the necessary
conditions give the unique minimum cost value, if a feasible optimal solution
exists. Hence the same conditions are both necessary and sufficient for all
convex cost functions.

Finally, I would like to give optimality criteria for the most general
problem in optimal plastic design, taking some of Professor Save's recent
work [R. 5] into consideration.

The problem can be formulated as follows:

$$\min_{Q_{hn}} \Phi = \int_D \sum_{l=1}^{t} \psi_l d\xi \tag{1}$$

such that $D = \bigcup_{\alpha \in I} R_\alpha$ with $R_z \cap R_w = 0$ for all $z \in I$, $w \in I$; on R_α, $\psi_l = \theta_{l\alpha}\gamma_{l\alpha}(\xi)$ such that

$$\psi_l \geq \bar{\Psi}_l(Q_{h,\eta}) \quad (h = 1, 2, \ldots, r), (l = 1, 2, \ldots, t), (\eta \in D_h \subseteq D) \tag{2}$$

where $Q_{h,\eta}$ is a state of stress that equilibrates a moving load system h,
the position of which is defined by the vector $\eta \in E^n$,

Φ is the total cost,
D is the structural domain,
ψ_l are design values of cost components,
R_α are subdomains over which the strength distributions are defined by
the 'shape function' $\gamma_{l\alpha}(\xi)$,
$\psi_l(Q)$ is a specific cost component function,
$\xi \in E^n$ where E^n is n-dimensional real Euclidean space.
The optimal solution is then associated with
(a) statically admissible stress fields $Q_{h,\eta}(\xi)$ and
(b) kinematically admissible strain fields $q_{h,\eta}(\xi)$ such that
(c) the strain vector at a point ξ for load h in position η is

$$q_{h,\eta} = \sum_{l=1}^{t} d\lambda_{lh\eta} G_{lh\eta}(\xi) \tag{3}$$

where $G_{lh\eta}$ is the cost gradient vector associated with load h in position η
and $d\lambda_{lh\eta}$ satisfies the following conditions:
(i) $d\lambda_{lh\eta}(\xi)$ is non-zero only if the value of the cost component function
$\bar{\Psi}_l(Q_{h,\eta})$ equals the design value ψ_l of such component at ξ, and

(ii) for each subdomain R_α and for each component $l = 1, 2, \ldots, t$

$$\int_{R_\alpha} \gamma_{l\alpha}(\xi)\mathrm{d}\xi = \sum_{h=1}^{r} \int_{R_\alpha} \int_{D_h} \gamma_{l\alpha}\mathrm{d}\lambda_{h,\eta}\mathrm{d}\xi \tag{4}$$

where

$$\mathrm{d}\lambda_{h,\eta} = \sum \frac{\partial \lambda_{h,\eta}}{\partial \eta_i} \, \mathrm{d}\eta_i$$

and $\eta = (\eta_1, \eta_2, \ldots, \eta_i, \ldots, \eta_n)$

The *RHS* of Eqn. (1) and Eqn. (2) give an *upper bound* on the minimum cost for statically admissible stress fields $Q_{h,\eta}$.

Denoting the *h*-th load system in position η by $p_{h,\eta}$ and the displacement fields corresponding to strains $q_{h,\eta}$ by $\mathrm{d}u_{h,\eta}$

$$\Phi = \sum_{h=1}^{r} \int_{D} \int_{D_h} p_{h,\eta}\mathrm{d}u_{h,\eta}\mathrm{d}\xi \tag{5}$$

gives a *lower bound* on the minimum cost if kinematic admissibility is fulfilled by $\mathrm{d}u_{h,\eta}$ with strain satisfying (3) and (4) in which $G_{lh\eta}(\xi)$ corresponds to *any* state of stress and the cost function $\bar{\psi}$ satisfies

$$\bar{\psi}(kQ) = k\bar{\psi}(Q) \text{ for all } \geq 0$$

All optimality criteria for various classes of problems of optimal plastic design and the upper and lower bound theorems of plastic limit analysis can be derived from the above statements.

Although Professor Szczepinski's contribution is not really an optimization problem, it appears to be an interesting method for the plastic analysis of complex systems.

[R. 1] Kaliszky, S., Economic Design by the Ultimate Load Method, *Conc. Constr. Engrg. 60* (1965) 365–372.

[R. 2] Charrett, D. E. and G. I. N. Rozvany, On Minimal Reinforcement in Concrete Slabs, *Archives of Mechanics, 24* (1972) 80–103.

[R. 3] Mróz, Z., Optimal design criteria for reinforced plates and shells, this volume, 425–429.

[R. 4] Rozvany, G. I. N., Optimal Plastic Design for Partially Preassigned Strength Distribution, *J. Optimiz. Theory Appl. 11* (June 1973).

[R. 5] Save, M. A., A Unified Formulation of the Theory of Optimal Plastic Design with Convex Cost Function, *J. Struct. Mech. 1* (1972) 247–276.

E. Shiratori, K. Ikegami and K. Kaneko

Subsequent yield surface determined in consideration of the Bauschinger effect

Foundations of Plasticity
pp. 477–489

DISCUSSION

H. Lippmann.[1]) It is a remarkable result of the authors to have experimentally determined yield surface in the three-dimensional stress-space. On the other hand, it seems to be somewhat puzzling that these surfaces already after small pre-strains do not include the origin (i.e., the stress state 'zero'). Therefore, the following questions arise:

1) How can this fact be interpreted? 2) Is this effect due to the uncertainty, mentioned by the authors themselves, regarding the definition of the yield point, especially after reversing the stress? 3) Can conclusions be drawn, or predictions be made, as regards the conditions encountered in metal forming where strains and pre-strains reach the magnitude of about 50% or even more?

O. A. Shishmarev.[2]) The authors have made an extensive (the most extensive I have ever met) study of the yield surfaces. They have to be congratulated for proposing an outstanding method of representing the surfaces in the deviatoric stress space. What is basically new here is that the yield surfaces are obtained and studied in a three-dimensional space.

I would like to draw attention to the following points still not sufficiently clear to me.

The authors use nominal stresses, which gives rise to an error in deter-

[1]) University of Karlsruhe, Karlsruhe, G.F.R.
[2]) Riazan, U.S.S.R.

510

mining the position of yield point on the surface. Under the largest obtained pre-strains of 5.1 per cent of equivalent strain the error amounted to 5 per cent in the normal stress and to 7 to 8 per cent in the circumferential stress. The form of the subsequent yield loci specified in the $\sigma_x - \tau_{xy}$ plane will therefore be somewhat distorted.

All points in the subsequent yield surface were obtained by means of a single specimen. This approach simplifies the experimental procedure and eliminates errors due to variable specimen dimensions and material non-uniformity. On the other hand, however this method introduces errors in determining the points of the yield locus. The errors are due to previous plastic deformation. I employed the foregoing method, only to become convinced of its imperfection. In Fig. A the curve a denotes the yield locus obtained by using the nickel specimens subjected to plastic strain of 1.1 per cent under tension. Each point on the locus was obtained from a different specimen. The curve b is associated with a locus, where all points were specified by means of a single specimen with the same amount of pre-straining. The yield points were determined at a fairly small reference strain (0.0006 per cent) but even then the curves diverge considerably as it can be seen in Fig. A. The experiment specifically designed to reveal the above effect led the authors to the conclusion that the effect is more pronounced on the opposite side. The effect is small in the vicinity of the point of loading and in my opinion this fact could be explained comparing the curves c and d

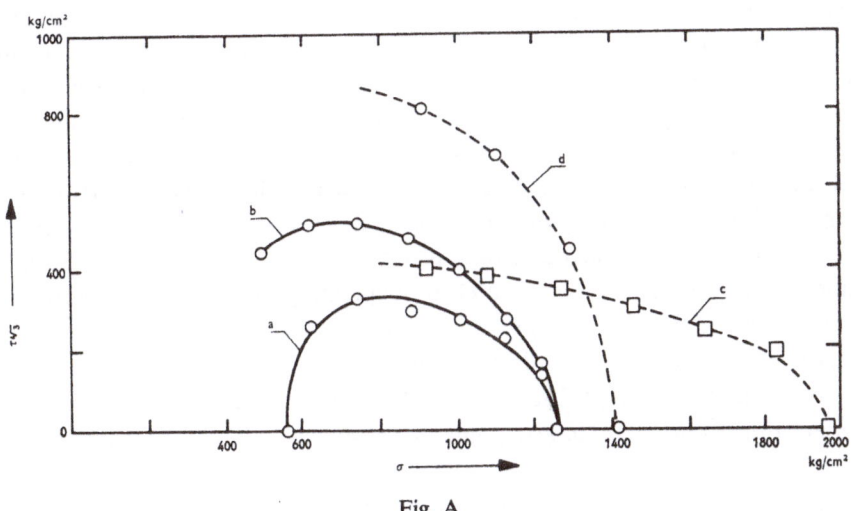

Fig. A

of Fig. A. The curves were obtained at the above specified reference strain using nickel specimens subjected to plastic pre-strain of 3.8 per cent in tension. The curve *c* was determined immediately after the initial loading, the curve *d* after a specimen had partially been unloaded and kept in this condition for 12 hr. Reloading the specimen in the direction of the initial prestrain the plastic strains occur much earlier then at the initial loading. Thus for some materials ageing results in weakening in the initial loading direction and in hardening in the perpendicular directions. For brass, having similar properties as those of nickel, the yield points specified using a single specimen will be influenced both by hardening resulting from the previous yielding and by the effect of ageing. This might well affect the accuracy of determining the position of the yield points even though the yield point definition sequence technique used by the authors is applied.

In studying the correlation between the shape of the yield locus and the reference strain chosen for determining the yield points the authors arrived at a somewhat paradoxical result: the yield point in loading moves in the opposite direction with respect to the prestrain. Since all points on the initial loading curve were obtained by means of a single specimen, partial unloading followed the determination of each point. The resulting inconsistency could be therefore attributed to the effect of ageing.

E. Shiratori.[1]) The authors thank Professor Lippmann for his discussion of this paper. The magnitude of the proof strain chosen when defining the yield point and not the uncertainty in determining the yield point, is responsible for the fact that the subsequent yield surface does not include the origin after small prestrain. For the material used in this experiment the stress-strain curve in reversing stress after prestrain is not a straight line up to the stress state 'zero', but the curved line as shown in Fig. B because of the Bauschinger effect. The authors regard this deviation from the straight line as yielding under the reverse stress, and define the stress point *K* corresponding to a given definite proof strain associated with the straight line *LM* as the yield point. Therefore, depending on the magnitude of the proof strain, the yield point in a reversing stress state lies above or below the 'zero' stress state (cf. Fig. 5 in the paper).

The behavior of the subsequent yield surface after large prestrain is very interesting from the practical point of view. Hence it will be necessary to

[1]) Tokyo Institute of Technology, Japan.

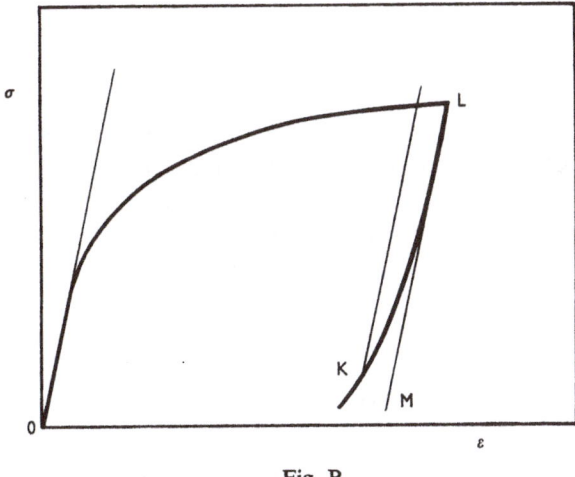

Fig. B

investigate the subsequent yield surface after large prestrain. The authors have not yet obtained experimental results for a large prestrain. A definite conclusion can not be drawn for such a case at present.

The authors wish to thank Professor Shishmarev for his useful comments.

A little difference may be observed between the shape of the subsequent yield surface represented by nominal stresses and that represented by true stresses. But, as Prof. Shishmarev points out, the effect is at most $7 \sim 8\%$ under the largest prestrain. Hence the authors do not consider that the difference has an important influence on the conclusion of the paper.

The subsequent yield surfaces shown in Fig. 5 were obtained by using one specimen to determine one yield point. There was little difference in shape between the subsequent yield surface for 0.02% proof strain in Fig. 5(a) and the 4th subsequent yield surface in Fig. 3(a), which have the same amount of prestrain.

In the authors' experiment, one yield surface was obtained within few hours. Hence the results will be free from the aging effect.

The dependence of the value of the yield stress on the proof strain in a preloading direction is more pronounced in the initial yield stress than in the subsequent yield stress. This causes that the ratio of the yield stress in the preloading direction moves in the opposite sense to the prestress point. This travelling of the point has no relation to the aging effect whatsoever.

O. A. Shishmarev

Experimental study on one type of plastic anisotropy not considered in simplified flow theories

Foundations of Plasticity
pp. 491–505

DISCUSSION

E. Shiratori.[1]) This paper presents an interesting method of modification of the flow theory in plasticity. But, in the author's experiment, the specimens were unloaded and then reloaded at the different angles to the first prestraining. Hence, the author neglects the effect of the plastic deformation during unloading.

The author has defined the length of restoration parth ΔS_{i0} from the point where the magnitude of modulus K reaches a stationary value. In other words, the value ΔS_{i0} is considered to be the length of the stress path beyond which the effect of the first prestrain can be neglected. Therefore the length ΔS_{i0} may be closely related to the so-called trace of delay.

The writer considers that the disagreements between the theoretical and experimental values in the author's paper were mainly caused by the incompleteness of theoretical representation of the shape of a subsequent yield locus. To derive a more precise flow theory by the author's method, the subsequent yield loci having an abrupt change in the prestraining direction may be necessary. Some experimental results obtained in the writer's laboratory are shown in Fig. A. It may be seen that the actual subsequent yield loci can not be predictee by a simple kinematic theory or an isotropic hardening theory.

O. A. Shishmarev.[2]) I have often studied the shape of yield loci obtained using a small reference strain as a basis for determining yield stress. The

[1]) Tokyo Institute of Technology, Tokyo, Japan.
[2]) Riazan, U.S.S.R.

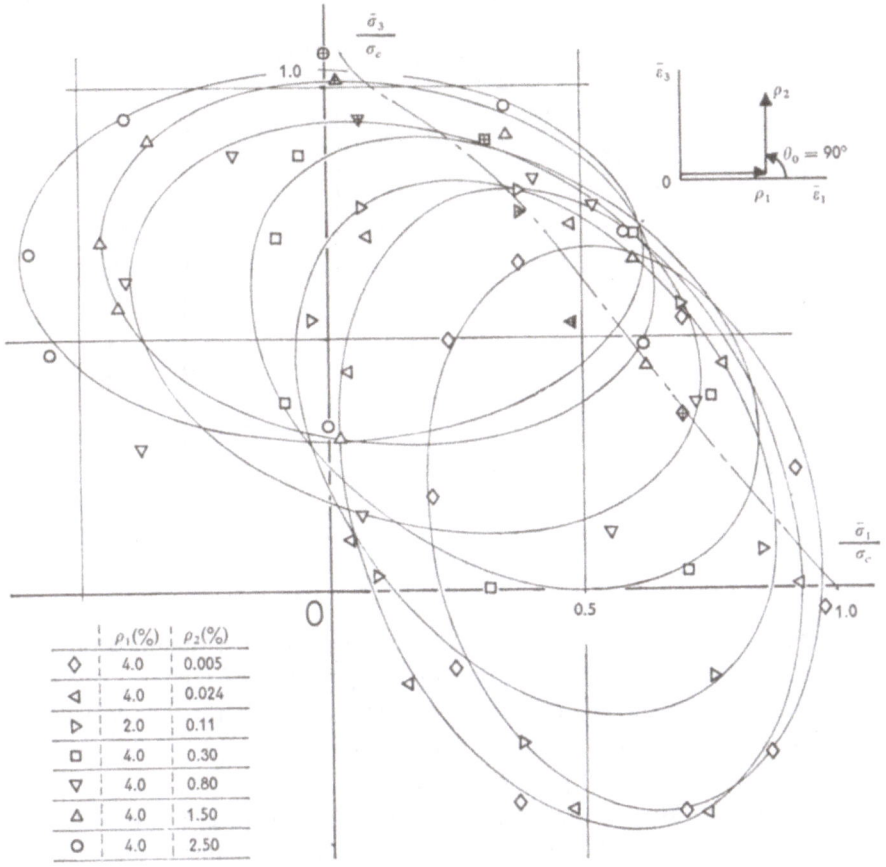

Fig. A

reference strains I used represented from 0,0006 to 0,01 per cent of the equivalent plastic strain, [S. 1–S. 4].

The loci for such reference strains proved, however too sensitive to the time elapsed between the end of prestraining and the moment of determining the yield points, and to the influence of creep and other effects as well. The shape of the yield surface corresponding to a small reference strain changes strongly with reference strain. Even a slight deviation in the reference value may cause appreciable errors in the position of the evaluated point of the yield surface. On the other hand the loci obtained at large reference strains (about one tenths percent) are more stable in this respect. Confirmation of

515

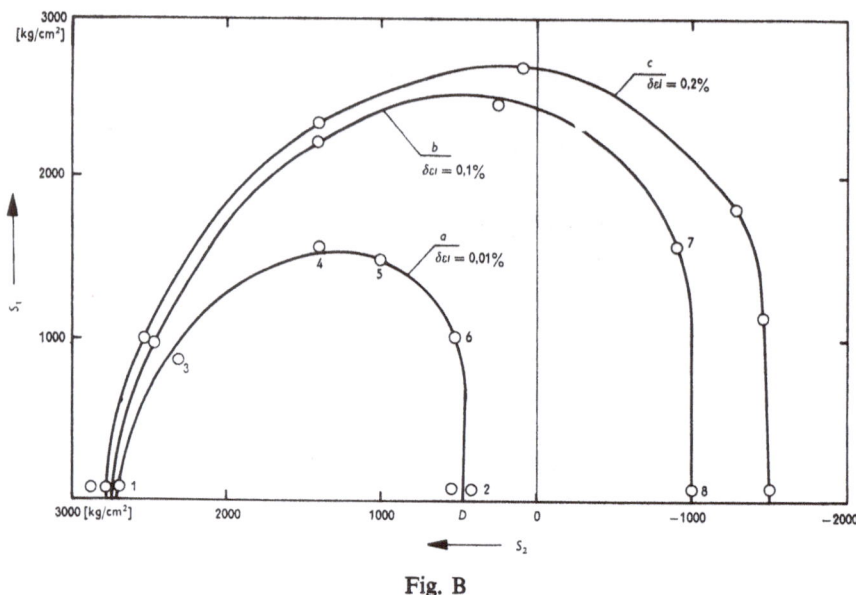

Fig. B

the above statement is provided in Fig. B where the loci obtained on steel specimens subjected to prestraining of 2.5 per cent of equivalent plastic strain in shear are shown. The values of reference strain for the plotted loci were chosen to be 0.01, 0.1 and 0.2 per cent (the curves *a*, *b*, *c*, respectively). The results explain why I gave up using small reference strains.

Professor Shiratori is right in observing that I disregard plastic strains caused by unloading when investigating the yield locus through secondary paths of loading after total unloading. However the influence of the error on the shape of the locus specified at a large reference strain of 0.2 per cent is minute. Specifying the yield points related to the opposite side of the locus (points 7, 8 of Fig. **B**) I took in fact into account plastic strains associated with unloading, they being included into the reference strain. As for the yield points lying on the locus within the first quadrant of the $S_1 - S_2$ plane, determined after full unloading on different paths, the unloading has no effect on the reference strain chosen. For small reference strains the effect of unloading may be more important (section *OD* in Fig. **B**). For the loci corresponding to a large reference strain this influence is rather small.

I cannot agree with Professor Shiratori that the discrepancy revealed between theoretical results and experimental data is largely due to an inaccuracy of a theoretical prediction as to the form of the subsequent yield loci. Such inaccuracies exist and lead to appreciable errors. But even if the loci obtained experimentally and theoretically coincide the discrepancy will still exist due to the neglect in flow theories the variation of the modulus K over the yield locus. The above mentioned variation disclosed along the *experimentally obtained* locus is very pronounced.

[S. 1] Yagn, Y. I. and O. A. Shishmarev, Doklady A.N. S.S.S.R., 119 (1958).

[S. 2] Shishmarev, O. A., Izvestiya Akad. Nauk S.S.S.R., OTN, Mekhanika i Mashi-nostroenie, N 4, (1962).

[S. 3] Shishmarev, O. A., Inzh. Zhurn., vol. 3., (1963).

[S. 4] Shishmarev, O. A., Inzh. Zhurn., Mekh. Tverd. Tela, 2 (1968).

A. J. M. Spencer and J. E. Ferrier

Some solutions for a class of plastic-elastic solids

Foundations of Plasticity
pp. 9–24

DISCUSSION

E. H. Lee.[1])   You stated, 'If plastic-rigid theory is to have value plastic-rigid solutions must, in some sense, be asymptotic solutions of the plastic-elastic equations. Such behavior is often assumed, but seems never to have been proven'. It seems to me that this question has received considerable attention in the literature for classical elastic-plastic theory. For example bending of beams and torsion of circular shafts, for which a residual 'elastic' region of infinitesimal width occurs for plastic-rigid theory. For torsion of other sections, the combined soap-film sand-hill analogy illustrates the same limiting situation along the roof ridges. The paper 'On Stress Discontinuities in Plane Plastic Flow', E. H. Lee, Proc. 3rd Symp. in Applied Math. of the American Math. Soc., 213–228, McGraw Hill, 1950, discusses such situations.

H. Lippmann.[2])   In the lecture, two problems have been discussed: The influence of different measures of stress-rate was examined first, and secondly, the difference between elastic-plastic and rigid-plastic solutions was considered. Thus, the question arises whether the error introduced by neglecting elastic properties, is smaller or larger than the scatter due to different stress-rates. For, if the first is true it obviously makes no sense to introduce elasticity when dealing with plastic deformations.

[1]) Stanford University, Stanford, California, USA.
[2]) University of Karlsruhe, Karlsruhe, G.F.R.

A. J. M. Spencer.[1]) Professor Lee is of course correct in stating that comparisons between plastic-elastic and plastic-rigid solutions have been made in classical plasticity theory. However, the examples which Professor Lee quotes employ the classical Prandtl-Reuss equations with a non-objective measure of stress-rate, and so the value of these is questionable when large deformations are concerned. We consider, therefore, that it is a useful exercise to repeat the solutions of some of the problems which Professor Lee mentions, but making use of suitable stress-rate measures.

Professor Lippmann raises a very interesting question. In reply, it seems necessary to distinguish between the cases of large and small plastic-elastic deformations. In the problems discussed in the paper, the effect of the elastic part of the deformation is shown through the influence of the parameter k/μ, with this parameter tending to zero in the limit of plastic-rigid behaviour, and the effect of variation of stress-rate measure is indicated by varying the parameter α. In our examples, for small deformations the influence of k/μ is important, for example in determining the strain at initial yield, and that of α less so. For large deformations the inclusion of elasticity gives rise only to second-order effects, and although k/μ determines the magnitude of these effects, their sign is determined by α. Hence in the absence of firm knowledge of the appropriate value of α it does indeed appear that there is no advantage in introducing elasticity in this case. However, for brevity and simplicity the questions of unloading and strain-hardening were not considered in the paper. Elasticity will certainly have an important influence in unloading, and may also assume greater importance if account is taken of strain-hardening.

Professor Dubey in his discussion note [S. 1] raises two points. The first gives arguments in favour of the choice of the co-rotational (in our terminology, the Jaumann) stress-rate tensor. We agree that when a choice of a particular stress-rate has to be made, there are good reasons for selecting the co-rotational stress-rate. However, the paper shows that, in two examples at least, a change from one objective stress-rate tensor to another has only a second-order effect on the solution. Although it is obviously not justified to draw firm conclusions from two simple examples, this suggests tentatively to us that in practice the choice of stress-rate may *in some problems* not have a large effect on the solution, and it may be possible to make a choice on grounds of mathematical convenience without introducing serious error.

[1]) University of Nottingham, England.

As his second point, Professor Dubey draws attention to the fact that where questions of stability are involved, the choice of stress-rate measure may be critical. We are in full agreement with this comment.

[S. 1] Dubey, R. N., On objective stress-rate in the constitutive relations for elastic-plastic solids, this volume, 382–385.

P. Stutz

Comportement elasto-plastique des milieux granulaires

Foundations of Plasticity
pp. 37–49

DISCUSSION

G. Gudehus.[1])   Dr. Stutz has used the formal concept of hypo-elasticity. As granular materials are not purely elastic the stress path will not be followed backward during the reversed strain path as in a hypo-elastic material. Therefore the theory presented can only serve as an approximation for certain monotonous deformation histories. For such histories, however, it will be simpler to use finite (quasi elastic) stress-strain laws; this was done by Goldscheider, e.g. (Diss. Univ. Karlsruhe, 1972). Only in the critical state that allows certain deformations without stress changes the hypo-elastic approximation is more powerful than a finite law. The following remarks only refer to this aspect of the theory.

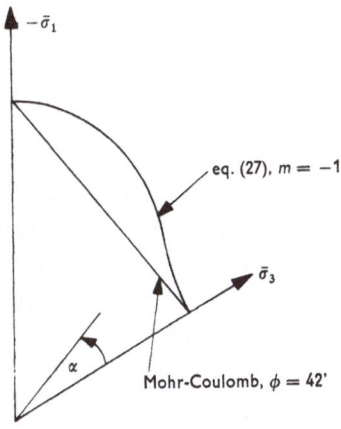

Fig. A

[1]) University of Karlsruhe, Karlsruhe, G.F.R.

The limiting surface eq. (27) contains equations proposed by Geniev ($m = -1$) and the author ($m = -1/2$) as special cases. It must be noted, however, that for $-4 < m < 1$ fitting of eq. (27) to the Mohr-Coulomb condition does not always lead to limiting surfaces which are not convex everywhere (Fig. A shows the envelope as calculated from eq. (27) for $\phi = 42°$ and $m = -1$ as an example: convexity does not hold for $\alpha = 0$). Convexity should hold for theoretical and experimental reasons.

Eq. (15) implies proportionality of deviator strain increment and deviator stress. This is at variance with the author's deviatoric normality condition as applied to eq. (27) and with results of cuboidal deformation tests.

A. Sawczuk.[1]) In connection with Professor Gudehus' first comment it might be worthwhile to mention that constitutive laws involving stress, objective stress rate and strain rates can describe more than hypo-elastic response. Various models of material behavior can be obtained on specifying the scalar functions entering the tensor equations in a quite similar way as it is done in 'infinitesimal' theories. Although, for example, both visco-elastic and elastic-plastic relations in 'infinitesimal' formulation involve rates of stress and strain they describe different materials.

P. Stutz.[2]) La relation (1) proposée dans notre communication permet de décrire l'évolution du matériau par un formalisme unique, quand il est soumis à un processus de chargement continu qui l'amène à l'état critique où son comportement peut être assimilé à celui d'un matériau parfaitement plastique. La loi proposée doit évidemment être complétée pour décrire les trajets correspondant à une décharge, afin de pouvoir suivre l'évolution du matériau quand on le soumet à des sollicitations plus complexes qui comprennent charge et décharge.

Si on remarque que la grandeur des déformations élastiques est négligeable, la déformation du matériau étant due presque uniquement aux glissements des grains, on est conduit à restreindre le domaine d'application de la relation (1) par la condition de dissipation positive. Le comportement du matériau est alors décrit par deux relations, (ayant des expressions formelles semblables à (1), mais avec des coefficients différents), le choix de l'une ou l'autre expression étant dicté par la condition de dissipation positive.

[1]) Institute of Fundamental Technological Research, Warsaw, Poland.
[2]) Université de Grenoble, St. Martin d'Hères, France.

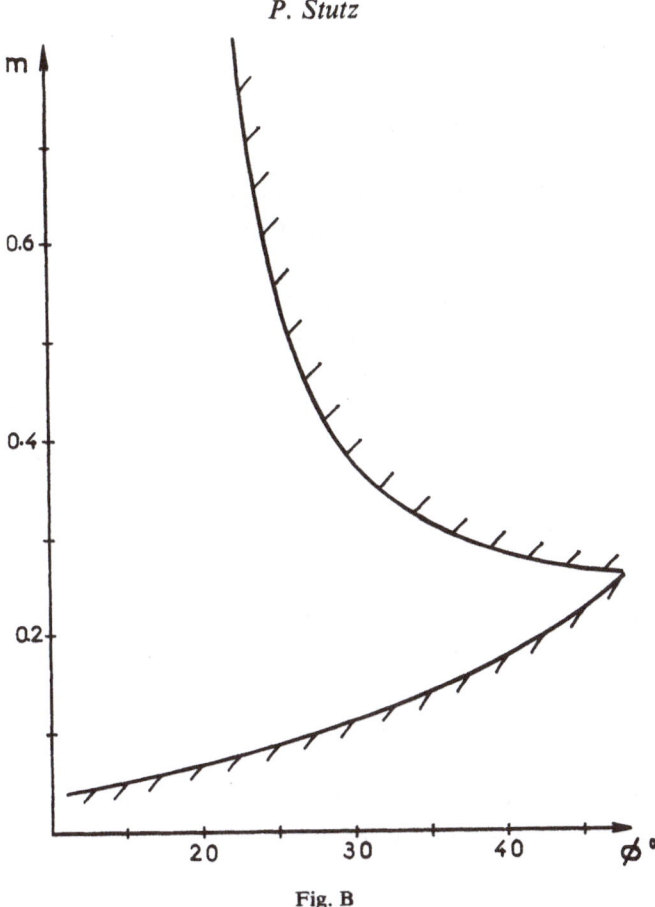

Fig. B

L'expression (30) de la surface limite dépend des paramètres ϕ et m. Si on veut imposer à cette surface d'être convexe, il en résulte évidemment des restrictions pour ces paramètres. Pour les valeurs usuelles de ϕ, les valeurs possibles pour m, assurant la convexité de la surface limite sont illustrées sur la figure B.

La relation (1) implique bien, à l'état critique, la proportionalité des déviateurs des tenseurs contrainte et vitesse de déformation. Si les resultats expérimentaux cités dans la communication de M. Gudehus (*AMS, 24* (1972) 395–402) représentent vraiment le travail d'un sable *à l'état critique*, il faut abandonner l'hypothèse de la linéarité de la loi par rapport au tenseur vitesse ou déformation. L'étude expérimentale du comportement d'un sable par de véritables essais triaxiaux, est donc d'une extrême importance. L'analyse de l'état critique n'est reste pas moins délicate, puisque le matériau, avant d'atteindre cet état, subit des déformations importantes qui favorisent une déformation non homogène de l'échantillon et une anisotropie du matériau.

T. Tokuoka

Fundamental relations of plasticity derived from hypo-elasticity

Foundations of Plasticity
pp. 1–8

DISCUSSION

S. Nemat-Nasser.[1])   Assuming that one accepts the concept of yield surface as a useful tool for dealing with plasticity, it is the yielding process and the relation between the kinematical and dynamical quantities which characterize it, that are of significance. The yield surface itself, when divorced from the corresponding transition processes, is of little significance. The basic difficulty in dealing with plasticity in the framework of the so-callled hypo-elasticity theory, is that this theory without further assumptions is incapable of dealing with the transition processes involved in elastoplastic deformations.

P. Germain.[2])   1. What is the physical significance (if any) of normal yield and shear yield? 2. Is the yield surface obtained by this method convex, or do we need to impose further assumptions in order to get a convex yield surface?

T. Tokuoka.[3])   The proposed plasticity theory in the framework of hypo-elasticity deals but the yield criterion and the flow rule, and the transition process from the unyield state to the yield state is not treated here. Generally we have two possible devices to treat this transition problem in our scope. *One* is that we assume another appropriate constitutive equation, e.g., an

[1]) Northwestern University, Evanston, Illinois, U.S.A.
[2]) Université Paris VI, Paris, France.
[3]) Kyoto University, Kyoto, Japan.

elastic stress-strain relation, in the unyield region of the same material; if so, we have no answer for the presented question. *Another* is the proposition that the hypo-elastic constitutive equation governs not only the yield state but also the unyield state. T. Y. Thomas (*Proc. Nat. Acad. Sci. U.S.A.*, *40* (1955) 720–726, 908–910) and C. Truesdell (*J. Appl. Phys. 27* (1956) 441–447) analyzed the transition process for a normal yield type hypo-elastic material, which has the v. Mises yield criterion as its singularity relation. Recently Tokuoka (*Int. J. Non-Linear Mech. 7* (1972) 609-620) treated a shear yield type hypo-elastic material, having the Tresca yield criterion, in the simple shear, and I found that for large angle of shear the maximum difference of the principal stresses approaches a Tresca yield value. Although we have no analysis for the transition process in the general case of hypo-elasticity, I now conjecture that according as the deformation increases the stress state approaches asymptotically the state, which satisfies the yield criterion defined as the singularity relation of the hypo-elastic constitutive relation.

With reference to the first question raised by Professor Germain I wish to add that with respect to the principal axes of stress we have, in general, non-zero normal components of plastic stretching in normal yield and non-zero shear component or components of it in shear yield. Then, if the work in plastic flow is defined by $W = \text{tr}(T_P D)$, where $_P D$ is the plastic stretching, we may have the non-zero work and no work in normal and shear yield, respectively.

For his second question I add that the yield surface is defined by the two-dimensional manifold, which is the set of stress vector t for det $H(t) = 0$, in a three-dimensional principal stress space. Then the yield surface is determined by the form of H. The hypo-elasticity H may have any form except from the condition that H is an isotropic tensor. Therefore the yield surface may be convex or concave. For example the surfaces (23)–(29) may express concave surfaces for appropriate combinations of material constants.

K. C. Valanis

Observed plastic behavior of metals vis-a-vis the endochronic theory of plasticity

Foundations of Plasticity
pp. 235–255

DISCUSSION

K. S. Havner.[1]) Unfortunately, I have not had opportunity to study Professor Valanis' earlier papers (*Arch. Mech. Stos. 23*, 1971) upon which the present theory is based, for judging from the results given here they certainly warrant study. The agreement with various experiments, obtained through particularizing the general equations (1.1)–(1.6) derived in those papers, is quite impressive.

I wish to ask several questions, beginning with one concerning the title of the paper. What is the definition (or your meaning) of the word 'endochronic'? I have searched in several dictionaries, including the 'Oxford English Dictionary', but have been unable to find it or to understand the combination of 'chronic' with the prefix 'endo' in the context of the paper.

My second question pertains to Sect. 2, Figs. 3 and 4, and the experiments which were conducted. It is stated that only *one* transverse strain of a uniaxially stressed thin flat bar was monitored, yet in the figures are shown plots of σ_{kk} versus ε_{kk}. Were indeed *two* transverse strains (as well as the axial strain) monitored, or was it merely *assumed* that the bar remained isotropic and the second transverse strain set equal to the first? If the latter, then the rather surprising results indicated would seem to require further investigation.

As a passing comment, I think the statement which ends Sect. 3 is a noteworthy example of scholarly restraint. Rather than being 'remarkable',

[1]) North Carolina State University, Raleigh, N.C., USA.

the agreement between theory and experiment in Fig. 6 is absolutely astounding.

In regard to the prediction of the 'unloading-loading loop' of Fig. 11, I would like Professor Valanis' opinion on the possible contribution of large deformation effects at this strain level (approximately 5 percent). Since the rate of work-hardening is apparently small, it may be that these effects are not negligible during incremental, predominantly plastic response. Of course they should be insignificant within the loop.

In Figs. 13 and 15 is it not intended that the vertical scale be the *critical* shear stress (or shear strength) of the crystal rather than the resolved shear stress? From Fig. 12, the latter repetitively cycles through zero during continued straining. Moreover, it should be mentioned that the data of Figs. 13 and 15 could be roughly predicted by the famous and quite simple hardening law proposed by Sir Geoffrey Taylor many years ago [H. 1, H. 2]. Of course, the close prediction in Fig. 15 is outstanding.

In closing, it seems to me that Professor Valanis' work represents an important new direction for macroscopic plasticity theory through its deemphasis of an elastic range and a precisely defined yield surface. In an extension of the numerical studies of f.c.c. crystalline aggregates reported in [H. 3], I have found that after a tensile strain of 4 to 5 times the proportional-limit strain the macroscopic elastic range is negligible or vanishes entirely. Similar results have been obtained by Hutchinson [H. 4]. The unloading response initially *appears* to be elastic until a sufficient number of new slip systems are activated to produce discernible nonlinearity. I believe Professor Valanis' macroscopic theory is consistent with these results.

[H. 1] Taylor, G. I., The distortion of aluminum crystals under compression, Part II, *Proc. R. Soc., A116* (1927) 16–38.

[H. 2] Taylor, G. I., Plastic strain in metals, *J. Inst. Metals 62* (1938) 307–324.

[H. 3] Havner, K. S. and R. Varadarajan, A quantitative study of a crystalline aggregate model, *Int. J. Solids Struct., 9* (1973) to appear.

[H. 4] Hutchinson, J. W., Elastic-plastic behavior of polycrystalline metals and composites, *Proc. R. Soc., A319* (1970) 247–272.

V. Kafka.[1]) In the paper under discussion some experimental evidence is given to support the assumption of constant Poisson's ratio as a more realistic one than the assumption of elastic hydrostatic response.

[1]) Institute of Theoretical and Applied Mechanics, Prague, Czechoslovakia.

I would like to point out that the literature dealing with the volume changes of metals undergoing plastic deformation is not so scarce as it may seem from the discussed paper (page 238). For example in [K. 1], there are ten more quotations concerning the subject and some of them rather old.

The results of all these works to the best of my knowledge differ very substantially from those presented in the discussed paper as to the magnitude of the volume change. To give an example it can be mentioned that discussing the paper [K. 2], where aluminium tests were reported, Batdorf called in question the accuracy of the method used by the authors, for one per cent volume change, improbably large was recovered. In their answer the authors referred to [K. 3] in which a volume change of approximately equal magnitude was revealed by a very precise method. I have found that in this reference such a volume change was really observed, but only in the neck region.

Contrary to all this evidence the volume change referred to in the discussed paper is as large as 8 per cent for aluminium (Fig. 4) and as large as 3 per cent for copper (Fig. 3).

I believe that there must be some misinterpretation. One possibility is that the plastic deformation was not homogeneously distributed and in the point, where the transverse strain was measured, the plastic deformation was much smaller than the average value. Calculating then the volume change from this measurement and from the elongation, measured on a large basis an unrealistic increase of volume may be obtained. Such a heterogeneous development of plastic deformation is really common, cf. [K. 1].

I would conclude that the assumption of constant Poisson's ratio can hardly be considered as proved to be better than the assumption of elastic hydrostatic response.

[K. 1] Kafka, V. and R. Novotny, A contribution to the solution of the question of volume changes of polycrystalline materials in plastic deformation. *Acta Technica* ČSAV, 1 (1971) 41.

[K. 2] Marin, J. and L. W. Hu, On the validity of assumptions made in theories of plastic flow for metals. *Trans. of the ASME*, 75 (1953) 1181.

[K. 3] Thomsen, E. G., J. Cornet, J. Lotze, and J. E. Dorn, Investigation of the validity of an ideal theory of elastic-plasticity for wrought aluminium alloys, NACA, TN 1552 (1948).

O. C. Zienkiewicz, G. C. Nayak and D. R. J. Owen

Composite and 'overlay' models in numerical analysis of elasto-plastic continua

Foundations of Plasticity
pp. 107–123

DISCUSSION

M. Kleiber.[1])   My question is of a fundamental nature and concerns the form the yield condition should take if formulated in the general form $F = F(\{\sigma\}, x) = 0$. I think there are many possibilities of assuming the explicit form of this condition. One of the basic problems in the plastic finite element analysis is to decide when a given element of the body can be considered as plastic and when it is still elastic. Would you be so kind to give an explicit form of the yield condition used in your numerical computations?

O. C. Zienkiewicz.[2])   In reply to Dr. Kleiber's question: the yield condition used in this paper is simply that of the von Mises type.

It represents the second stress invariant and can be written in the component form. Other yield criteria has been used such as the Mohr – and for the explicit form which we find very convenient I would like to refer him to a publication entitled:

'Convenient form of stress invariant for plasticity' by myself and Dr. G. C. Nayak which appeared in the *Journal of Structural Division*, American Society of Civil Engineers (April 1972). Clearly the form is very much problem dependent.

---

[1]) Institute of Fundamental Technological Research, Warsaw, Poland.
[2]) University of Wales, Swansea, U.K.

REFEREES

Referees

N. C. Lind, Waterloo, Canada
H. Lippmann, Karlsruhe, GFR
G. Maier, Milan, Italy
L. E. Malvern, Gainsville, USA
P. V. Marçal, Providence, USA
J. B. Martin, Providence, USA
Ch. Massonnet, Liège, Belgium
J. P. Miles, Manchester, Great Britain
A. J. A. Morgan, Los Angeles, USA
Z. Mróz, Warsaw, Poland
T. Mura, Evanston, USA
S. Murakami, Nagoya, Japan
J. Murzewski, Cracow, Poland
J. Najar, Warsaw, Poland
L. B. Nikitin, Moscow, USSR
V. N. Nikolaevskii, Moscow, USSR
J. Ostrowska, Warsaw, Poland
D. R. Owen, Pittsburgh, USA
A. C. Palmer, Cambridge, Great Britain
P. R. Paslay, Providence, USA
R. K. Penny, Liverpool, Great Britain
P. Perzyna, Warsaw, Poland
A. Phillips, New Haven, USA
T. H. H. Pian, Cambridge, Mass. USA
A. C. Pipkin, Providence, USA
A. R. S. Ponter, Leicester, Great Britain
E. P. Popov, Berkeley, USA
H. L. D. Pugh, East Kilbridge, Great Britain
D. Radenkovic, Paris, France
J. R. Rice, Providence, USA
R. D. Rivlin, Bethlehem, USA
G. I. N. Rozvany, Clayton, Australia
G. Sacchi, Pavia, Italy

J. Salençon, Paris, France
R. Sankaranarayanan, Bangalore, India
M. Save, Mons, Belgium
A. Schofield, Manchester, Great Britain
W. J. Shack, Cambridge, Mass. USA
E. Shiratori, Tokyo, Japan
O. A. Shishmarev, Riazan, USSR
O. M. Sidebottom, Urbana, USA
J. A. Simmons, Washington, USA
R. Sowerby, Hamilton, Canada
A. J. M. Spencer, Nottingham, Great Britain
B. Storåkers, Stockholm, Sweden
P. Stutz, Grenoble, France
J. L. Swedlow, Pittsburgh, USA
P. S. Symonds, Providence, USA
W. Szczepiński, Warsaw, Poland
Z. R. Tamuzh, Riga, USSR
T. C. T. Ting, Chicago, USA
T. W. Ting, Urbana, USA
T. Tokuoka, Kyoto, Japan
A. A. Vakulenko, Leningrad, USSR
K. C. Valanis, Hoboken, USA
C. C. Wang, Houston, USA
T. Wierzbicki, Warsaw, Poland
A. S. Wineman, Ann Arbor, USA
R. de Wit, Washington, USA
S. Yaghmai, Teheran, Iran
R. N. Yong, Montreal, Canada
S. Zahorski, Warsaw, Poland
J. Zarka, Paris, France
J. Zawidzki, Warsaw, Poland
O. C. Zienkiewicz, Swansea, Great Britain

SYMPOSIUM PARTICIPANTS

B. O. ALMROTH, Dept. 52-33, B/205, Lockheed Palo Alto Research Laboratory, 3251 Hanover Street, Palo Alto, Calif. 94304, USA

M. ARCISZ, Institute of Fundamental Technological Research, Swiętokrzyska 21, Warsaw, Poland

G. AUGUSTI, University of Florence, Via di St. Marta 3, 50139 Florence, Italy

K. B. E. AXELSSON, Chalmers University of Technology Fack, S-40220, Göteborg 5, Sweden

G. BACKHAUS, Techn. Universität Dresden, 8027 Dresden, Mommsenstr., GDR

Z. BACZYŃSKI, Institute of Fundamental Technological Research, Swiętokrzyska 21, Warsaw, Poland

M. BALARIN, DAW Zentralinstitut für Festkörperphysik und Werkstofforschung, 8027 Dresden, GDR

A. BALTOV, Institute of Technical Mechanics, Sofia, Oboriste 21, Bulgaria

N. V. BANICHUK, Institute of Problems of Mechanics, USSR Academy of Sciences, Moscow A-40, Leningradskii Prospekt 7, USSR

J. BAUER, Institute of Fundamental Technological Research, Swiętokrzyska 21, Warsaw, Poland

J. BEJDA, Institute of Fundamental Technological Research, Swiętokrzyska 21, Warsaw, Poland

A. BERIO, University of Cagliari Ist. Scienza Construzioni, Piazza d'Armi, Cagliari, Italy

J. BIATKIEWICZ, Technical University of Cracow, Cracow, Poland

T. Z. BLAZYNSKI, University of Leeds, Dept. of Mechanical Engineering, Leeds LS2 9JT, Great Britain

O. BRUHNS, Ruhr-Universität Bochum, Institut für Mechanik, 463 Bochum, Postfach 2148, GFR

K. BUCHACEK, Institut of Theoret. and Appl. Mechanics, Phaha 2, Vysehradska 49, Czechoslovakia

H. F. BUECKNER, General Electric Co., 1184 Bellemead Court, Schenectady, N.Y., USA

A. Shiu Lau Chan, Dept. of Aeronautics, Imperial College of Science and Technology, Prince Consort Road, London SW 7, Great Britain

F. L. Chernous'ko, Institute of Problems of Mechanics, USSR Academy of Sciences, Moscow A-40, Leningradskii Prospect 7, USSR

M. Como, University of Naples, Corso Vittorio Emanuele 656, Naples, Italy

N. Coutris, Comité National Français de Mécanique, 11, rue Mansard, 92 Vannes, France

N. Cristescu, Institutul de Matematica, Calea Grinta 21, Bucuresti 12, Rumania

Y. D'escatha, Laboratorie de Mécanique de l'Ecole Polytechnique, 17, rue Descartes, Paris 5, France

L. Dietrich, Institute of Fundamental Technological Research, Swiętokrzyska 21, Warsaw, Poland

A. Drescher, Institute of Fundamental Technological Research, Swiętokrzyska 21, Warsaw, Poland

E. Drescher, Institute of Fundamental Technological Research, Swiętokrzyska 21, Warsaw, Poland

D. C. Drucker, University of Illinois, College of Engineering, Urbana, Ill., USA

J. L. Duncan, Dept. of Mechanical Engineering, McMaster University, Hamilton, Ontario, Canada

M. K. Duszek, Institute of Fundamental Technological Research, Swiętokrzyska 21, Warsaw, Poland

Cz. Eimer, Institute of Fundamental Technological Research, Swiętokrzyska 21, Warsaw, Poland

M. I. Erkhov, Institute for Structural Engineering (CNIKI), Moscow K-25, USSR

J. E. Ferrier, Hunting Engineering Ltd., Ampthill, Bedfordshire, Great Britain

J. T. Fong, U.S. National Bureau of Standards, Washington, D.C.20234, USA

H. Fukuoka, Faculty of Engineering Science Osaka University, 1-1, Michikaneyamacho, Toyonaka-shi, Osaka, Japan

I. M. Fyfe, University of Washington, Seattle, Washington, USA

P. Germain, Faculté des Sciences, Université Paris VI, 9, quai Saint-Bernard, Tour 66, Paris 5e, France

J. J. Gilman, Allied Chemical Corporation, Morristown, N.J. USA

P. Glockner, University of Calgary, Calgary, Alberta, Canada

B. GOWDA, University of Oxford, Dept. of Metallurgy, Parks Road, Oxford, Great Britain

W. GRABCZYŃSKA, Institute of Fundamental Technological Research, Świętokrzyska 21, Warsaw, Poland

A. GRIMALDI, University of Naples, Parco Margherita 24, Naples, Italy

P. GRUNDY, Dept. of Civil Eng., Monash University, Clayton, 3168, Victoria, Australia

G. GUDEHUS, Institut für Bodenmechanik und Felsmechanik, Universität Karlsruhe, 7500 Karlsruhe, GFR

W. GUTKOWSKI, Institute of Fundamental Technological Research, Świętokrzyska 21, Warsaw, Poland

V. GUTTMANN, Euratom, Petten, The Netherlands

K. S. HAVNER, North Carolina State University, Raleigh, N.C., USA

P. G. HODGE, JR., University of Minnesota, Aeronautical Engineering Building, Minneapolis, Minn., 55455, USA

H. G. HOPKINS, Dept. of Mathematics, The University of Manchester, Sackville Street, Manchester M60 1QD, Great Britain

Y. HORIE, North Carolina State University, Raleigh, N.C. 27607, USA

M. JANAS, Institute of Fundamental Technological Research, Świętokrzyska 21, Warsaw, Poland

N. JONES, Massachusetts Institute of Technology, Dept. of Ocean Engineering, Cambridge, Mass. 01239, USA

V. KAFKA, Institute of Theor. and Appl. Mechanics, Praha 2, Vysehradska 49, Czechoslovakia

S. KAJFASZ, Institute of Fundamental Technological Research, Świętokrzyska 21, Warsaw, Poland

S. KALISZKY, Technical University, Müegyetem rpt. 3, Budapest XI, Hungary

K. KÄMMEL Bergakademie Freiberg, Sektion Maschinen- u. Energietechnik, Freiberg, GDR

I. KISIEL, Technical University of Wrocław, Wrocław, Poland

M. KLEIBER, Institute of Fundamental Technological Research, Świętokrzyska 21, Warsaw, Poland

J. KLEPACZKO, Institute of Fundamental Technological Research, Świętokrzyska 21, Warsaw, Poland

P. KNOLL, Institut für Bergbausicherheit, 703 Leipzig, Friedrikenstr. 60, GDR

W. KOITER, Technische Hogeschool Delft, Laboratorium voor Technische Mechanica, Delft, Mekelweg, The Netherlands

J. A. KÖNIG, Institute of Fundamental Technological Research, Świętokrzyska 21, Warsaw, Poland

W. KOSIŃSKI, Institute of Fundamental Technological Research, Świętokrzyska 21, Warsaw, Poland

R. KOWALCZYK, Institute of Fundamental Technological Research, Świętokrzyska 21, Warsaw, Poland

J. KRATOCHVÍL, Czechoslovak Acad. Sci., Inst. Solid State Physics, Prague 6, Czechoslovakia

J. KRAVTCHENKO, Laboratoire de Mécanique des Fluides, Domaine Univ. de Grenoble, 38 St. Martin d'Hères, France

E. KRÖNER, Universität Stuttgart, 7 Stuttgart, GFR

V. J. KUFNER, Czech Polytechnic Institute of Prague, Civil Engineering Faculty, Prague 6 - Dejvice, Czechoslovakia

M. KWIECIŃSKI, Technical University of Warsaw, Warsaw, Poland

G. LANDGRAF, Technische Universität Dresden, 8027 Dresden, Liebigstr. 24. GDR

R. W. LARDNER, 3066 Butternut, Port Coquitlam, B.C., Canada

F. A. LECKIE, Dept. of Engineering, University of Leicester, Leicester LE1 7RH, Great Britain

E. H. LEE, Dept. of Applied Mechanics, Stanford University, Stanford, Calif. 94305, USA

L. H. N. LEE, University of Notre Dame, Dept. of Aerospace and Mechanical Eng., Notre Dame, Indiana, USA

TH. LEHMANN, Ruhr-Universität Bochum, Institut für Mechanik, 463 Bochum, Postfach 2148, GFR

T. LESER, Moores Hill Road, Bel Air, Maryland 21014, USA

T. H. LIN, Mechanics and Structural Dept. University of California, Los Angeles, Calif. 90024, USA

N. C. LIND, Faculty of Engineering, Dept. of Civil Engineering, University of Waterloo, Waterloo, Ontario, Canada

J. LITOŃSKI, Institute of Fundamental Technological Research, Świętokrzyska 21, Warsaw. Poland

H. LIPPMANN, University of Karlsruhe, Institut für Technische Mechanik und Festigkeitslehre, Karlsruhe, GFR

J. MANDEL, Ecole Polytechnique, Paris, 16, rue Colonel Bonnet, Paris 16e, France

G. MAIER, Politecnico di Milano, Instituto di Scienza e Tecnica delle Costruzioni, 20133 Milano, Italy

N. N. MALININ, Moscow Technological University, Moscow B. 5., USSR

Z. MARCINIAK, Technical University of Warsaw, Warsaw, Poland

F. A. McCLINTOCK, Dept. of Mechanical Engineering, Massachusetts Institute of Technology, Cambridge, Mass. 02139, USA

J. MIASTKOWSKI, Institute of Fundamental Technological Research, Swiętokrzyska 21, Warsaw, Poland

M. MICUNOVIC, Masinski Fakultet, S. Janjica 6, 34000 Kragujevac, Yugoslavia

Z. MRÓZ, Institute of Fundamental Technological Research, Swiętokrzyska 21, Warsaw, Poland

T. MURA, Northwestern University, Evanston, Ill. 60201, USA

S. MURAKAMI, Nagoya University, Dept. of Mechanical Engineering, Chikusa-ku, Nagoya, Japan

J. NAJAR, Institute of Fundamental Technological Research, Swiętokrzyska 21, Warsaw, Poland

S. NEMAT-NASSER, Nothwestern University, Evanston, Ill. 60201, USA

T. NONAKA, Disaster Prevention Research Inst., Kyoto University, Kyoto, Japan

H. NOVOTNA, Katedra stavebne mechaniky, CVUT, Praha 6, Czechoslovakia

W. NOWACKI, Polish Academy of Sciences, Warsaw, Poland

F. K. G. ODQVIST, Terstensonsvägen 7D, Djursholm, Sweden

W. NOWACKI, Polish Academy of Sciences, Warsaw, Poland and International Centre for Mechanical Sciences, Udine, Italy

E. T. ONAT, Dept. of Engineering and Applied Science, Becton Center, Yale University, New Haven, Conn. 06520, USA

J. OSTROWSKA-MACIEJEWSKA, Institute of Fundamental Technological Research, Swiętokrzyska 21, Warsaw, Poland

J. ORKISZ, Technical University of Cracow, Cracow, Poland

D. R. OWEN, Dept. of Mathematics, Carnegie-Mellon University, Pittsburgh, Penn. 15232, USA

B. PEGEL, Deutsche Akademie der Wissenschaften zu Berlin, Zentralinstitut für Festkörperphysik und Werkstofforschung, 8027 Dresden, GDR

O. PEJCOCH, Faculty of Metallurgy VSB, Ostrava 1, Osvoboditelu 33, Czechoslovakia

P. PERZYNA, Institute of Fundamental Technological Research, Swiętokrzyska 21, Warsaw, Poland

A. PHILLIPS, Dept. of Engineering and Applied Science, Yale University, New Haven, Conn.06520, USA

M. PIAU, Université Paris VI, Mécanique Théorique, Tour 66, 9 quai St. Bernard, 75 Paris 5e, France

W. PRAGNER, Brown University, Providence, R.I. 02912, USA

D. RADENKOVIC, Laboratoire de Mécanique des Solides, Ecole Polytechnique 17, rue Descartes, Paris 5e France

M. REINER, Technion, Haifa, Israel

G. ROMANO, University of Naples, via Brigata Bologna 16, 80125 Naples, Italy

M. ROMANO, University of Naples, via Brigata Bologna 16, 80125 Naples, Italy

G. I. N. ROZVANY, Monash University, Clayton, Victoria 3168, Australia

J. RYCHLEWSKI, Institute of Fundamental Technological Research, Swiętokrzyska 21, Warsaw, Poland

G. SACCHI, Istituto di Scienza e Technica delle Costruzioni, Universita di Pavia, 27100 Pavia, Italy

J. SALENÇON, Laboratoire de Mécanique des Solides, Ecole Polytechnique, 17, rue Descartes, Paris 5e, France

M. SAVE, Faculté Polytechnique, Mons, Belgium

A. SAWCZUK, Institute of Fundamental Technological Research, Swiętokrzyska 21, Warsaw, Poland

V. SCHMIDT, Deutsche Akademie der Wissenschaften zu Berlin, Zentralinstitut für Festkörperphysik und Werkstofforschung, 401 Halle/Saale, Weinberg 2, GDR

H. D. W. SCHUMANN, Pädagogische Hochschule Dresden, 806 Dresden, Wigardstraße 17, GDR

B. R. SETH, Birla Institut of Technology, Mesra, Ranchi, India

A. H. SHABAIK, Materials Dept. University of California, Los Angeles, Calif. 90024, USA

F. SIDOROFF, Université Paris VI, 124, rue Hoche, 93100 Montreuil, France

J. A. SIMMONS, Institute for Material Research, US Dept. of Commerce, National Bureau of Standards, Washington, D.C.20234, USA

A. J. M. SPENCER, Dept. of Theoretical Mechanics, University of Nottingham, Nottingham NG7 2RD, Great Britain

U. STAHLBERG, The Royal Inst. of Technology, Brinellvagen 23, Stockholm, Sweden

J. STASTNA, Technical University of Prague, Stavebni fakulta CVUT, Praha 6, Czechoslovakia

B. STORÅKERS, The Royal Institute of Technology, S-10044 Stockholm 70, Sweden

D. C. STOUFFER, Engineering Analysis Dept., University of Cincinnati, Cincinnati, Ohio 45221, USA

H. STRIFORS, The Royal Institute of Technology, S-10044 Stockholm 70, Sweden

P. STUTZ, Université de Grenoble, 32, rue Champ-Rochas, 38 Meylan, France

W. SYNOLD, Kantstr. 8, 50 Erfurt, GDR

W. SZCZEPIŃSKI, Institute of Fundamental Technological Research, Świętokrzyska 21, Warsaw, Poland

Cz. SZYMAŃSKI, Institute of Fundamental Technological Research, Świętokrzyska 21, Warsaw, Poland

R. TAKSERMAN-KROZER, Technion, Haifa, Israel

J. J. TELEGA, Technical University of Silesia, Gliwice, Poland

C. TEODOSIU, Centre of Mechanics of Solids, Str. Constantin Mille 15, Bucuresti, Roumania

T. C. T. TING, Dept. of Materials Engineering, Chicago, Ill. 60680, USA

K. C. VALANIS, Stevens Institute, Hoboken, N.J. 07030, USA

C. VIVANET, University of Cagliari, Istuto di Scienza delle Construzioni, Piazza d'Armi, Gagliari, Italy

Z. WASIUTYŃSKI, Institute of Fundamental Technological Research, Świętokrzyska 21, Warsaw, Poland

Z. WASZCZYSZYN, Technical University of Cracow, Cracow, Poland

M. P. WHITE, University of Massachusetts, Amherst, Mass., USA

T. WIERZBICKI, Institute of Fundamental Technological Research, Świętokrzyski 21, Warsaw, Poland

R. DE WIT, National Bureau of Standards, Metallurgy Division NBS, Washington, D.C. 20234, USA

E. WŁODARCZYK, Military Academy, Warsaw, Lazurowa 239

M. P. WNUK, Mechanical Engineering Dept. South Dakota State University, Brookings, S. Dakota 57006, USA

W. WOJNO, Institute of Fundamental Technological Research, Świętokrzyska 21, Warsaw, Poland

S. ZAHORSKI, Institute of Fundamental Technological Research, Świętokrzyska 21, Warsaw, Poland

A. ZAOUI, Ecole Polytechnique, 17, rue Descartes, Paris 5e, France

J. ZARKA, Ecole Polytechnique, 17, rue Descartes, Paris 5e France

J. ZAWIDZKI, Institute of Fundamental Technological Research, Swięto-krzyska 21, Warsaw, Poland

O. C. ZIENKIEWICZ, University of Wales, Singleton Park, Swansea SA2 8PP, Great Britain

H. ZORSKI, Institute of Fundamental Technological Research, Swiętokrzyska 21, Warsaw, Poland

M. ŻYCZKOWSKI, Technical University of Cracow, Cracow, Poland